Springer Series on Atomic, Optical, and Plasma Physics

Volume 106

The Springer Series on Atomic, Optical, and Plasma Physics covers in a comprehensive manner theory and experiment in the entire field of atoms and molecules and their interaction with electromagnetic radiation. Books in the series provide a rich source of new ideas and techniques with wide applications in fields such as chemistry, materials science, astrophysics, surface science, plasma technology, advanced optics, aeronomy, and engineering. Laser physics is a particular connecting theme that has provided much of the continuing impetus for new developments in the field, such as quantum computation and Bose-Einstein condensation. The purpose of the series is to cover the gap between standard undergraduate textbooks and the research literature with emphasis on the fundamental ideas, methods, techniques, and results in the field.

More information about this series at http://www.springer.com/series/411

Xin Tu • J. Christopher Whitehead
Tomohiro Nozaki
Editors

Plasma Catalysis

Fundamentals and Applications

Springer

Editors
Xin Tu
Department of Electrical Engineering
and Electronics
University of Liverpool
Liverpool, UK

J. Christopher Whitehead
School of Chemistry
The University of Manchester
Manchester, UK

Tomohiro Nozaki
Department of Mechanical Engineering
Tokyo Institute of Technology
Tokyo, Japan

ISSN 1615-5653 ISSN 2197-6791 (electronic)
Springer Series on Atomic, Optical, and Plasma Physics
ISBN 978-3-030-05188-4 ISBN 978-3-030-05189-1 (eBook)
https://doi.org/10.1007/978-3-030-05189-1

This Springer imprint is published by the registered company Springer Nature Switzerland AG
The registered company address is: Gewerbestrasse 11, 6330 Cham, Switzerland

Preface

The combination of nonthermal plasma and heterogeneous catalysis, known as plasma catalysis, is regarded as a promising and emerging technology for environmental cleanup, energy conversion, and the synthesis of fuels and chemicals at ambient pressure and low temperatures. In nonthermal plasmas, the working gas is activated to create highly energetic electrons and a range of reactive species (free radicals, excited atoms, ions, and molecules) for the initiation of plasma-assisted physical and chemical reactions. Nonthermal plasma has a distinct non-equilibrium character, which means the overall gas kinetic temperature in the plasma can be as low as room temperature, while the electrons are highly energetic with a typical energy of 1–10 eV. As a result, nonthermal plasma can easily break most chemical bonds and overcome the disadvantage of high temperature and/or high pressure required by conventional thermal catalysis and enable thermodynamically unfavorable chemical reactions to proceed at low temperatures under ambient conditions. The high reaction rates and fast attainment of steady state in plasma chemical reactions allows rapid start-up and shutdown of the process compared to thermal treatment technologies, which significantly reduces the overall energy cost and offers a promising route for chemical energy storage using renewable energy sources such as solar and wind power.

The coupling of nonthermal plasma with suitable catalysts has great potential to reduce the activation energy of reactions, enhance the conversion of reactants, and improve selectivity toward the desired products. These all contribute in different ways to enhancing the energy efficiency of the plasma process, as well as the catalyst stability by reducing poisoning, coking, and sintering, due to a synergy that occurs between the plasma and the catalyst. Plasma catalysis has been investigated for the removal of low-concentration environmental pollutants in high-volume waste gas and liquid streams, for the synthesis of platform chemicals and synthetic fuels from a range of sources (e.g., methane, carbon dioxide, and biomethanol), and for the synthesis of carbon nanomaterials. Different plasma sources and plasma-catalyst configurations have been designed and developed to enhance and optimize the performance of plasma-catalytic processes. Plasma catalysis has also been extended to the synthesis, preparation, and treatment of catalysts (e.g., supported metal catalysts) at low temperatures as an attractive alternative to the thermal processes (e.g., thermal calcination and reduction) for catalyst preparation. As we know, the interactions between plasma and catalyst are very complex in a hybrid plasma catalysis system, especially when using a single-stage plasma catalysis configuration where the catalyst is placed directly in the plasma. Both the properties of the plasma (e.g., local and global electric field) and catalyst (e.g., surface structure, morphology, metal active sites) can be modified by the presence of each other, which in turn affects the physicochemical interactions between the plasma and catalyst, consequently changing the performance of the chemical reaction. In situ diagnostics and plasma modeling have been identified as the key tools to better understand the reaction mechanisms on the catalyst surface under plasma environment. There have been considerable and increasing research activities in this emerging area in recent years.

Plasma catalysis is a broad, highly interdisciplinary, and still evolving field. Covering even the most important aspects of plasma catalysis in a single book that reaches readers ranging from students to active researchers in academia and industry is a significant challenge. This book provides a unique opportunity for a group of world-leading researchers active in the field of plasma catalysis to work together and to contribute to the first book specified in plasma catalysis, covering both fundamentals and applications of plasma catalysis. This book is divided into three sections. The first section (Chaps. 1, 2, 3, and 4) of the book provides a broad overview of plasma catalysis history, plasma catalysis systems, plasma-catalyst interactions, and plasma catalysis modeling. The second section (Chaps. 5, 6, 7, 8, 9, and 10) of the book covers a wide range of plasma-catalytic chemical processes including the removal of NO_x and volatile organic compounds (VOCs), ammonia decomposition for hydrogen production, methane activation, carbon dioxide conversion, and alcohol reforming. The last section (Chap. 11) discusses the challenges and future perspectives to the advance of this emerging field.

We believe the broad range of topics included in this book will provide the reader with an insight into the realm of possibilities that now exist in plasma catalysis research and development, and bridge the gap between plasma and catalysis community. It is also our hope that this book will stimulate further research in this exciting area.

We would like to express our deepest appreciation to all the authors who shared our excitement in preparing this book and contributed unique chapters that illustrate the depth and breadth of this emerging research topic. We gratefully acknowledge the assistance of our students Yaolin Wang, Michael Craven, and Jonathan Harding.

Liverpool, UK — Xin Tu
Tokyo, Japan — Tomohiro Nozaki
Manchester, UK — J. Christopher Whitehead

Contents

Contributors

Bryony Ashford Department of Electrical Engineering and Electronics, University of Liverpool, Liverpool, UK

Annemie Bogaerts Research group PLASMANT, Department of Chemistry, University of Antwerp, Antwerp, Belgium

Pieter Cools Research Unit Plasma Technology, Department of Applied Physics, Ghent University, Ghent, Belgium

Michael Craven Department of Electrical Engineering and Electronics, University of Liverpool, Liverpool, UK

Patrick Da Costa Sorbonne Université, Institut Jean Le Rond d'Alembert, Saint Cyr l'école, France

Nathalie De Geyter Research Unit Plasma Technology, Department of Applied Physics, Ghent University, Ghent, Belgium

Hongchen Guo State Key Laboratory of Fine Chemicals, School of Chemical Engineering, Dalian University of Technology, Dalian, Liaoning, China

Seigo Kameshima Department of Mechanical Engineering, Tokyo Institute of Technology, Tokyo, Japan

Ahmed Khacef GREMI, UMR 7344 CNRS – Université d'Orléans, Orléans, France

Hyun-Ha Kim National Institute of Advanced Industrial Science and Technology (AIST), Tsukuba, Ibaraki, Japan

Dae Hoon Lee Plasma Engineering Laboratory, Korea Institute of Machinery and Materials, Daejeon, South Korea

Department of Environment & Energy Engineering, University of Science and Technology, Daejeon, South Korea

Akira Mizuno Department of Environmental and Life Sciences, Graduate School of Engineering, Toyohashi University of Technology, Toyohashi, Aichi, Japan

Rino Morent Research Unit Plasma Technology, Department of Applied Physics, Ghent University, Ghent, Belgium

Erik Neyts Research group PLASMANT, Department of Chemistry, University of Antwerp, Antwerp, Belgium

Tomohiro Nozaki Department of Mechanical Engineering, Tokyo Institute of Technology, Tokyo, Japan

Atsushi Ogata National Institute of Advanced Industrial Science and Technology (AIST), Tsukuba, Ibaraki, Japan

Zunrong Sheng Department of Mechanical Engineering, Tokyo Institute of Technology, Tokyo, Japan

Keishiro Tamura Department of Mechanical Engineering, Tokyo Institute of Technology, Tokyo, Japan

Yoshiyuki Teramoto National Institute of Advanced Industrial Science and Technology (AIST), Tsukuba, Ibaraki, Japan

Xin Tu Department of Electrical Engineering and Electronics, University of Liverpool, Liverpool, UK

Li Wang State Key Laboratory of Fine Chemicals, School of Chemical Engineering, Dalian University of Technology, Dalian, Liaoning, China

College of Environmental Sciences and Engineering, Dalian Maritime University, Dalian, Liaoning, China

Department of Electrical Engineering and Electronics, University of Liverpool, Liverpool, UK

Yaolin Wang Department of Electrical Engineering and Electronics, University of Liverpool, Liverpool, UK

J. Christopher Whitehead School of Chemistry, The University of Manchester, Manchester, UK

Takumi Yamazaki Department of Mechanical Engineering, Tokyo Institute of Technology, Tokyo, Japan

Yanhui Yi State Key Laboratory of Fine Chemicals, School of Chemical Engineering, Dalian University of Technology, Dalian, Liaoning, China

Chapter 1
Plasma Catalysis: Introduction and History

J. Christopher Whitehead

1.1 Historical Introduction

Plasma catalysis (sometimes called plasma-enhanced catalysis, plasma-catalyst coupling, plasma-assisted catalysis or plasma-driven catalysis) is a hybrid technique where a catalytic material is used in conjunction with a gas discharge yielding a viable technique which gives enhanced performance for a range of gas processing applications such as removal of pollutants such as NO_x, SO_x and volatile organic compounds (VOCs) and production of a range of chemicals such as ammonia from N_2 and H_2, hydrogen and oxygenates by the reforming of hydrocarbons and a range of added-value chemicals from the conversion of CO_2 [1–9]. Perhaps in order to properly set the scene for discussing the mechanism and impact of plasma catalysis, it is necessary to separately examine the two techniques that were brought together to form this hybrid process.

Heterogeneous catalysis in which a solid catalytic material is placed into a stream of reactive gases can provide alternative reaction pathways with lowered energy barriers. These can increase the rates of the reactions involved, increasing the overall yield of the process at a given temperature and thereby improving the efficiency of the process. Mechanistically, these alternative reaction pathways may also improve the yield or selectivity for particular products, and using different catalysts can favor particular outcomes. This technique, often called thermal catalysis, has been used for hundreds of years and now forms the basis of many large-scale industrial processes particularly in the oil, gas and chemical industries. Catalysis was discovered by Sir Humphry Davy in 1817 who noted that heated platinum gauze or foil would bring about the slow combustion of vapors such as alcohol, ether, coal gas and methane below their ignition temperature [10]. Davy conducted his research at the Royal Institution in London and is probably most remembered for his invention of a

J. C. Whitehead (✉)
School of Chemistry, The University of Manchester, Manchester, UK
e-mail: j.c.whitehead@manchester.ac.uk

X. Tu et al. (eds.), *Plasma Catalysis*, Springer Series on Atomic, Optical, and Plasma Physics 106, https://doi.org/10.1007/978-3-030-05189-1_1

miners' safety lamp. (See the very readable account of the life of Davy by Thomas [11] for more information.) The Swedish chemist, Berzelius, acknowledged Davy's discovery in his treatise on catalysis where that word was first used. Berzelius recognized the existence of a catalytic force that '*is reflected in the capacity that some substances have, by their mere presence and not by their own reactivity, to awaken activities that are slumbering in molecules at a given temperature*' [12]. We now most commonly use the definition given by Ostwald: "*Catalysis is the acceleration of a chemical reaction, which proceeds slowly, by the presence of a foreign substance which is not in itself necessary for the reaction*" [13]. It is a question of the kinetics rather than the energetics of the overall reaction.

The study of gas discharges where a high voltage is applied to electrodes in a glass tube at reduced pressure had its origins at the start of the eighteenth century when Francis Hauksbee, an English draper, observed a glow in an evacuated tube containing mercury when the tube became electrically charged: electroluminescence [14]. It took another century before significant advances in gas discharges were obtained. This in part came from the development of better vacuum pumps such as that of Geissler in 1855. This allowed Geissler to make evacuated tubes with platinum electrodes containing different gases and show that discharges of different colors could be produced that were characteristic of the gas in the tube. These tubes were initially regarded as scientific novelties [15]. Michael Faraday, who worked at the Royal Institution in London, as Humphry Davy's assistant, looked at the structures or strata of light that were produced in gaseous discharges noting that the glow was not uniform but was brighter close to the electrodes with a dark space in between [16]. In time, such observations led to an understanding of the identity of the species responsible for luminescence and conductivity in gaseous discharges.

In 1834, Faraday also laid down the foundations for the mechanism of heterogeneous catalysis based on his observation of the spontaneous combination of oxygen and hydrogen on a platinum surface at room temperature [17]. He proposed that the two gases were condensed on the surface of the metal and that "*the approximation of the particles to those of the metals may be very great*". He stated that there would be a continual interchange of particles between the adsorbed layer and the gas in contact with it. The water product was less attracted by the platinum than the reacting gases and so could evaporate. He emphasized the importance of the forces between the different species stating: "*I admit . . . that the sphere of action of the particles extends beyond those other particles with which they are immediately and evidently in union, and in many cases produces effects arising into considerable importance*" [18]. It could be said that the work of Davy and Faraday at the Royal Institution played a key role in establishing the foundations of the two components that would come together a century later to form the basis of plasma catalysis.

Plasma is an ionized gas containing an equal number of positive ions as of electrons and negative ions. It is thus electrically neutral and its degree of ionization can range from very low, i.e. partially ionized, to 100%, or fully, ionized. In the laboratory or factory, plasma can be created by a variety of discharge techniques involving chambers with electrodes energized by direct dc, pulsed or ac currents including electrodes screened by a dielectric material or electrode-less systems such as inductively or capacitively coupled radio frequency radiation and microwaves. These discharges bring about a breakdown of the gas and create a range of species such as electrons, ions, dissociated and excited species giving plasma reactive properties that can potentially bring about chemical transformations, and which might be enhanced when a catalyst is present.

Additionally, we can characterize plasma by the pressure regime in which it operates which can be either low, being less than atmospheric, typically ca. 100 mbar or high pressure which is atmospheric or higher. Working at pressures less than atmospheric requires the provision of vacuum pumps which adds to the complexity and cost of the system, but low pressure conditions favor surface collisions making catalytic effects easier to notice, and they also minimize the deactivation of the excited states produced in the plasma through gas-phase collisions but at the price of reduced gas throughput. Another distinction concerns the degree of thermal equilibrium achieved in the plasma. In thermal plasma, all the degrees of freedom including the electrons, ions and neutral species are equilibrated and have the same temperature as the bulk gas (typically >1000 K). Nonthermal plasma has a high degree of non-equilibrium between the light electrons and the heavier particles such as the ions, radical and molecules. The disparity in mass between the electron and the gaseous species means that little kinetic energy is transferred between them and the heavy atomic and molecular species remain close to their ambient temperature. Thus, a nonthermal discharge can create excited and reactive species that can only be produced in an equilibrium system such as an arc or flame at very high temperatures. This means that we can have an ionized, excited and reactive gas interacting with the catalyst at temperatures at which conventional thermal catalysis would be inactive. Typically, much of the research in plasma catalysis has focused on the use of atmospheric pressure, nonthermal plasma because of the simplification in engineering systems without the need for vacuum systems and its operation at low temperatures that minimize corrosion and deterioration of the catalyst through sintering or coking.

It is unclear whether the discovery of the effect of combining a catalyst with plasma was a deliberate act or just a serendipitous observation. Kim [1] has given an extensive history of the development of plasma catalysis which pinpoints the first experiment where a catalyst was intentionally combined with an electrical discharge as being performed in 1921 by Ray and Anderegg [19]. They attempted to oxidize carbon monoxide to carbon dioxide by first using a silent discharge in an atmospheric pressure mixture of oxygen and carbon monoxide and then passing these gases over a silver catalyst. They noted that: "*It is obvious that part of the oxidation takes place in the ozonizer while the gases are under the influence of the silent discharge and that part takes place while the gases are under the influence of the catalyst*". The use of a catalyst downstream gave enhanced oxidation of the CO compared to experiments performed in 1879 by Berthelot [20] who used a similar configuration but without a catalyst and attributed the incomplete CO oxidation to decomposition of the CO_2 product back to CO [13].

Over the next three or four decades, reports were relatively infrequent and focused on systems of industrial interest where potential improvement in efficiency

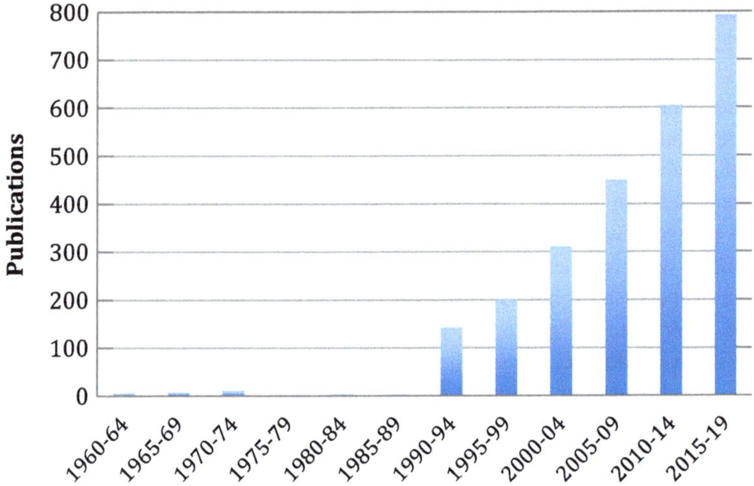

Fig. 1.1 The number of articles describing "plasma catalysis" published during different 5-year periods from 1960. Note that the period 2015–2019 only contains publication up to June 2018. (The data has been obtained by analysis using the Web of Science databases)

could be transformational. These included gas to liquid conversion of hydrocarbons, ammonia production and decomposition and the removal of VOCs. Experiments were performed both at atmospheric and reduced pressures where the latter has more potential for identifying any possible influence of the catalyst as the ratio of surface to gas-phase collisions increases as the gas pressure decreases. If we use the number of publications as an indicator of activity in the field of plasma catalysis, it would seem from Fig. 1.1 that research in the field took off in the 1990s, and, while still relatively modest, it is increasing rapidly.

1.2 The Plasma Catalysis Effect

There are two main configurations in which a catalyst can be combined with plasma. In one, the catalyst is place directly in contact with the plasma (variously called a one-stage arrangement or in-plasma catalysis, IPC) and, in the second, the catalyst is placed downstream of the discharge (a two-stage arrangement or post-plasma catalysis, PPC). These arrangements can be extended in a complex way by using a sequence of catalysts each with a different role. For example, the destruction of low concentrations toluene in air was achieved by initially treating the gas stream using a series of packed-bed discharges followed by two downstream catalysts in sequence [21]. These were MnO_2 followed by a MnO_2/CuO mixed catalyst. In the absence of the catalysts, the plasma only destroys a proportion of the toluene, but the addition of either of the catalysts results in complete removal. Oxidation of toluene gives CO_2, but CO is also produced due to incomplete oxidation. The MnO_2/CuO catalyst promotes the conversion of CO to CO_2. The MnO_2 catalyst is effective in totally removing any excess ozone that is produced in the discharge from the breakdown of air and not used in the catalytic oxidation of the toluene.

Compared to a thermal catalysis experiment where the energy needed to promote reaction comes from supplied heat, in plasma-activated catalysis the energy comes from the electrical discharge. This produces reactive and energetic species in the gas phase that may then come into contact with the surface of the catalyst. The primary species produced in plasma are generally short-lived and consist of electrons, photons, ions, excited-state atoms or molecules and radicals. In the one-stage plasma-catalyst arrangement, it will be these species that will interact with the catalyst. But in a two-stage plasma-catalyst configuration, it will only be the relatively long-lived species that exit from the plasma that are in contact with the downstream catalyst. These will be the end-products, by-products and long-lived reactive intermediates from the plasma processing (of which ozone is an important and common example, as noted above) and, possibly, vibrationally excited species. Vibrational energy can be a significant mode for efficiently activating surface adsorption and desorption processes.

The effectiveness of plasma catalysis processing can be assessed in several ways. Commonly, the products of the processing are determined by analysis of the downstream gas either in real time or by taking samples for remote analysis. A variety of spectroscopic and analytical techniques are employed such as Fourier-transform infrared spectroscopy (FTIR), gas chromatography (GC) and mass spectrometry. The formation of liquid products can also be monitored by downstream condensation and collection. To be useful, the analysis of the end products should be performed in a quantitative manner although the extent of removal of the input gas can be achieved by a relative measurement. Qualitative measurement of the minor end products or intermediates, which aids their identification, may be of value in determining the reaction mechanism. The primary quantities that are measured include the degree of conversion or destruction, *X*, of the input species, AB, usually expressed as a percentage:

$$X(\%) = \frac{\text{No. of moles of AB removed}}{\text{Total no. of moles of AB}} \times 100$$

Another important quantity is the selectivity, *S*, for the formation of a particular product, CD,

$$S(\%) = \frac{\text{No. of moles of CD formed}}{\text{No. of moles of AB removed}} \times 100$$

where appropriate allowance must be made for the stoichiometry of the reaction.

The prime consideration in the assessment of the effectiveness of the combination of plasma with catalyst is that there should be some improvement in either of these two parameters as a consequence of hybridizing the two techniques. One measure of this is to determine whether any synergy results from the combination. This can be evaluated using a *synergy factor* where the measured effect of the processing (e.g. the conversion or selectivity) using the combination of plasma and catalyst must be greater than the sum of the effect of processing using plasma alone and of catalyst alone under equivalent conditions [22]. The synergy factor will be greater than one when the combination is synergistic. This is illustrated in Fig. 1.2 from the work of Wang et al. [23] on the decomposition of NH_3 in a dielectric barrier discharge reactor packed with a range of metal catalysts.

Unfortunately, far too many publications mistakenly assume that the combination of plasma and catalysis will always be synergistic, but that is not necessarily true, and only the calculation of a synergy factor that is greater than unity will provide the evidence of synergy. The addition of the catalyst to a discharge may improve the conversion or product selectivity compared to using the discharge on its own, but that does not necessarily demonstrate a synergistic effect. Vandenbroucke et al. studied the plasma catalysis decomposition of the decomposition of trichloroethylene (TCE) and found that the synergy factor for the degree of dissociation of TCE ranges from 0.78 to 4.78 in a range of experiments with changing parameters demonstrating that synergy may only exist under certain experimental conditions in a particular plasma catalysis system [22].

Another very important criterion for judging the effectiveness of plasma catalysis is to consider the energy efficiency of the processing. This is of vital importance in objectively comparing one plasma catalysis process with another that may use a different catalyst or plasma configuration. It is also a way in which the technology of plasma catalysis can be compared with other competitive technologies and is a key way in which to judge the economics and hence the commercial viability of plasma catalysis as a processing technology for a particular application. Essentially, the energy efficiency tells us how much material we can convert or produce for a defined amount of energy. It can be expressed in a bewildering array of units, e.g. kg/J,

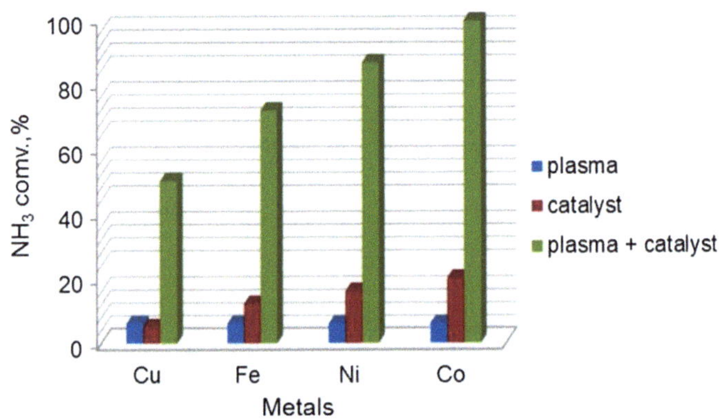

Fig. 1.2 Demonstration of the synergistic effect of plasma catalysis for the conversion of NH_3 at a specific energy density of 28 kJ/L using a dielectric barrier reactor packed with a range of supported metal catalysts at a temperature of 450 °C. Results are presented for plasma alone, catalyst alone and the plasma-catalyst combination. (Reproduced from Ref. [23] with permission. Copyright 2015, the American Chemical Society)

mol/J, molecule/eV, mol/kWh, etc., but they can be easily interconverted! The reciprocal of the energy efficiency is sometimes used and called the specific energy. Gutsol et al. have provided a very useful review with summaries of different ways of describing energy efficiency in plasma systems [24]. Alternatively, the energy efficiency can be expressed by comparing the actual energy required to create (or destroy) a given amount of material with the theoretical energy required for the same process. This theoretical energy for the formation or destruction process can be identified with the reaction enthalpy for that process if a clear reaction mechanism has been identified.

In considering the performance of a plasma catalysis system, a balance often has to be struck between different outcomes. For example, it is not always possible to maximize conversion or yield and energy efficiency simultaneously. It is commonly found that as the energy supplied to the system (commonly represented as the specific energy density, *SED*, or the specific energy input, *SEI*, which is the energy input per unit volume of the reactor) increases, the conversion increases, but the energy efficiency decreases. Figure 1.3 demonstrates this for the decomposition of CO_2 in a packed bed plasma reactor filled with $BaTiO_3$. It can be seen that the optimal energy efficiency is obtained at the lowest *SEI* value, while the magnitude of the dissociation increases with increasing energy input.

In Fig. 1.4, this effect is demonstrated by a plot that shows the optimum regions in terms of energy efficiency and the obtained decomposition of CO_2 for a range of different plasma sources with and without catalysts [26]. Such a plot can be used as an aid to evaluate which plasma or plasma-catalyst combination will be best to achieve a particular performance. This methodology has been extended to a wider range of plasma systems for the conversion of pure CO_2 and the dry reforming of CH_4 with methane CO_2 and demonstrates that certain plasma systems are more effective for certain processes. (See Figs. 24 and 32 in [8].)

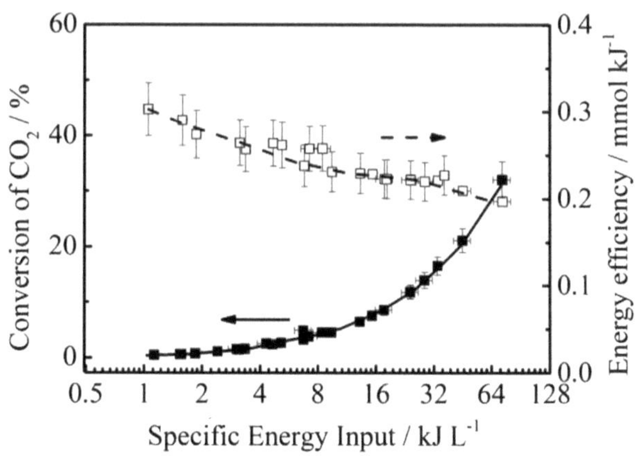

Fig. 1.3 A comparison of the variation of % CO_2 conversion and the energy efficiency as a function of the *SEI* for the treatment of an atmospheric pressure stream of pure CO_2 in a DBD reactor packed with $BaTiO_3$ beads. (Drawn using data taken from Ref. [25])

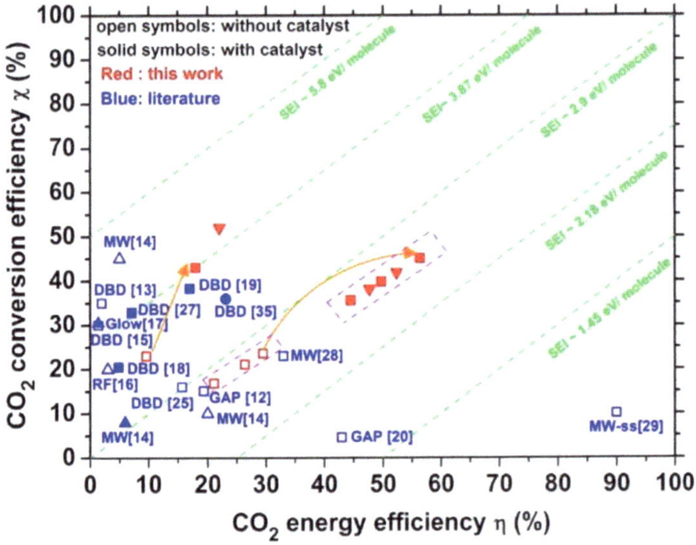

Fig. 1.4 Comparison of conversion and energy efficiency for CO_2 decomposition with different types of nonthermal plasmas. Square symbols represent pure CO_2; triangles, CO_2-Ar mixture; inverted triangle, CO_2-H_2O mixture; diamond, CO_2-H_2O-Ar mixture. MW microwave, RF radio frequency, GAP gliding arc plasmatron, MW-ss microwave (supersonic flow), DBD dielectric barrier discharge. The *SEI* is given for reference. (Reproduced from Ref. [26] with permission. Copyright 2017, Elsevier)

1.3 How Does Plasma Catalysis Work?

1.3.1 What Does the Plasma Do to the Catalyst?

It is not the aim of this section to give an answer to this question as that will be the focus of many of the chapters that follow. Instead, I will just illustrate the way in which researchers have attempted to rationalize the observation of a plasma catalysis effect. Essentially there are two basic ways in which plasma catalysis works as mentioned before. First, it could be that placing a catalyst either directly in a plasma or downstream from it changes the operation of the discharge in a physical or chemical way that enhances the processing. Alternatively, using plasma to activate a catalyst changes the behavior of the catalytic process in a beneficial way. Basic theories of catalysis tell us that there are several stages involved in a catalytic reaction [27]. Firstly, species in the gas phase must collide with the surface and become adsorbed. Following adsorption, they might migrate on the surface to a reactive site that might be located within a pore inside the catalyst. At this stage, surface reaction can take place between species adsorbed on the surface, which is called a Langmuir-Hinshelwood mechanism, or a species still in the gas phase may react with one that is adsorbed, which is designated an Eley-Rideal process. Then, the surface-bound reaction product must desorb from the surface and make its way back into the bulk gas.

In thermal catalysis, the surface is heated to supply energy to overcome any barriers to adsorption. Species can be strongly adsorbed onto the catalyst with bonds that are comparable to chemical called chemisorption, or the bonding can be weaker in magnitude comparable to physical forces such as van der Waals and that is called physisorption. The nature of binding to the catalyst depends on its chemical composition giving the active sites for binding and the nature of the adsorbed species. Adsorption may be of the species in the same form as it is when gaseous or it might involve the dissociation into more reactive fragments, e.g. the dissociation of molecular hydrogen into two hydrogen atoms when adsorbed on Ni, a process called dissociative chemisorption. Adsorption of species onto a surface can improve reactivity by increasing their interaction time with the plasma, and this goes someway to explaining why enhanced performance can be obtained in plasma catalysis systems when using materials such as alumina, Al_2O_3 or titania, TiO_2, that in conventional catalysis would be regarded as being inert support materials. It is important to realize that the rate of progress of a plasma catalysis process can be determined not only by the adsorption of the reagents, but also if a reaction product (or intermediate) is strongly adsorbed onto the surface and the reaction slows down as fewer surface sites become available for adsorption. The relative binding energies to the surface (enthalpies of adsorption) for the reagents, intermediates and products are key in determining the effectiveness of the catalytic process. As an example, using infrared in situ probing of the surface loading of a TiO_2 catalyst during plasma catalysis, Barakat et al. [28] have investigated the oxidation of isopropanol by ozone in a downstream configuration and find that the rate of formation of gaseous end-products is limited by the rate of oxidation of an acetone intermediate, which saturates the surface.

As we have previously mentioned, there is a rich wealth of species in the plasma that includes many species that are potentially reactive including ions, excited atomic and molecular species and molecular and atomic free radicals created from gas-phase dissociation that are all created in the discharge. Once created, these species can undergo gas-phase collisions that can create further unstable species by reaction or excitation. Alternatively, the collisions can bring about deactivation by quenching or by recombination. Some short-lived excited species will decay by emitting a photon, a process called fluorescence that in the case of electronically excited species yields ultraviolet or visible emission giving the characteristic color to the discharge as was observed by Hauksbee and Faraday in their nineteenth century experiments with gaseous discharges.

In determining how these discharge-produced species might interact with the catalyst, it is not just necessary to consider how energetic or reactive they might be, but we must also consider their lifetimes in the gas phase that result from the collisional processes of activation, reaction and deactivation, just described and also how far they can travel during this lifetime. In other words, do these plasma-created species make it to the catalyst? Typically at atmospheric pressure and room temperature, they travel less than 100 nm and the time between collisions is less than 1 ns. The lifetime of a ground state oxygen atom, $O(^3P)$, in an atmospheric pressure plasma in dry air is about 14 μs [29] and that of a ground state OH radical depends on the gas composition but is typically 100 μs in an air plasma although this reduces to about 20 μs when a potential reagent such as trichloroethylene is added [30]. This means that there will only be a narrow region immediately above the catalyst surface from which the short-lived plasma-created species can be adsorbed. Kim et al. [31] have calculated that for ground state oxygen atoms or OH radicals, this layer will have a thickness of ~50 μm. Outside of this layer, species produced by the plasma will react or be deactivated in the gas phase in the same way that they do in the absence of the catalyst. This suggests that to gain the maximum effect from plasma catalysis, the reactor must be designed in a way that produces the reactive species very close to the surface of the catalyst.

Many catalysts consist of a metal coated onto a support material, and the effectiveness of the catalyst depends not only on the identity of the metal but also on the fractional coverage of the support by the metal and the size of the deposited metal particles. In many experiments, it is found that using nanoscale particles increases their effectiveness. At a molecular level, the dynamics of the reactive processes involved in the plasma-catalytis processes are of many types. In Fig. 1.5, some possible processes for the formation of ammonia in hydrogen/nitrogen dielectric barrier plasma with a Ru metal catalyst deposited on an alumina support are illustrated [32]. These indicate the interplay between gas-phase excitation and dissociation processes and adsorption onto both the metal particle and the alumina support, spillover onto the alumina from the Ru and migration of species across the alumina surface, followed by surface reaction and desorption of products. Gaseous excited N_2 molecules created in the discharge adsorb dissociatively onto the alumina, while H_2 is either dissociated in the gas phase or on the Ru metal. Diffusion of N and H atoms from the metal across the surface of the alumina creates NH_3 in a stepwise manner by two different reactions where the H atoms created on the Ru metal react more quickly with the N atoms on the Al_2O_3 than do the H atoms adsorbed directly from the gas phase. This of course is a mechanism proposed to explain kinetic measurements without any direct identification of species on the catalyst surface.

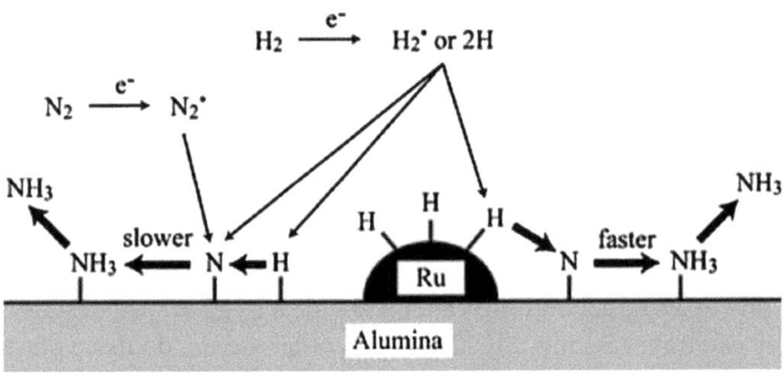

Fig. 1.5 Schematic diagram of the reaction pathways in the formation of NH_3 from a N_2/H_2 plasma on a Ru/Al_2O_3 catalyst. (Taken from Ref. [32] by permission. Copyright 2007, Springer)

Another important effect that the plasma may have on the catalyst is that of surface modification. Indeed, plasma is increasingly being used as a method of pre-treating catalysts for conventional thermal processing [33, 34]. For example, using a low pressure glow discharge or radio frequency plasma with Ar, He, N_2 or air as a source of energetic electrons, it is possible to reduce a metal in a more environmentally sustainable way than with conventional thermal or chemical processing [35, 36]. Plasma preparation can change the dispersion of metals on a catalyst, allowing for the treatment of sensitive low temperature materials and the production of novel classes of catalysts involving selective doping and the production of nanoparticles. It seems clear that if plasma can be used to prepare a catalyst in isolation under controlled conditions, then with the normal operating conditions of a one-stage plasma catalysis process, some of the plasma activity may result in a continual dynamic modification of the catalytic surface. Marinov et al. [37] have noted this for a SiO_2 surface under exposure to a low pressure nitrogen plasma commenting that the surface under contact with the plasma is not static and suggesting that this type of behavior is expected for a range of surfaces and plasma chemistries. This dynamic interaction between plasma and catalyst may give a unique benefit in terms of reduced poisoning or coking and increased stability and activity.

Finally, we should consider how the catalyst and its adsorbed species will behave when exposed to the electrons produced by the discharge. Electron-stimulated desorption is a well-known process in surface science. This can arise due to local heating of the surface by electron impact leading to thermal desorption. It is also possible that the addition of electrons can affect the bonding between a molecular adsorbate and the surface creating repulsive interactions leading to desorption. Dissociative electron attachment of a species on the surface giving an anion that then dissociates may be an important process when dealing with electronegative species. It has been shown [38] that electron attachment to a CO_2 molecule that is adsorbed onto an oxygen vacancy on a TiO_2 surface brings about dissociation of the CO_2^- anion to give CO and the filling of the oxygen vacancy by the O^- ion. The CO may then desorb migrate to another region of the surface. Mei et al. [39] have developed this mechanism to describe the dissociation of CO_2 on photocatalytic materials such as $BaTiO_3$ and TiO_2 where plasma-produced electrons of sufficient energy can create electron-hole pair states in an analogous method to the excitation of these materials by photons of the same energy in photocatalysis. This process is shown schematically in Fig. 1.6.

1.3.2 What Does the Catalyst Do to the Plasma?

In general terms, placing a catalytic material into plasma will affect the physical and electrical properties of the gas discharge that might thereby cause some modification to the outcomes of the plasma-catalytic processes. The addition of packing material might change the nature of the discharge from a filamentary form that propagates as

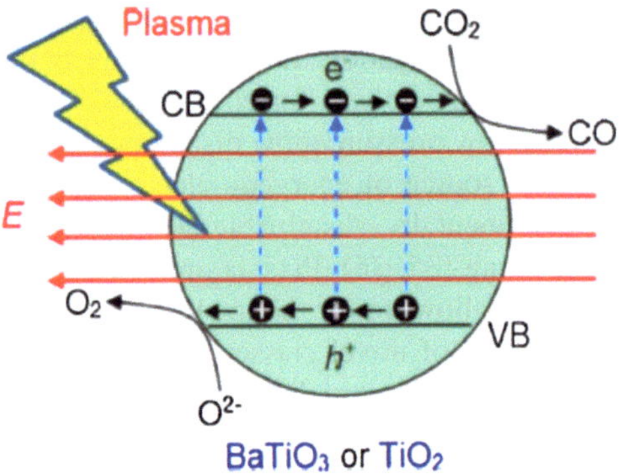

Fig. 1.6 A schematic mechanism for the plasma catalysis conversion of CO_2 on the surface of a photocatalyst. (Reproduced from Ref. [39] with permission. Copyright 2017, Elsevier)

microdischarges or streamers through the gas between the electrodes to that of a surface discharge that moves on the surface of the material. Such surface discharges on a catalytic surface have recently been imaged for the first time [40]. This change in discharge type may also be associated with an alteration of the electron energy distribution that may result in changes to the yields and identities of the species formed in the discharge. Neyts and Bal [41] has recently suggested that the addition of a catalytic material to a discharge can introduce an additional electric field influencing the retention time of the species at the catalyst surface and altering the charge distribution at the surface increasing reactivity. The efficiency of the processing will also depend on the structural features of the catalyst such as its morphology, porosity and chemical activity as these may interact with the nature of the plasma. The use of dielectric materials as packing can have a profound effect on the electrical properties of the discharge by changing the capacitance of the reactor. This can influence the breakdown voltage and the charge that can be transferred between the electrodes. In general, more electrical energy can be deposited into a discharge when materials of higher dielectric constant are used resulting in the formation of more electrons of higher energy giving increased yield of ionized and excited species. However, the overall effect of using high dielectric materials on the efficiencies and selectivities for the plasma catalysis processing is by no means clear and is found to vary between systems suggesting that there may be a subtle interplay between several effects that is not completely understood.

Interesting observations have been made where a dielectric barrier discharge reactor is filled with spherical beads of a high dielectric material such as $BaTiO_3$ or ZrO_2. Such materials are often ferroelectric, and polarization effects can cause enhanced electric fields at the points of contact between the beads giving higher concentrations of excited species. The discharge transitions from streamers propagating in the gas when there is no packing to a mixture of gaseous and surface discharge when there is packing present. Increased conversion of CO_2 is observed [25, 42–44] in these arrangements, and simulations [45, 46] confirm that there are enhanced electric fields where the beads touch and the electron density is higher in these regions. This causes stronger and faster development of the plasma. The bead size is also important in affecting the conversion of the CO_2 and increases as the bead size decreases as long as it is possible to sustain a discharge as the breakdown voltage for plasma action increases strongly as the size decreases [44]. Michielsen et al. [43] suggest that all these observations have wider implications for plasma catalysis as the conversion obtained is not just a function of the active sites on the catalyst which is deposited on a support material but also on how that support materials is packed into the reactor. There is a positive contribution of the packing on the conversion obtained, due to electric field enhancement at the contact points, but a negative contribution due to the lower residence time of the gases because of the volume reduction at a given flow rate and a positive or negative effect depending on the nature of material inserted into the voids. This conclusion that the important role played by how a reactor is packed with catalyst is supported by observations that using a dielectric barrier discharge (DBD) reactor that is only partially filled with a catalyst doubles the CH_4 conversion and H_2 yield during the dry reforming of CH_4 with CO_2 compared to a filled reactor [47].

The nature of the catalyst surface can have an effect on the electric field in the discharge, and irregularities in the surface such as roughness or the presence of pores can create local variations in the electric field and local regions of field intensification that can become a source of high-energy electrons. The islands of nanosized dispersed metals on a support will also encourage enhanced electrical fields as will edges, steps or other irregularities in the crystalline structure of the catalyst. Pores can selectively adsorb molecules depending on their size increasing their reactivity, and penetration of plasma into the pores may specifically excite such species. Experimental studies on the removal of pollutants from air [48] show that for nanoporous materials (pore size <0.8 μm), the electrical discharge does not penetrate into the pores, but for mesoporous materials (≥ 15 μm), it is possible for microdischarges to penetrate into the pores and that excited species can be stabilized in the pores giving enhanced reaction times. This is supported by modeling studies [49] that suggest that plasma species can be created within pores on the μm scale which may interact with the catalyst surface and affect the plasma catalytic process. Kim [31] has demonstrated that estimates of the rates of diffusion indicate that penetration into pores of nanometer dimension (<2 nm) is unfeasibly slow but that on the mesoscale (2–50 nm) it could be possible.

1.4 How Can We Learn More About Plasma Catalysis?

Having given an introduction and brief background to our current progress and understanding in the field of plasma catalysis, it is appropriate to outline the directions in which research might be focused in order to increase our understanding. The following chapters of this book contain detailed accounts, expertly written, on many of these areas. Much of our understanding so far has come from a very wide-ranging survey of many types of plasma-catalyst experiments applied to a wide range of systems involved in the fields of pollution control, energy and environmental aspects of reuse and conversion of materials and the production of high-value chemicals from different feedstocks. While many of these experiments are largely phenomenological in type, there are many examples of systematic studies where the grossly complicated, interacting, multiparameter monster that is plasma catalysis has been tamed to a degree by isolating different aspects and divining what the key parameters are. In this way, much progress has been made in understanding different aspects of the process including the relative contributions of chemical, physical and engineering principles. We now understand much about the physics of gas discharges and how this impinges upon the nature of the electrical inputs in terms of how the electrical power can be supplied in terms of the spectrum used (from microwave, radiofrequency, ac, pulsed to dc) and how these combine with the construction to determine the electrical field strengths and how the plasma will develop and propagate. We know much about how a wide range of ionizations, excitations and fragmentations results from collisions of the electrons produced in the discharge with the gaseous species and can calculate densities under certain conditions. We can also make deductions about the kinetics and reaction mechanisms by comparing the results of modeling with experiment particularly in terms of being able to predict the yields of reaction products and how these change with processing parameters. To a limited degree, we can model using realistic 2- and 3-D representations of our reactors and take account of mass and heat transport in the system. As we explore more, we become aware of areas where we do not have detailed experimental observations of key phenomena nor adequate models to describe our results. Most of our experimental observations are on steady state systems or involve processes with relatively long time constants. We know very much more about the outcomes of the processing than we do about the intermediate stages in the process. We can identify the major end products but know nothing about the intermediates other than the prediction of computation.

Kim [1] has recently discussed the timescales over which some of the different aspects of plasma catalysis take place and has presented an excellent summary in pictorial form displayed here in Fig. 1.7.

It can be seen that the different processes cover timescales ranging from nanosecond to second. Herein lie many of difficulties and limitations that prevent a complete description and experimental study of plasma processes. We need to cover this range, but no single technique will allow this, and different techniques are needed for different timescales. In the plasma, the electron-induced processes that create the reactive species take place on the nanosecond scale whereas the subsequent chemical reactions in the gas phase and processes on the surface are considerably slower (microsecond to millisecond and longer) and compete with deactivation processes that may decrease yields by removing reagents and encouraging back-reactions converting products back to reagents and reduce selectivities by allowing for the formation of unwanted by-products.

Another aspect of our lack of understanding is also apparent from the diagram that concerns heating, heat flow, thermodynamics and temperature. In using nonthermal plasma, we operate in a regime of nonthermal equilibrium where the use of a single temperature to describe different properties of the system is not valid. The electrons will have a very large mean energy (1–10 eV, typically several thousand degrees

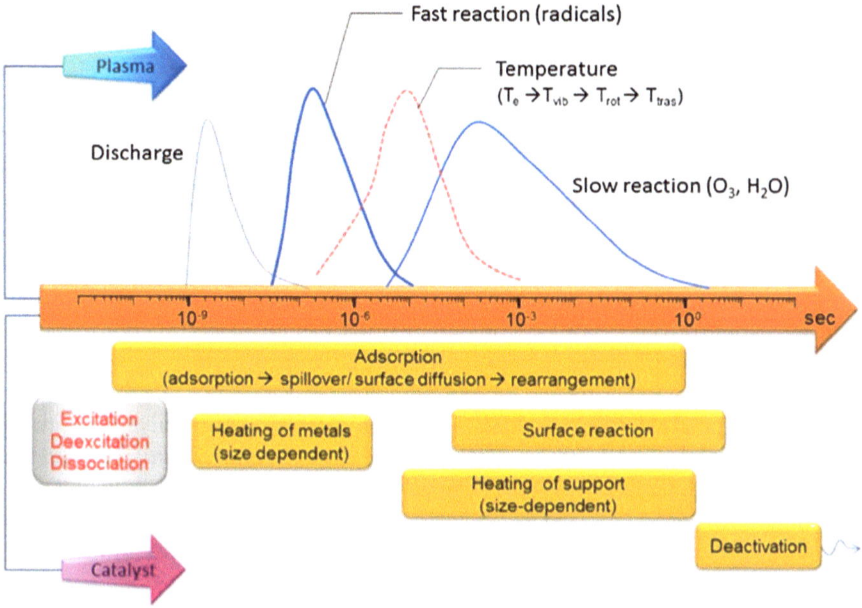

Fig. 1.7 Characteristic timescales in plasma and catalytic processes. (Taken from Kim et al. [1], by permission. Copyright 2016, Springer)

Kelvin, if they can be described by Boltzmann statistics). In the gas, there will be a disequilibrium between the different degrees of freedom (electronic, translation and vibrational and rotational for a molecular species) which means that there is no single temperature that can describe the gas and in general $T_{\text{vibrational}} > T_{\text{rotational}} \approx T_{\text{translational}}$. This trend reflects the relative rates of relaxation, which depend on the identity of the gas and the pressure. The gas temperature in a plasma experiment is often equated to the rotational temperature of some gas, e.g. N_2 which is determined spectroscopically by Raman or optical emission spectroscopy. The retention of vibrational excitation by some molecules or radicals is recognized as being of particular importance in plasma catalysis systems because such species may be easier to dissociate and also may adsorb more easily on a surface. It was recognized at an early stage of the development of plasma catalysis that vibrational excitation could provide enhanced adsorption onto catalyst surfaces and increase the reactivity for endothermic processes such as the production of NO from N_2 and O_2 on WO_3 and MoO_3 catalysts [50]. It has also been shown that vibrationally excited CH_4 chemisorbs with more than three orders of magnitude greater efficiency on a Ni surface than it does when in its ground vibrational state [51]. The extent of electronic excitation depends on the radiative lifetime, τ_0, of the state and the appropriate quenching rates. Species commonly encountered in plasma include atomic and molecular states of oxygen ($O(^1D)$, $E = 1.967$ eV, $\tau_0 = 150$ s: O_2 ($a^1\Delta$), $E = 0.977$ eV, $\tau_0 = 64.6$ min) and nitrogen ($N(^2D)$, $E = 2.38$ eV, $\tau_0 \sim 17$ h: $N(^2P)$, $E = 3.576$ eV, $\tau_0 \sim 12$ s: N_2 ($A\ ^3\Sigma_u$), $E = 6.224$ eV, $\tau_0 = 2.0$ s). The high reactivity of $O(^1D)$, for example, reduces its lifetime to ~1 ns in a humid atmospheric pressure, air plasma due to quenching with air molecules and reaction with water to form the radical OH. The high-energy metastable nitrogen state, N_2 ($A\ ^3\Sigma_u$), plays an important role in plasma chemistry as it can collisionally dissociate many molecules [7].

In describing the operation of the catalyst during plasma catalysis, the concept of a single temperature is also problematic. In the experiment, the plasma may be externally heated but a more common configuration is to just place the catalyst into the discharge. It would however be a mistake to assume that the temperature of the catalyst surface will be the same as that of the surrounding gas. The impact of the plasma on the catalyst may bring about some heating due to electron or dielectric heating and the effects of the exo- or endothermicity of the reactions taking place on the surface that will either add or remove heat from the catalyst, respectively. Thermal imaging experiments by Nozaki and Okasaki [52] clearly show the non-uniformity of temperature along the surface of a Ni/Al_2O_3 catalyst mounted in a dielectric barrier discharge reactor during the exothermic steam reforming of methane with the temperature increasing along the direction of the gas flow. Correspondingly, a temperature decrease in the catalyst bed can be seen during the endothermic dry reforming reaction of CH_4 with CO_2 [53]. Further difficulties in using a single temperature to describe the catalyst come from the enhanced electric fields associated with surface irregularities or the points of contact formed between the catalytic particles as discussed above. These can give rise to local hotspots where a disproportionally high amount of the plasma catalysis processing may take place [9]. A final comment relates to the timescales noted in Fig. 1.7 relating this time to the timescale for heat transfer where the catalyst may be in the form of nanoscale metal particulates on a support material such as alumina. Because the metal particles (10 nm) are much smaller than the size of support particle (2 mm), as energy is supplied to the catalyst by the plasma, the metal will be heated at a much more rapid rate than the support giving heating times of 10^{-12} and 5 s, respectively. Thus, the metal particles will heat and cool much more rapidly than the support.

All of this goes to show that believing that the plasma catalysis process takes place at a single temperature could be a very big mistake as there are multiple temperatures needed to describe the species in the gas phase and on the catalyst that may vary with position within the reactor and also with time. In essence, there is a high degree of inhomogeneity with respect to time and position for many of the properties association with the plasma catalysis process which include the concentrations of the different species (including the electrons), the electron energies, the electric fields and the temperatures.

References

1. Kim, H. H., Teramoto, Y., Ogata, A., Takagi, H., & Nanba, T. (2016). Plasma catalysis for environmental treatment and energy applications. *Plasma Chemistry and Plasma Processing, 36*, 45–72.

2. Patil, B. S., Wang, Q., Hessel,V., & Lang, J. (2016). Plasma-assisted nitrogen fixation reactions. In G. Stefanidis & A. Stankiewicz (Eds.), *Alternative energy sources for green chemistry* (pp. 296–338). Cambridge, UK: Royal Society of Chemistry.
3. Yu, Q., Wang, H., Liu, T., Xiao, L., Jiang, X., & Zheng, X. (2012). High-efficiency removal of NO_x using a combined adsorption-discharge plasma catalytic process. *Environmental Science & Technology, 46*, 2337–2344.
4. Tao, X., Bai, M., Li, X., Long, H., Shang, S., Yin, Y., & Dai, X. (2011). CH_4-CO_2 reforming by plasma – challenges and opportunities. *Progress in Energy and Combustion Science, 37*, 113–124.
5. Whitehead, J. C. (2010). Plasma catalysis: A solution for environmental problems. *Pure and Applied Chemistry, 82*, 1329–1336.
6. Whitehead, J. C. (2014). Plasma catalysis for volatile organic compounds abatement. In *Handbook of advanced methods and processes in oxidation catalysis* (pp. 155–172). London: Imperial College Press.
7. Whitehead, J. C. (2016). Plasma-catalysis: The known knowns, the known unknowns and the unknown unknowns. *Journal of Physics D-Applied Physics, 49*, 243001.
8. Snoeckx, R., & Bogaerts, A. (2017). Plasma technology – A novel solution for CO_2 conversion? *Chemical Society Reviews, 46*, 5805–5863.
9. Neyts, E. C. (2016). Plasma-surface interactions in plasma catalysis. *Plasma Chemistry and Plasma Processing, 36*, 185–212.
10. Davy, H. (1817). Some new experiments and observations on the combustion of gaseous mixtures, with an account of a method of preserving a continuous light in mixtures of inflammable gases and air without flame. *Philosophical Transactions of the Royal Society A, 107*, 77–85.
11. Thomas, J. M. (2015). Sir Humphry Davy and the coal miners of the world: A commentary on Davy (1816) 'An account of an invention for giving light in explosive mixtures of fire-damp in coal mines'. *Philosophical Transactions – Royal Society. Mathematical, Physical and Engineering Sciences, 373*, 20140288.
12. Berzelius, J. J. (1836). Considerations respecting a New Power which acts in the Formation of Organic Bodies. *Edinburgh New Philosophical Journal, 21*, 223.
13. Ostwald, W. (1894). *Zeitschrift für Physikalische Chemie, 15*, 705–706.
14. Hauksbee, F. (1709). *Physico-mechanical experiments on various subjects; containing an account of several surprizing phenomena touching light and electricity, producible on the attrition of bodies, with many other remarkable appearances... together with the explanations of all the machines... and other apparatus us'd in making the experiments.* London: For the author; pr. R. Brugis.
15. Maia, E., Serra, I., & Peres, M. (2010). The gas discharges in history and teaching of physics and chemistry. *Travaux de laboratoire institut Rocha Cabral Lisbonne, 100*, 1–14.
16. Faraday, M. (1838). Experimental researches in electricity. –Thirteenth series. *Philosophical Transactions. Royal Society of London, 128*, 125–168.
17. Faraday, M. (1834). On the power of metals and other solids to induce the combination of gaseous bodies. *Philosophical Transactions of the Royal Society A, 124*, 55–76.
18. Robertson, A. J. B. (1983). The development of ideas on heterogeneous catalysis: Progress from Davy to Langmuir. *Platinum Metals Review, 27*, 31–39.
19. Ray, A. B., & Anderegg, F. O. (1921). The oxidation of carbon monoxide by passage with oxygen or air through the silent discharge and over ozone decomposing catalysts. *Journal of the American Chemical Society, 43*, 967–978.
20. Berthelot, P. E. M. (1879). Recherches sur l'ozone et sur l'effuse électrique. *Comptes rendus de l'Académie des Sciences, 88*, 50–52.
21. Harling, A. M., Glover, D. J., Whitehead, J. C., & Zhang, K. (2009). The role of ozone in the plasma-catalytic destruction of environmental pollutants. *Applied Catalysis B-Environmental, 90*, 157–161.
22. Vandenbroucke, A. M., Mora, M., Jimenez-Sanchidrian, C., & Romero-Salguero, F. J. (2014). TCE abatement with a plasma-catalytic combined system using MnO_2 as catalyst. *Applied Catalysis B-Environmental, 156*, 94–100.

23. Wang, L., Yi, Y., Zhao, Y., Zhang, R., Zhang, J., & Guo, H. (2015). NH_3 decomposition for H_2 generation: Effects of cheap metals and supports on plasma-catalyst synergy. *ACS Catalysis, 5*, 4167–4174.
24. Gutsol, A., Rabinovich, A., & Fridman, A. (2011). Combustion-assisted plasma in fuel conversion. *Journal of Physics D-Applied Physics, 44*, 274001.
25. Xu, S., Whitehead, J. C., & Martin, P. A. (2017). CO_2 conversion in a non-thermal, barium titanate packed bed plasma reactor: The effect of dilution by Ar and N_2. *Chemical Engineering Journal (Amsterdam, Netherlands), 327*, 764–773.
26. Chen, G., Godfroid, T., Britun, N., Georgieva, V., Delplancke-Ogletree, M., & Snyders, R. (2017). Plasma-catalytic conversion of CO_2 and CO_2/H_2O in a surface-wave sustained microwave discharge. *Applied Catalysis B-Environmental, 214*, 114–125.
27. Froment, G. F., Bischoff, K. B., & De Wilde, J. (1990). *Chemical reactor analysis and design* (Vol. 2). New York: Wiley.
28. Barakat, C., Gravejat, P., Guaitella, O., Thevenet, F., & Rousseau, A. (2014). Oxidation of isopropanol and acetone adsorbed on TiO_2 under plasma generated ozone flow: Gas phase and adsorbed species monitoring. *Applied Catalysis B-Environmental, 147*, 302–313.
29. Holzer, F., Roland, U., & Kopinke, F. D. (2002). Combination of non-thermal plasma and heterogeneous catalysis for oxidation of volatile organic compounds part 1. Accessibility of the intra-particle volume. *Applied Catalysis B-Environmental, 38*, 163–181.
30. Hibert, C., Gaurand, I., Motret, O., & Pouvesle, J. M. (1999). OH(X) measurements by resonant absorption spectroscopy in a pulsed dielectric barrier discharge. *Journal of Applied Physics, 85*, 7070–7075.
31. Kim, H. H., Teramoto, Y., Negishi, N., & Ogata, A. (2015). A multidisciplinary approach to understand the interactions of nonthermal plasma and catalyst: A review. *Catalysis Today, 256*, 13–22.
32. Mizushima, T., Matsumoto, K., Ohkita, H., & Kakuta, N. (2007). Catalytic effects of metal-loaded membrane-like alumina tubes on ammonia synthesis in atmospheric pressure plasma by dielectric barrier discharge. *Plasma Chemistry and Plasma Processing, 27*, 1–11.
33. Glonek, K., Wroblewska, A., Makuch, E., Ulejczyk, B., Krawczyk, K., Wrobel, R. J., Koren, Z. C, & Michalkiewicz, B. (2017). Oxidation of limonene using activated carbon modified in dielectric barrier discharge plasma. *Applied Surface Science, 420*, 873–881.
34. Liu, L., Zheng, C., Wu, S., Gao, X., Ni, M., & Cen, K. (2017). Manganese-cerium oxide catalysts prepared by non-thermal plasma for NO oxidation: Effect of O_2 in discharge atmosphere. *Applied Surface Science, 416*, 78–85.
35. Liu, C., Li, M., Wang, J., Zhou, X., Guo, Q., Yan, J., & Li, Y. (2016). Plasma methods for preparing green catalysts: Current status and perspective. *Chinese Journal of Catalysis, 37*, 340–348.
36. Wang, W., Wang, Z., Wang, J., Zhong, C., & Chang, J. (2017). Highly active and stable Pt-Pd alloy catalysts synthesized by room-temperature electron reduction for oxygen reduction reaction. *Advanced Science, 4*, 1600486.
37. Marinov, D., Guaitella, O., de los Arcos, T., von Keudell, A., & Rousseau, A. (2014). Adsorption and reactivity of nitrogen atoms on silica surface under plasma exposure. *Journal of Physics D-Applied Physics, 47*, 475204.
38. Lee, J., Sorescu, D. C., & Deng, X. (2011). Electron-induced dissociation of CO_2 on $TiO_2(110)$. *Journal of the American Chemical Society, 133*, 10066–10069.
39. Mei, D., Zhu, X., Wu, C., Ashford, B., Williams, P. T., & Tu, X. (2016). Plasma-photocatalytic conversion of CO_2 at low temperatures: Understanding the synergistic effect of plasma-catalysis. *Applied Catalysis B-Environmental, 182*, 525–532.
40. Kim, H.-H., Teramoto, Y., & Ogata, A. (2016). Time-resolved imaging of positive pulsed corona-induced surface streamers on TiO_2 and gamma-Al_2O_3-supported Ag catalysts. *Journal of Physics D-Applied Physics, 49*, 415204.
41. Neyts, E. C., & Bal, K. M. (2017). Effect of electric fields on plasma catalytic hydrocarbon oxidation from atomistic simulations. *Plasma Processes and Polymers, 14*, 1600158.

42. Mei, D., Zhu, X., He, Y., Yan, J. D, & Tu, X. (2015) Plasma-assisted conversion of CO_2 in a dielectric barrier discharge reactor: Understanding the effect of packing materials. *Plasma Sources Science & Technology, 24*, 015011.
43. Michielsen, I., Uytdenhouwen, Y., Pype, J., Michielsen, B., Mertens, J., Reniers, F., Meynen, V., & Bogaerts, A. (2017). CO_2 dissociation in a packed bed DBD reactor: First steps towards a better understanding of plasma catalysis. *Chemical Engineering Journal, 326*, 477–488.
44. Butterworth, T., Elder, R., & Allen, R. (2016). Effects of particle size on CO_2 reduction and discharge characteristics in a packed bed plasma reactor. *Chemical Engineering Journal, 293*, 55–67.
45. Zhang, Y., Wang, H., Jiang, W., & Bogaerts, A. (2015). Two-dimensional particle-in cell/ Monte Carlo simulations of a packed-bed dielectric barrier discharge in air at atmospheric pressure. *New Journal of Physics, 17*, 083056.
46. Koen Van, L., & Annemie, B. (2016). Fluid modelling of a packed bed dielectric barrier discharge plasma reactor. *Plasma Sources Science and Technology, 25*, 015002.
47. Tu, X., & Whitehead, J. C. (2012). Plasma-catalytic dry reforming of methane in an atmospheric dielectric barrier discharge: Understanding the synergistic effect at low temperature. *Applied Catalysis B-Environmental, 125*, 439–448.
48. Roland, U., Holzer, F., & Kopinke, F. D. (2002). Improved oxidation of air pollutants in a non-thermal plasma. *Catalysis Today, 73*, 315–323.
49. Zhang, Y., Van Laer, K., Neyts, E. C., Bogaerts, A. (2016). Can plasma be formed in catalyst pores? A modeling investigation. *Applied Catalysis B-Environmental, 185*, 56–67.
50. Gicquel, C., Cavadias, S., & Amouroux, J. (1986). Heterogeneous catalysis in low-pressure plasmas. *Journal of Physics D: Applied Physics, 19*, 2013–2042.
51. Juurlink, L. B. F., McCabe, P. R., Smith, R. R., DiCologero, C. L., & Utz, A. L. (1999). Eigenstate-resolved studies of gas-surface reactivity: CH_4 (ν_3) dissociation on Ni(100). *Physical Review Letters, 83*, 868–871.
52. Nozaki, T., & Okazaki, K. (2013). Non-thermal plasma catalysis of methane: Principles, energy efficiency, and applications. *Catalysis Today, 211*, 29–38.
53. Kameshima, S., Tamura, K., Mizukami, R., Yamazaki, T., & Nozaki, T. (2017). Parametric analysis of plasma-assisted pulsed dry methane reforming over Ni/Al_2O_3 catalyst. *Plasma Processes and Polymers, 14*, 1600096.

Chapter 2
Plasma Catalysis Systems

Akira Mizuno and Michael Craven

2.1 Introduction

Plasmas constitute the forth state of matter after solids, liquids, and gases. A plasma is an ionized gas composed of ions and electrons that is electrically conductive and highly responsive to electromagnetic fields. A plasma may be formed by heating a gas to exceedingly high temperatures, or by exposing a gas to a large potential difference between two electrodes, that can separate electrons from the gas molecules to form ions. The number of positively charged and negatively charged particles present in a plasma is equal, meaning plasmas exhibit quasi-neutrality. A plasma can be completely ionized, whereby practically all the neutral gas species have been converted to ions (the degree of ionization, or density ratio of the major charged particles to neutral gas species, is close to unity), or weakly ionized, where the degree of gas ionization is between 10^{-7} and 10^{-4} and a mixture of neutral atoms and molecules, ions, electrons, excited species, and radicals are formed in the plasma. Fully ionized plasmas are present at very high temperatures in thermonuclear systems involving nuclear fusion, such as those in the core of the sun, in stars and in solar corona, whereas weakly ionizing plasmas are formed at much lower temperatures. Due to their low gas temperatures and the fact that they can generate a variety of chemically active species (e.g., ions, radicals, and excited species), weakly ionizing plasmas can be used for plasma chemistry.

Weakly ionizing plasmas can be divided into two categories: thermal plasmas and nonthermal plasmas (NTP), that are defined by their electron temperatures. In thermal plasmas, the bulk gas molecules and electrons are in thermal equilibrium

A. Mizuno (✉)
Department of Environmental and Life Sciences, Graduate School of Engineering, Toyohashi University of Technology, Toyohashi, Aichi, Japan
e-mail: mizuno@ens.tut.ac.jp

M. Craven (✉)
Department of Electrical Engineering and Electronics, University of Liverpool, Liverpool, UK
e-mail: m.j.craven@liverpool.ac.uk

X. Tu et al. (eds.), *Plasma Catalysis*, Springer Series on Atomic, Optical, and Plasma Physics 106, https://doi.org/10.1007/978-3-030-05189-1_2

at temperatures in the range of 11,600 to 23,200 K (1 to 2 eV). In nonthermal plasmas, the temperature of the electrons is significantly higher (11,600 to 116,00K) than that of the bulk gas molecules (~273 K), and so they are said to be "non-equilibrium" or "cold" plasmas. Electrons at this temperature have sufficient energy (1–10 eV) to break chemical bonds and to excite both atomic and molecular species. Compared to thermal plasmas, NTPs can be generated using relatively low input energy and they can produce high concentrations of chemically active species at temperatures that do not alter the thermodynamic equilibrium of a reaction, which is why they have found application as promoters for a variety of different chemical reactions. The performance of NTP-promoted chemical reactions can be enhanced by combining the plasmas with catalysts, which results from synergetic effects from interactions between the catalyst and the plasma. Although these effects are not completely understood, they have been exploited by combining different NTPs and catalysts using a plethora of reactor configurations for an even greater number of reactions.

In this chapter, we will outline some of the underlying principles of plasma chemistry, introduce some of the most important NTPs used in plasma-chemical and plasma-catalytic reactions, and explore some of the different reactor configurations used for plasma-catalytic reactions.

2.2 Chemical Reactions Promoted by Nonthermal Plasmas

NTPs generate high-energy electrons (between 1 and 10 eV), while the energy levels of ions and neutrals remains low [1–4]. This energy difference results from the different size of the particles and the relative distances that they travel between the electrodes in a plasma reactor. The electrons have a much smaller mass than the ions and so the distance they travel is much larger, relatively, than that travelled by the ions; therefore, when a voltage is applied (particularly when using short pulses), the electrons are driven across a larger voltage drop, so most of the input electrical energy is used to energize the electrons. These electrons can excite gaseous molecules and atoms which can enhance the speed of reactions without increasing the gas temperature, and so heat loss due to excitation of ions and neutral particles can be reduced.

For a chemical reaction to proceed, activation energy must be provided to the system. This energy excites the reactant molecules and provides them with enough energy to effectively interact with one another, where they enter a transitional state and a thermodynamic equilibrium is established, as shown in Eq. 2.1. A change in temperature induces a change in the position of the equilibrium, in accordance with Le Chatelier's principle. For exothermic reactions, more product, *C*, is produced at lower temperatures and heat is released (ΔH <0).

$$\mathrm{A} + \mathrm{B} \rightleftarrows C + \Delta H \ (\Delta H : \text{enthalpy, negative for exothermic reactions}) \qquad (2.1)$$

In thermochemical reactions, the activation energy is typically provided to the reaction by heating the mixture, but the increase in temperature favors the endothermic reaction ($\Delta H > 0$) and the equilibrium shifts to produce more of the reactants, A and B. NTPs can provide the activation energy through highly energetic electrons without significantly increasing the gas temperature, thus favoring the exothermic reaction and production of the product, *C*. This is an important advantage of NTPs to promote chemical reactions [5, 6]. For endothermic reactions, such as dry reforming of CO_2 and the dissociation of stable molecules, vibrational excitation has been found to play an important role in plasma-promoted reactions [7]. Gliding arc and microwave plasmas, having higher ionizing rates and higher transient temperatures, are more suitable to improve the conversion efficiency of these stable molecules. These plasmas are sometimes known as "warm plasma."

Collisions of the high-energy electrons with gas molecules in plasmas generate a plethora of active species that can take part in chemical reactions. These species are excited molecules and atoms, ions, radicals, and even new stable molecules. A summary of some of the main chemical reactions that can occur in plasmas are listed in Table 2.1.

Radicals generated by NTPs are very reactive at low temperatures [8–10] and are useful for various environmental applications such as decomposition of volatile

Table 2.1 The main chemical reactions that occur in plasmas

Electron/molecular reactions	
Excitation	$e^- + A_2 \rightarrow A_2^* + e^-$
Dissociation	$e^- + A_2 \rightarrow 2A + e^-$
Attachment	$e^- + A_2 \rightarrow A_2^-$
Dissociative attachment	$e^- + A_2 \rightarrow A^- + A$
Ionization	$e^- + A_2 \rightarrow A_2^+ + 2e^-$
Dissociative ionization	$e^- + A_2 \rightarrow A^+ + A + e^-$
Recombination	$e^- + A_2^+ \rightarrow A_2$
Detachment	$e^- + A_2^- \rightarrow A_2 + 2e^-$
Atomic/molecular reactions	
Penning dissociation	$M + A_2 \rightarrow 2A + M$
Penning ionization	$M^* + A_2 \rightarrow A_2^+ + M + e^-$
Charge transfer	$A^{\pm} + B \rightarrow B^{\pm} + A$
Ion recombination	$A^- + B^+ \rightarrow AB$
Neutral recombination	$A + B + M \rightarrow AB + M$
Decomposition	
Electronic	$e^- + AB \rightarrow A + B + e^-$
Atomic	$A^* + B_2 \rightarrow AB + B$
Synthesis	
Electronic	$e^- + A \rightarrow A^* + e^-$, $A^* + B \rightarrow AB$
Atomic	$A + B \rightarrow AB$

Here, A and B are atoms, A_2 and B_2 are molecules, *e* is an electron, M is a temporary collision partner, + and – indicate ions with their respective charge symbols, and $*$ indicates an excited or radical species [2]

organic compounds (VOCs) and odors [11–13], reducing gaseous pollutants such as diesel emissions [4, 14–16], enhancement of combustion [17, 18], sterilization, and treatment of drinking and waste water. Conversion of CO_2 and CH_4 to syngas or liquid fuel [19–23] and production of hydrogen carriers (e.g., H_2 and NH_3) by NTP [24, 25] are also important technological challenges for the development of renewable energy storage for a greener future without fossil fuels.

2.3 Generation of Nonthermal Plasma in Atmospheric Pressure

The simplest way of creating a nonthermal plasma is to apply an external electric field between two electrodes positioned inside a volume of gas. The minimum voltage required to dissociate a gas into electrons and ions to form a plasma discharge is called the breakdown voltage (V_b). The breakdown voltage is described by Paschen's Law (Eq. 2.2) and is controlled by the distance between the electrodes (d) and the gas pressure (p), where a is the saturation ionization in the gas at a particular E/p (electric field/pressure) and b is related to the excitation and ionization energies (i.e., a and b are both constants that depend on the type of gas used for the plasma) [26].

$$V_b = \frac{a(pd)}{ln(pd) + b} \tag{2.2}$$

NTPs typically operate at low pressures (1–10^3 Pa) or atmospheric pressure and are comprised of "cold," nonthermal discharges, such as avalanches, Townsend, streamer, and glow discharges. Reactors that generate NTPs are typically designed to avoid the formation of arc and spark discharges which lead to plasma thermalization.

Some of the most commonly used NTPs in plasma catalysis are discussed in this section.

2.3.1 Corona Discharge and Pulsed Streamer Corona

In a non-uniform electrode system, such as needle-to-plate or wire-to-cylinder electrodes, streamer microdischarges are produced from a sharp electrode. The streamers expand from the tip of the electrode into space where the electric field decreases and can eventually break down. This is called a corona discharge, and the area in which the discharge occurs is called the corona region. The name "corona" is derived from the crown-like appearance of the many streamers initiated from the sharp electrode. In the narrow channels produced by the streamers, energetic electrons dissociate the surrounding gas molecules, resulting in ionization [27]. The corona can be sustained, and the electrical breakdown of the discharges avoided, by using short, ramped voltage pulses. The voltage in the corona region is dependent on the distance between the electrodes, the duration of the pulses, and the gas composition in the discharge area. The discharge can operate at atmospheric pressure and high pressures. The characteristics of the discharges vary depending on the configuration of the electrodes and the polarity of the field.

Negative corona: In a negative corona, the high electric field is centered around the cathode. At first, a negative corona starts intermittently as a Trichel pulse. As the voltage is increased, a pulseless corona is produced and finally flashover occurs with a spark, and the discharge bridges the electrodes to form an arc. The flashover voltage of a negative corona is normally around two times larger than that of a positive corona.

Positive corona: In a positive corona, the high electric field is centered around the anode. Upon initiation, a burst pulse corona is formed. As the voltage is increased, brush-like streamers are produced, followed by a uniform glow discharge. Further increase in the applied voltage causes the streamers to extend, and a flashover takes place with a spark. Occasionally a stable Hermstein glow forms before flashover. These transitions are affected by the geometry of the electrodes, especially with the radius and curvature of the sharp corona electrode.

Pulsed streamer corona: Streamers are generated by applying a short (typically < microsecond) voltage pulse to an electrode [1]. Due to the short duration of the pulses, very high peak voltages can be applied while ensuring that no transition to spark discharge takes place, thus avoiding flashover. In addition, space charge does not accumulate during or between pulses, so there is no shielding of the electric field around the needle tip. Due to these reasons, streamers can propagate over a longer distance, increasing the corona region, which can, therefore, ionize a larger volume of gas, as indicated in Fig. 2.1 [28]. Indeed, both negative and positive pulsed coronas have been successfully used for the production of O^* radicals [29] and for the removal of NO from gas streams [30]; moreover, negative coronas were found to produce twice as much O^* compared to positive coronas, and positive coronas were more effective in NO removal.

Corona discharges have been used in plasma chemistry to prepare synthetic fibers, for electrostatic precipitation, for the destruction of toxic compounds, to control static in semi-conductor industries, to generate ozone, and to control the amount of acid gases (e.g., NO_x and SO_x) produced in combustion reactions [31].

2.3.2 DBD (Dielectric Barrier Discharge) and Surface Discharge

Originally known as a silent discharge due to a lack of noisy spark formation, the dielectric discharge is a nonthermal plasma that can be operated in a wide range of pressures (mbar up to atmospheric pressure) [32]. Figure 2.2a shows an example of a

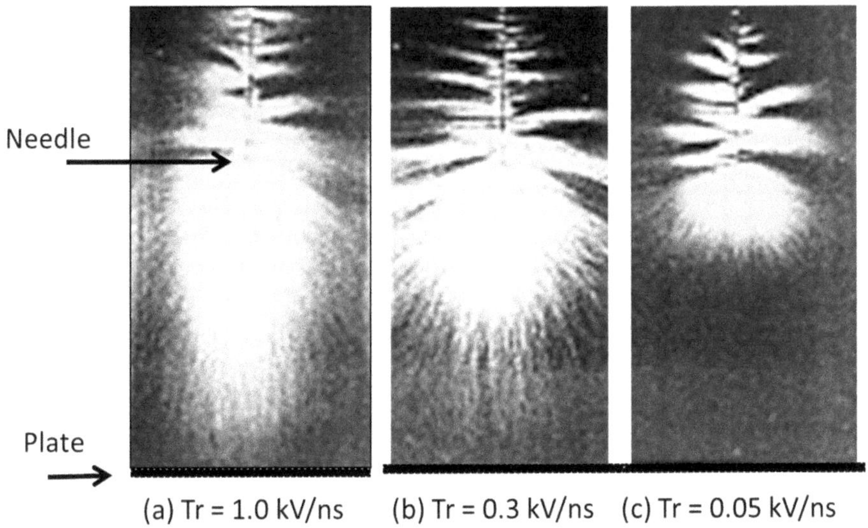

Fig. 2.1 Effect of raising pulse voltage rate of a positive pulsed streamer corona. Needle (1.0 mm diameter, 20 mm length)-to-plate (40 mm diameter) with 40 mm separation; peak voltage +50 kV [28]

DBD reactor with parallel plate electrodes, although it is important to note that coaxial cylindrical configurations are also commonly used [2]. An insulating plate made of a dielectric material (i.e., materials with high relative permittivity such as quartz, glass, and ceramics) is inserted in the discharge gap between the electrodes as a barrier to avoid flashover. The spacing in the discharge gap can vary from 0.1 mm to a few centimeters depending on the applied voltage and the gas occupying the discharge gap. The electrode system can be regarded as a capacitor with an air gap. Application of AC or pulsed voltages charges the capacitor, but when the field strength of the gap in the capacitor exceeds its breakdown strength, discharge takes place across the gap [12]. The plasma produced is non-uniform and consists of many microdischarges or filaments that extend across the discharge gap and cover the entire surface of the insulating plate. The build-up of charge on the surface of the insulating plate limits the flow of current, which extinguishes the filaments and leaves a significant charge deposited on its surface; this occurs within a few ns after breakdown. When the polarity of the applied voltage is reversed, again the electric field across the gap increases, and discharge takes place to lower the voltage difference of the surface and the counter electrode. As the polarity of the electrodes is rapidly changing, the breakdown voltage is reached during consecutive voltage half-waves, resulting in microdischarges forming at a frequency which is twice that of the applied frequency, which can range between 50 Hz and 1 MHz [32]. The short duration of the discharges leads to limited transport of heavier charged particles such as ions, so there is very little gas heating, making this a NTP. A large portion of the electron energy can be used to excite atoms or molecules of gas in the discharge gap, which can be used to initiate chemical reactions [33]. Indeed, DBD has been used for

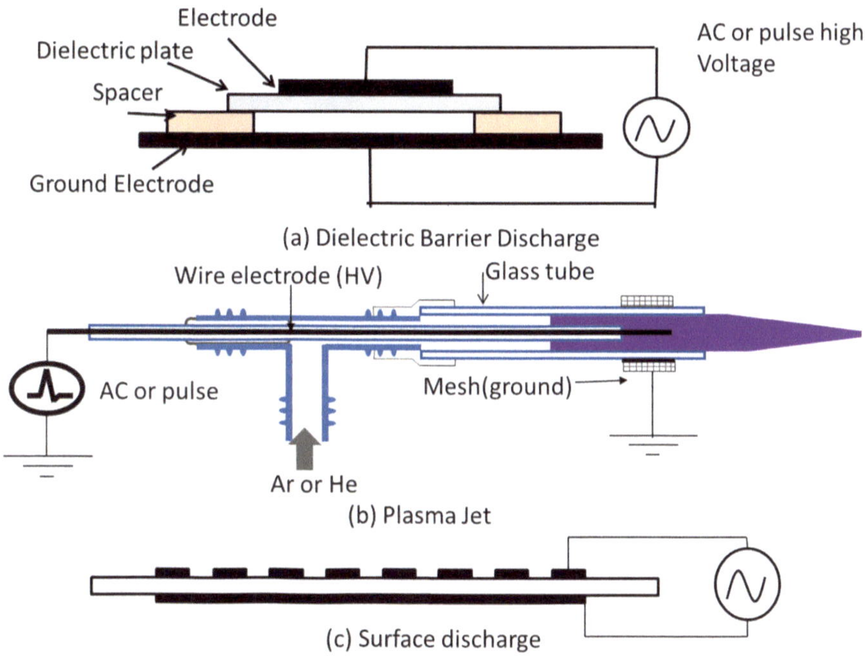

Fig. 2.2 Electrode systems for generation of NTPs [2, 36, 41]

ozone generation, surface treatment, fuel synthesis, and various other applications [34]. He is often used as a working gas in the discharge gap as this can produce a uniform discharge, designated as an atmospheric pressure glow discharge [35], which is important for material processing.

An example of an atmospheric plasma jet, or atmospheric pressure glow discharge, is shown in Fig. 2.2b [36]. A concentric cylinder electrode system is used, with a high-voltage inner electrode surrounded an insulating tube made of a dielectric material (e.g., glass) and an outer electrode made of mesh situated at the nozzle. Ar or He are used as a feed gas. A DBD is generated in the nozzle, and the energetic electrons excite molecules of the feed gas. Due to meta-stable energy levels with longer lifetimes, the excited Ar or He atoms are transported outside the electrode system through the nozzle and mixed with ambient air to produce a plasma jet. Excitation, ionization, and dissociation of air molecules in the jet can produce excited species and radicals. Plasma jets have found applications in the production of microelectronic devices, surface treatment, and microbial decontamination and for the pre-treatment of catalysts [37–40].

Figure 2.2c shows surface discharge [41]. Surface discharge is another configuration of DBD, with two electrodes set either side of an insulating plate made of a dielectric material. On one side of the electrode is a series of strip electrodes and on the other side is a parallel, high-voltage electrode. With application of an AC voltage or pulse, a discharge takes place around the edges of each electrode in the gas space

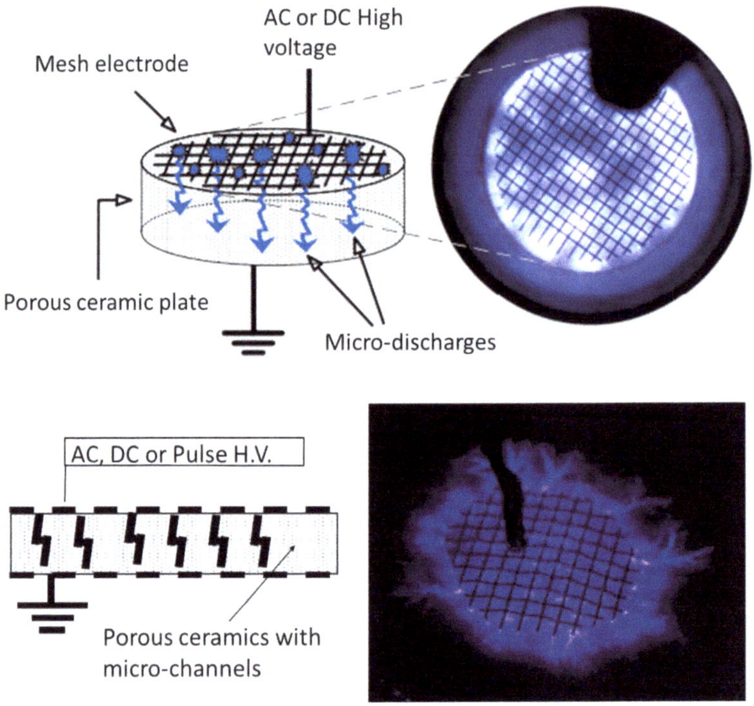

Fig. 2.3 Microdischarges in porous ceramic plate [48, 49]

above the insulating plate, propagating along the surface [42]. Surface DBD has been shown to be effective for the treatment of VOCs, ozone synthesis and nitric oxide conversion from air, and the oxidation of mercury [43–45].

Figure 2.3 shows a reactor in which microdischarges are generated using porous ceramics [46, 47]. Discharges were found to occur in the pores of the ceramics, and the type of discharge was found to be influenced by the pore sizes. In small pores (0.8–15 μm), the discharge developed on the surface of the dielectric (surface discharge), and no discharge was observed inside the pores. In larger pores (15–90 μm), microdischarges were generated. The optimum generation and distribution of the microdischarges were observed with pore sizes of 50 and 80 μm. This system was proposed as a novel way to generate stable, large volume atmospheric pressure plasmas in hybrid plasma-catalyst reactors as a method for pollutant abatement from car exhausts.

2.3.3 Gliding Arc Plasma

Gliding arc plasma can be generated between two or more flat electrodes with curved knife edges in a transverse gas flow [50]. The electrodes are positioned with the knife edges pointing toward each other to create a diverging discharge gap, with the gas

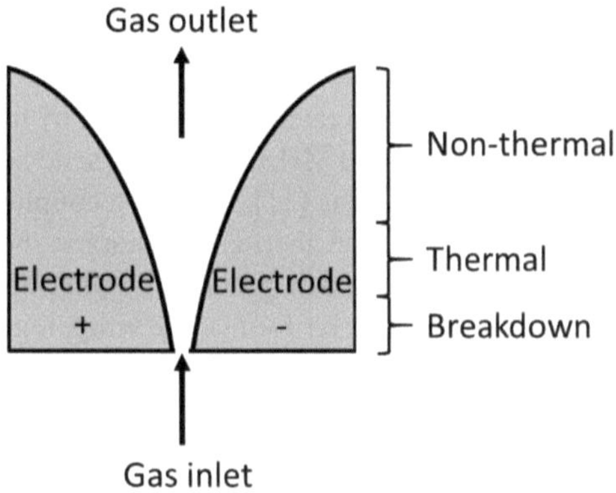

Fig. 2.4 Schematic diagram of a gliding arc plasma reactor. (Adapted from Nijdam et al. [51])

inlet situated upstream of the narrowest end of the gap (Fig. 2.4). A high-voltage generator (AC or DC) provides the electric field to ionize the gas. The arc discharge is initiated at the shortest point between the electrodes (a few millimeters) and then proceeds to glide upward through the widening electrode gap, directed by the transverse gas flow. The length of the discharge grows as it moves along the discharge axis until it can no longer sustain itself with the supplied power. The arc is then extinguished and a new arc is generated at the narrowest point within a 10 ms timescale. High gas flow rates (several L/min) are required to sustain the discharges.

Gliding arc plasma is an intermediate plasma between low and high temperature and is sometimes referred to as a "warm plasma." As such, gliding arcs have properties of both thermal plasmas (large power, currents, and electron densities) and nonthermal plasmas (low gas temperature). Discharges in the lower part of the gliding arc close to the breakdown tend to have thermal characteristics, which gradually evolve into nonthermal discharges as they move up the electrode and their power decreases. Breakdown voltages are usually a few kilovolts. The properties of the discharge tend to depend on the gas flow rate and the input power, which can range from tens of W to several tens of kW.

Arc discharges are widely used in welding devices [52] and have found applications in plasma chemistry for CO_2 conversion [53], tar removal from gas streams [54], hydrophobization of cotton fabric [55], hydrogen production from methanol [56], and dry reforming of methane to generate syngas and carbon nanomaterials [57]. In addition to the standard bi-plate set-up, several other types of gliding arc plasma reactors have been developed including rotating gliding arcs co-driven by magnetic field and tangential flow [58], cone-shaped rotating arc reactors [59], and reverse vortex flow gliding arc reactors [60].

2.3.4 Radio-Frequency (RF) Plasma

RF plasmas are generated and sustained by high-frequency electromagnetic fields in the range of 0.1–100 MHz, with a frequency of 13.56 MHz most commonly used for industrial applications [31]. Inductive coupling of the electromagnetic field with the plasma will generate thermal discharges (between 6000 and 10,000 K), whereas capacitive coupling will generate nonthermal discharges [61]. Inductively coupled plasmas (ICPs) can be formed by wrapping an inductive solenoid coil around the discharge chamber or by placing a planar coil adjacent to it. A strong magnetic field is produced in the discharge gap as a current is passed through the induction coil, which induces a low electric field that can generate high-density plasmas. These plasmas are often used to generate ions and excited species for sample analysis when coupled with analytical techniques such as atomic emission spectroscopy (ICP-AES) or mass spectrometry (ICP-MS), or for reactive-ion etching of surfaces (ICP-RIE).

Capacitively coupled plasmas (CCPs) are formed by applying an RF voltage between two planar electrodes that are spaced a few centimeters apart either inside or outside of the discharge chamber. Application of the RF voltage establishes a strong electric field which enables formation of a plasma discharge. In response to the electric field, the electrons rapidly oscillate back and forth within a cloud of positively charged ions. Low pressures (1–10^3 Pa) are typically used to maintain the nonthermal plasma. Nonthermal RF plasmas have been used with metal (Fe, Cu, Pd, Ag, and Au) mesh catalysts for ammonia synthesis [62], and for the synthesis of biofuels from lignocellulosic biomass in a hydrogen background gas [63]. They have also been in polymer treatment [64] and in the microelectronics industry for film deposition and etching [65].

2.3.5 Microwave Plasmas

Microwave plasmas are produced inside of a waveguide structure that directs microwaves from a power source toward the plasma load with the aid of an applicator, which optimizes the energy transfer to the plasma and minimizes the reflected power [32]. A circulator is also used to protect the power source from reflected microwave energy. A continuous flow of gas occupies the region where the plasma is formed, which can be contained inside a tubular discharge chamber that is often made of a dielectric material such as quartz.

The operating frequency range for microwave discharges is between 300 MHz to 10 GHz, which is much higher than that of RF discharges. Indeed, the most commonly used frequency is 2.45 GHz. At these frequencies, the electrons are the only charged particles with a mass that is small enough to oscillate with the electric field; therefore, a nonthermal plasma is produced. They can be operated over a wide pressure range, from about 100 Pa to atmospheric pressure, and electron densities between 10^8 and 10^{15} cm^{-3} have been reported [2].

Atmospheric plasmas have been used to produce hydrogen via methane reforming with a hydrogen production rate of 255 g[H_2]/h and 85 g[H_2]/kWh energy efficiency [25]. A recent study has investigated the feasibility of using nonthermal microwave plasma for nuclear waste management. Up to 92% of radioactive wastes could be removed, and oxygen in the plasmas was found to play a significant role in the reactions.

2.3.6 Electron Beam Irradiation Plasma

Plasma can be generated by irradiating a gaseous medium with a high-energy electron beam. The beam is created by accelerating electrons using a high voltage inside a vacuum system. The accelerated electrons are passed through a thin foil Ti window at one end of the system that also functions as a vacuum seal. Usually, electrons with several keV energy are used to irradiate the gas [66].

This system allows plasmas with a high degree of ionization to be formed in low-pressure environments; plasmas with electron densities of 10^{10} to 10^{12} cm^{-3} and 11,600 K (1 eV) electron temperature can be obtained with working gas pressures between 1 and 100 Pa [32]. It is also a very energy-efficient method for plasma generation as up to 70% of the beam energy can be transferred to the plasma.

Large flows of ions with energies in the range of 1 to 5 eV can be produced in electron beam-generated plasmas, which can be used for atomic layer etching and engineering of surface chemistries and morphologies [67]. Moreover, electron beam-generated plasmas have been used for the processing of graphene [68], polymer treatment for improved Cu and Al film adhesion, and stainless steel surface nitriding [69] and for the decomposition of volatile organic compounds such as ethyl chloride and vinyl chloride [70].

2.4 NTP-Catalyst Systems

The chemistry in NTPs is rather complex, owing to the formation of many different active species (e.g., vibrationally excited species, radicals, etc.) in the plasma discharge, and so the selectivity of the desired products in any given reaction is typically low. Combining NTP with catalysts in a well-designed reactor system is an effective method to improve the selectivity and energy efficiency of a reaction as it can merge the high selectivity of catalytic reactions with the low temperature performance of plasma reactions in a compact reactor [24, 71]. Furthermore, the use of catalysts can also reduce the energy consumption of a process, which can be large for some reactions when NTP is used alone.

As shown in Fig. 2.5, NTP can be combined with catalysts in either a single stage, two-stage, or multiple-stage (multi-stage) configuration. In a single-stage configuration (Fig. 2.5a), the catalyst is situated inside the plasma discharge area, where it can

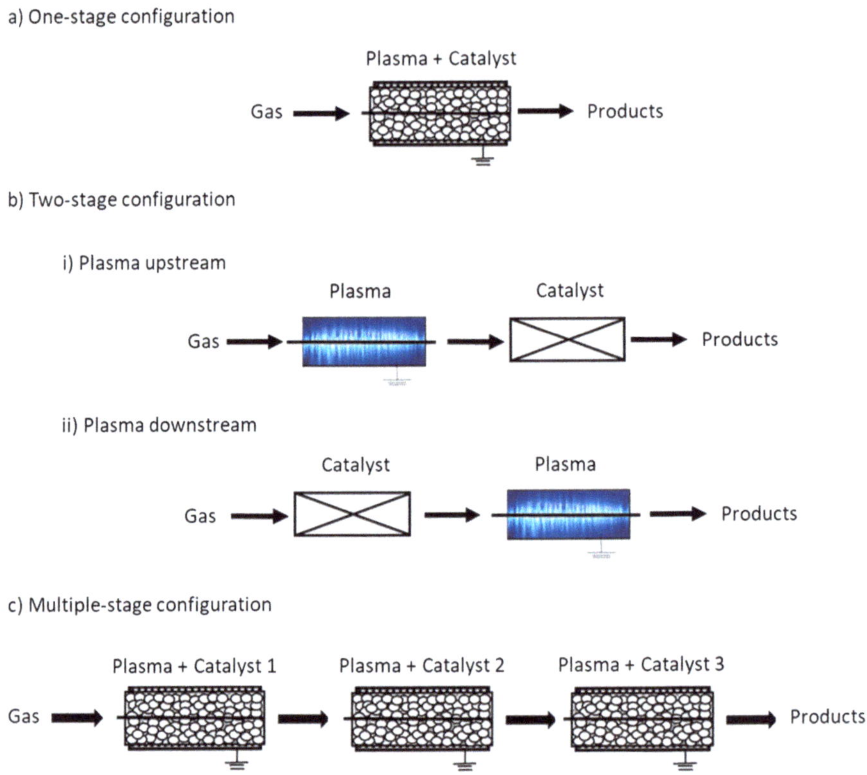

Fig. 2.5 Schematic diagrams of plasma catalysis system configurations [24]

either partially or completely fill the discharge area. This enables the plasma and the catalyst to directly interact with each other, which can lead to synergistic enhancements in both the plasma and catalyst performances. Indeed, the degree of catalyst packing in the discharge area has been found to have a strong effect on the discharge mode [57]. Work by Tu and Whitehead demonstrated that a fully packed configuration altered the discharge mode from filamentary microdischarges (in the absence of catalyst) to predominantly surface discharges on the catalyst surface, whereas, although filamentary discharges were still observed in a partially packed configuration, interactions between the plasma and catalyst were improved significantly [57]. The effect of plasma-catalyst interactions is discussed in more depth in Chap. 3. Reactions in a one-stage configuration involve short-lived active species such as radicals, excited atoms and molecules, electrons, and photons [72].

In a two-stage configuration (Fig. 2.5b), the plasma is situated either upstream or downstream of the catalyst bed. The catalyst bed is typically heated in these systems to promote thermal catalytic reactions. Plasma pre-processing occurs when the plasma is positioned upstream of the bed, where the plasma converts the reactants into more chemically active species to accelerate the catalytic reaction. The active species formed must have fairly long lifetimes in order for them to pass from the plasma downstream to the catalyst bed. These are relatively stable intermediates and are unlikely to be vibrationally excited species, which only have short lifespans (1–100 ns) at atmospheric pressure and are, therefore, unlikely to survive the transition from the plasma zone to the catalyst bed. Plasma post-processing occurs when the plasma is positioned downstream of the catalyst bed and is most commonly used to destroy undesirable by-products from the thermal catalytic reaction, or to convert any residual reactants into other more valuable products.

Multi-stage configurations (Fig. 2.5c) enable different plasma-catalyst reactions to be used in a single reactor set-up. The novelty of this configuration is that each catalyst can perform a different function in the reaction, in accordance with its position in the gas stream relative to the other catalysts and the anticipated reaction, enabling a variety of different processing combinations to be achieved, for instance, breakdown of the initial precursor → oxidation of the resulting intermediates → destruction of excess O_3 in the product stream [24, 73, 74].

Plasma-catalyst systems have been used for a wide variety of reactions, including VOCs and tar decomposition, NO_x and SO_x removal, ozone generation, and NH_3 synthesis, to name but a few. Figure 2.6 depicts some of the different NTP-catalyst systems that been used to remove NO_x from gas streams [24]. In a one-stage process (Fig. 2.6a), additives such as C_2H_4 and NH_3 can be injected into the gas stream prior to processing in a plasma-driven catalytic reactor to improve NO removal. In some cases, deposited carbon on the surfaces in the reactor, formed by decomposition of carbon-containing compounds in the plasma, functions as a reducing agent for NO_2, thus aiding with removal.

In a two-stage process (Fig. 2.6b), NO is oxidized to more reactive NO_2 in the first stage by plasma catalysis. Additives such as C_2H_4 are used to enhance oxidation. The NO_2-rich gas is then introduced to the second catalyst-containing reactor, where selective catalytic reduction (SCR) occurs with the addition of NH_3 [75, 76]. As oxidation of NO to NO_2 in the first stage requires a higher temperature, the use of a second reactor ensures that a lower temperature (around 400 K) can be used in order for the reduction reaction to take place. The optimum ratio of NO:NO_2 for SCR with NH_3 is ~ 1:1 [77, 78]. As the power of NTP can be controlled quickly (typically through use of a variac), the ratio of NO and NO_2 can be easily maintained – if the gas composition is monitored in real time – to improve the efficiency of the reaction. This is an advantage of using NTP for the oxidation process.

The indirect radical injection system (Fig. 2.6c) is a variation of the two-step process. Here, O_3 is generated by plasma in the first step and then introduced into the gas stream for NO removal. As with the reaction in Fig. 2.6b, NO is oxidized to NO_2 and subsequently reduced to N_2.

In a batch system (Fig. 2.6d), or cycled process system, an absorbent-containing reactor removes pollutants from the gas stream. The reactor is then separated from the main gas stream, and the absorbed pollutant gases can be either oxidized or reduced with plasma processing in the presence of appropriate reactants [79].

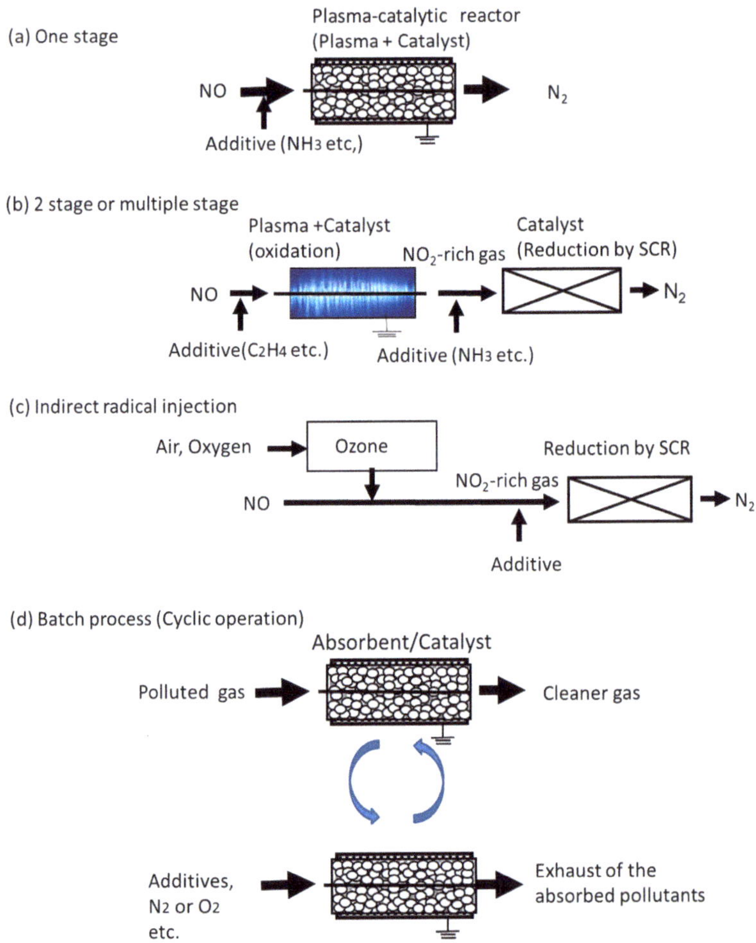

Fig. 2.6 Plasma catalysis system for deNO_x processes [24]

2.4.1 *Packed Bed DBD Reactors*

A packed bed reactor is a variation of a DBD reactor that is often used for one-stage catalytic reactions. The reactor is packed with pellets made of dielectric materials such as glass, quartz, aluminate, or ferroelectrics (e.g., $BaTiO_3$). Examples of packed bed reactor designs are illustrated in Fig. 2.7 [80]. With application of an AC or pulsed voltage, discharges form between the contact points of the pellets due to the enhanced electric field produced by concentrated electrostatic lines of flux in the dielectric material. In a typical reactor, there are a large number of contact points in close proximity to each other, which can amplify the electric field strength and increase the electron energies: a phenomenon that is called *the packed bed effect*. Figure 2.8 shows the discharge power of a packed bed discharge inside a glass tube

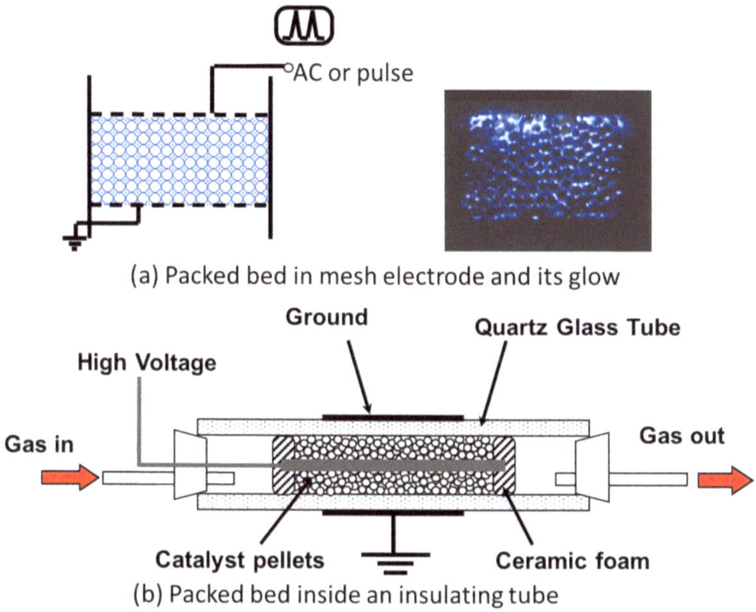

Fig. 2.7 Packed bed discharge (packed with dielectric catalyst pellets) [71, 80]

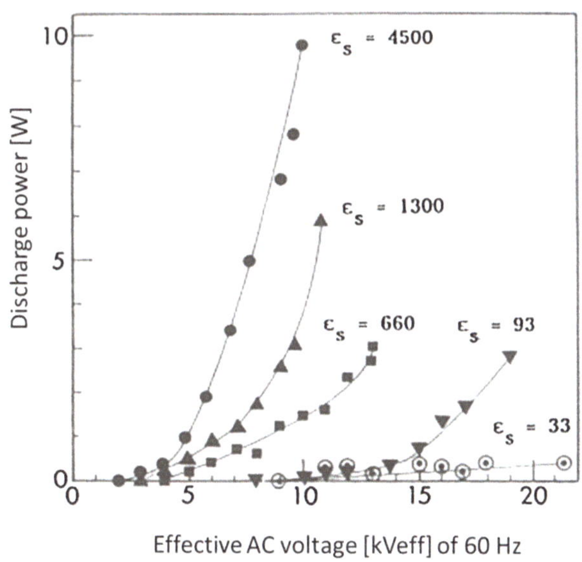

Fig. 2.8 Discharge power of a packed bed reactor. (Size of a packed bed: 40 mm diameter and 20 mm thickness. Both sides of the bed are attached to a metal mesh electrode, and an AC 60 Hz high voltage is applied. Packing material: spherical pellets of 3.3 mm diameter with different specific dielectric constants) [80]

with a 40 mm inner diameter and 20 mm thickness. The bed was packed with pellets of 3 mm diameter made of different materials with specific dielectric constants (εs) in the range of 33–4500. AC high voltage of 60 Hz frequency was applied. The results show that more intense discharges can be generated using materials with

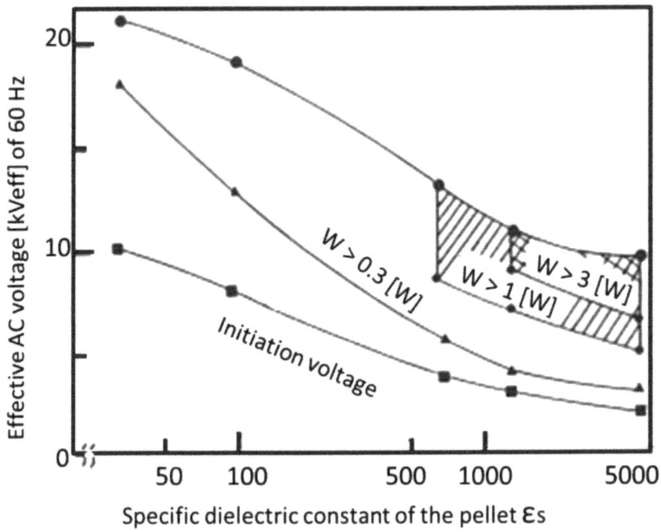

Fig. 2.9 Stable region of a packed bed discharge. (Size of a packed bed: 40 mm diameter and 20 mm thickness. Both sides of the bed are attached to a metal mesh electrode, and an AC 60 Hz high voltage is applied. Packing material: spherical pellets of 3.3 mm diameter with different specific dielectric constants) [80]

larger dielectric constants, which in turn produces higher energy electrons. The frequency of the applied AC voltage also affects the associated current pulses, as shown in Fig. 2.9. The lower curve indicates the initiation voltage of the discharge, and the upper curve is the sparking voltage. Between these two lines, a stable region of the partial discharge can be obtained. Another configuration for a cylindrical packed bed reactor is shown in Fig. 2.7b. Inclusion of the insulating wall in this design can prevent sparking.

As well as effecting the discharge properties of the plasma reactor, inclusion of the packing material can also reduce the costs of using large volumes of expensive catalysts to fill the discharge area (if a completely packed reactor is needed), with the added benefit of providing quasi-homogeneous dispersion of the catalyst in the discharge area to maximize reactant-catalyst and plasma-catalyst interactions. Indeed, with careful selection of catalysts and packing materials, packed bed reactors show promising performances for various reactions, including fuel reforming and CO_2 conversion [81]. Compared to a reaction conducted with plasma only, Ni/Al_2O_3 pellets in a packed bed reactor were shown to improve H_2 selectivity and increase the H_2/CO molar ratio of syngas from 0.84 to 2.53 in the dry reforming of CH_4. The improvement was attributed to the formation of spatially limited discharges and surface discharges on the catalyst surface, coupled with improved charge transfer from the conductive Ni active sites. It has been noted in other works that propagation of surface discharge streamers along the surface of a catalyst pellet in a packed bed reactor is affected by the chemical composition of the catalyst material and the presence of metal active sites on the catalyst surface [82]. With careful selection of catalyst materials and metal sites, the properties of the surface discharges could be better controlled to improve performance.

The packing materials can also function as catalysts themselves. For example, a packed bed reactor using zirconia beads was found to increase CO_2 conversion and energy efficiency by factors of up to 1.9 and 2.2, respectively, compared to the corresponding unpacked DBD reactor. This was explained by a lower breakdown voltage and higher electron energy and local electric fields near the contact points of the beads, which enabled more effective splitting of CO_2 in the packed bed reactor. Later reports investigated the influence of specific packing materials and reactor properties on the conversion and energy efficiency of CO_2 dissociation in a packed DBD reactor [83]. The results showed that $BaTiO_3$ gave the highest conversion and energy efficiency, with 25% and 4.5%, respectively, with the conversion twice as much as a DBD with no packing material. Moreover, the findings indicated that three effects were found to influence the reaction: (i) electric field enhancement at the contact points improved conversions and efficiency; (ii) the presence of packing materials reduced the residence time of CO_2 in the plasma, leading to a decrease in conversion; and (iii) the voids between the particles can have a positive or negative effect on the reaction depending on the size of the beads and the material they are made of.

The chemical and physical properties of the catalysts (e.g., chemical composition, pellet size, topography, and morphology) greatly influence the performance of the reactor for any given reaction and, therefore, must be of the utmost consideration when designing new reactors if the reactors are to perform successfully [84–86].

2.4.2 Honeycomb Discharge Reactors

Some commercial NTP systems, especially those used for control of NO_x in vehicle emissions, require low-pressure drops and high catalyst stability against mechanical vibrations. Ceramic honeycomb catalysts, with many narrow pore channels in which plasma discharges can be produced, are suitable candidates as they can enable good gas flow rates while providing large surface areas for production of a uniform plasma discharge [46, 47]. Unlike pellet-type catalysts in packed bed reactors, the honeycomb channels are fixed in place inside the catalyst and will not move around inside a reactor with mechanical vibrations like pellets can, which can alter the mode of the discharges during use and thus affect performance. However, the generation of a stable, homogenous electrical discharge through the channels in a honeycomb is difficult. One way to overcome this problem is to use a sliding discharge [87, 88] inside the honeycomb.

Sliding discharges can be generated through honeycomb catalyst channels by using a packed bed discharge with an AC voltage in front of the honeycomb and applying a DC electric field across the honeycomb. The sliding discharges are produced when the AC-driven barrier discharges "slide" along the dielectric material in the honeycomb channels with application of the DC voltage. This electrode configuration enables the ionization of the fine channels (1 mm^2) of a honeycomb made of cordierite. Figure 2.10 illustrates the configuration of a honeycomb plasma reactor that uses a bundle of transparent capillaries to simulate a honeycomb catalyst in order to better understand the generation of honeycomb discharges. The packed bed portion of the reactor constitutes γ-Al_2O_3 pellets (3 mm diameter) set in a quartz glass tube (inner diameter of 26 mm) with a stainless steel rod (6 mm in diameter) at the center as a discharge electrode. The outer surface of the tube is covered with an aluminum sheet (15 mm length) as a ground electrode. Bundled thin quartz glass capillary tubes are used to simulate a honeycomb catalyst and to observe light emission from the discharges. The inner diameter of the individual quartz glass capillaries is 1 or 2 mm with a length of 20 mm. The bundle (26 mm diameter) is inserted into the outer quartz glass tube. A stainless steel mesh (16 mesh) is placed on the upper end of the capillaries where a DC power supply is connected to form the honeycomb discharge in the capillary. Discharges are generated in the packed bed at contacting points of the γ-Al_2O_3 pellets (3 mm) using an AC voltage. When a high DC voltage is applied between the bundle of quartz glass capillaries, the volume inside of the capillaries can be ionized. If the AC (packed bed) discharge occurs, the sliding discharge fills the volume between capillaries by the effect of the DC electric field.

Figure 2.11 shows images of the discharges created when different combinations of DC and AC voltages (−11 kV and 25 kVp-p, respectively) are applied to the reactor. No discharge is observed in the absence of an applied voltage (Fig. 2.11a). Figure 2.11b shows that only a packed bed discharge is observed when only the AC

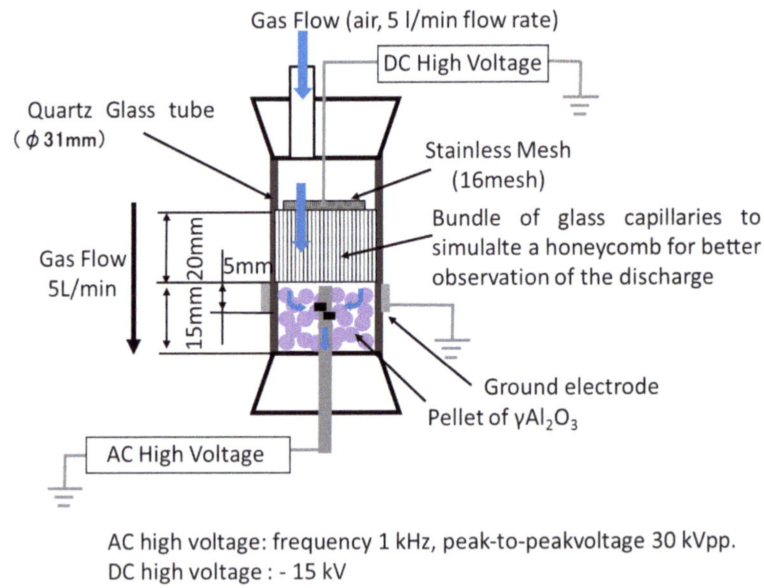

Fig. 2.10 Set-up for honeycomb discharge [89]

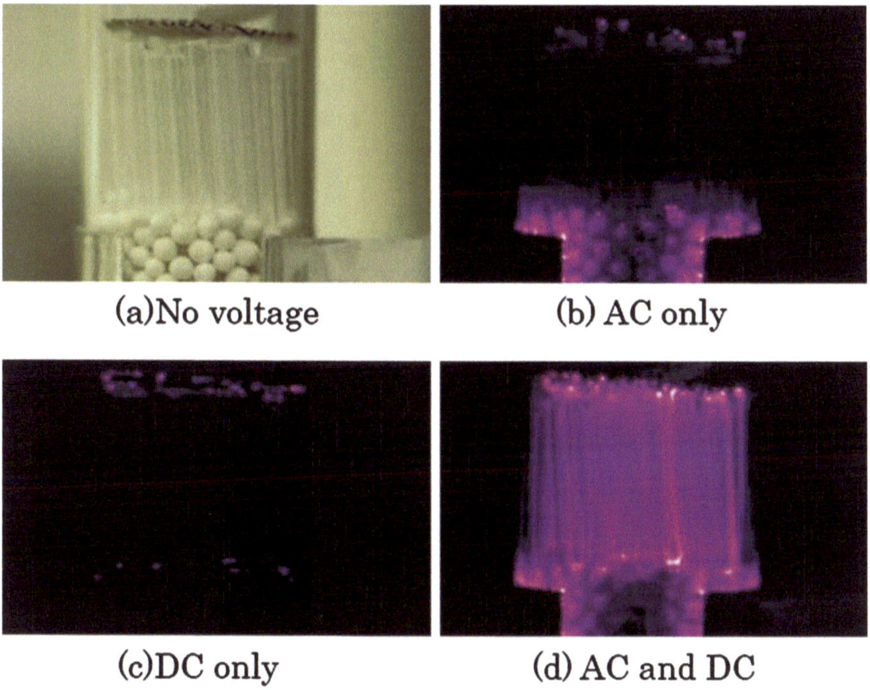

Fig. 2.11 Images of discharges created in a bundle of glass capillaries simulating a honeycomb with application of AC and DC voltages. (Quartz capillary: 1 mm inner diameter, 20 mm in length, AC: 25 kVpp, 1 kHz, DC: −11 kV) [89]

voltage is applied. Application of a DC voltage only (Fig. 2.11c) shows that corona discharges are generated in the upper part of the quartz glass tube at the edges of the mesh electrode. With application of both AC and DC voltages (Fig. 2.11d), a honeycomb discharge occurs in the bundle of capillaries, and uniform light emission along the capillary tube is observed. The intensity of the light emissions from the plasma increases with increasing DC voltage until sparking occurs.

The relationship between DC and AC high voltages used to generate uniform honeycomb discharges in 1 and 2 mm quartz glass capillaries is shown in Fig. 2.12. The starting voltage (*V*i) of the honeycomb discharge and the maximum voltage (flashover voltage, *V*s) of the DC voltages are shown when the AC voltage is changed. The DC voltage was increased slowly from zero until sparking took place. When this was measured, the AC high voltage and the frequency were fixed. Vi decreased with increasing the AC voltage. Both positive and negative DC voltages were used. With negative DC, (*V*s – *V*i) is larger than the same with positive DC. With negative DC, both *V*s and *V*i values are lower than those with positive DC [89]. Increasing the inner diameter of the capillaries from 1 to 2 mm decreased *V*i and *V*s. This may have been due to easier transition of the surface barrier discharge mode into capillary microdischarge mode through the larger capillary pores [47].

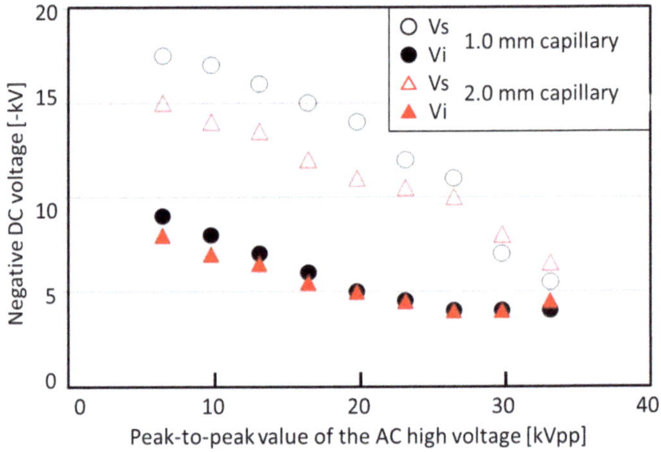

Fig. 2.12 Relationship of the AC and negative DC high voltage for generation of uniform discharge in the glass capillary (Capillary: length 20 mm, inner diameter of 1.0 mm and 2.0 mm; *Vi* is the initiation voltage of the discharge inside the glass capillary, *Vs* is the sparking voltage) [89]

It is well known that positive streamers tend to extend further than negative streamers. In this honeycomb discharge, ionized gases in the packed bed discharge act as an electrode, which supply electrons, ions, and photons to the capillaries more easily than metal electrodes. Therefore, more effective streamer generation and propagation could be achieved. The honeycomb discharge takes place with the average DC electric field in the capillary of 3–8.5 kV/cm, depending on the AC voltage value.

Instead of using a packed bed discharge, a flat ceramic plate electrode can be used to initiate a honeycomb discharge as shown in Fig. 2.13 [90]. An induction electrode is embedded inside the ceramic plate, and the other electrode is printed on the surface of the honeycomb. With AC application between the induction and the printed electrode, surface discharge takes place at the inner wall of each hole. This can be used to initiate a honeycomb discharge. With the application of negative DC to the outer electrode, a honeycomb discharge can be generated. This honeycomb discharge is practical because the pressure drop can be kept low. It is also noted that the electrode system is stable to vibrations; therefore, it has the potential to be used in vehicles to remove noxious emissions from exhaust gasses. It should also be noted that the surface treatment of honeycomb catalysts by plasma exposure can be carried out using this technique.

2.5 Conclusions

NTP is a versatile processing technology with various environmental applications. Making plasmas competitive requires informed modifications in the process design that includes selection of the optimal plasma parameters and the best-matched

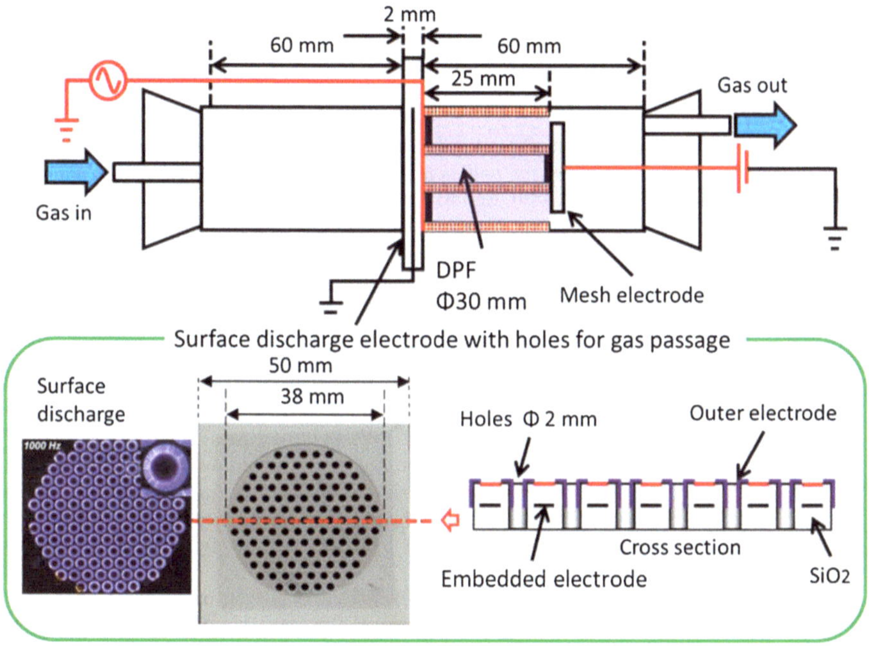

Fig. 2.13 Honeycomb discharge initiated by surface discharge [90]

catalyst. Physical aspects of the interaction between plasma and catalyst, such as propagation of discharge on surfaces and electrohydrodynamics (EHD) gas flow, should also be considered in the design of plasma-catalyst reactors.

For further improvement of the process efficiency of NTP-catalyst reactors, intensive studies are necessary to understand the synergetic effect of the combination of plasma with catalyst, which includes precise measurement and modeling of excited species, ionized species and radicals in plasma, the time-course of reactions, and their interaction with the catalyst and electrode surfaces. This can only be achieved using a multidisciplinary approach that brings together expertise in engineering, physics, and chemistry.

Acknowledgments A. Mizuno is grateful for the researchers and engineers collaborating together for the development of the plasma processes. The author also thanks the members of the laboratory and the graduated students who have worked for the plasma processes at Toyohashi University of Technology.

References

1. Mizuno, A., Clements, J. S., & Davis, R. H. (1986). A method for the removal of sulfur dioxide from exhaust gas utilizing pulsed streamer corona for electron energization. *IEEE Transactions on Industry Applications, IA-22*, 516–522.

2. Eliasson, B., Member, S., & Kogelschatz, U. (1991). Nonequilibrium volume plasma chemical processing. *IEEE Transactions on Plasma Science, 19*, 1063–1077.
3. Kim, H.-H., Teramoto, Y., Negishi, N., & Ogata, A. (2015). A multidisciplinary approach to understand the interactions of nonthermal plasma and catalyst: A review. *Catalysis Today, 256*, 13–22.
4. Penetrante, M. B. (1993). Plasma chemistry and power consumption in non-thermal $DeNO_x$. In *Non-thermal plasma techniques for pollution control. Part A. Overview, fundamentals and supporting technologies* (pp. 65–89). Berlin: Springer.
5. Nozaki, T., & Okazaki, K. (2013). Non-thermal plasma catalysis of methane : Principles , energy efficiency , and applications. *Catalysis Today, 211*, 29–38.
6. Okazaki, K., Kishida, T., Ogawa, K., & Nozaki, T. (2002). Direct conversion from methane to methanol for high efficiency energy system with exergy regeneration. *Energy Conversion and Management, 43*, 1459–1468.
7. Snoeckx, R., Aerts, R., Tu, X., & Bogaerts, A. (2013). Plasma-based dry reforming: A computational study ranging from the nanoseconds to seconds time scale. *Journal of Physical Chemistry C, 117*, 4957–4970.
8. Bruggeman, P., Iza, F., Lauwers, D., & Gonzalvo, Y. A. (2010). Mass spectrometry study of positive and negative ions in a capacitively coupled atmospheric pressure RF excited glow discharge in He–water mixtures. *Journal of Physics D: Applied Physics, 43*, 12003.
9. Ono, R., & Oda, T. (2003). Dynamics of ozone and OH radicals generated by pulsed corona discharge in humid-air flow reactor measured by laser spectroscopy. *Journal of Applied Physics, 93*, 5876–5882.
10. Ikawa, S., Kitano, K., & Hamaguchi, S. (2010). Effects of pH on bacterial inactivation in aqueous solutions due to low-temperature atmospheric pressure plasma application. *Plasma Processes and Polymers, 7*, 33–42.
11. Kim, H. H., Kim, J. H., & Ogata, A. (2008). Adsorption and oxygen plasma-driven catalysis for total oxidation of VOCs. *International Journal of Environmental Science and Technology, 2*, 106–112.
12. Vandenbroucke, A. M., Morent, R., De Geyter, N., & Leys, C. (2011). Non-thermal plasmas for non-catalytic and catalytic VOC abatement. *Journal of Hazardous Materials, 195*, 30–54.
13. Van Durme, J., Dewulf, J., Leys, C., & Van Langenhove, H. (2008). Combining non-thermal plasma with heterogeneous catalysis in waste gas treatment: A review. *Applied Catalysis B: Environmental, 78*, 324–333.
14. Sato, S., Kawada, Y., Sato, S., Hosoya, M., & Mizuno, A. (2011). The study of NO_x reduction using plasma-assisted SCR system for a heavy duty diesel engine. *SAE Technical Paper*. 2011-01-0310, 2007.
15. Kuwahara, T., Yoshida, K., Kannaka, Y., Kuroki, T., & Okubo, M. (2011). Improvement of NO_x reduction efficiency in diesel emission control using nonthermal plasma combined exhaust gas recirculation process. *IEEE Transactions on Industry Applications, 47*, 2359–2366.
16. Mizuno, A., Chakrabarti, A., & Okazaki, K. (1993). Application of corona technology in the reduction of greenhouse gases and other gaseous pollutants. *Non-Thermal Plasma Techniques for Pollution Control, 34*, 165–185.
17. Starikovskii, A. Y., Anikin, N. B., Kosarev, I. N., Mintoussov, E. I., Nudnova, M. M., Rakitin, A. E., Roupassov, D. V., Starikovskaia, S. M., & Zhukov, V. P. (2008). Nanosecond-pulsed discharges for plasma-assisted combustion and aerodynamics. *Journal of Propulsion and Power, 24*, 1182–1197.
18. Fridman, A. G., & A.R. and A. (2011). Combustion-assisted plasma in fuel conversion. *Journal of Physics D: Applied Physics, 44*, 274001.
19. Tu, X., Gallon, H. J., Twigg, M. V., & Gorry, P. A. (2011). Dry reforming of methane over a Ni/Al_2O_3 catalyst in a coaxial dielectric barrier discharge reactor. *Journal of Physics D: Applied Physics, 44*, 274007.
20. Jiang, T., Li, Y., Liu, C., Xu, G., Eliasson, B., & Xue, B. (2002). Plasma methane conversion using dielectric-barrier discharges with zeolite A. *Catalysis Today, 72*, 229–235.

21. Okumoto, M., Su, Z., Katsura, S., & Mizuno, A. (1999). Dilution effect with inert gases in direct synthesis of methanol from methane using nonthermal plasma. *IEEE Transactions on Industry Applications, 35*, 1205–1210.
22. Paulussen, S., Verheyde, B., Tu, X., De Bie, C., Martens, T., Petrovic, D., Bogaerts, A., & Sels, B. (2010). Conversion of carbon dioxide to value-added chemicals in atmospheric pressure dielectric barrier discharges. *Plasma Sources Science and Technology, 19*, 34015.
23. Jogan, K., Mizuno, A., Yamamoto, T., & Chang, J. (1993). The effect of residence time on the CO2 reduction from combustion flue gases by an AC ferroelectric packed bed reactor. *IEEE Transactions on Industry Applications, 29*, 876–881.
24. Kim, H.-H., Teramoto, Y., Ogata, A., Takagi, H., & Nanba, T. (2016). Plasma catalysis for environmental treatment and energy applications. *Plasma Chemistry and Plasma Processing, 36*, 45–72.
25. Jasiński, M., Dors, M., & Mizeraczyk, J. (2008). Production of hydrogen via methane reforming using atmospheric pressure microwave plasma. *Journal of Power Sources, 181*, 41–45.
26. Fridman, A., & Kennedy, L. A. (2011). *Plasma physics and engineering*. Hoboken: CRC Press.
27. Loeb, L.B. (1965). Electrical coronas: their basic physical mechanisms. Berkeley: University of California Press, pp. 694
28. Mizuno, A., Kamase, Y., Tsugawa, H., Shibuya, A., & Yamamoto, K. (1987). Effect of pulse rising rate on current emission and ozone formation in a pulsed streamer corona discharge in air. *8th International symposium on plasma chemistry* (pp. 2216–2221). Tokyo, Japan.
29. van Heesch, E. J. M., Winands, G. J. J., & Pemen, A. J. M. (2008). Evaluation of pulsed streamer corona experiments to determine the O^* radical yield. *Journal of Physics D: Applied Physics, 41*, 234015.
30. Shang, K., & Wu, Y. (2010). Effect of electrode configuration and corona polarity on NO removal by pulse corona plasma. *2010 Asia-Pacific power energy engineering conference*, Asia Pacific, p. 1–4.
31. Fridman, A. (2008). *Plasma chemistry*. Cambridge, MA: Cambridge University Press.
32. Conrads, H., & Schmidt, M. (2000). Plasma generation and plasma sources. *Plasma Sources Science and Technology, 9*, 441.
33. Kogelschatz, U., Eliasson, B., & Egli, W. (1999). From ozone generators to flat television screens: history and future potential of dielectric-barrier discharges. *Pure and Applied Chemistry, 71*, 1819.
34. Wang, L., Yi, Y., Wu, C., Guo, H., & Tu, X. (2017). One-step reforming of CO_2 and CH_4 into high-value liquid chemicals and fuels at room temperature by plasma-driven catalysis. *Angewandte Chemie International Edition, 56*, 13679–13683.
35. Okazaki, S., Kogoma, M., Uehara, M., & Kimura, Y. (1993). Appearance of stable glow discharge in air, argon, oxygen and nitrogen at atmospheric pressure using a 50 Hz source. *Journal of Physics D: Applied Physics, 26*, 889.
36. Schutze, A., Jeong, J. Y., Babayan, S. E., Park, J., Selwyn, G. S., & Hicks, R. F. (1998). The atmospheric-pressure plasma jet: a review and comparison to other plasma sources. *IEEE Transactions on Plasma Science, 26*, 1685–1694.
37. Ehlbeck, E., Schnabel, U., Polak, M., Winter, J., von Woedtke, T., Brandenburg, R., Hagen, T. v. d., & Weltmann, K.-D. (2011). Low temperature atmospheric pressure plasma sources for microbial decontamination. *Journal of Physics D: Applied Physics, 44*, 013002.
38. Liu, G., Li, Y., Chu, W., Shi, X., Dai, X., & Yin, Y. (2008). Plasma-assisted preparation of Ni/SiO_2 catalyst using atmospheric high frequency cold plasma jet. *Catalysis Communications, 9*, 1087–1091.
39. Shang, S., Liu, G., Chai, X., Tao, X., Li, X., Bai, M., Chu, W., Dai, X., Zhao, Y., & Yin, Y. (2009). Research on $Ni/\gamma\text{-}Al_2O_3$ catalyst for CO_2 reforming of CH_4 prepared by atmospheric pressure glow discharge plasma jet. *Catalysis Today, 148*, 268–274.
40. Cheng, D.-G. (2008). Plasma decomposition and reduction in supported metal catalyst preparation. *Catalysis Surveys from Asia, 12*, 145–151.

41. Masuda, S., Akutsu, K., Kuroda, M., Awatsu, Y., & Shibuya, Y. (1988). A ceramic-based ozonizer using high-frequency discharge. *IEEE Transactions on Industry Applications, 24*, 223–231.
42. Pietsch, G. J. (2001). Peculiarities of dielectric barrier discharges. *Contributions to Plasma Physics, 41*, 620–628.
43. Malik, M. A., Schoenbach, K. H., & Heller, R. (2014). Coupled surface dielectric barrier discharge reactor-ozone synthesis and nitric oxide conversion from air. *Chemical Engineering Journal, 256*, 222–229.
44. Affonso Nobrega, P., Blin-Simiand, N., Bournonville, B., Jorand, F., Lacour, B., Pasquiers, S., Rohani, V., Cauneau, F., & Fulcheri, L. (2017). Comparison between performances of surface and volume nanosecond pulsed dielectric barrier discharges for the treatment of volatile organic compounds. *23rd International Symposium on Plasma Chemistry*, at Montréal.
45. Zhao, Y., Shang, K., Duan, L., Shang, K., Tang, S., & Lu, N. (2013). Oxidation efficiency of elemental mercury in two DBD plasma reactors Oxidation efficiency of elemental mercury in two DBD plasma reactors. *Journal of Physics Conference Series, 418*, 012118.
46. Hensel, K., Katsura, S., & Mizuno, A. (2005). DC microdischarges inside porous ceramics. *IEEE Transactions on Plasma Science, 33*, 574–575.
47. Hensel, K., Martišovitš, V., Machala, Z., Janda, M., Leštinský, M., Tardiveau, P., & Mizuno, A. (2007). Electrical and optical properties of AC microdischarges in porous ceramics. *Plasma Processes and Polymers, 4*, 682–693.
48. Hensel, K., Katsura, S., & Mizuno, A. (2005). DC microdischarges inside porous ceramics. *IEEE Transactions on Plasma Science, 33*, 574–575.
49. Hensel, K., Martisovits, V., Machala, Z., Janda, M., Lestinsky, M., Tardiveau, P., & Mizuno, A. (2007). Electrical and optical properties of AC microdischarges in porous ceramics. *Plasma Processes and Polymers, 4*, 682–693.
50. Locke, B. R., Sato, M., Sunka, P., Hoffmann, M. R., & Chang, J.-S. (2006). Electrohydraulic discharge and nonthermal plasma for water treatment. *Industrial and Engineering Chemistry Research, 45*, 882–905.
51. Nijdam, S., van Veldhuizen, E., Bruggeman, P., & Ebert, U. (2012) An introduction to nonequilibrium plasmas at atmospheric pressure. In: *Plasma chemistry and catalysis in gases and liquids*. Weinheim, Germany: Wiley-VCH Verlag GmbH.
52. Chen, S., Zhang, R., Jiang, F., & Dong, S. (2018). Experimental study on electrical property of arc column in plasma arc welding. Weinheim, Germany: Wiley-VCH Verlag GmbH.
53. Indarto, A., Yang, D., Choi, J.-W., Lee, H., & Keun Song, H. (2007). Gliding arc plasma processing of CO2 conversion. *Journal of Hazardous Materials, 146*, 309–315.
54. Liu, S., Mei, D., Wang, L., & Tu, X. (2017). Steam reforming of toluene as biomass tar model compound in a gliding arc discharge reactor. *Chemical Engineering Journal, 307*, 793–802.
55. Cerny, P., Bartos, P., Olsan, P., & Spatenka, P. (2019). Hydrophobization of cotton fabric by Gliding Arc plasma discharge. *Current Applied Physics, 19*, 128–136.
56. Lu, Y., Yan, W., Hu, S., & Wang, B. (2012). Hydrogen production by methanol decomposition using gliding arc gas discharge. *Journal of Fuel Chemistry and Technology, 40*, 698–706.
57. Tu, X., & Whitehead, J. C. (2012). Plasma-catalytic dry reforming of methane in an atmospheric dielectric barrier discharge: Understanding the synergistic effect at low temperature. *Applied Catalysis B: Environmental, 125*, 439–448.
58. Zhang, H., Li, X. D., Zhang, Y. Q., Chen, T., Yan, J. H., & Du, C. M. (2012). Rotating Gliding arc codriven by magnetic field and tangential flow. *IEEE Transactions on Plasma Science, 40*, 3493–3498.
59. Lee, D. H., Kim, K.-T., Cha, M. S., & Song, Y.-H. (2010). Plasma-controlled chemistry in plasma reforming of methane. *International Journal of Hydrogen Energy, 35*, 10967–10976.
60. Kalra, C. S., Cho, Y. I., Gutsol, A., & Fridman, A. (2005). Gliding arc in tornado using a reverse vortex flow. *The Review of Scientific Instruments, 76*, 025110.

61. Shunko, E. V., Stevenson, D. E., & Belkin, V. S. (2014). Inductively coupling plasma reactor with plasma electron energy controllable in the range from ~6 to ~100 eV. *IEEE Transactions on Plasma Science, 42*, 774–785.
62. Shah, J., Wang, W., Bogaerts, A., & Carreon, M. L. (2018). Ammonia synthesis by radio frequency plasma catalysis : revealing the underlying mechanisms. *ACS Applied Energy Materials, 1*, 4824–4839.
63. Gao, Y., Uner, N. B., Thimsen, E., & Foston, M. B. (2018). Accessing unconventional biofuels via reactions far from local equilibrium. *Fuel, 226*, 472–478.
64. Okubo, M., Saeki, N., & Yamamoto, T. (2008). Development of functional sportswear for controlling moisture and odor prepared by atmospheric pressure nonthermal plasma graft polymerization induced by RF glow discharge. *Journal of Electrostatics, 66*, 381–387.
65. Schäfer, J., Foest, R., Sigeneger, F., Loffhagen, D., Weltmann, K.-D., Martens, U., & Hippler, R. (2012). Study of thin film formation from silicon-containing precursors produced by an RF non-thermal plasma jet at atmospheric pressure. *Contributions to Plasma Physics, 52*, 872–880.
66. Penetrante, B. M., & Schultheis, S. E. (Eds.). (1993). *Non-thermal plasma techniques for pollution control*. Berlin Heidelberg: Springer.
67. Walton, S. G., Boris, D. R., Hernandez, S. C., Lock, E. H., Petrova, T. B., Petrov, G. M., & Fernsler, R. F. (2015). Electron beam generated plasmas for ultra low T e processing. *ECS Journal of Solid State Science and Technology, 4*, 5033–5040.
68. Walton, S. G., Hernández, S. C., Boris, D. R., Petrova, T. B., & Petrov, G. M. (2017). Electron beam generated plasmas for the processing of graphene. *Journal of Physics D: Applied Physics, 50*, 354001.
69. Leonhardt, D., Muratore, C., & Walton, S. G. (2005). Applications of electron-beam generated plasmas to materials processing. *IEEE Transactions on Plasma Science, 33*, 783–790.
70. Vitale, S. A., Hadidi, K., Cohn, D. R., & Bromberg, L. (1997). Decomposition of ethyl chloride and vinyl chloride in an electron beam generated plasma reactor. *Radiation Physics and Chemistry, 49*, 421–428.
71. Mizuno, A., Shimizu, K., Yanagihara, K., Kinoshita, K., Tsunoda, K., Kim, H. H., & Katsura, S. (1996). Effect of additives and catalysts on removal of nitrogen oxides using pulsed discharge. *IAS '96. Conference record 1996 IEEE industry application conference: Thirty-first IAS annual meeting,* San Diego, CA, USA, *3,* 1808–1812.
72. Whitehead, J. C. (2010). Plasma catalysis: A solution for environmental problems. *Pure and Applied Chemistry, 82*, 1329.
73. Yoshida, H., Marui, Z., Aoyama, M., Sugiura, J., & Mizuno, A. (1989). Removal of odor gas component utilizing plasma chemical reactions promoted by the partial discharge in a ferro-electric Pellet Layer. *J. Inst. Electrost. Japan, 13*, 425–430.
74. Kim, H.-H., Sugasawa, M., Hirata, H., Teramoto, Y., Kosuge, K., Negishi, N., & Ogata, A. (2013). Ozone-assisted catalysis of toluene with layered ZSM-5 and Ag/ZSM-5 zeolites. *Plasma Chemistry and Plasma Processing, 33*, 1083–1098.
75. Djéga-Mariadassou, G., Baudin, F., Khacef, A., Da Costa, P. (2012). NO_x abatement by plasma catalysis. In: *Plasma chemistry and catalysis in gases and liquids*. Weinheim, Germany: Wiley-VCH Verlag GmbH.
76. Penetrante, B. M., Brusasco, R. M., Merritt, B. T., & Vogtlin, G. E. (1999). Environmental applications of low-temperature plasmas. *Pure and Applied Chemistry, 71*, 1829–1835.
77. Mok, Y. S., Koh, D. J., Kim, K. T., & Nam, I.-S. (2003). Nonthermal plasma-enhanced catalytic removal of nitrogen oxides over V_2O_5/TiO_2 and Cr_2O_3/TiO_2. *Industrial and Engineering Chemistry Research, 42*, 2960–2967.
78. Bröer, S., & Hammer, T. (2000). Selective catalytic reduction of nitrogen oxides by combining a non-thermal plasma and a V_2O_5-WO_3/TiO_2 catalyst. *Applied Catalysis B: Environmental, 28*, 101–111.
79. Kim, H., Ogata, A., & Futamura, S. (2008). Oxygen partial pressure-dependent behavior of various catalysts for the total oxidation of VOCs using cycled system of adsorption and oxygen plasma. *Applied Catalysis B: Environmental, 79*, 356–367.

80. Mizuno, A., & Ito, H. (1990). Basic performance of an electrostatically augmented filter consisting of a packed ferroelectric pellet layer. *Journal of Electrostatics, 25*, 97–107.
81. Sentek, J., Krawczyk, K., Młotek, M., Kalczewska, M., Kroker, T., Kolb, T., Schenk, A., Gericke, K., & Schmidt-szałowski, K. (2010). Environmental Plasma-catalytic methane conversion with carbon dioxide in dielectric barrier discharges. *Applied Catalysis B: Environmental, 94*, 19–26.
82. Van Laer, K., & Bogaerts, A. (2015). Improving the conversion and energy efficiency of carbon dioxide splitting in a Zirconia-packed dielectric barrier discharge reactor. *Energy Technology, 3*, 1038–1044.
83. Michielsen, I., Uytdenhouwen, Y., Pype, J., Michielsen, B., Mertens, J., Reniers, F., Meynen, V., & Bogaerts, A. (2017). CO_2 dissociation in a packed bed DBD reactor: First steps towards a better understanding of plasma catalysis. *Chemical Engineering Journal, 326*, 477–488.
84. Zhang, Y., Van Laer, K., Neyts, E. C., & Bogaerts, A. (2016). Can plasma be formed in catalyst pores ? A modeling investigation. *Applied Catalysis B: Environmental, 185*, 56–67.
85. Babaeva, N. Y., & Kushner, M. J. (2010). Intracellular electric fields produced by dielectric barrier discharge treatment of skin. *Journal of Physics D: Applied Physics, 43*, 185206.
86. Wada, N. (2013). High efficiency and high concentration ozone generation. *J. Soc. Electr. Mater. Eng., 22*, 26–34.
87. Moreau, E. (2007). Airflow control by non-thermal plasma actuators. *Journal of Physics D: Applied Physics, 40*, 605.
88. Hensel, K., Sato, S., & Mizuno, A. (2008). Sliding discharge inside glass capillaries. *IEEE Transactions on Plasma Science, 36*, 1282–1283.
89. Sato, S., Hensel, K., Hayashi, H., Takashima, K., & Mizuno, A. (2009). Honeycomb discharge for diesel exhaust cleaning. *Journal of Electrostatics, 67*, 77–83.
90. Benard, N., Mizuno, A., & Moreau, E. (2015). Manipulation of a grid-generated mixing with an active honeycomb dielectric barrier plasma discharge. *Applied Physics Letters, 107*, 233508.

Chapter 3
Plasma-Catalyst Interactions

Hyun-Ha Kim, Yoshiyuki Teramoto, and Atsushi Ogata

Low-temperature catalytic reactions induced by nonthermal plasma (NTP) have been of great interest in plasma chemistry. The presence of plasma can alter the elementary steps in catalysis. The type of catalysts also provides substantial changes in plasma properties. This chapter reviews the current progress in the experimental observations and the understanding of bilateral interactions of NTP with catalysts. Among the various types of combinations, this chapter will mostly focus on the single-stage plasma-driven catalysis (PDC). Several lines of experimental evidence on the interaction are introduced: type of reactor, propagation of discharge streamer, key components of plasma, and catalyst durability.

3.1 From Plasma Alone to Plasma Catalysis

3.1.1 Electrical Discharges and Plasma

Electrical discharges can form plasma over a wide range of conditions: pressure, gas mixture, and temperature. In addition to various electrical discharges, electron-beam irradiation and gamma or X-ray also have a long history of being used over the past centuries. The latter is also referred to as radiation chemistry. Ozone generation by dielectric barrier discharge (DBD) has been a landmark of plasma applications in industry. From the 1980s, atmospheric pressure non-equilibrium plasma has gathered attention compared to the low-pressure (vacuum) plasmas. Pulsed corona discharges and DBD have been the two representative tools for generating NTP under mild conditions (i.e., ambient temperature and atmospheric pressure). Both

H.-H. Kim (✉) · Y. Teramoto · A. Ogata
National Institute of Advanced Industrial Science and Technology (AIST),
Tsukuba, Ibaraki, Japan
e-mail: hyun-ha.kim@aist.go.jp

X. Tu et al. (eds.), *Plasma Catalysis*, Springer Series on Atomic, Optical, and Plasma Physics 106, https://doi.org/10.1007/978-3-030-05189-1_3

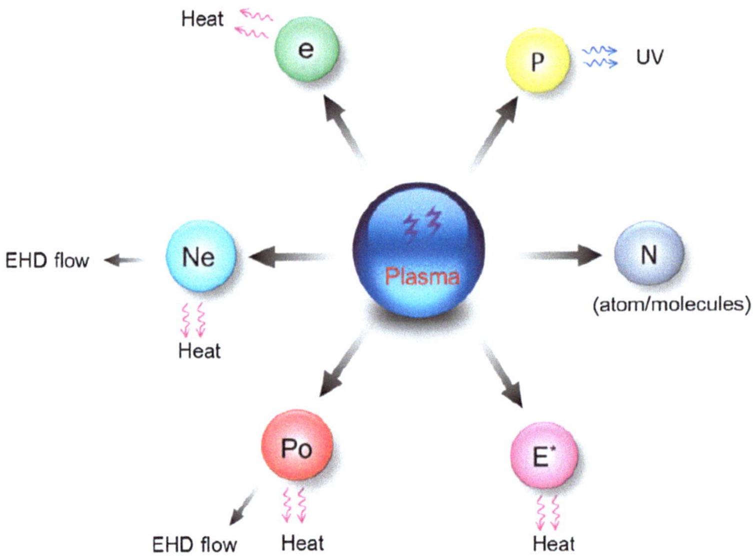

Fig. 3.1 Basic six particles in plasma: e electron, P photon, N neutrals (reactive species), E∗ excited molecules, Po positive ions, Ne negative ions [1]

electrical discharges and ionizing irradiation can generate plasma, which consists of six major particles, as shown in Fig. 3.1. Electron temperature (T_e) is often used to classify plasmas into "thermal" and "nonthermal." In NTP, T_e is considerably higher than that of the ions (T_i) and ambient gas (T_g): $T_e \gg T_i \approx T_g$. These intrinsic advantages of NTPs attracted researchers to look at environmental remediation [2, 3], such as removal of NO_x and SO_x from flue gases [4–6], odor removal, and destruction of volatile organic compounds (VOCs) [7] or hazardous air pollutants (HAPs). Extensive studies in the 1990s shed light on the merits and demerits of using NTP processes for environmental applications [4, 7, 8]. The demerits include high energy consumption and the formation of unwanted byproducts [9, 10]. Plasma catalysis has been proposed as one of possible alternatives to pure plasma processes. The interaction of these six particles and the catalyst will be discussed in Sect. 3.3.

Early reports on the catalytic effect of electrode materials in corona discharges may date back to the 1920s [11]. Figure 3.2 illustrates the scheme on how O_3 formation is suppressed by the catalytic action of electrode materials. Silver was found to be most prominent in O_3 suppression. Later, these effects were found to be related with the electrode oxidation/reduction of electrode surface (aging condition) and catalytic activity toward O_3 decomposition [12–14]. For example, it has been reported that an O atom fixed on Ag (Ag-O) can catalytically convert O_3 to O_2 [15], which may explain the O_2 suppression with an Ag electrode. Sporadic studies have been reported over several decades [16], and studies on plasma catalysis became widespread after 2000. Surface-plasma interactions have also been found in plasma-enhanced chemical vapor deposition (PE-CVD), sputtering, and plasma etching, but

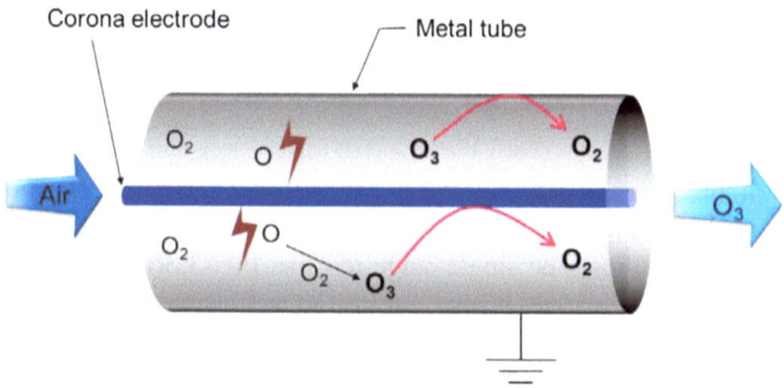

Fig. 3.2 Catalytic decomposition of O_3 on the surface of electrodes [12]

these processes are usually operated at low pressure or vacuum, whereas the pressure range of plasma catalysis is basically atmospheric pressure.

3.1.2 Beyond the Incompatibility

In principle, low temperature and high activity are completely incompatible in conventional catalysis. The most important issue in plasma catalysis is the low working temperature below or near the light-off temperature in catalyst-alone processes. This temperature isolation can make the two processes cooperative rather than having a competitive relationship. The combination of NTP and catalyst aims to achieve a synergistic effect, which enables better removal efficiency, selectivity, and carbon balance [17–20]. Plasma can generate highly reactive chemical species (both oxidative and reductive) even at ambient temperature and pressure. This complementary combination is the essence of a plasma-catalyst. Plasma enables catalysts to be effective at lower temperatures. Catalysts provide a better selectivity by providing new chemical pathways via surface reactions. There is a large body of literature reporting synergistic effects in plasma catalysis [1]. However, some of the results were obtained under extremely high specific input energy, where the heating is not negligible (see also Sect. 3.3.4). The number of catalysts (combinations between active metals and supports) tested for plasma catalysis is rapidly increasing, but there still exists tremendous number of catalysts hitherto unstudied. Even for a given combination, catalytic activity changes depending on the loading amount, size, and redox condition of active metals. A number of reviews and references are available for the applications of plasma catalysis, which include steam reforming [21], dry reforming with CO_2 [22, 23], CO/CO_2 methanation [24], hydrogenation of CO_2 [25], de-NO_x [26–31], NO_x synthesis [32], NH_3 synthesis [33–35], volatile organic compound (VOC) removal [36], and water purification [37]. This chapter will mostly discuss the lateral interactions between plasma and catalyst to draw the best guideline available as of now.

3.2 Plasma Catalysis Reactors

3.2.1 Reactor Classification

Classification of plasma catalysis reactors is usually based on the position and number of catalysts. Figure 3.3 schematically illustrates the typical reactors used in plasma-catalyst studies. The single-stage reactor (Fig. 3.3(i)), where catalysts are located inside the plasma zone, is also referred to as plasma-driven catalysis (PDC) or in-plasma catalysis (IPC). The term "PDC" was first used in 1999 to describe the combined process of pulsed corona and TiO_2 catalyst for the removal of NO_x [38]. The presence of a dielectric barrier (glass, ceramic, quartz, etc.) can prevent a transition into a spark discharge. Bilateral interactions between plasma and in PDC reactor impose difficulties in understanding the elementary processes. Each elementary property becomes interdependent when catalysts are placed in the plasma zone. In contrast, catalysts are located downstream of the plasma reactor in the two-stage reactor (Fig. 3.3(ii)). In chemical viewpoint, short-lived radicals ($O(^3P)$, $O(^1D)$, OH, N) cannot be utilized in the two-stage configuration. Ozone-assisted catalysis is one of the most well-known two-stage reactors. A multi-stage process (Fig. 3.3(iii)) could be an interesting option for the industrial application. The most important aspect is that each catalyst has a different function along the flow direction, which can be fine-tuned for specific purposes [39–42]. Further oxidation of byproducts or intermediates and reduction of O_3 could be optimized by using multi-stage configurations.

The VOC oxidation in single-stage PDC was found to be accelerated by increasing O_2 content in the gas mixture [43, 44] while this trend is not observable in either pure plasma or catalyst processes. The cycled system for VOC oxidation (Fig. 3.3 (iv)) is based on this unique O_2-dependent behavior. The operation of the cycled system is quite similar to the conventional adsorption, except for the catalyst regeneration using oxygen plasma. Once the catalyst bed reaches adsorption saturation, the catalyst bed is purged with pure O_2 and then plasma is turned on to decompose adsorbed VOCs. The primary advantages are the fast and energy-efficient reduction/oxidation under highly controlled environment (i.e., pure O_2 or N_2). Many recent studies confirmed the advantage of the cycled system for the removal of benzene [45–47], toluene [48–51], and HCHO [52] using a variety of adsorbent/catalyst combinations. Adsorption followed by treatment with air plasma can also enhance the energy efficiency of VOC removal [53–55]. However, NO_x formation is inevitable in air, so the input power (i.e., treating capacity) should be carefully considered to meet the regulation of NO_x emission. Oxygen plasma instead completely suppresses NO_x formation during the oxidation of adsorbed VOCs. TiO_2 is a highly sensitive catalyst to O_2 content, but the low adsorption capacity limited its potential for the cycled system. Despite the large surface area and the well-established performance, activated carbon is not suitable for the cycled system due to its electrical conducting nature (electrical resistivity of ~10^{-1} Ω·cm) [56]. Up to

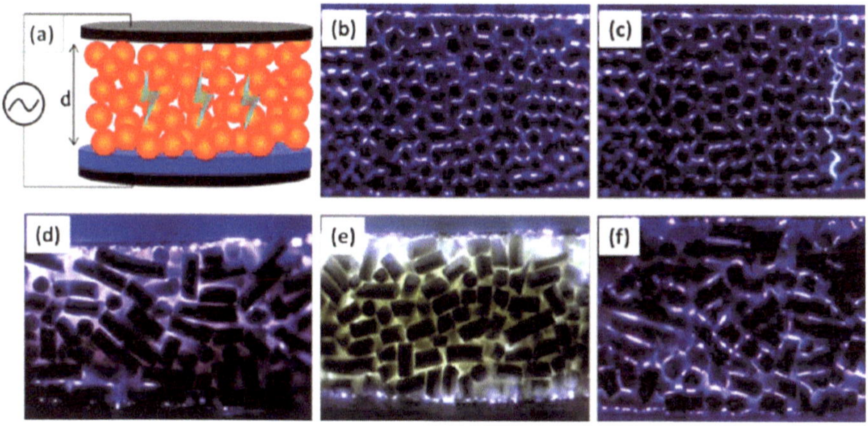

(i)

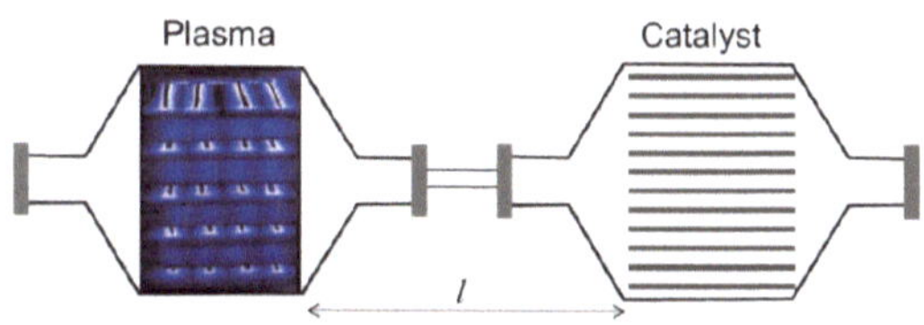

(ii)

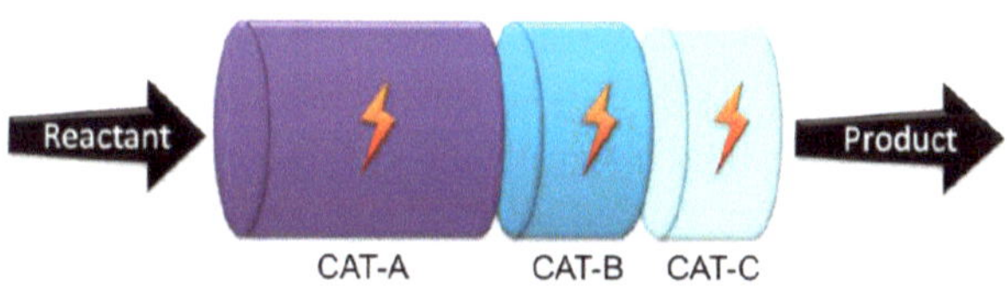

(iii)

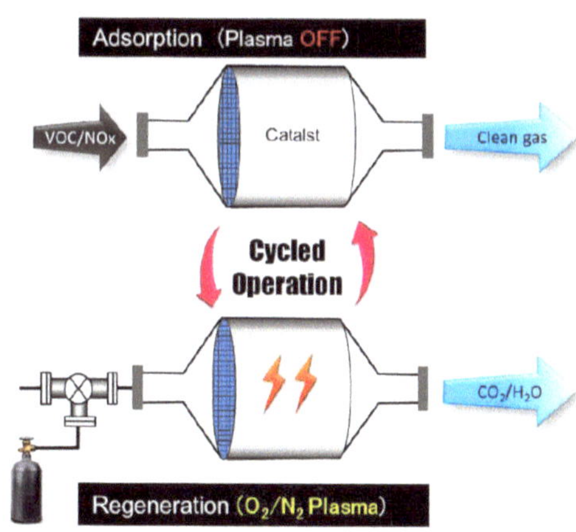

(iv)

Fig. 3.3 Plasma-catalyst reactors. (i) Single-stage plasma-catalyst reactor. (**a**) Typical plane-type packed bed reactor, (**b**) $BaTiO_3$ (plasma on), (**c**) $BaTiO_3$ (plasma with spark discharges), (**d**) bare HY, (**e**) 2 wt% Ag/HY, (**f**) 5 wt% Ag/Ms-13X (**b~f**, mesh-to-mesh electrode) (ii) Two-stage plasma-catalyst reactor. (iii) Multi-stage plasma-catalyst reactor [1]. (iv) Cycled system (adsorption ↔ regeneration using O_2 or N_2 plasma)

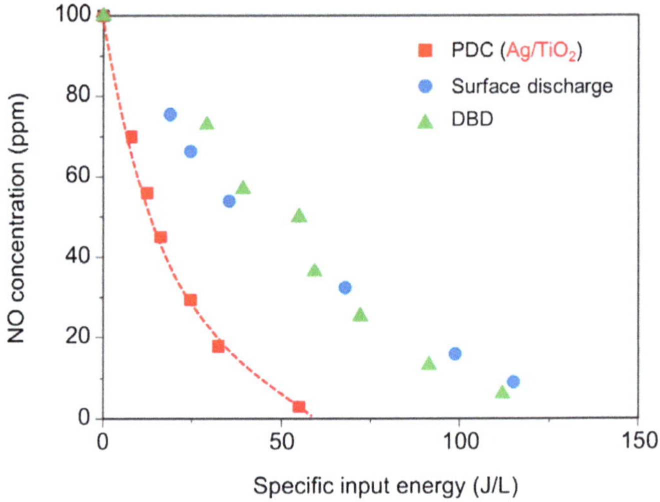

Fig. 3.4 Comparison of PDC and plasma alone (surface discharge, DBD) in NO removal in N_2 at 373 K. Energy constant k_E for the PDC and plasma-alone reactors were $49 \pm 1 \times 10^{-3}$ L/J and $19 \pm 4 \times 10^{-3}$ L/J, respectively [60]

date, metal nanoparticle-loaded zeolites satisfy the requirements of the cycled system: adsorption capacity and high catalytic activity under an O_2 environment.

A cycled system has been studied for NO_x reduction into N_2 [57–59]. In these cases, N_2 plasma is used to form N atoms for NO_x reduction. It is interesting to see that PDC is also superior to plasma-alone process as shown in Fig. 3.4 for the removal of 100 ppm of NO in N_2. Comparison of the energy constant k_E revealed [60] that the PDC reactor packed with Ag/TiO_2 catalyst showed 2.6 times more efficiency than the pure plasma reactor. Fixation of radicals on the surface of catalyst and its chemical reaction are further discussed in Sect. 3.3.3.

3.2.2 *Equivalent Electrical Circuit*

The physicochemical parameters of catalysts (i.e., loading material and their size distribution, supporter, pore size, void volume, surface area, etc.) are important parameters in determining the catalytic activity. Except for activated carbon, most of catalytic materials are electrically insulating. In contrast to a two-stage reactor, resistivity, capacitance, and dielectric constant of catalysts are key components in a single-stage plasma-catalyst reactor. Figure 3.5 illustrates the electrical equivalent circuit of (a) dielectric-barrier discharge and (b) single-stage plasma-catalyst reactor. In general, DBD is represented by the bipolar Zener diode [62] or time-varying resistance $R(t)$ [63]. In the case of a single-stage plasma-catalyst reactor, the resistance (R_C) and capacitance (C_C) of the catalyst are connected in series to the air gap.

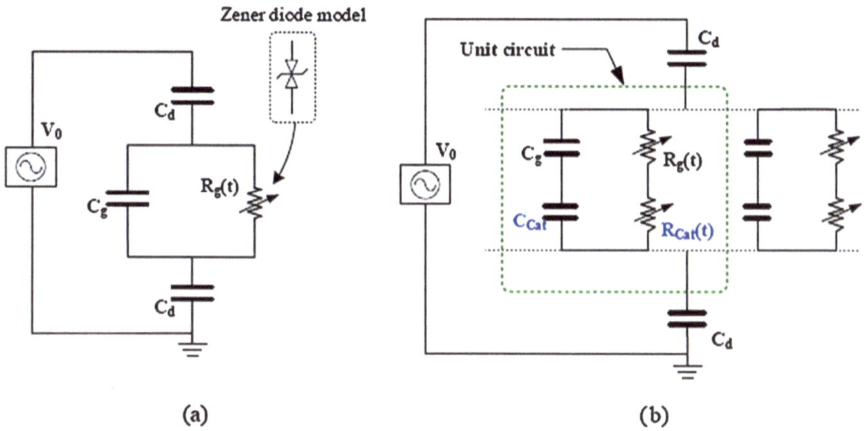

Fig. 3.5 Equivalent electrical circuit of plasma-catalyst reactor [61] (**a**) DBD reactor, (**b**) Single-stage plasma-catalyst reactor

This capacitance in series divides the applied voltage on the air gap more, which leads to a favorable condition for partial discharges at the contact points of catalysts. Plasma onset voltage is usually lowered when catalysts are packed in DBD reactors [64]. This can be simply explained by the reduced gap distance between the contact point of catalyst pellets. Dielectric constant of packing materials also influences the discharge power and plasma onset voltage [65]. In other words, changing packing material from one to another can alter the equivalent circuit and the plasma generation. A Lissajous figure and current waveforms are handy methods to monitor the change in the plasma properties [66]. When Ag was impregnated on a given framework of HY zeolites, the oxidation state of Ag substantially changed according to the Si/Al ratio. The higher the Si/Al ratio, the larger the drop of electrical resistivity, in turn leading to a diminished propagation of surface streamers [67]. This observation clearly indicated that the electrical parameters of each of the packing materials are highly important in determining not only the propagation of surface streamers but also the catalytic activity under oxygen plasma.

3.2.3 *Use of Plasma for Catalyst (Preparation, Reaction, and Regeneration)*

Plasma can be utilized for catalysis in three different ways: preparation, reaction, and regeneration. Firstly, plasma is used for catalyst preparation such as precursor decomposition [68], template removal [69], reduction of active metal nanoparticles [70, 71], and stable dispersion [72]. A group from Tianjin University studied the effects of glow plasma in the preparation of various catalysts and reported enhanced dispersity and higher activity [68, 73]. Menard et al. compared thermal and ozone treatments for the removal of the protecting ligand in Au cluster catalysts [74]. They reported that the ozone-treated Au/TiO_2 showed smaller mean diameter, higher activity, and better thermal stability. Residual chloride in the prepared catalyst is a poison of active site, so subsequent treatment is a crucial step before use. Secondly, plasma can enhance catalytic reactions, which is the key aspect in single-stage PDC reactor. Details of the plasma effect on catalysis will be discussed in Sect. 3.3.

Thirdly, regeneration of the catalyst could be also achieved by NTP. Plasma regeneration strongly depends on the type of deactivation. When the coverage by organic substances is the main cause of deactivation, there is a possibility for plasma to be effective. Low-pressure glow discharge in Ar-O_2 mixture was found to be effective for the decoking of catalysts [75, 76]. Nano-sized Au catalysts are well-known for high activity toward CO and H_2 oxidation at room temperature, or even down to below 200 K. The high catalytic activity for CO oxidation quickly drops when Au/TiO_2 is exposed to 100 ppm of toluene or C_3H_6 [77]. Heating is not suitable for a nano-sized Au catalyst due to the thermal agglomeration of Au particles that eventually leads to irreversible deactivation. However, oxygen plasma can regenerate Au/TiO_2 catalyst at ambient temperature [77, 78]. Interestingly, plasma can promote CO oxidation over the fully deactivated Au/TiO_2 catalyst by the adsorption of hydrocarbons (propylene and toluene). Plasma can directly supply reactive oxygen species to the surface of Au from the gas phase leading to the oxidation of CO. We referred to this type of reaction as a "reverse Eley-Rideal"-type reaction [1]. In this case, plasma not only regenerates the deactivated surface but also promotes the chemical reaction.

The shape of supported metal catalysts is known to change according to reaction conditions. Recently, Yoshida et al. visualized the shape change of Au on CeO_2 by means of aberration-corrected ETEM (environmental transmission electron microscopy) [79]. Plasma can form strongly oxidative or reducing conditions depending on the gas mixture. This transient behavior will be an interesting subject worth studying in the future.

3.2.4 Elementary Steps in Plasma Catalysis

Both conventional and plasma catalysis consist of the common elementary steps: adsorption and surface diffusion of reactants, surface reaction, and desorption of products. The characteristic difference between them is the driving force, i.e., heat vs plasma. Each of these elementary steps influences the overall reaction rate. In many cases, catalytic reactions are kinetically expressed by the Langmuir-Hinshelwood (L-H) mechanism as shown in Eq. (3.1). The reaction rate coefficient k can be expressed by the Arrhenius Eq. (3.2)

$$r = -\frac{dC}{dt} = \frac{k\mathrm{KC}}{1 + \mathrm{KC}} \tag{3.1}$$

$$k = A\,exp\left(-\frac{E_a}{RT}\right) \tag{3.2}$$

E_a and K denote the activation energy and equilibrium constant, respectively. The smaller the E_a and the higher the T, the larger the reaction rate coefficient k becomes. The presence of plasma can alter these four factors (k, K, E_a, and T), which eventually leads to synergistic effect in plasma catalysis. Clearly, plasma provides new elementary steps to catalysis, which in turn leads to either a larger k by decreasing E_a or a better selectivity. It should be noted that the Arrhenius equation is based on a state in thermal equilibrium while plasma catalysis is in a non-equilibrium state. In this sense, careful attention should be paid before implementing an Arrhenius plot to the data analysis for plasma catalysis.

3.3 Players in Catalyst Activation

3.3.1 Time-Scale Matching

Figure 3.6 compares the characteristic time scales of the elementary processes in plasma and catalytic reactions. Microdischarges in DBD and streamers in pulsed corona have a typical duration of 10^{-8} s, which can be easily monitored by current waveforms. Excitation and electron-impact dissociation lead to formation of reactive

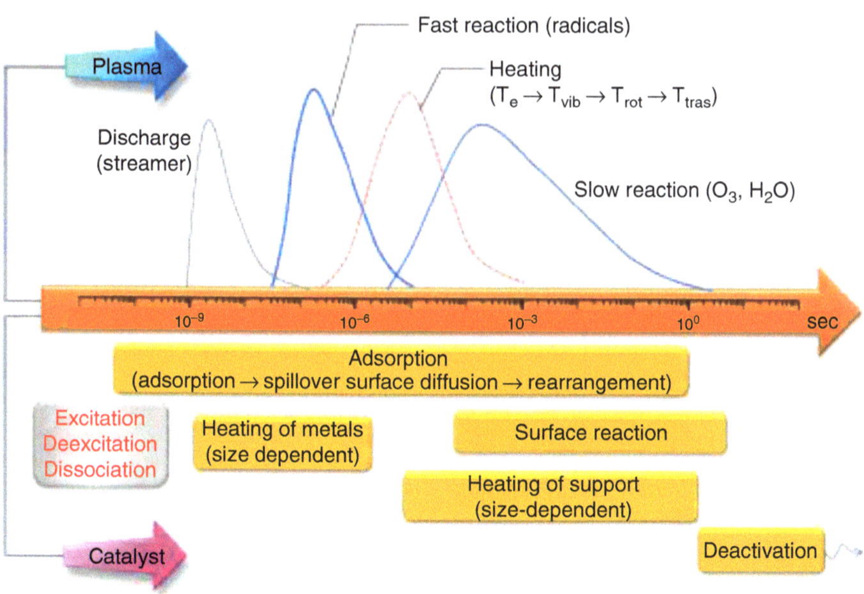

Fig. 3.6 Characteristic time scales in plasma and catalytic reaction [1]

species. Radical reactions have typical time scales of 10^{-5} s, and recombination of radicals yields relatively stable species (O_3 and H_2O_2) which contributes to slower reaction times. In a catalytic reaction, the initial chemisorption step occurs within 10^{-9} s. However, the total adsorption process proceeds over a significantly longer time scale (up to 100 s), as a result of surface rearrangement via spillover and surface diffusion, along with gas-phase diffusion into the micropores. The catalytic turnover time scale is known to be in the range of 10^{-2} to 10^2 s [80], which is slower than the fast radical reactions by a factor of 2–3 orders of magnitude. This different time scale suggests that the rate-limiting step is a surface reaction rather than a plasma-involved process in the gas phase.

3.3.2 Less Important Particles (Electrons, Photons, and Ions)

As illustrated in Fig. 3.1, plasma chemistry is the combined actions of the six major particles. Among them, neutral atoms (i.e., radicals) and molecules (O_3, H_2O_2) are generally believed to play a key role in most of the chemical reactions. The electron density and its energy distribution have been studied intensively in order to understand the physical properties of plasmas. Ion chemistry (ion molecules) proceeds even faster than radical reactions, but their direct contributions to the overall reaction are considered to be negligible due to the dilute concentration of pollutants [21]. In a typical gas composition, i.e., air containing pollutants in ppm level, most of electron energy is preferentially transferred to the dominant gases (O_2 and N_2), which leads to the formation of reactive species.

Various methods of optical diagnostics have been developed and noticeable advancements have been made during the last decade. Optical emission spectroscopy (OES) has indicated that plasma in an air-like mixture produces UV light, which has potential to activate catalysts. This qualitative information led researchers to speculate the contribution of UV light as a possible reason for synergistic effects in plasma catalysis [38]. The solar flux on the Earth's surface is about 100 mW/cm^2 (137 mW/cm^2 at the top of the atmosphere, solar constant), and the UV flux occupies about 7 mW/cm^2. Recent quantitative measurements of UV photon flux in air revealed that it is only on the order of several μW/cm^2 [81–83], which is about three orders of magnitude less than that of the outdoor solar UV flux. The contribution of UV light in plasma catalysis is a matter of the UV photon flux, but not the mechanism itself. When external UV with a photon flux of several tens mW/cm^2 was used to irradiate a plasma-catalyst reactor, a measurable enhancement was observed [84, 85].

3.3.3 Important Particles (Radicals, Excited Molecules)

The formation and contribution of various radicals ($O(^3P)$, $O(^1D)$, OH, N) and reactive molecules (O_3, H_2O_2) have been considered key factors in determining the plasma chemistry. Laser diagnostics have been implemented to measure time- and space-resolved formation and decay of these reactive species [86].

Radical fixation on the surface of a catalyst is a unique pathway in plasma catalysis. Generally, the lifetime of radicals in gas phase lies in the order of 10^{-5} s, but the surface-fixed species can survive over several tens of minutes [87–89] or even hours [90]. Guaitella et al. studied adsorption/reaction kinetics of C_2H_2 on the surface of plasma-treated SiO_2 and TiO_2 catalysts and found that O_2 or air plasma-pretreated TiO_2 can remove C_2H_2 while no removal occurred with the SiO_2 catalyst [87]. The NO measurement using in situ tunable diode laser absorption spectroscopy (TDLAS) revealed that RF plasma pretreatment can fix N atoms (in N_2) [88] or O atoms (in synthetic air) [91] on Pyrex which were stable over several tens of minutes.

Carbon monoxide (CO) is one of the major byproducts in VOC removal. CO oxidation in gas phase is limited due to the slow reaction rate coefficients.

$$CO + OH \rightarrow CO_2 + H \ ; \quad k = 1.4 \times 10^{-13} cm^3/(molecule \cdot s) \tag{3.R1}$$

$$CO + O \rightarrow CO_2; \quad k = 1.4 \times 10^{-17} cm^3/(molecule \cdot s) \tag{3.R2}$$

$$CO + O_3 \rightarrow CO_2 + O_2 \ ; \quad k = 4.0 \times 10^{-25} cm^3/(molecule \cdot s) \tag{3.R3}$$

Ozone has quite poor reactivity in gas phase with most of the chemicals, except for hydrocarbons containing a C=C double bond. However, the long lifetime of ozone enables a two-stage configuration. Furthermore, ozone-metal interactions can modify its reactivity. For example, Au-O is formed when nanoporous Au is exposed to O_3, even at room temperature. Biener et al. demonstrated that ozone-induced oxygen uptake is only limited to a monolayer rather than fully oxidizing the nanoporous Au [92]. The Au-O successfully oxidized CO into CO_2.

$$Au - O + CO \rightarrow CO_2 + Au \tag{3.R4}$$

Excited molecules exhibit a much larger reactivity when compared to the ground state. The contribution of vibrationally excited molecules has also been discussed in surface chemistry [93]. Nozaki et al. reported that vibrationally excited methane could be used to improve dissociative chemisorption on Ni/SiO_2 catalyst [21, 94].

3.3.4 Additional Effect (Heat)

In principle, plasma activates catalysts in plasma catalysis in the same way as heat, when looking at conventional catalysis. The electrical energy injected into plasma reactors is eventually converted into heat, so the temperature increase is a function of how much energy has been injected into a given reactor. Heat released from plasma can also play a role in plasma catalysis. The sources of heat in plasma catalysis can be divided into electrical and chemical factors. Electrical factors include both dielectric loss in barrier and catalysts and joule heating by the current flow. Chemical effects are a combination of vibrational-vibrational (V-V) and a vibrational-transitional (V-T) relaxation, as well as the heat of reactions (i.e., exothermal reactions). The heat of a reaction is in the range of several thousand kJ/mole. However, the concentration of organic pollutants is usually in several hundred ppmv level, so its contribution to heating process is not significant in treating dilute VOCs. The degree of heating is primarily dependent on the power input into a plasma reactor. The measured rates of temperature increase were determined to be approximately 30~40 K per 100 JL^{-1} [61]. The gas temperature and gas composition change along the flow direction, so the main reactions can be sectionalized from inlet to outlet. The inlet section should be effective for the initiation reaction, while the middle and end parts are for intermediates and unwanted byproducts (CO or O_3). In the case of VOC oxidation, the usual intermediates are aldehydes and CO. Some catalysts exhibit even higher activity toward these intermediates than the reactants. It is therefore expected that contribution of thermal catalytic reaction can vary with the position.

$$\text{Reactant(inlet)} \rightarrow \text{Intermediates(middle)} \rightarrow \text{Products(outlet)} \qquad (3.R5)$$

For example, decomposition of aromatic hydrocarbons requires temperatures of about 473 K. Specific input energy of 200 J/L in air leads to a temperature increase of about 75~95 K. This temperature is far less than that required for the thermal catalysis of most of aromatic compounds but sufficient for the light-off of CO oxidation (threshold temperature for CO oxidation). The Au catalyst on the oxide supports (TiO_2, CeO_2, and Fe_2O_3) is a well-known catalyst for CO oxidation, even at temperatures as low as 200 K [95]. Jiang et al. reported the complete oxidation of C_2H_4 over a Pt/MCM-41 catalyst at room temperature [96]. These results clearly suggest that the heating effect in plasma catalysis should be considered as depending not only on the substances under reaction but also on the type of catalysts. From the perspective of reactor design, a heat-tight configuration will be highly important for endothermic reactions, while cooling will be beneficial for exothermic reactions [1].

3.3.5 Probing Interaction of Plasma and Catalyst

Despite the various experimental approaches, a clear understanding of the mechanism is still elusive. The development of in situ measurement techniques has played an important role both in catalysis and NTP studies. However, the use of these well-established in situ techniques in catalyst study (such as in situ X-ray powder diffraction (XRD) with heating and gas flow, diffuse reflectance infrared Fourier transform spectroscopy (DRIFTS), and in situ Raman spectroscopy) is hampered in plasma catalysis due to the presence of high-voltage and electromagnetic noise. Recently, DRIFTS has been used in measuring plasma-induced surface reactions, such as hydrocarbon-assisted selective catalytic reduction of NO_x [97] and VOC decomposition over three metal oxides (γ-Al_2O_3, TiO_2, and CeO_2) [98]. Plasma-induced fixation of oxygen species over Ag-supported TiO_2 and MS-13X was studied using isotopically labeled oxygen ($^{18}O_2$) and an online mass spectrometer [89]. The amount of surface-fixed oxygen species was proportional to the amount of Ag loaded, and their lifetimes were found to be up to 10 h, based on the NO titration method [90].

Laser-induced fluorescence (LIF) has been used in the nondestructive diagnostics of reactive species with a high sensitivity [86]. Su et al. used a planar LIF imaging and near-IR thermography for high-throughput screening of active catalysts for naphthalene oxidation by detecting fluorescence signals using a CCD camera [99, 100]. A total of 15 catalyst samples were mounted in a flow cell reactor heated at 503–643 K; the fluorescence emissions of 515–546 nm from the reaction product (naphthoquinone) were measured just above the catalysts. Application of these techniques in plasma catalysis may accelerate the progress of plasma catalysis by providing fast and high-throughput screening of catalysts.

3.4 Discharge Characteristics and Surface Streamers

3.4.1 Surface Streamers

Two characteristic discharge modes are found in a catalyst-packed plasma reactor: partial discharge (PD) and surface streamers [101, 102]. Partial discharge consists of numerous microdischarges at the contact points of catalysts where the electric field is augmented due to large dielectric constants of the packing materials [65, 103–105]. The PD appears as a spot, and its size becomes larger as the applied voltage increases. The second type of discharge is surface streamers. Propagation of surface streamers was found to highly correlate with performance of plasma catalysis [101]. This kind of correlation is a new insight hitherto never considered. Investigation of a surface streamer requires an intensified CCD camera because of its low intensity. Figure 3.7 compares three random images of plasmas in air in the absence (a) and the presence of Ag nanoparticles (b and c). In the case of bare MS-13X (a),

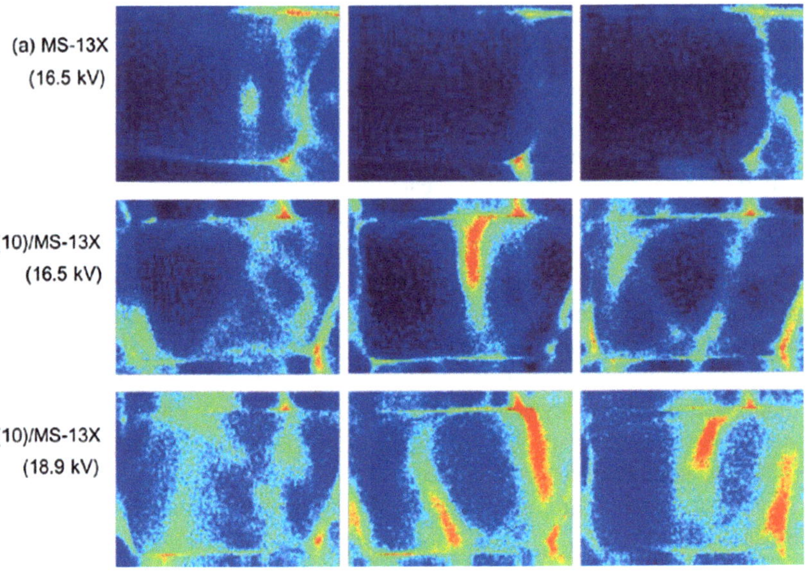

Fig. 3.7 Three random images of plasma generation for (**a**) bare MS-13X (16.5 kV) and (**b** and **c**) Ag(10)/MS-13X. Frequency and ICCD gate time were 50 Hz and 10 ms, respectively

PD was dominant at contact point of pellets, and surface streamers appeared in limited number and position. On the other hand, surface streamers became dominant for the Ag(10)/MS-13X (b and c). Of course, the bare MS-13X exhibits poor VOC removal, CO_2 selectivity, and carbon balance. A considerable enhancement can be made when Ag nanoparticles are loaded on the MS-13X. This physical interaction is a completely new category determining the catalytic activity.

Figure 3.8 shows the ns-gated images of surface streamers propagating on the bare γ-alumina. Total emission (b) and $N_2(C)$ emission (c) obtained with a band-pass filter showed a similar pattern. $N_2(C)$ emission, which is also referred to as second positive band of $N_2(C^3\Pi_u) \rightarrow N_2(B^3\Pi_g)$ transition, requires threshold energy of 11 eV. In an air-like discharge, the $N_2(C)$ emission at 337.1 nm is the most prominent line regardless of the type of plasma reactor. A distinctive feature of the surface streamers was a sheet-like flat structure with a width of about 500 μm, which is different from the gas-phase streamer. Gas-phase streamer appearing at the gap of neighboring pellets in (b) had a diameter of about 150 μm, which is consistent with literature values for barrier discharge [106, 107].

Figure 3.9 shows a sequence of surface streamers on a 2 wt% Ag/TiO_2 catalyst up to 14 ns. The primary streamer (PS) started not only from the tip of the point electrode but also from the catalyst contact points along the needle electrode. When the PS reached a contact point with the next bead, the discharge intensity became stronger by merging the PS1 and PD at 6 ns. These observations clearly indicated that PD can be a staging point during the propagation of surface streamers over multiple catalyst beads. Propagation velocity of PS was about 250 km/s at 7.1 kV and increased linearly with the applied voltage. However, it is much slower than those reported values for PS (500–3000 km/s) in gas phase [107, 109–111].

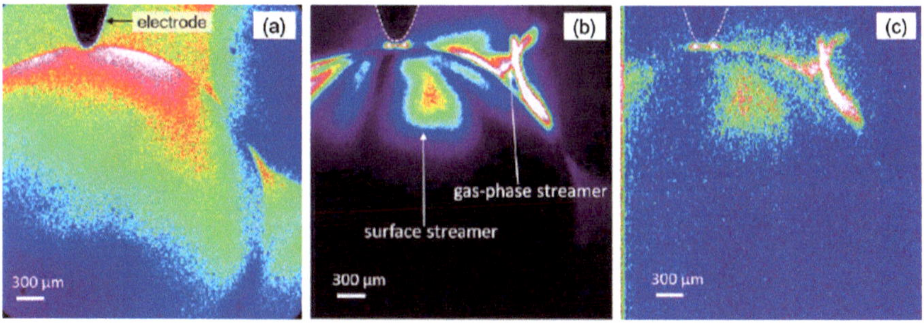

Fig. 3.8 Surface streamers on bare alumina with a point-to-plate configuration: (**a**) observation area (without discharge), (**b**) total emission, and (**c**) excited N_2 emission of $N_2(C_3\Pi_u) \rightarrow N_2(B_3\Pi_g)$ transition at 337.1 nm (the second positive band). Applied voltage was 9.3 kV

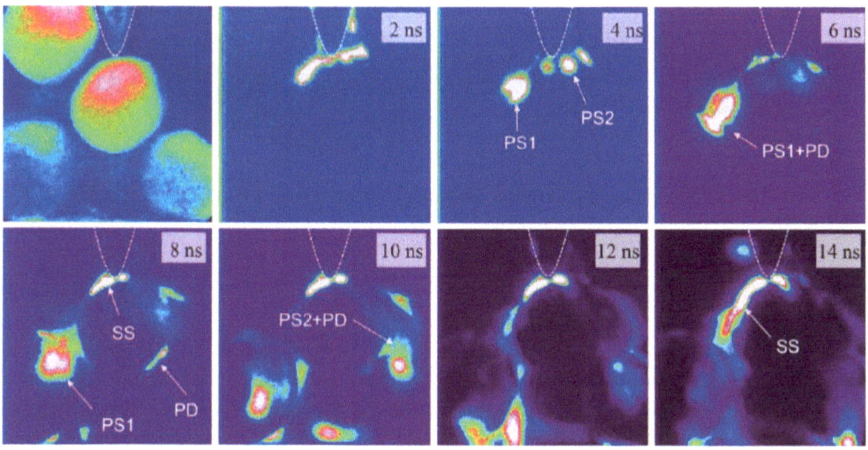

Fig. 3.9 Time-resolved images of surface streamer on Ag(2)/TiO_2 catalyst (at 7.1 kV, 50 scans with 3-ns gate time) [108]

3.4.2 Criteria for Interaction

As long as a catalyst is used below its light-off temperature, radical transfer from plasma to surface is the essence for interaction. Propagation of surface streamer can be a measure testing compatibility of the catalytic materials in plasma catalysis. A poor propagation of surface streamers indicates less interaction area that is detrimental to catalytic reactions. Figure 3.10 schematically illustrates the conditions necessary for the interactions between plasma-induced active species and catalyst surface. The dimensionless parameter Λ describes the criteria for the interaction between reactive species and the catalyst [61].

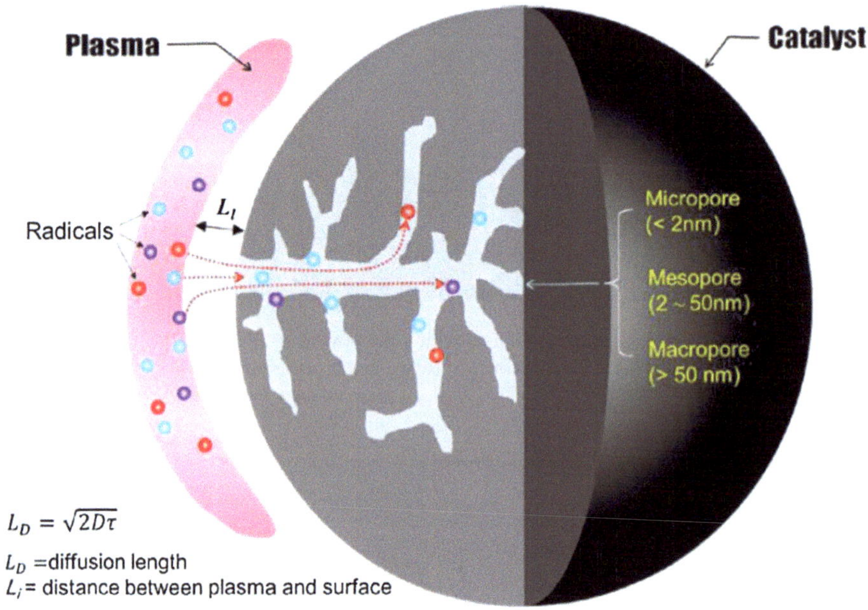

Fig. 3.10 Criteria for interaction between plasma and catalyst based on diffusion length (L_D) and distance between plasma and surface (L_l)

$$\Lambda = \frac{L_l}{L_D} \leq 1 \tag{3.3}$$

The L_D and L_l indicate the diffusion length of reactive species and distance between the plasma and catalyst. The diffusion lengths of radicals under atmospheric pressure are about 60 μm [61]. These simple criteria indicate that the L_l should be shorter than 60 μm to guarantee the interactions. The parameter Λ can also be applied to other plasma processes such as sterilization, plasma medicine, and surface treatment. In this regard, the closer the distance L_l, the higher the chance for interaction between plasma and target materials (i.e., catalyst in the case of plasma catalysis).

It is clear that the presence of surface streamers and its characteristics are important factors determining good catalytic activity in plasma catalysis. Streamer imaging can be used as rough but fast tool for catalyst screening whether it is active in plasma catalysis.

3.5 Conclusion Remarks

In summary, this chapter addressed some multidisciplinary characteristics regarding interactions of NTP and catalysts possess. The effect of plasma on catalysis was discussed based on the six major particles as well as the heat. The electrical characteristics of catalysts can greatly change the properties of plasma generation. It is therefore necessary to look at the process from multidisciplinary viewpoints to draw out useful insights into the fundamental mechanism. Especially, the recent progress on the surface streamers is presented including ns-gated time-resolved ICCD imaging. Further studies are necessary to understand the working mechanism of surface streamers and the intrinsic interactions between surface streamer and catalysts.

Acknowledgments The authors would like to acknowledge financial support from the JSPS KAKENHI Grant Number JP26400539.

References

1. Kim, H. H., Teramoto, Y., Ogata, A., Takagi, H., & Nanba, T. (2016). Plasma catalysis for environmental treatment and energy applications. *Plasma Chemistry and Plasma Processing, 36*, 45–72.
2. Penetrante, B. M., & Schultheis, S. E. (1993). *Non-thermal plasma techniques for pollution control: Part a: Overview, fundamentals and supporting technologies*. New York: Springer.
3. Penetrante, B. M., & Schultheis, S. E. (1993). *Non-thermal plasma techniques for pollution control: Part B: Electron beam and electrical discharge processing*. New York: Springer.
4. Mizuno, A., Clements, J. S., & Davis, R. H. (1986). A method for the removal of sulfur dioxide from exhaust gas utilizing pulsed streamer corona for electron energization. *IEEE Transactions on Industry Applications, 22*, 516–522.
5. Dinelli, G., Civitano, L., & Rea, M. (1990). Industrial experiments on pulsed corona simultaneous removal of NO_x and SO_2 from flue gas. *IEEE Transactions on Industry Applications, 26*, 535–541.
6. Masuda, S., & Nakao, K. (1990). Control of NO_x by positive and negative pulsed corona discharges. *IEEE Transactions on Industry Applications, 26*, 374–383.
7. Nunez, C. M., Ramsey, G. H., Ponder, W. H., Abbott, J. H., Hamel, L. E., & Kariher, P. H. (1993). Corona destruction: An innovative control technology for VOCs and air toxics. *Journal of the Air & Waste Management Association, 42*, 242–247.
8. Clements, J. S., Mizuno, A., Finney, W. C., & Davis, R. H. (1989). Combined removal of SO_2, NO_x, and fly ash from simulated flue gas using pulsed streamer corona. *IEEE Transactions on Industry Applications, 25*, 62–69.
9. Penetrante, B. M., Bardsley, J. N., & Hsiao, M. C. (1997). Kinetic analysis of non-thermal plasmas used for pollution control. *Japanese Journal of Applied Physics, 36*, 5007–5017.
10. Yamamoto, T. (1999). Optimization of nonthermal plasma for the treatment of gas streams. *Journal of Hazardous Materials, B67*, 165–181.
11. Anderegg, F. O. (1923). Surface complications in the corona discharge. *Trans. Am. Electrochemical Soc., 44*, 203–214.
12. Yehia, A., & Mizuno, A. (2008). Suppression of the ozone generation in the positive and negative dc corona discharges. *International Journal of Plasma Environmental Science and Technology, 2*, 44–49.
13. Newsome, P. T. (1926). A study of the influence of the electrodes on the formation of ozone at low pressures in the electrical discharge. *Journal of the American Chemical Society, 48*, 2035–2045.
14. Petrov, A. A., Amirov, R. H., & Samoylov, I. S. (2009). On the nature of copper cathode erossion in negative corona discharge. *IEEE Transactions on Plasma Science, 37*, 1146–1149.
15. Dhandapani, B., & Oyama, S. T. (1997). Gas phase ozone decomposition catalysts. *Applied Catalysis B: Environmental., 11*, 129–166.
16. Henis, J. M. (1976). *Nitrogen oxide decomposition process*. USA: Monsanoto Company.

17. Park, M. K., Ryu, S. G., Park, H. B., Lee, H. W., Hwang, K. C., & Lee, C. H. (2004). Decomposition of cyanogen chloride by using a packed bed plasma reactor at dry and wet air in atmospheric pressure. *Plasma Chemistry and Plasma Processing, 24*, 117–135.
18. Chang, C. L., & Lin, T. S. (2005). Elimination of carbon monoxide in the gas streams by dielectric barrier discharge systems with Mn catalyst. *Plasma Chemistry and Plasma Processing, 25*, 387–401.
19. Holzer, F., Kopinke, F. D., & Roland, U. (2005). Influence of ferroelectric materials and catalysts on the performance of non-thermal plasma (NTP) for the removal of air pollutants. *Plasma Chemistry and Plasma Processing., 25*, 595–611.
20. Magureanu, M., Mandache, N. B., Parvulescu, V. I., Subrahmanyam, C., Renken, A., & Kiwi-Minsker, L. (2007). Improved performance of non-thermal plasma reactor during decomposition of trichloroethylene: Optimization of the reactor geometry and introduction of catalytic electrode. *Applied catalysis. B, Environmental, 74*, 270–277.
21. Nozaki, T., Muto, N., Kado, S., & Okazaki, K. (2004). Dissociation of vibrationally excited methane on Ni catalyst part 1. Application to methane steam reforming. *Catalysis Today, 89*, 57–65.
22. Kraus, M., Eliasson, B., Kogelschatz, U., & Wokaun, A. (2001). CO_2 reforming of methane by the combination of dielectric-barrier discharges and catalysis. *Physical Chemistry Chemical Physics, 3*, 294–300.
23. Wang, Q., Yan, B. H., Jin, Y., & Cheng, Y. (2009). Dry reforming of methane in a dielectric barrier discharge reactor with Ni/Al_2O_3 catalyst: Interaction of catalyst and plasma. *Energy & Fuel, 23*, 4196–4201.
24. Jwa, E., Lee, S. B., Lee, H. W., & Mok, Y. S. (2013). Plasma-assisted catalytic methanation of CO and CO_2 over Ni-zeolite catalysts. *Fuel Processing Technology, 108*, 89–93.
25. Scapinello, M., Martini, L. M., & Tosi, P. (2014). CO_2 hydrogenation by CH_4 in a dielectric barrier discharge: Catalytic effects of nickel and Copper. *Plasma Processes and Polymers, 11*, 624–628.
26. Hammer, T., & Broer, S. (1998). Plasma enhanced selective catalytic reduction of NO_x for diesel cars. Society of Automotive Engineers: Paper No. 982428.
27. Hoard, J., & Balmer, M. L. (1998). Analysis of plasma-catalysis for diesel NO_x remediation. *Society of Automotive Engineers, 982429*, 13–19.
28. Penetrante, B. M., Brusasco, R. M., Merritt, B. T., & Vogtlin, G. E. (1999). Environmental application of low-temperature plasmas. *Pure and Applied Chemistry, 71*, 1829–1835.
29. Kim, H. H., Takashima, K., Katsura, S., & Mizuno, A. (2001). Low-temperature NO_x reduction processes using combined systems of pulsed corona discharge and catalysts. *Journal of Physics D: Applied Physics, 34*, 604–613.
30. Miessner, H., Francke, K.-P., Rudolph, R., & Hammer, T. (2002). NO_x removal in excess oxygen by plasma-enhanced selective catalytic reduction. *Catalysis Today, 75*, 325–330.
31. Mok, Y. S., Dors, M., & Mizerazcyk, J. (2004). Effect of reaction temperature on NO_x removal and formation of ammonium nitrate in nonthermal plasma process combined with selective catalytic reduction. *IEEE Transactions on Plasma Science, 32*, 799–807.
32. Patil, B., Cherkasov, N., Lang, J., Ibhadon, A., Hessel, V., & Wang, Q. (2016). Low temperature plasma catalytic NO_x synthesis in a packed DBD reactor: Effect of support materials and supported active metal oxides. *Applied Catalysis B: Environmental, 194*, 123–133.
33. Mizushima, T., Matsumoto, K., Sugoh, J.-I., Ohkita, H., & Kakuta, N. (2004). Tubular membrane-like catalyst for reactor with dielectric-barrier discharge plasma and its performance in ammonia synthesis. *Applied Catalysis A: General, 265*, 53–59.
34. Hong, J., Aramesh, M., Shimoni, O., Seo, D. H., Yick, S., Greig, A., Charles, C., Prawer, S., & Murphy, A. B. (2016). Plasma catalytic synthesis of ammonia using functionalized-carbon coatings in an atmospheric-pressure non-equilibrium discharge. *Plasma Chemistry and Plasma Processing., 36*, 917–940.
35. Kim, H. H., Teramoto, Y., Ogata, A., Takagi, H., & Nanba, T. (2017). Atmospheric-pressure nonthermal plasma synthesis of ammonia over ruthenium catalysts. *Plasma Processes and Polymers, 14*, 1600157.

36. Vandenbroucke, A. M., Morent, R., Geyter, N. D., & Leys, C. (2011). Non-thermal plasmas for non-catalytic and catalytic VOC abatement. *Journal of Hazardous Materials, 195*, 30–54.
37. Kusic, H., Koprivanac, N., & Locke, B. R. (2005). Decomposition of phenol by hybrid gas/liquid electrical discharge reactors with zeolite catalysts. *Journal of Hazardous Materials, B125*, 190–200.
38. Kim, H. H., Tsunoda, K., Katsura, S., & Mizuno, A. (1999). A novel plasma reactor for NO_x control using photocatalyst and hydrogen peroxide injection. *IEEE Transactions on Industry Applications, 35*, 1306–1310.
39. Yoshida, H., Marui, Z., Aoyama, M., Sugiura, J., & Mizuno, A. (1989). Removal of odor gas component utilizing plasma chemical reactions promoted by the partial discharge in a ferroelectric pellet layer. *Journal of the Institute of Electrostatics Japan, 13*, 425–430.
40. Harling, A. M., Glover, D. J., Whitehead, J. C., & Zhang, K. (2008). Novel method for enhancing the destruction of environmental pollutants by the combination of multiple plasmadischarges. *Environmental Science & Technology, 42*, 4546–4550.
41. Park, S. Y., Deshwal, B. R., & Moon, S. H. (2008). NO_x removal from the flue gas of oil-fired boiler using a multistage plasma-catalyst hybrid system. *Fuel Processing Technology, 89*, 540–548.
42. Hübner, M., Guaitella, O., Rousseau, A., & Röpcke, J. (2013). A spectroscopic study of ethylene destruction and by-product generation using a three-stage atmospheric packed-bed plasma reactor. *Journal of Applied Physics, 114*, 033301.
43. Kim, H. H., Oh, S. M., Ogata, A., & Futamura, S. (2005). Decomposition of gas-phase benzene using plasma-driven catalyst reactor: Complete oxidation of adsorbed benzene using oxygen plasma. *Journal of Advanced Oxidation Technologies, 8*, 226–233.
44. Kim, H. H., Ogata, A., & Futamura, S. (2008). Oxygen partial pressure-dependent behavior of various catalysts for the total oxidation of VOCs using cycled system of adsorption and oxygen plasma. *Applied Catalysis. B: Environmental, 79*, 356–367.
45. Kim, H. H., Ogata, A., & Futamura, S. (2007). Complete oxidation of volatile organic compounds (VOCs) using plasma-driven catalysis and oxygen plasma. *International Journal of Plasma Environmental Science and Technology, 1*, 46–51.
46. Fan, H. Y., Shi, C. S., Li, X. S., Zhao, D. X., Xu, Y., & Zhu, A. M. (2009). High-efficiency plasma catalytic removal of dilute benzene from air. *Journal of Physics D: Applied Physics, 42*, 225105.
47. Fan, H. Y., Li, X. S., Shi, C., Zhao, D. Z., Liu, J. L., Liu, Y. X., & Zhu, A. M. (2011). Plasma catalytic oxidation of stored benzene in a cycled storage-discharge (CSD) process: Catalysts, reactors and operation conditions. *Plasma Chemistry and Plasma Processing, 31*, 799–810.
48. Mok, Y. S., & Kim, D. H. (2011). Treatment of toluene by using adsorption and nonthermal plasma oxidation process. *Current Applied Physics, 11*, S58–S62.
49. Dang, X., Huang, J., Cao, L., & Zhou, Y. (2013). Plasma-catalytic oxidation of adsorbed toluene with gas circulation. *Catalysis Communications, 40*, 116–119.
50. Wang, W., Wang, H., Zhu, T., & Fan, X. (2015). Removal of gas phase low-concentration toluene over Mn, Ag and Cemodified HZSM-5 catalysts by periodical operation of adsorption and non-thermal plasma regeneration. *Journal of Hazardous Materials, 292*, 70–78.
51. Qin, C., Dang, X., Huang, J., Teng, J., & Huang, X. (2016). Plasma-catalytic oxidation of adsorbed toluene on Ag-Mn/γ-Al_2O_3: Comparison of gas flow-through and gas circulation treatment. *Chemical Engineering Journal, 299*, 85–92.
52. Zhao, D. Z., Li, X. S., Shi, C., Fan, H. Y., & Zhu, A. M. (2011). Low-concentration formaldehyde removal from air using a cycled storage- discharge (CSD) plasma catalytic process. *Chemical Engineering Science, 66*, 3922–3929.
53. Oh, S. M., Kim, H. H., Einaga, H., Ogata, A., Futamura, S., & Park, D. W. (2006). Zeolite-combined plasma reactor for decomposition of toluene. *Thin Solid Films, 506-507*, 418–422.
54. Kuroki, T., Hirai, K., Matsuoka, S., Kim, J. Y., & Okubo, M. (2011). Oxidation system of adsorbed VOCs on adsorbent using non thermal plasma flow. *IEEE Transactions on Industry Applications, 47*, 1916–1921.

55. Saulich, K., & Muller, S. (2013). Removal of formaldehyde by adsorption and plasma treatment of mineral adsorbent. *Journal of Physics D: Applied Physics, 46*, 045201.
56. Subrenat, A., Baléo, J. N., Cloirec, P. L., & Blanc, P. E. (2001). Electrical behaviour of activated carbon cloth heated by the joule effect: Desorption application. *Carbon, 39*, 707–716.
57. Okubo, M., Inoue, M., Kuroki, T., & Yamamoto, T. (2005). NO_x reduction aftertreatment system using nitrogen nonthermal plasma desorption. *IEEE Transactions on Industry Applications, 41*, 891–899.
58. Yoshida, K., Kuwahara, T., Kuroki, T., & Okubo, M. (2012). Diesel NO_x aftertreatment by combined process using temperature swing adsorption, NO_x reduction by nonthermal plasma, and NO_x recirculation: Improvement of the recirculation process. *Journal of Hazardous Materials, 231-232*, 18–25.
59. Yu, Q. Q., Wang, H., Liu, T., Xiao, L. P., Jiang, X. Y., & Zheng, X. M. (2012). High-efficiency removal of NO_x using a combined adsorption-discharge plasma catalytic process. *Environmental Science & Technology, 46*, 2337–2344.
60. Kim, H. H., Prieto, G., Takashima, K., Katsura, S., & Mizuno, A. (2002). Performance evaluation of discharge plasma process for gaseous pollutant removal. *Journal of Electrostatics, 55*, 25–41.
61. Kim, H. H., Teramoto, Y., Negishi, N., & Ogata, A. (2015). A multidisciplinary approach to understand the interactions of nonthermal plasma and catalyst: A review. *Catalysis Today, 256*, 13–22.
62. Francke, K. P., Rudolph, R., & Miessner, H. (2003). Design and operation characteristics of a simple and reliable DBD reactor for use with atmospheric air. *Plasma Chemistry and Plasma Processing, 23*, 47–57.
63. Kogelschatz, U. (2003). Dielectric-barrier discharges: Their history, discharge physics, and industrial applications. *Plasma Chemistry and Plasma Processing, 23*, 1–46.
64. Kim, H. H., & Ogata, A. (2012). Interaction of nonthermal plasma with catalyst for the air pollution control. *International Journal of Plasma Environmental Science and Technology, 6*, 43–48.
65. Mizuno, A., & Ito, H. (1990). Basic performance of an electrostatically augmented filter consisting of a packed ferroelectric pellet layer. *Journal of Electrostatics, 25*, 97–107.
66. Harling, A. M., Kim, H. H., Futamura, S., & Whitehead, J. C. (2007). Temperature dependence of plasma-catalysis using a nonthermal, atmospheric pressure packed bed; the destruction of benzene and toluene. *Journal of Physical Chemistry C, 111*, 5090–5095.
67. Kim, H. H., Teramoto, Y., Sano, T., Negishi, N., & Ogata, A. (2015). Effects of Si/Al ratio on the interaction of nonthermal plasma and Ag/HY catalysts. *Applied Catalysis B: Environmental, 166-167*, 9–17.
68. Kuai, P.-Y., Liu, C.-J., & Huo, P.-P. (2009). Characterization of CuO-ZnO catalyst prepared by decomposition of carbonates using dielectric-barrier discharge plasma. *Catalysis Letters, 129*, 493–498.
69. Liu, Y., Pan, Y.-X., Kuai, P., & Liu, C.-J. (2010). Template removal from ZSM-5 zeolite using dielectric-barrier discharge plasma. *Catalysis Letters, 135*, 241–251.
70. Tu, X., Gallon, H. J., & Whitehead, J. C. (2013). Plasma-assisted reduction of a NiO/Al_2O_3 catalyst in atmospheric pressure H_2/Ar dielectric barrier discharge. *Catalysis Today, 211*, 120–125.
71. Kim, T., Lee, D. H., Jo, S. K., Pyun, S. H., Kim, K. T., & Song, Y. H. (2016). Mechanism of the accelerated reduction of an oxidized metal catalyst under electric discharge. *ChemCatChem, 8*, 685–689.
72. Liu, X., Mou, C. Y., Lee, S., Li, Y., Secrest, J., & Jang, B. W. L. (2012). Room temperature O_2 plasma treatment of SiO_2 supported Au catalysts for selective hydrogenation of acetylene in the presence of large excess of ethylene. *Journal of Catalysis, 285*, 152–159.
73. Liu, C. J., Zhao, Y., Li, Y., Zhang, D. S., Chang, Z., & Bu, W. H. (2013). Perspectives on electron-assisted reduction for preparation of highly dispersed noble metal catalysts. *ACS Sustainable Chemistry & Engineering, 2*, 3–13.

74. Menard, L. D., Xu, F., Nuzzo, R. G., & Yang, J. C. (2006). Preparation of TiO_2-supported Au nanoparticle catalysts from a Au_{13} cluster precursor: Ligand removal using ozone exposure versus a rapid thermal treatment. *Journal of Catalysis, 243*, 64–73.
75. Khan, M. A., & Al-Jalal, A. A. (2004). Enhanced decoking of a coked zeolite catalyst using a glow discharge in Ar-O_2 gas mixture. *Applied Catalysis A: General, 272*, 141–149.
76. Al-Jalal, A. M., & Khan, M. A. (2010). Optical emission and raman spectroscopy studies of reactivity of low-pressure glow discharges in Ar-O_2 and He-O_2 gas mixtures with coked catalysts. *Plasma Chemistry and Plasma Processing, 30*, 173–182.
77. Kim, H. H., Tsubota, S., Daté, M., Ogata, A., & Futamura, S. (2007). Catalyst regeneration and activity enhancement of Au/TiO_2 by atmospheric pressure nonthermal plasma. *Applied Catalysis A: General, 329*, 93–98.
78. Fan, H. Y., Shi, C. A., Li, X. S., Zhang, S., Liu, J. L., & Zhu, A. M. (2012). In-situ plasma regeneration of deactivated Au/TiO_2 nanocatalysts during CO oxidation and effect of N_2 content. *Applied Catalysis B: Environmental, 119*, 49–55.
79. Yoshida, H., Kuwauchi, Y., Jinschek, J. R., Sun, K., Tanaka, S., Kohyama, M., Shimada, S., Haruta, M., & Takeda, S. (2012). Visualizing gas molecules interacting with supported nanoparticulate catalysts at reaction conditions. *Science, 335*, 317–320.
80. Somorjai, G. A. (1992). The experimental evidence of the role of surface restructuring during catalytic reactions. *Catalysis Letters, 12*, 17–34.
81. Sano, T., Negishi, N., Sakai, E., & Matsuzawa, S. (2006). Contributions of photocatalytic/catalytic activities of TiO_2 and γ-Al_2O_3 in nonthermal plasma on oxidation of acetaldehyde and CO. *Journal of Molecular Catalysis A: Chemical, 245*, 235–241.
82. Rajasekaran, P., Mertmann, P., Bibinov, N., Wandke, D., Viol, W., & Awakovicz, P. (2010). Filamentary and homogeneous modes of dielectric barrier discharge (DBD) in air: Investigation through plasma characterization and simulation of surface irradiation. *Plasma Processes and Polymers, 7*, 665–675.
83. Ochiai, T., Nakata, K., Murakami, T., Morito, Y., Hosokawa, S., & Fusishima, A. (2011). Development of an air-purification unit using a photocatalysis-plasma hybrid reactor. *Electrochemistry, 79*, 838–841.
84. Huang, H. B., Ye, D. Q., Fu, M. L., & Feng, F. D. (2007). Contribution of UV light to the decomposition of toluene in dielectric barrier discharge plasma/photocatalysis system. *Plasma Chemistry and Plasma Processing, 27*, 577–588.
85. Maciuca, A., Batiot-Dupeyrat, C., & Tatibouet, J. M. (2012). Synergetic effect by coupling photocatalysis with plasma for low VOCs concentration removal from air. *Applied Catalysis B: Environmental, 125*, 432–438.
86. Ono, R. (2016). Optical diagnostics of reactive species in atmospheric-pressure nonthermal plasma. *Journal of Physics D: Applied Physics, 49*, 083001.
87. Guaitella, O., Lazzaroni, C., Marinov, D., & Rousseau, A. (2010). Evidence of atomic adsorption on TiO_2 under plasma exposure and related C_2H_2 surface reactivity. *Applied Physics Letters, 97*, 011502.
88. Marinov, D., Guaitella, O., Rousseau, A., & Ionikh, Y. (2010). Production of molecules on a surface under plasma exposure: Example of NO on pyrex. *Journal of Physics D: Applied Physics, 43*, 115203.
89. Kim, H. H., Ogata, A., Schiorlin, M., Marotta, E., & Paradisi, C. (2011). Oxygen isotope ($^{18}O_2$) evidence on the role of oxygen in the plasma-driven catalysis of VOC oxidation. *Catalysis Letters, 141*, 277–282.
90. Teramoto, Y., Kim, H. H., Ogata, A., & Negishi, N. (2013). Study of plasma-induced surface active oxygen on zeolite-supported silver nanoparticles. *Catalysis Letters, 143*, 1374–1378.
91. Guaitella, O., Hubner, M., Welzel, S., Marinov, D., Ropcke, J., & Rousseau, A. (2010). Evidence for surface oxidation on pyrex of NO into NO_2 by adsorbed O atoms. *Plasma Sources Science and Technology, 19*, 045206.
92. Biener, J., Wittstock, A., Zepeda-Ruiz, L. A., Biener, M. M., Zielasek, V., Kramer, D., Viswanath, R. N., Weissmüller, J., Bäumer, M., & Hamza, A. V. (2009). Surface-chemistry-driven actuation in nanoporous gold. *Nature Materials, 8*, 47–51.

93. Utz, A. L. (2009). Mode selective chemistry at surfaces. *Current Opinion in Solid State & Materials Science, 13*, 4–12.
94. Nozaki, T., Muto, N., Kadio, S., & Okazaki, K. (2004). Dissociation of vibrationally excited methane on Ni catalyst part 2. Process diagnostics by emission spectroscopy. *Catalysis Today, 89*, 67–74.
95. Haruta, M., Yamada, N., Kobayashi, T., & Iijima, S. (1989). Gold catalysts prepared by coprecipitation for low-temperature oxidation of hydrogen and of carbon monoxide. *Journal of Catalysis, 115*, 301–309.
96. Jiang, C., Hara, K., & Fukuoka, A. (2013). Low-temperature oxidation of ethylene over platinum nanoparticles supported on mesoporous silica. *Angewandte Chemie, International Edition, 52*, 6265–6268.
97. Stere, C. E., Adress, W., Burch, R., Chansai, S., Goguet, A., Graham, W. G., & Hardacre, C. (2015). Probing a non-thermal plasma activated heterogeneously catalyzed reaction using in situ DRIFTS-MS. *ACS Catalysis, 5*, 956–964.
98. Rodrigues, A., Tatibouet, J. M., & Fourre, E. (2016). Operando DRIFT spectroscopy characterization of intermediate cpecies on catalysts surface in VOC removal from air by non-thermal plasma assisted catalysis. *Plasma Chemistry and Plasma Processing, 36*, 901–915.
99. Su, H., & Yeung, E. S. (2000). High-throughput screening of heterogeneous catalysts by laser-induced fluorescence imaging. *Journal of the American Chemical Society, 122*, 7422–7423.
100. Su, H., Hou, Y., Houk, R. S., Schrader, G. L., & Yeung, E. S. (2001). Combinatiorial screening of heterogeneous catalysts in selective oxidation of naphthalene by laser-induced fluorescence imaging. *Analytical Chemistry, 73*, 4434–4440.
101. Kim, H. H., Kim, J. H., & Ogata, A. (2009). Microscopic observation of discharge plasma on the surface of zeolites supported metal nanoparticles. *Journal of Physics D: Applied Physics, 42*, 135210.
102. Kim, H. H., & Ogata, A. (2011). Nonthermal plasma activates catalyst: From current understanding and future prospects. *European Physical Journal Applied Physics, 55*, 13806.
103. Mizuno, A., Yamazaki, Y., Ito, H., & Yoshida, H. (1992). Ac energized ferroelectric pellet bed gas cleaner. *IEEE Transactions on Industry Applications, 28*, 535–540.
104. Mizuno, A., Yamazaki, Y., Obama, S., Suzuki, E., & Okazaki, K. (1993). Effect of voltage waveform on partial discharge in ferroelectric pellet layer for gas cleaning. *IEEE Transactions on Industry Applications, 29*, 262–267.
105. Takaki, K., Chang, J. S., & Kostov, K. G. (2004). Atmospheric pressure of nitrogen plasmas in a ferro-electric packed bed barrier discharge reactor part I: Modeling. *IEEE Transactions on Dielectrics and Electrical Insulation, 11*, 481–490.
106. Nozaki, T., Unno, Y., Miyazaki, Y., & Okazaki, K. (2001). Optical diagnostics for determining gas temperature of reactive microdischarges in a methane-fed dielectric barrier discharge. *Journal of Physics D: Applied Physics, 34*, 2504–2511.
107. Hoft, H., Kettlitz, M., Weltmann, K.-D., & Brandenburg, R. (2014). The bidirectional character of O_2 concentration in pulsed dielectric barrier discharges in O_2/N_2 gas mixtures. *Journal of Physics D: Applied Physics, 47*, 455202.
108. Kim, H. H., Teramoto, Y., & Ogata, A. (2016). Time-resolved imaging of positive pulsed corona-induced surface streamers on TiO_2 and γ-Al_2O_3-supported Ag catalysts. *Journal of Physics D: Applied Physics, 49*, 415204.
109. Marode, E. (1975). The mechanism of spark breakdown in air at atmospheric pressure between a positive point and a plane. 1. Experimental: Nature of the streamer track. *Journal of Applied Physics, 46*, 2005–2015.
110. Namihira, T., Wang, D., Katsuki, S., Hackam, R., & Akiyama, H. (2003). Propagation velocity of pulsed streamer discharges in atmospheric air. *IEEE Transactions on Plasma Science, 31*, 1091–1094.
111. Huiskamp, T., Pemen, A. J. M., Hoeben, W. F. L. M., Beckers, F. J. C. M., & Heesch, E. J. M. V. (2013). Temperature and pressure effects on positive streamers in air. *Journal of Physics D: Applied Physics, 46*, 165202.

Chapter 4
Plasma Catalysis Modeling

Annemie Bogaerts and Erik Neyts

4.1 Introduction

Plasma catalysis is gaining increasing interest for various applications, as is obvious from the other chapters in this book. However, the underlying mechanisms are very complex and not yet fully understood. A better insight in these mechanisms can be obtained by experiments, but also computer modeling can be very useful for this purpose. Indeed, a model allows us to unravel the individual processes, before combining them into the overall picture, which is not always possible in experiments. This should allow us to obtain detailed insight into the underlying mechanisms.

Although modeling is quite well established for describing the plasma behavior for various kinds of plasmas and diverse applications and computer simulations are also extensively used for describing (thermal) catalytic processes, the modeling attempts reported in literature for describing the combination of both, in plasma catalysis, are very limited up to now. The reason for this is probably the enormous degree of complexity of the entire process.

This chapter gives an overview of the modeling work performed already for plasma catalysis. We start this chapter with a brief overview of plasma-catalyst interactions, to identify the various aspects that need to be considered by modeling.

A. Bogaerts (✉) · E. Neyts (✉)
Research group PLASMANT, Department of Chemistry, University of Antwerp, Antwerp, Belgium
e-mail: annemie.bogaerts@uantwerpen.be; erik.neyts@uantwerpen.be

X. Tu et al. (eds.), *Plasma Catalysis*, Springer Series on Atomic, Optical, and Plasma Physics 106, https://doi.org/10.1007/978-3-030-05189-1_4

4.2 Plasma-Catalyst Interactions

The major difference between plasma catalysis and thermal catalysis is that the source gas will already be activated by the plasma, causing among others dissociation, excitation, and ionization of the molecules. Indeed, plasma does not only contain molecules but also various radicals, excited species and ions originating from these molecules, as well as electrons and photons. All these species might interact with the catalyst surface, in addition to electric fields that are also characteristic of a plasma, all contributing to the complexity of plasma catalysis. Vice versa, the catalyst will also affect the plasma behavior in various respects. In the following we will give a brief summary of the effects of plasma on a catalyst, as well as the effects of a catalyst on the plasma.

4.2.1 Effects of the Plasma on a Catalyst

Plasma may affect the catalyst and catalysis mechanisms in various ways. It may cause:

(a) Changes in the physicochemical properties of the catalyst, e.g., a higher adsorption probability [1]; a higher surface area [2, 3]; a change in the oxidation state [4–9]; reduced coke formation, thus preventing deactivation of the catalyst [5]; and a change in the work function due to the presence of a voltage and current (or charge accumulation) at the catalyst surface [10, 11], affecting the catalytic activity [12].
(b) The formation of hot spots [13], possibly modifying the local plasma chemistry [14]. These hot spots might even thermally activate the catalyst locally [15], or they can also deactivate the catalyst due to plasma-induced damage [16, 17].
(c) Lower activation barriers, due to the existence of short-lived active species, such as radicals and vibrationally excited species [4].
(d) Activation by photon irradiation [18, 19], although this effect is still under debate, because other studies reported no effect [20, 21]. It is indeed claimed [18, 22, 23] that the UV light originating from the plasma is not intensive enough. Indeed, the UV dose in typical photocatalytic processes should be in the order of several mW/cm^2, whereas in typical (air) plasmas it is only in the order of $\mu W/cm^2$ [22]. However, it is definitely possible that photocatalysts are activated by other (energetic) plasma species, like ions, metastables, or electrons with suitable energy [13, 23, 24].
(e) A change in the reaction pathways, because the plasma contains not only gas molecules, like in thermal catalysis, but also many more species, such as radicals, ions, electrons, and vibrationally and electronically excited species, which can undergo other types of reactions at the catalyst surface. For instance, besides the more common Langmuir-Hinshelwood mechanism, where two adsorbed species undergo a chemical reaction, in plasma catalysis the

Eley-Rideal mechanism, where a plasma-produced radical reacts with an adsorbed species, can also occur more readily, while this process occurs in thermal catalysis only at very high temperatures [25].

4.2.2 *Effects of the Catalyst on the Plasma*

Vice versa, a catalyst may have a profound effect on the plasma. It may cause:

(a) Enhancement of the local electric field in the plasma, because the catalyst is mostly present in a structured packing (e.g., pellets, beads, honeycomb, etc., so-called packed bed reactor), or simply due to the porosity of the catalyst surface [18, 26–28]. This enhanced electric field consequently results in an increase of the high energy tail of the electron energy distribution [14, 29, 30].
(b) A change of the discharge type from streamers inside the plasma to streamers along the catalyst surface [6, 31–37], which might result in more intense plasma around the contact points [13, 38]. The latter will affect the plasma chemistry.
(c) The formation of microdischarges in the catalyst pores [29, 39–43]. This might lead to a strong electric field inside the pores, which will also affect the plasma chemistry.
(d) The adsorption of plasma species on the catalyst surface, affecting the residence time and hence the concentration of species in the plasma [13], while new reactive species might be formed at the catalyst surface. The adsorption will increase with the porosity of the catalyst surface [44].

A schematic overview of some plasma-catalyst interaction processes is presented in Fig. 4.1. More details on these interaction processes can be found in many excellent reviews on this topic (e.g., [20, 22, 25, 32, 45–56]) and in Chap. 3 of this book. In the next section, we will illustrate how computer modeling can contribute to obtain a better insight in these interaction mechanisms.

4.3 Modeling the Effects of a Plasma on the Catalyst and on the Catalytic Surface Reactions

Modeling the effects of the plasma on the catalyst is quite a difficult task. This is due on one hand to the wide variation in time and length scales in plasma-surface interactions and to the complexity of the interactions on the other hand. While chemical surface reactions of plasma species at the catalyst surface can nowadays be accurately simulated (see examples below), modeling plasma effects, such as plasma-induced morphological changes, surface charging, the effects of photons, etc., is highly challenging. Likewise, simulating adsorption probabilities and coke formation may be feasible, while the enlargment of the surface area or (global) changes in surface oxidation state are much more complicated to simulate.

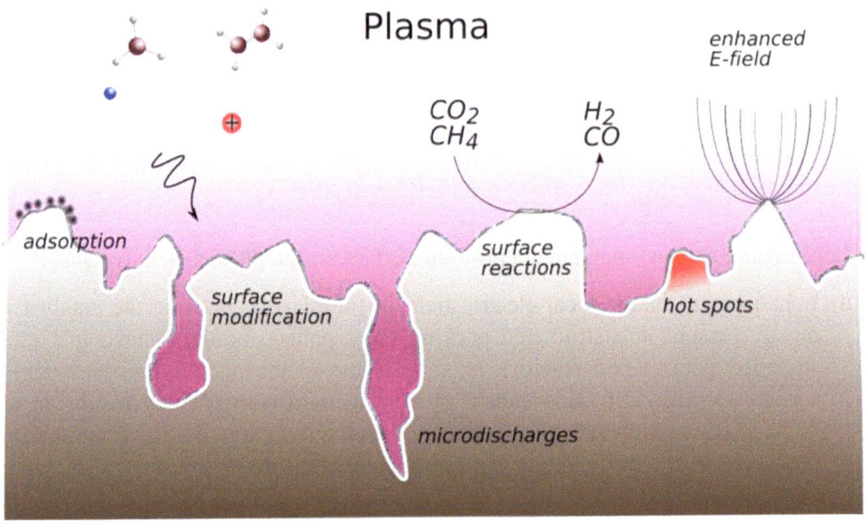

Fig. 4.1 Schematic overview of some plasma-catalyst interactions. (Modified from [46])

So far, very little work has been performed in most of these areas. On the other hand, extensive literature exists on simulations of chemical reactions of a source gas at a catalyst surface, without plasma effects. Such simulations can be carried out at the atomistic level, based on either a classical molecular dynamics (MD) or quantum-mechanical approach (mostly density functional theory or DFT); on a mesoscale level, based on a kinetic Monte Carlo (kMC) approach; or on a global level, based on rate equations. Here we give a few examples, being most relevant for the study of plasma catalysis.

In plasma catalysis, as mentioned above, the plasma activates the source gas, creating among others radicals (besides electrons, ions, and excited species). At the molecular level [57], the interaction of radicals with a catalyst surface has been studied by both classical MD simulations (e.g., [58–60]) and DFT calculations (e.g., [61–63]). Again a few examples will be given in this section. Although the link with plasma catalysis is still weak in most of these studies as they only consider radicals, some simulation works recently appeared in which also other plasma effects, like the effect of electrons [64, 65] and the electric field [66], were addressed.

Some plasma effects, however, were previously considered for another (closely related) application, i.e., carbon nanotube (CNT) growth. The effect of the electric field was investigated in [67], while the effect of ion bombardment was investigated in [68]. The same principle should also apply to plasma catalysis.

The effect of excited species on plasma catalytic processes has also been investigated, more specifically for vibrationally excited CH_4 [69]. This, however, is not straightforward, because of the required highly accurate description of the interatomic interactions and forces. This will also be explained below.

To the best of our knowledge, the interaction of photons with catalyst surfaces has not yet been studied in the context of plasma catalysis, and we shall therefore not elaborate on this topic. Finally, the future needs to model plasma-induced effects on a catalyst surface will be discussed.

4.3.1 Changes in the Physicochemical Properties of the Catalyst

The plasma may modify the physicochemical properties of the catalyst, including a change in both the morphological surface properties (e.g., surface structure and surface area) and electronic surface properties (e.g., work function and surface oxidation state). In Sects. 4.3.3 and 4.3.5 we shall discuss plasma-induced changes in catalyst surface reactivity.

So far, however, there are no simulations or model results in the literature available describing the plasma-induced modification of such catalyst physicochemical properties. The effect of such changes, on the other hand, has been described. For instance, Ni et al. [70] developed a heat balance model accounting for temperature gradients resulting from a change in catalyst oxidation state (see Sect. 4.3.2). This model, however, does not elucidate how the plasma may have induced this change in the catalyst oxidation state.

In contrast, there are various reports on models for catalyst restructuring in thermal catalysis, e.g., based on KMC modeling. One example of such a model was developed by Zhdanov et al. [71], to simulate the reconstruction of a Pt(100) surface in the CO-NO reaction. Reconstruction was enabled by considering two types of Pt atoms, i.e., stable and metastable. Although the site density was kept constant, the reconstruction could grow by virtue of accounting for the interactions between the stable and metastable sites. The standard Metropolis algorithm was employed to switch between both states, while accounting for adsorbate-catalyst interactions. An example of the resulting catalyst structure is shown in Fig. 4.2, along with the simulated oscillatory behavior of the reactants.

4.3.2 Hot Spots and Other Thermal Effects

As the plasma contains ions and photons and plasma species which react exo- or endothermically at the catalyst surface, it is natural to imagine that the plasma will affect the thermal balance of the surface. An excellent review paper on this topic can be found in [72]. Several models have been developed to describe the thermal balance of the catalyst in plasma catalysis.

A heat balance model for a packed bed dielectric barrier discharge (DBD) under natural convection was developed by Li et al. [73]. The model was developed in order to understand and explain the experimental observation that the temperature of the plasma phase is significantly higher than the temperature of the pellet phase. The model

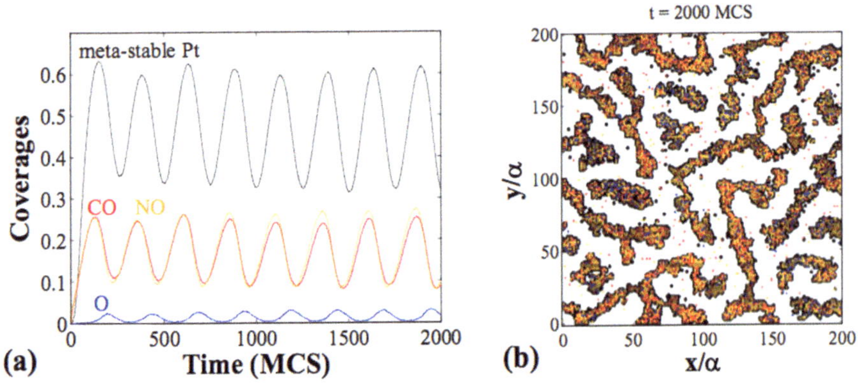

Fig. 4.2 (**a**) Simulated oscillatory surface coverage of CO and NO reactants in their reaction on a Pt (100) surface. (**b**) Snapshot of the reconstructed catalyst structure at the end of the KMC simulation, where the metastable Pt states are shown as black dots. The red, orange, and blue dots denote the CO, NO, and O adsorbates, respectively. The lattice constant is α. (Reproduced with permission from [71])

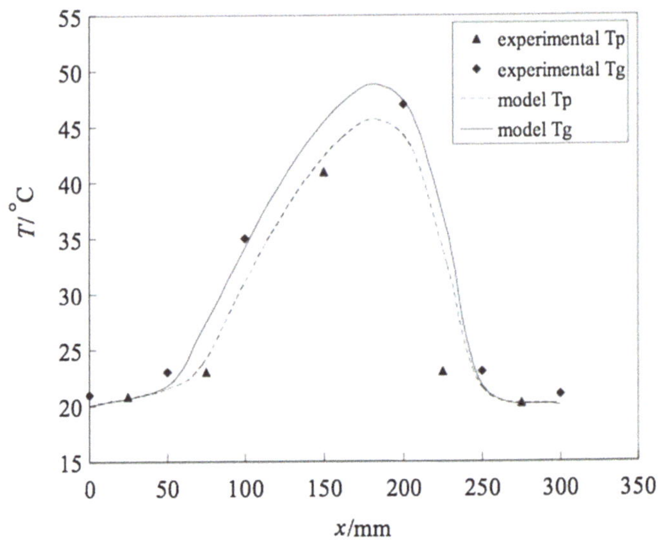

Fig. 4.3 Comparison between the experimental gas phase (Tg) and pellet phase (Tp) temperature and the respective simulated temperatures, as function of position along the length of the DBD. (Reproduced with permission from [73])

assumes that no chemical surface reactions (which may be either endothermic or exothermic) occur. Another important limiting feature of the model is that forced convection is not accounted for. Nevertheless, reasonably good agreement with the experimental data is demonstrated. In particular, it can be seen from Fig. 4.3 that the model indeed predicts a gas phase temperature exceeding the pellet phase temperature. Based on the model results, the authors explain that this results from a higher energy consumption in the plasma compared to the pellet phase, while at the same time the heat capacity of the plasma is much less than that of the pellet phase. A second important conclusion of the model from the plasma catalytic point of view is that as the pellet size becomes smaller, the temperature difference between plasma and packed bed increases.

In relation to morphological changes in the catalyst, Ni et al. studied oxidative methane coupling as a function of oxygen concentration [70]. It was experimentally found that the formation of ethane and ethylene especially was selectively dependent on the oxygen concentration, which was attributed to temperature inhomogeneities in the catalyst bed due to the microwave heating. In order to investigate this temperature inhomogeneity, a heat transfer model was developed. For a thin catalyst bed of 1 cm length and based on an estimated thermal conductivity of 3 W/m/K, a temperature difference of 85 K exists between the center of the catalyst and its periphery when the gas mixture contains 12.5% of oxygen. In the absence of oxygen, the temperature of the catalyst at the center was calculated to be 168 °C above the catalyst temperature at its periphery. According to the authors, this difference in temperature gradient results from a change in the catalyst properties and in particular a change in catalyst oxidation state from CeO_2 to Ce_2O_3. Therefore, the model succeeds in predicting a temperature gradient as a function of a change in dielectric loss, in agreement with their experiments.

4.3.3 Modification of Catalyst Reaction Kinetic Parameters

It is well known that the reaction rate of any surface reaction is not only determined by the catalyst element but also by the precise geometrical and electronic structure of the active site. Thus, when the plasma modifies, e.g., the catalyst surface morphology, kinetic parameters are likely to be modified as well.

Nozaki et al. carried out a numerical integration of the rate equation for methane steam reforming, in order to obtain the rate coefficient [74]. By calculating the rate coefficient for a plasma-catalyst system and for a regular catalytic system, it was found that while the energy barrier is unchanged, the pre-exponential factor is enhanced by a factor of 50 in the DBD.

Jiang et al. performed a very large number (~18,000) of DFT calculations to explicitly map out a nine-dimensional potential energy surface for the dissociative adsorption of CO_2 on Ni(100) [75]. In particular in relation to plasma catalysis, the effect of vibrational excitation on the CO_2 dissociation was investigated. From these calculations, the authors found that the vibrational efficacity strongly depends on the translational energy, as can be seen in Fig. 4.4. In particular, below a translational energy of 0.4 eV, the symmetric and asymmetric stretch vibrations show a vibrational efficacity of only 0.3 and 0.2, respectively. In this regime, a small increase in translational energy yields a significant change in dissociation probability. For higher translational energies, however, these efficacities increase to about 2.2, demonstrating the dominant role of vibration in the dissociative chemisorption at these higher translational energies.

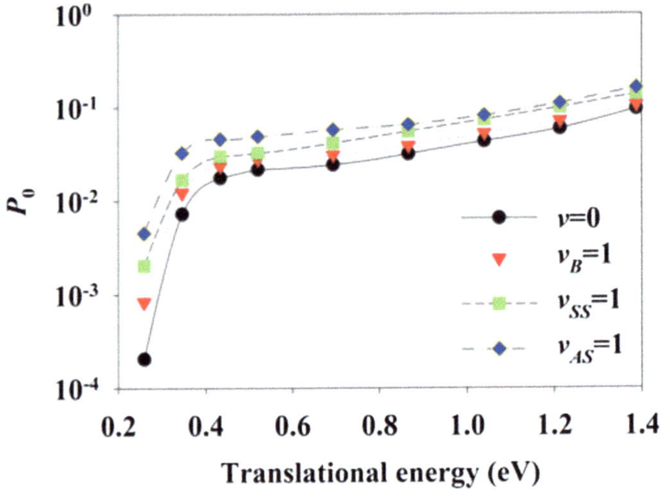

Fig. 4.4 Calculated normal incident dissociative sticking probability P_0 of CO_2 as function of translational energy and for several vibrationally excited modes. (Reproduced with permission from [75])

4.3.4 Reaction Pathways

As the plasma may modify the kinetic parameters, it may of course also affect the actual reaction pathways. For instance, Neyts et al. employed long time scale atomistic simulations based on a classical force field to study the methanol-to-formaldehyde conversion process in the presence and absence of an external electric field [66]. Although the field was not taken into account self-consistently (i.e., the charge distribution at the surface was not modified as a function of the E-field), including the field did demonstrate a significant effect on the surface retention time of the reactants. Moreover, it was found that this residence time is also a function of the polarity of the field: the retention time was decreased by 6% if the field was pointing toward the surface, while it increased by 13% if the field had opposite polarity. However, the time to first conversion (which is representative for the surface reaction time at the pristine surface) did not show any dependence on the electric field polarity nor on its magnitude.

Also employing a classical force field, Somers et al. performed a series of molecular dynamics simulations to investigate the interaction of CH_x radicals with various Ni surfaces in a wide temperature range, aiming mainly at methane reforming and H_2 formation [58–60]. The simulations demonstrate that the consecutive impingement of plasma radicals quickly renders the initially crystalline surface amorphous, altering the surface catalytic properties. This, in turn, is seen to be a result of C diffusion into the (sub-)surface layers, which can be controlled to a certain extent by controlling the catalyst temperature. Thus, reducing the temperature as is typically done in plasma catalysis and allowing the plasma to dissociate CH_4 into CH_3 and lower radicals enable to maintain a well-controlled catalyst surface and catalyst reactivity.

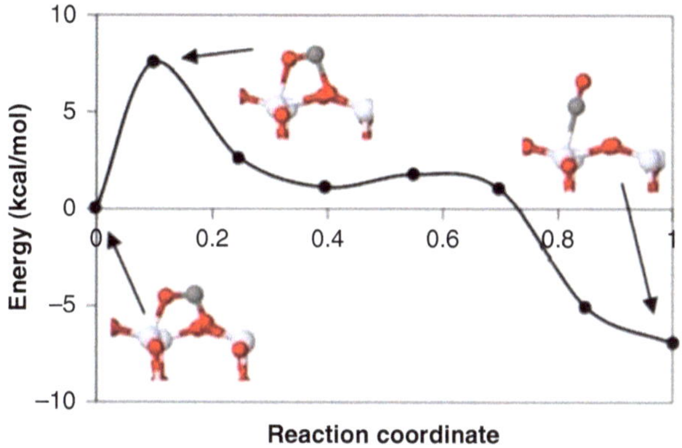

Fig. 4.5 Minimal CO_2 dissociation path on anatase (001) in the presence of a surface oxygen vacancy. (Reproduced with permission from [61])

Often, the accuracy of such studies depends strongly on the accuracy of the available force fields. While various rather accurate force fields have been developed, each new elemental combination requires the fitting of a new parametrization. In order to overcome this issue, DFT calculations may be employed instead.

As an example, Huygh et al. performed extensive DFT calculations to investigate how oxygen vacancies affect the reactivity of the TiO_2 anatase (001) surface toward CO_2 dissociation [61]. While CO_2 was shown not to dissociate at any appreciable rate on the stoichiometric surface, CO_2 dissociation does occur on a surface containing oxygen vacancies (see Fig. 4.5). While these calculations do not simulate the plasma directly, they demonstrate that if the plasma can create the required oxygen vacancies, the plasma may indirectly enhance the catalytic CO_2 dissociation process.

The same authors also investigated the adsorption of plasma-generated CH_x radicals on the same surface, again in relation to the presence of oxygen vacancies. From their DFT calculations, it turns out that the precise location of the vacancies greatly affects the adsorption interaction. This in turn affects the possible surface reactions [62].

Dry reforming of methane on a Ni(111) surface was studied at low temperature, relevant for plasma catalysis, by Shirazi et al. [63], as a function of H coverage. It was found that there is a clear (albeit non-linear) relation between the increase in H coverage and the decrease in energy barriers for methanol and ethane formation. In particular, it was found that the increase in H coverage may sufficiently lower the barriers to make the relevant reactions in the dry reforming process thermally accessible at the low temperatures relevant for plasma catalysis. This is shown in Fig. 4.6, schematically representing the decrease in activation barrier as a function of H coverage for the hydrogenation of CH_2OH to methanol.

The importance of surface charging for adsorbate/catalyst interactions was recently convincingly demonstrated by Bal et al. using DFT calculations [64, 65]. By adding a fixed proton to the vacuum space above a slab of catalytic material and enforcing charge neutrality in the simulation cell, a negatively charged surface is obtained with a surface charge density that can be tuned by the size of the

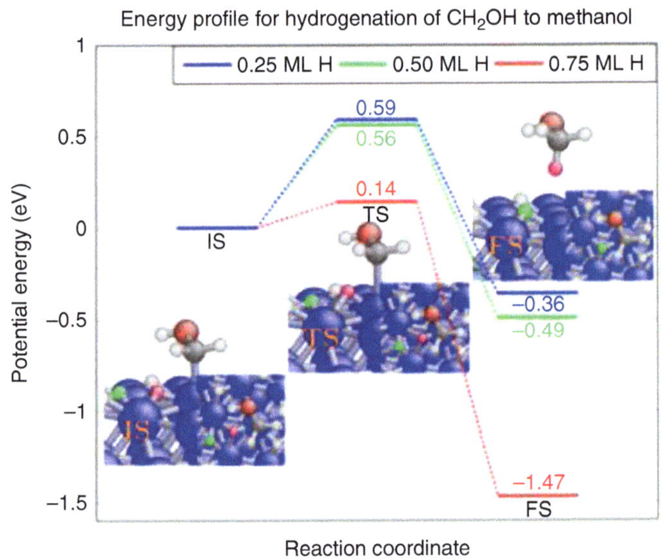

Fig. 4.6 Influence of H coverage of the Ni(111) surface on the energy profile for hydrogenation of CH_2OH to methanol, as obtained from DFT calculations. (Reproduced with permission from [63])

simulation cell. Additionally, an electric field arises naturally in this setup. The positive counter charge in the vacuum space can then be thought of as the positive space charge in the sheath of the plasma region in contact with the surface.

Using this methodology, Bal et al. investigated the adsorption and splitting of CO_2 on charged and neutral surfaces of alumina-supported single atom catalysts (Ti, Ni, and Cu). It was found that the surface charge significantly enhances the reductive power of the catalyst, strongly promoting the adsorption and dissociation of CO_2 to CO and O as shown in Fig. 4.7. Furthermore, it was also found that the relative catalytic activity of the investigated materials is modified by the surface charge. The origin of this effect lies in the increase of the reductive power (or the Lewis basicity) of the surface by the negative surface charge. Overall, the CO_2 splitting reaction was found to become less endothermic, and furthermore, also the splitting barriers were found to be reduced by the presence of the surface charge. It was therefore concluded that surface charge might play a highly important role in the possible synergism observed in plasma catalysis [64, 65].

Although DFT is sufficiently general enough to model essentially any reactive system relevant in plasma catalysis, it is highly computationally expensive and is limited to small system sizes. Moreover, if dynamics are required, the accessible time scale is typically very limited as well (order of pico- or nanoseconds). To access long time scale dynamics, a novel generic method was recently developed, termed collective variable-driven hyperdynamics (CVHD) [76]. While CVHD was shown to allow the extension of the time scale by up to nine orders of magnitude in classical molecular dynamics, i.e., into the μs-ms-sec region (see, e.g., [66]), it has

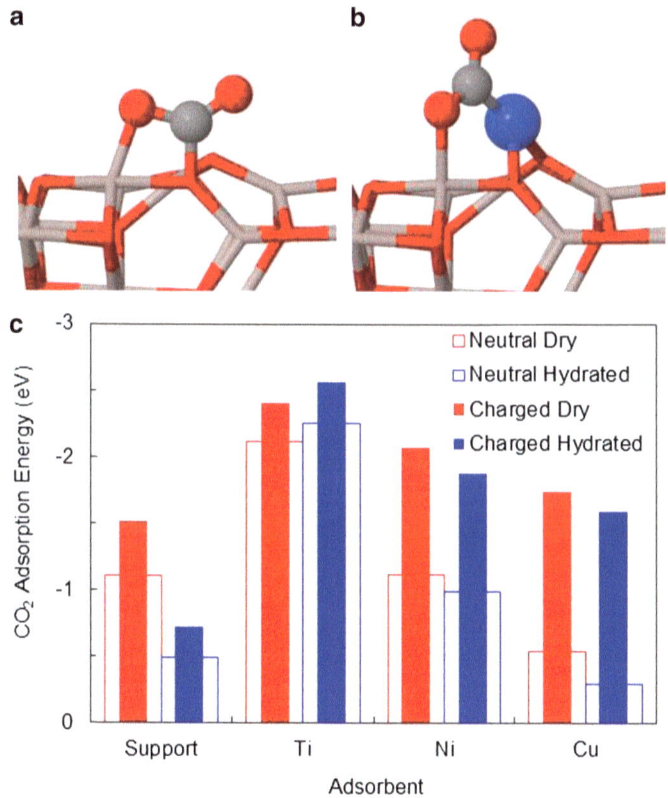

Fig. 4.7 Effect of surface charging on CO_2 adsorption. (**a**) and (**b**) Most favorable adsorption configuration on the support and supported transition metal atom. (**c**) Adsorption energies on all sites, with and without extra charge. (Reproduced with permission from [64])

not been applied in the context of plasma catalysis except for the CO_2 splitting process described above [76].

Therefore, various other types of models have been developed and used, which, incidentally, allow for including actual plasma effects somewhat easier than atomistic models. In microkinetic models, a set of elementary reactions is specified. Each elementary reaction is characterized by its rate constant, which can be determined from experiments or from DFT calculations. Solving the set of reactions yields the time evolution of the concentration of each of the species included. Alternatively, other models such as KMC (see also above) can be applied. Finally, simplified kinetic models can be constructed as well, typically to complement experimental data.

As a very recent example of Monte Carlo simulations, Guerra and Marinov developed and compared several dynamical Monte Carlo models to investigate surface reactions [77]. Although it is mentioned that these models were developed to study plasma-surface interactions, the authors acknowledge that only stable neutrals are taken into account, and the surface is considered to be of constant composition and morphology. Thus, ions, photons, electrons, excited species, and electromagnetic fields are not yet considered, and thus the link to actual plasma-surface interactions is still quite weak. Nevertheless, such models hold great promise, as they are much faster than atomistic simulations (for which the link to actual plasmas is also weak), they allow you to study long time scale evolution, and they can be easily coupled with gas phase chemistry. On the other hand, they are strongly dependent on a complete list of possible surface reactions and appropriate model parameters such as sticking coefficients and energy barriers. Very recently, the same authors also provided a description and comparison between KMC models and rate balance equation models [78].

Microkinetic modeling was performed by Blaylock et al. for steam methane reforming over multifaceted nickel catalysts [79, 80]. These authors first performed extensive DFT calculations to determine the kinetic and thermochemistry parameters for a considerable list of possible reactions. Interestingly, from such models the change in reaction pathway can be determined as a function of changing conditions. However, to the best of our knowledge, such models have not yet been performed in the context of plasma catalysis.

In contrast, simplified kinetic models have been developed for plasma catalysis. For instance, Tochikubo developed a combined plasma/catalytic rate equation model to describe the reduction of nitrogen oxides with hydrocarbon admixtures [81]. The catalyst model consists of a set of mass balanced equations for relevant surface reactions. The allowed surface reactions are shown in Fig. 4.8. Although the surface coverage is variable in the model, the number of active sites that are catalytically active is kept fixed in the model. The model is capable of identifying a number of important reactions, and the general trends of the dependency of NO_x removal on temperature and hydrocarbon admixture are reproduced. Still, more reliable surface reaction data is required to allow for a more quantitative understanding of the NO_x reduction chemistry. In particular, it is assumed that each reaction accounted for is described by the Arrhenius equation and, more importantly, that there are no

Number	Reaction	ε_a (eV)
(R1)	$2S + O_2 \rightarrow 2\,S\text{-}O$	0.96
(R2)	$2\,S\text{-}O \rightarrow 2\,S + O_2$	2.66
(R3)	$S + NO_2 \rightarrow S\text{-}NO_2$	0.50
(R4)	$S\text{-}NO_2 \rightarrow S + NO_2$	1.16
(R5)	$S\text{-}O + NO \rightarrow S\text{-}NO_2$	0.29
(R6)	$S\text{-}NO_2 \rightarrow S\text{-}O + NO$	0.70
(R7)	$S\text{-}O + S\text{-}NO_2 \rightarrow S\text{-}NO_3 + S$	0.88
(R8)	$S\text{-}NO_3 + S \rightarrow S\text{-}O + S\text{-}NO_2$	1.29
(R9)	$S\text{-}O + HC \rightarrow S\text{-}HC^*$	0.68
(R10)	$2\,S\text{-}NO_3 + 3HC \rightarrow 2\,S\text{-}HC^* + CO + H_2O + 2NO$	0.48
(R11)	$S\text{-}HC^* + S\text{-}O \rightarrow 2\,S + CO + H_2O$	1.84
(R12)	$S\text{-}HC^* + S\text{-}NO_3 \rightarrow S\text{-}NO_2 + S + CO + H_2O$	1.39
(R13)	$2\,S\text{-}HC^* + S\text{-}NO_3 \rightarrow 2\,S + S\text{-}NCO + CO_2 + 2H_2O$	1.45
(R14)	$S\text{-}NCO + S\text{-}NO_3 \rightarrow 2\,S\text{-}O + CO_2 + N_2$	1.67
(R15)	$S\text{-}NCO + S\text{-}NO_2 \rightarrow 2\,S\text{-}O + CO + N_2$	1.03
(R16)	$S\text{-}O + CO \rightarrow S + CO_2$	0.72
(R17)	$S\text{-}NO_3 + CO \rightarrow S\text{-}NO_2 + CO_2$	0.39

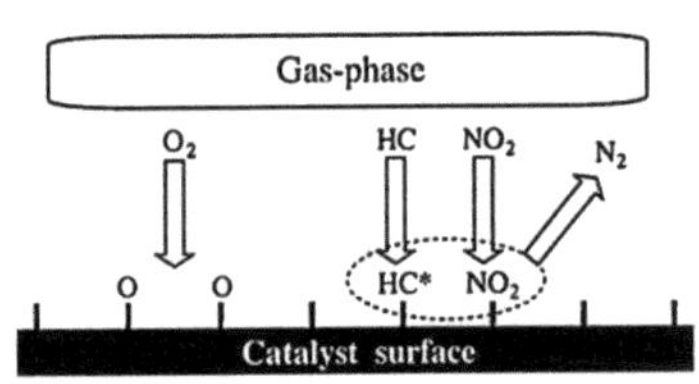

Fig. 4.8 List of surface reactions allowed in the rate equation model of Tochikubo et al. (Reproduced with permission from [81])

reactions besides those included in the model. Thus, while atomistic models are limited due to the required interaction model (classical force fields) and the accessible time and length scales, global models are limited due to their inevitably incomplete list of reactions and assumed reaction behavior.

Delagrange et al. performed a series of experiments on toluene oxidation in a plasma reactor containing manganese-based catalysts. It was concluded from these experiments that ozone is highly important in the oxidation process. This was confirmed by applying a simplified rate equation model, showing reasonable agreement with the experiments, especially for low amounts of toluene [82]. In particular, the authors concluded from their combined experimental and model results that the increase in toluene oxidation in the presence of the catalyst compared to the plasma alone is probably due to the importance of the surface reactions between plasma-generated ozone and toluene.

Finally, Kim et al. also studied the decomposition of several volatile organic compounds (VOCs) in a plasma-driven setup. Except for styrene, zero-order kinetics was invariably observed. A kinetic model was then constructed to represent these zero-order kinetics with respect to the specific energy input [83]. In this model, the plasma enters through the formation of active surface sites, which are assumed to be proportional to the discharge power density. From the model, a so-called energy constant for each VOC can be defined, whose value is indicative of the energy required to decompose each VOC to a certain degree in the plasma-catalyst setup. Overall, the model results were found to be in good agreement with the experimental results.

4.3.5 Future Needs

From the above, it is quite clear that a lot of work still needs to be done. Currently, there are no models which are capable of describing the influence of the various plasma effects on the surface reactions or on the catalyst properties. The development of such a model is, naturally, complicated by the inherent wide range in time and length scale of the processes and phenomena involved. Therefore, a multi-scale or multi-method approach seems to be most appropriate. In such a multi-method approach, DFT calculations could for instance be used to calculate either reaction probabilities of adsorbates on the catalyst, both for neutral, charged, and excited species, or kinetic parameters for surface reactions including these species. This information can then be used to construct, e.g., a Monte Carlo model, which is capable of handling longer time and length scales. The dynamics of the atomistic processes can be handled through advanced simulation techniques such as collective variable-driven hyperdynamics [76].

4.4 Modeling the Effects of a Catalyst on the Plasma

There are also not many attempts yet to model the effects of a catalyst on the plasma behavior. Here we will discuss the various aspects, more or less following the list presented in Sect. 4.2.2.

4.4.1 Enhancement of the Local Electric Field and Change of the Discharge Type

The most prominent effect of the presence of a catalyst on the plasma characteristics is the change in discharge behavior and the local electric field enhancement. Indeed, most plasma catalysis experiments are carried out in a dielectric barrier discharge, where the catalyst is typically present as pellets, or as a coating on (dielectric/ support) beads, in a packed bed DBD reactor. A schematic illustration of such a packed bed DBD reactor is shown in Fig. 4.9.

The packing beads (or pellets) are typically dielectric materials, and the applied electric field between both electrodes of the DBD reactor thus causes polarization of the dielectric beads. At the contact points between the beads, opposite charges are present, which might cause a strong local electric field enhancement in the plasma. This behavior has been reported based on experiments, as outlined in Sect. 4.2.2 above, but it has also been the subject of modeling.

Chang [27] applied a zero-dimensional (0D) plasma chemistry model for a $N_2/NF_3/O_2/H_2$ mixture in a $BaTiO_3$ packed bed plasma reactor, to predict which by-products were formed. The enhancement factor of the electric field in the voids between the ferroelectric pellets was simply deduced from the ratio of the dielectric constant of the pellets and the gas phase. Takaki et al. [85] developed a simplified time-averaged 1D numerical plasma model for N_2, based on solving the transport equations as well as the Poisson equation. They reported that all plasma parameters increased upon increasing the applied potential and dielectric constant of the packing beads. Kang et al. [26] developed a 2D model of a DBD reactor with two stacked ferroelectric beads inside, studying the propagation of the microdischarges during the first 20 ns. The behavior of electrons and ions was described by a set of fluid equations, but no plasma chemical species were taken into account. It was reported that the discharges evolve in three phases, avalanche, streamer, and decay, and also that the streamer discharges tend to be stabilized by the presence of dielectric materials. Russ et al. [86] applied a 2D fluid model to simulate transient microdischarges in a packed bed DBD reactor filled with dry exhaust gas, but only focusing on a short discharge (few tens of nanoseconds). Finally, although not directly applied to a packed bed reactor, Babaeva et al. also showed very interesting modeling results for the effect of dielectric spheres (dust particles) blocking a plasma streamer, using a 2D fluid model in humid air [87]. The effect of particle size, shape, and material properties of the dust particles on the streamer dynamics was

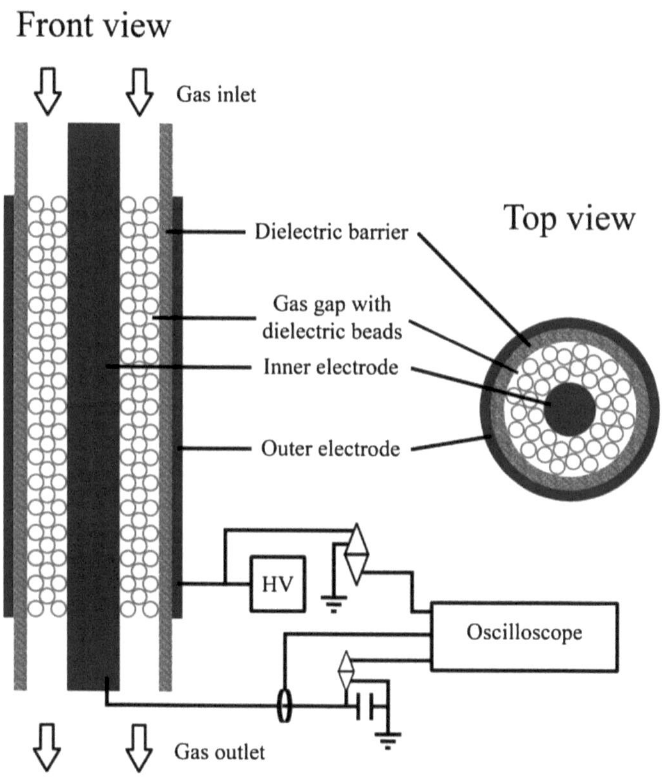

Fig. 4.9 Schematic illustration of a packed bed DBD reactor. (Reproduced with permission from [84])

investigated, and it was revealed that small dielectric particles (< tens of μm) are enveloped by the streamer, while larger particles can intercept and reinitiate streamers [87].

Kruszelnicki et al. [88] presented 2D fluid model simulations for a packed bed reactor, constructed out of dielectric rods, in humid air, studying in detail the mechanism of discharge propagation. They reported that overall, the discharges in a packed bed reactor can be classified in three modalities: positive restrikes, filamentary microdischarges, and surface ionization waves. Restrikes are formed, following breakdown in regions of high electric field. When the restrikes are confined between two dielectrics, they result in the formation of filamentary microdicharges that bridge the gap between the dielectrics. Eventually, surface charging near the feet of the microdischarges creates electric field components parallel to the dielectric surface, leading to the formation of surface ionization waves. The calculations revealed that the production of reactive species primarily takes place near the surfaces, as a result of restrikes and surface ionization waves. Hence, the production of reactants in a packed bed reactor is not a continuous process, but it rather results from the accumulation of individual, transient events.

In the same paper, the effect of separation between the dielectric rods and the rod orientation in the packed bed reactor was investigated, and it was found that the type of discharge dominating the production of reactive species depends on the dielectric facilitated electric field enhancement, which is a function of the topography and orientation of the dielectric lattice [88]. Whereas filamentary microdischarges and subcritical discharges and their follow-on negative streamers are stable and occupy relatively large volumes, they do not necessarily contribute to a large extent to the plasma chemical processes, because of their lower electron densities and temperatures. On the other hand, restrikes and surface ionization waves are characterized by higher electron densities and temperatures, and thus, in spite of their smaller volume and lifetime, they often produce larger amounts of reactive species. As the packed bed geometry affects the type of discharge that is favored, it will thus also affect the magnitude and reproducibility of reactant production. Finally, the authors also reported that photoionization plays an important role in discharge propagation through the dielectric lattice, as it seeds initial charge in regions of high electric field, which are difficult to access for electrons from the main streamer. This implies that knowledge of the UV spectral distribution is important for the propagation of discharges through packed bed reactors [88].

Figure 4.10 illustrates the calculated time-integrated densities of excited N_2 species, as well as experimental data, obtained by fast camera imaging of visible light emission in a packed bed reactor constructed also of dielectric rods, as considered in the model [88]. The formation of a cathode-seeking filamentary microdischarge (FM) between the rods is indicated both in the simulated and experimental results. Lack of plasma near the bottom pole of the top rod is in part due to the direction of the applied electric field, which points upward, toward the cathode. In addition to the filamentary discharges, surface discharges also occur, as is clear from this figure. Ions produced in the positive polarity filamentary microdischarges are accelerated toward the surface of the central rod, positively charging its surface. The latter produces an electric field component parallel to the surface, eventually leading to the development of a surface ionization wave (SIW), also indicated in the figure. This process takes about 4 ns in the model [88].

Recently, Kang et al. [89] also presented a 2D fluid model to study surface streamer propagation in a simplified packed bed reactor, in comparison with experimental data, obtained from time-resolved intensified charge coupled device (ICCD) imaging.

Within our group PLASMANT, we recently also performed a number of different modeling studies on a packed bed DBD reactor [84, 90–94], and some results of these models will be illustrated below.

(a) *Particle-in-Cell-Monte Carlo Collision (PIC-MCC) Model of a Packed Bed DBD Reactor*

Zhang et al. [90] developed a 2D particle-in-cell-Monte Carlo collision (PIC-MCC) model to describe the filamentary discharge behavior in a parallel-plate packed bed DBD reactor in air (N_2/O_2 = 80/20), comparing an unpacked (i.e., empty) and a packed bed DBD reactor, at an applied voltage of −20 kV. The simulations predict that the dielectric packing leads to a transition in discharge

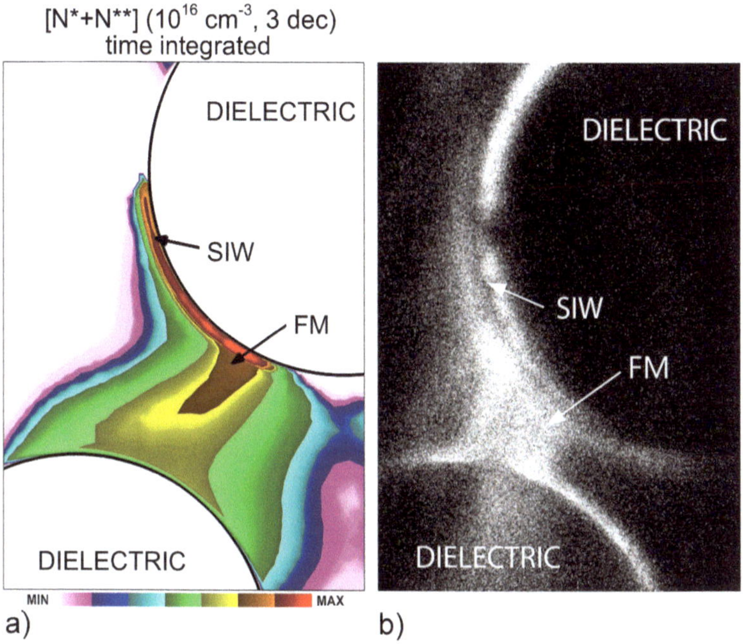

Fig. 4.10 Calculated time-integrated densities of excited N_2 species in a packed bed reactor in humid air, at an applied voltage of −30 kV and bead separation of 0.7 mm, as obtained from 2D fluid simulations by Kushner and coworkers (**a**), and measured visible light emission, obtained with an ICCD camera at an observation gate width of 0.5 μs, in the packed bed reactor with similar 2D geometry, constructed out of dielectric rods (**b**). (Reproduced with permission from [88])

behavior from a combination of negative streamers and unlimited surface streamers on the bottom dielectric surface, in case of an empty DBD reactor, to a combination of predominant positive streamers and limited surface streamers on the dielectric surfaces of the beads and plates, in the packed bed DBD reactor. The calculated electron density distribution is plotted in Fig. 4.11, for different moments in time, in a parallel-plate packed bed DBD reactor with five packing beads, assuming a dielectric constant (ε) of 22 for both the packing beads and the upper and lower dielectric barriers.

Furthermore, the electric field enhancement inside the dielectric material, as well as in the plasma near the contact points between the beads and the dielectric plates, was demonstrated in this study [90]. Calculations were performed for different dielectric constants of the packing beads and dielectric barriers (i.e., $\varepsilon = 4$, 9, and 22), and the corresponding maximum calculated electric field strengths and electron densities, at different times, for these three different dielectric constants, are listed in Table 4.1.

It is clear that the strength of the electric field and the maximum electron densities generally increase as a function of time, as well as with a rising dielectric constant, because the dielectric materials are more effectively polarized. The electric field

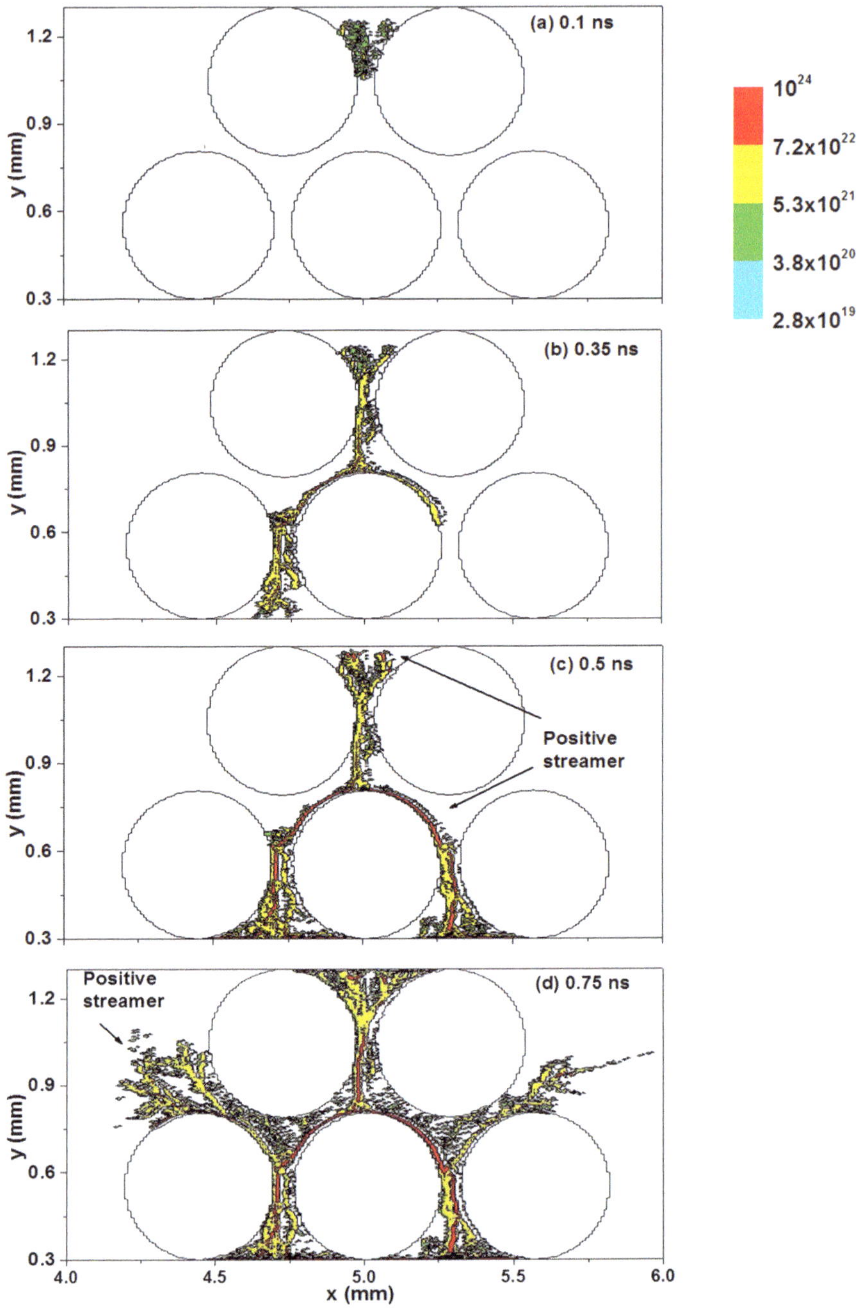

Fig. 4.11 Electron density (m^{-3}), calculated with the PIC-MCC model, at different times, i.e., (a) 0.1 ns, (b) 0.35 ns, (c) 0.5 ns, and (d) 0.75 ns, in a parallel-plate packed bed DBD reactor with five packing beads, assuming a dielectric constant of the packing beads and the dielectric barriers equal to 22, for an applied voltage of – 20 kV. (Reproduced with permission from [90])

Table 4.1 Calculated maximum values of the electric field strength and electron density, at four different times, for three different dielectric constants (ε) of the packing beads and dielectric barriers, for a parallel-plate packed bed DBD reactor, as obtained from a PIC-MCC model [90]

	Max. electric field (V/m)			Max. electron density (m^{-3})		
Time (ns)	$\varepsilon = 4$	$\varepsilon = 9$	$\varepsilon = 22$	$\varepsilon = 4$	$\varepsilon = 9$	$\varepsilon = 22$
0.1	1.9×10^7	3.1×10^7	5.5×10^7	1.1×10^{22}	1.4×10^{22}	3.8×10^{22}
0.35	4.9×10^7	6.0×10^7	1.3×10^8	8.5×10^{22}	1.1×10^{23}	2.6×10^{23}
0.5	6.0×10^7	1.4×10^8	1.8×10^8	2.2×10^{23}	5.2×10^{23}	1.4×10^{24}
0.75	1.5×10^8	3.4×10^8	4.0×10^8	2.5×10^{23}	5.6×10^{23}	1.1×10^{24}

values are also much higher than in an empty DBD reactor, resulting in a stronger and faster development of the plasma and thus also higher electron densities. More details about these modeling results can be found in [90].

Gao et al. also applied a PIC-MCC model to study the mode transition from volume to surface discharge in a packed bed DBD in O_2/N_2 mixtures, upon changing the applied voltage and O_2/N_2 mixing ratio [91]. A higher voltage yields a mode transition from combined volume + surface discharge to pure surface discharge, because the charged species can escape more easily to the beads due to the strong electric field and thus charge the bead surface. This surface charging will enhance the tangential component of the electric field along the dielectric bead surface, yielding SIWs. The latter give rise to a high concentration of reactive species on the surface, and thus they might enhance the surface activity of the beads, which is of interest for plasma catalysis. The SIWs were found to propagate more slowly with increasing O_2 content in the mixture, due to loss of electrons by attachment to O_2 molecules.

(b) *Fluid Model of a Packed Bed DBD Reactor*

Van Laer and Bogaerts developed a 2D fluid model for a packed bed DBD reactor [84]. In principle, such a reactor needs to be modeled in three dimensions, to fully account for the packing geometry, as there is no axial symmetry to reduce the geometry to 2D. However, the mesh size for modeling a packed bed DBD reactor needs to be very small, to account for the regions near the contact points between the beads. More specifically, the distance between two mesh points in the gas gap was around 10 μm, while it was around 1 μm near the contact points and at the material surfaces, yielding up to 100,000 mesh elements in a 2D geometry [84]. Therefore, modeling a packed bed DBD reactor in 3D is not yet feasible within a realistic calculation time. For this reason, Van Laer and Bogaerts have developed two complementary axisymmetric 2D fluid models, in order to approach the 3D geometry. These models are based on a 3D unit cell of a close-packed DBD reactor, i.e., a so-called "contact point" model and a "channel of voids" model; see Fig. 4.12 [84]. The combination of both models allows you to describe the two important features of a packed bed DBD reactor, i.e., (i) the contact between the beads, which is expected to lead to local electric field enhancement in the discharge due to polarization effects, and (ii) the fact that the voids between the beads are connected, allowing the plasma to travel from one side of the discharge gap to the other. The first

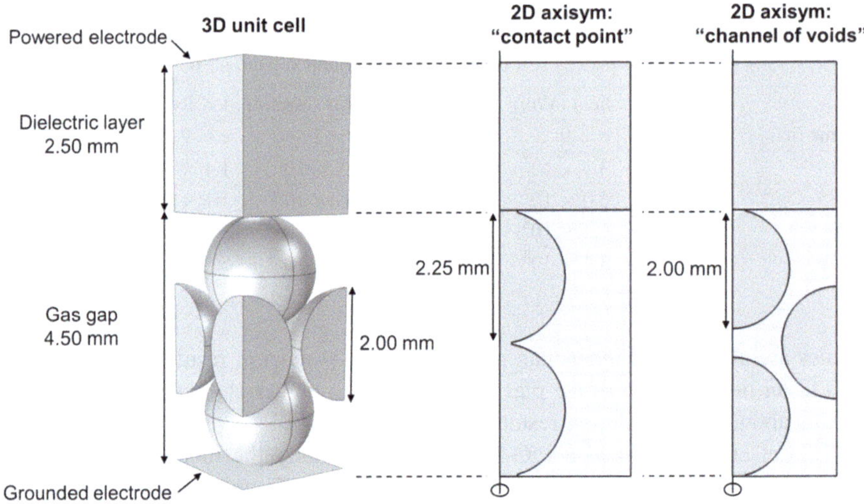

Fig. 4.12 3D unit cell of a packed bed DBD reactor (left) and its 2D representations used in the "contact point" model (middle) and "channel of voids" model (right). (Reproduced with permission from [84])

model considers two packing beads, which are slightly larger than in the real (3D) geometry, to allow them to be in direct contact with each other, while the second model describes three packing beads, with the same size as in reality, with a "channel of voids" in between them (see Fig. 4.12).

This fluid model was developed in helium, (i) because of the simplified plasma chemistry, thus reducing the calculation time, and (ii) because helium yields a homogeneous discharge in a DBD, which is easier to describe with a fluid model. The model solves conservation equations for the densities of the various plasma species (i.e., He atoms, He^+ and He_2^+ ions, metastable He* atoms, He_2* dimers, and the electrons) and for the average electron energy. The other plasma species, i.e., the so-called heavy particles, are assumed to be in thermal equilibrium with the gas, so that no extra energy equation needs to be solved for them. The conservation equations for the species densities are based on source and loss terms, defined by the chemical reactions, while the source of the electron energy is due to heating by the electric field, and the energy loss is again dictated by collisions. In addition, transport is included in the conservation equations, defined by diffusion and by migration in the electric field (for the charged species). These conservation equations are coupled with the Poisson equation for a self-consistent calculation of the electric field distribution from the charged species densities. The packing beads are treated as solid objects in the model, with zero space charge and certain dielectric properties inside the beads, as well as charge accumulation on their surface. The model was developed with the COMSOL Multiphysics Simulation Software. More details can be found in [84].

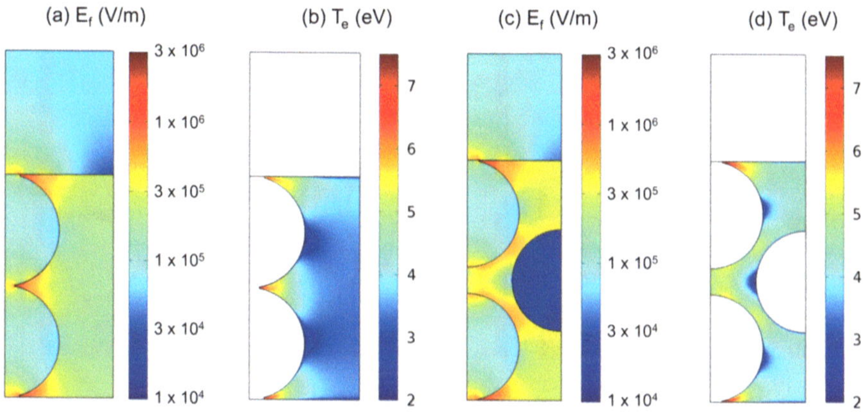

Fig. 4.13 Calculated time-averaged 2D profiles of the electric field (E_f) and electron temperature (T_e) in a packed bed DBD reactor, as obtained from a 2D fluid model for two complementary 2D geometries, reflecting the specific features of a packed bed DBD reactor, i.e., a "contact point" geometry (**a**, **b**) and a "channel of voids" geometry (**c**, **d**), at a peak-to-peak voltage of 4 kV and a frequency of 23.5 kHz

Figure 4.13 illustrates the calculated time-averaged electric field and electron temperature distributions, for a peak-to-peak voltage of 4 kV and a frequency of 23.5 kHz, in the 2D geometries of both the "contact point" model (a, b) and the "channel of voids" model (c, d). The "contact point" model clearly illustrates the local electric field enhancement near the contact points, due to polarization of the beads, both inside the material and in the gas gap (see Fig. 4.13a). The latter gives rise to more electron heating near the contact points, which is reflected by the higher electron temperature in Fig. 4.13b. The same behavior is also predicted by the "channel of voids" model (see Fig. 4.13c, d), although it is somewhat less pronounced, because the beads are not in direct contact with each other.

At this relatively low applied voltage of 4 kV, the plasma is initiated at the contact points and remains in this region, reflecting the properties of a Townsend discharge. At higher applied voltage, the discharge will spread out more into the bulk of the reactor, from one void space to the other. This is illustrated in Fig. 4.14 for the electron density, which is plotted at four consecutive moments of gas breakdown, at an applied voltage of 7.5 kV (peak-to-peak). The corresponding electrical current profile during one cycle of the applied voltage is plotted in Fig. 4.15. Four current peaks are observed in the first half cycle, which correspond to the electron density profiles illustrated in Fig. 4.14. The first three current peaks correspond to local glow-like discharges, taking place separately and consecutively, and they depend on the time needed for the electric field to reach high enough values to create a breakdown in the gas. The first peak corresponds to a local discharge between the two beads on the left (Fig. 4.14a), the second peak reflects the discharge between the upper left bead and the dielectric layer (Fig. 4.14b), and the third peak represents the simultaneous discharge between the dielectric layer and the top of the right bead and between the bottom of the right bead and the grounded electrode (Fig. 4.14c).

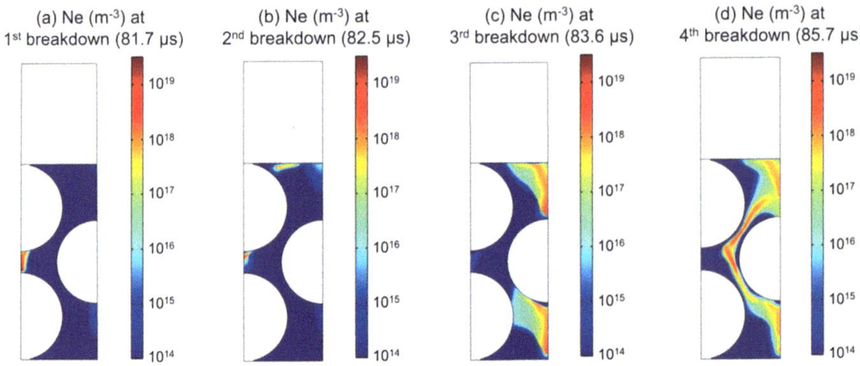

Fig. 4.14 Calculated 2D electron density profiles in the "channel of voids" geometry, at four consecutive moments of gas breakdown, i.e., (a) first, (b) second, (c) third, and (d) fourth breakdown, corresponding to the current peaks of Fig. 4.15, at a peak-to-peak voltage of 7.5 kV and a frequency of 23.5 kHz

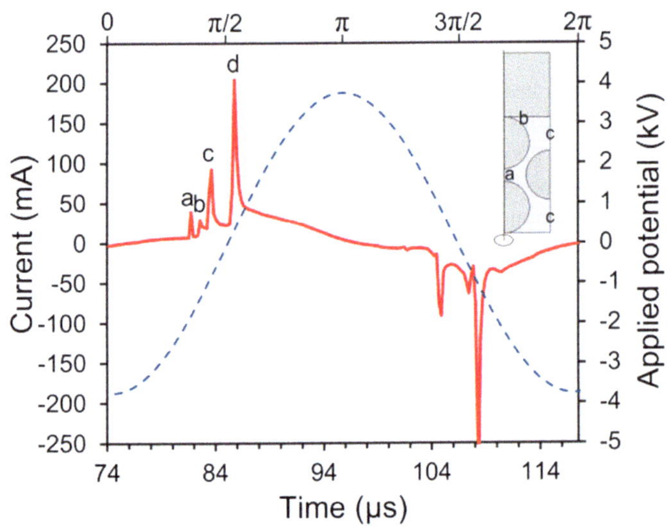

Fig. 4.15 Calculated electrical current profile (solid line, left y-axis) during one cycle of the applied voltage (dashed line, right y-axis), for a peak-to-peak voltage of 7.5 kV and a frequency of 23.5 kHz. The current peaks labeled with (**a–d**) in the first half cycle correspond to the electron density profiles (**a–d**) in Fig. 4.14. The inset shows where the discharge takes place during peaks (**a**), (**b**), and (**c**)

Subsequently, a discharge is formed over the whole gas gap, with plasma traveling through the "channel of voids," from one wall to the other (see Fig. 4.14d and the fourth peak in Fig. 4.15, which is most intense). A similar behavior occurs during the next half cycle, although the timing of the peaks relative to each other can change. Nevertheless, the strongest current peak always corresponds to the discharge travelling through the entire gas gap.

Hence, this example illustrates that modeling can reveal where exactly the discharge is created and how it expands inside the voids between the packing. This gives some idea on the discharge evolution (along the surface of the beads), but the actual change of discharge type (from streamers inside the plasma to streamers along the catalyst surface) is difficult to capture with this fluid model. More details of the discharge evolution, and how the discharge can travel through the gap, can also be found in [84].

Experimental measurements in a packed bed DBD reactor are not straightforward, among others, due to visual blocking of optical diagnostics by the packing beads. However, Kim and coworkers did measurements in a packed bed DBD reactor, using an ICCD camera [31–34], and they also observed that at low applied potential the discharge stays local at the contact points, while at higher potential, it spreads across the surface of the packing material [32, 33], in good agreement with the model results of Van Laer and Bogaerts [84]. Similar observations were also made by Tu et al. [35].

Van Laer and Bogaerts also studied the influence of the dielectric constant (ε) of the packing material on the plasma characteristics, for two different gap sizes, i.e., a mm-gap (4.5 mm) and a microgap (0.5 mm) [92]. The calculated time- and space-averaged electric field, electron temperature, and electron and ion densities are plotted as a function of ε in Fig. 4.16, for both gap sizes. Calculations were performed for $\varepsilon = 5$, 9, 25, 100, and 1000, which are representative for silica (SiO_2), alumina (Al_2O_3), zirconia (ZrO_2), titania (TiO_2), and barium titanate ($BaTiO_3$) materials, respectively.

The plasma behavior was found to be significantly different in both cases, even when applying the same voltage. Indeed, the electric field was found to be more enhanced in the microgap reactor, as is obvious from Fig. 4.16a, because the same voltage is applied over a shorter gap. When comparing both y-axes, it is clear that the electric field in the microgap reactor is at least a factor of 5 larger than in the mm-gap reactor, for the same applied voltage and dielectric constant. This stronger electric field results in more current peaks per half cycle, as was illustrated in [91], because the required electric field strength to cause a breakdown is more often reached.

In both the microgap and mm-gap reactor, the calculated electric field increases upon higher dielectric constant of the packing beads, but only up to a certain extent, according to the model. In the mm-gap reactor, the electric field does not increase anymore above $\varepsilon = 100$, while in the microgap reactor, the electric field only increases when ε rises from 5 to 9, but then it stays constant. The reason is that at higher dielectric constants, the model reveals that the electric field enhancement only takes place at the top part of the reactor, where the packing beads are in contact with the dielectric covering the powered electrode, while the electric field near the bottom (grounded) electrode is rather weak, due to less polarization between the lower beads and the grounded electrode. In the mm-gap reactor, this phenomenon only comes into play above $\varepsilon = 100$.

The calculated electron temperature shows a quite similar trend as the electric field strength upon rising dielectric constant; see Fig. 4.16b. In the mm-gap reactor, the electron temperature rises gradually with the dielectric constant, due to the

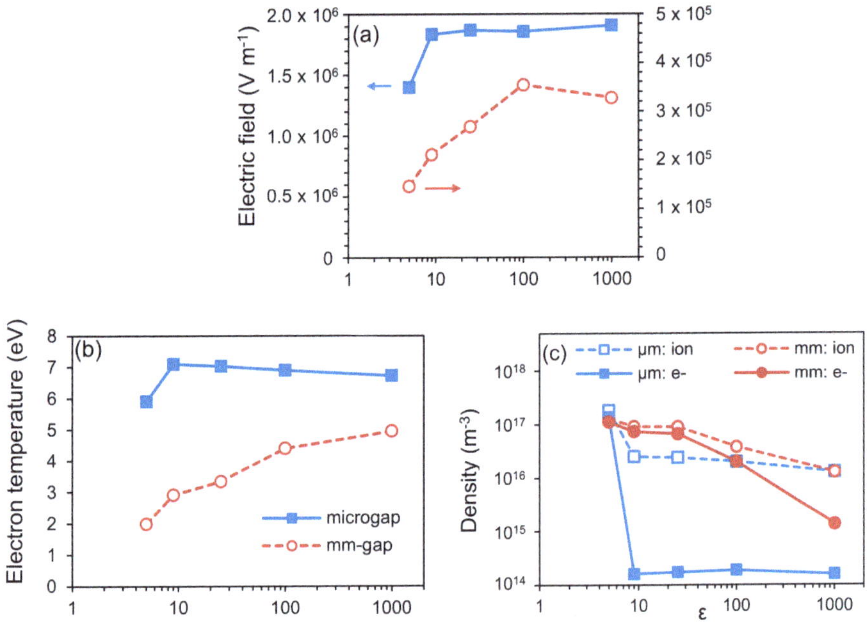

Fig. 4.16 Space- and time-averaged electric field strength (**a**), electron temperature (**b**), and electron and ion density (**c**), as a function of the dielectric constant of the packing beads, for both a microgap and a mm-gap packed bed DBD reactor

enhanced electric field. In the microgap reactor, the electron temperature only rises from $\varepsilon = 5$ to $\varepsilon = 9$, followed by a slight drop, which is attributed to the fact that the electric field enhancement mainly takes place at the top part of the reactor (see above).

Finally, the calculated electron and ion densities follow more or less the opposite trend as the electric field strength (see Fig. 4.16c). The ion density drops almost one order of magnitude, while the electron density drops by three orders of magnitude between $\varepsilon = 5$ and 9 in the microgap reactor, whereas a more gradual drop is observed between $\varepsilon = 25$ and 1000 for the mm-gap reactor, being again more pronounced for the electrons than for the ions. This drop is attributed to a change in discharge mechanism. For the mm-gap reactor, the plasma loses its ability to travel through the channels between the voids when ε rises from 25 to 1000, resulting in an overall lower electron and ion density. In the microgap reactor, our calculations reveal that the plasma can only travel through the channel between the voids when $\varepsilon = 5$. This was illustrated in detail in [92]. At higher dielectric constants, the electrons and ions get more easily absorbed on the walls and surfaces of the packing beads, due to the small dimensions and enhanced electric field, so the plasma loses its ability to travel through these small channels. This explains the much lower overall electron (and ion) density. The effect is more pronounced for the electrons, because of their smaller mass and thus their higher mobility. Hence, as is clear from Fig. 4.16c, the electron density is much lower than the ion density in the microgap reactor (except for $\varepsilon = 5$), and the same behavior becomes apparent for $\varepsilon \geq 100$ in the mm-gap reactor, thus resulting in a non-quasineutral plasma at these conditions.

Recently, Van Laer and Bogaerts also applied the above fluid models to study the effect of different bead sizes and dielectric constants of the packing beads, while keeping the interelectrode spacing constant [93]. They demonstrated that the discharge behavior changes considerably as a function of bead size and dielectric constant. At low dielectric constant, the plasma is spread out over the full discharge gap, with a significant density in the voids as well as in the connecting void channels. The electric current profile shows a strong peak during each half cycle. Upon increasing dielectric constant, the plasma becomes localized in the voids, and the current profile consists of many smaller peaks during each half cycle. For large bead sizes, the shift from full gap discharge to localized discharges was found to take place at a higher dielectric constant than for smaller beads. In addition, a higher breakdown voltage seems to be required to cause plasma formation for smaller beads or beads with a lower dielectric constant [93].

Although the above models are developed for helium, we expect a similar behavior in reactive gases that are typically used for environmental applications of plasma catalysis, except that streamer formation will occur in the latter cases. The higher electron temperature will result in more electron impact ionization, excitation, and dissociation of the gas molecules, for the same applied power, and this can explain why a packed bed DBD reactor gives higher pollutant decomposition or CO_2 conversion and higher energy efficiencies than an empty reactor, although it is suggested in [92] that this effect might also be (partially) compensated by the lower electron density.

To study in more detail the behavior of streamer propagation in a packed bed DBD, Wang et al. [94] recently developed a 2D fluid model for a packed bed DBD in air, and they investigated the behavior of positive restrikes, filamentary microdischarges, and surface discharges, as well as the transition in discharge modes upon changing the dielectric constant of the packing beads.

Positive restrikes between the dielectrics result in the formation of filamentary microdischarges. Surface charging creates electric field components parallel to the dielectric surface, leading to surface ionization waves. At a low dielectric constant of the packing (e.g., $\varepsilon_r = 5$), plasma ignition between the beads occurs directly as surface discharges or surface ionization waves, which can connect with the surface of the adjacent bead, as illustrated in Fig. 4.17. On the other hand, at high dielectric constants (e.g., $\varepsilon_r = 1000$), no surface streamer jumping toward the adjacent bead surface takes place and spatially limited filamentary microdischarges are generated between the beads, as shown in Fig. 4.18. At intermediate dielectric constants, a mixed mode of surface discharges and local discharges is observed [94]. The calculation results were in good qualitative agreement with experiments, as detailed in [94].

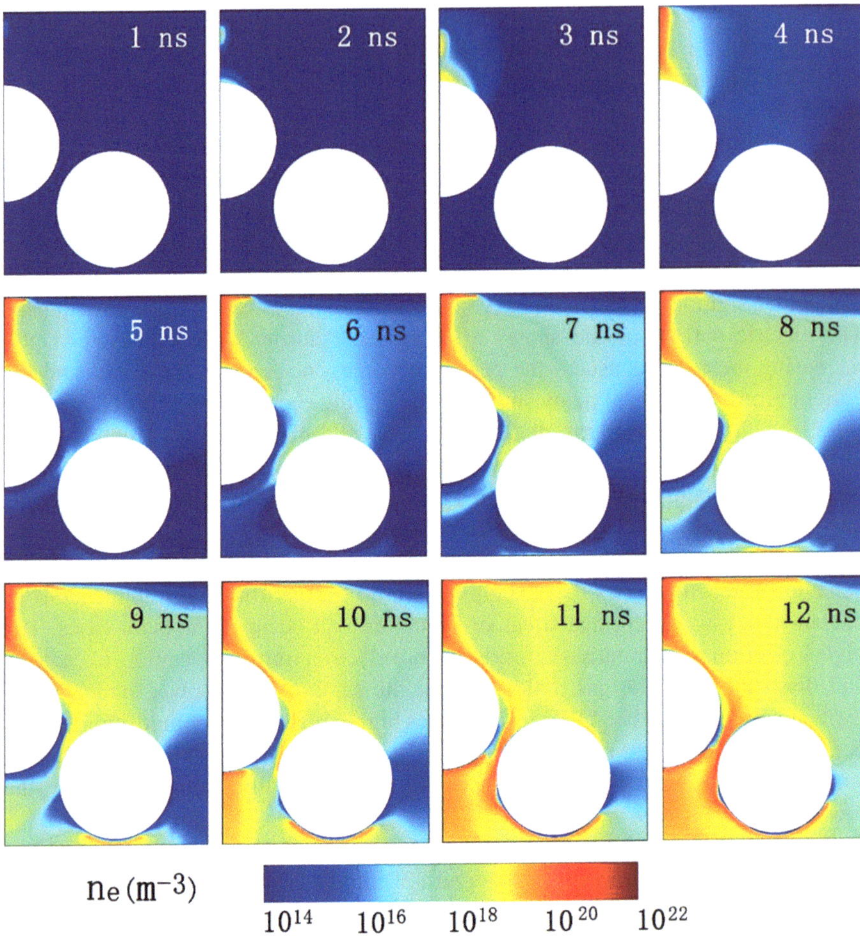

Fig. 4.17 Calculated electron number density distribution as a function of time, for a packed bed DBD reactor in dry air, with packing beads of $\varepsilon_r = 5$. (Adapted from [94] with permission)

Due to the enhanced electric field, energetic electrons are mainly created in the positive restrikes, local discharges, and surface discharges, causing the production of many reactive species. Our results indicate that a higher dielectric constant constrains the discharge to the contact points of the beads, which may limit the catalyst activation due to the limited catalyst surface area in contact with the discharge. This may have implications for the efficiency of plasma catalysis. Indeed, the best performance is not always reached for packing material with the highest dielectric constant [95, 96].

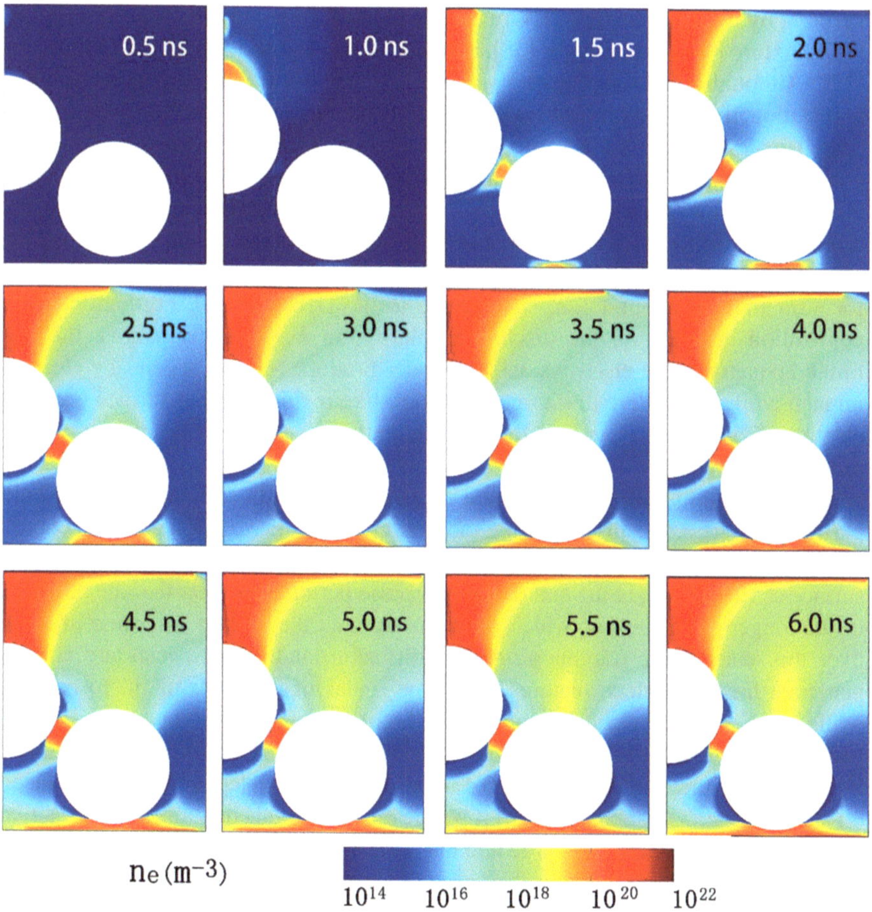

Fig. 4.18 Calculated electron number density distribution as a function of time, for a packed bed DBD reactor in dry air, with packing beads of $\varepsilon_r = 1000$. (Adapted from [94] with permission)

4.4.2 *Formation of Microdischarges in the Catalyst Pores*

It is important for plasma catalysis applications to obtain more insight in the occurrence and/or mechanism of microdischarge formation inside catalyst pores, as the presence of plasma species inside catalyst pores might affect the physical and chemical properties of the catalyst, and more importantly, it will increase the active surface area of the catalyst that is available for surface reactions. The formation of microdischarges inside catalyst pores was investigated experimentally by Holzer, Roland, and Kopinke [29, 39, 40], as well as by Hensel et al. [41–43], but again, the modeling attempts are more limited.

Bhoj and Kushner have developed a comprehensive multi-scale 2D fluid-surface kinetics model to describe the functionalization of rough polymer (polypropylene) surfaces by an atmospheric pressure dielectric barrier-corona discharge in a $He/O_2/H_2O$ mixture [97], in humid air [98], and in $He/NH_3/H_2O$ and $He/O_2/H_2O$ mixtures [99]. Although it was not a catalyst surface, the study is still relevant for this application, due to the similarities. The multi-scale model accounts for two different spatial scales, i.e., (i) a macroscopic scale (~ mm), defined by the spacing of the corona streamers, and (ii) a microscopic scale (~ μm), defined by the surface roughness, to investigate whether plasma species can penetrate into the microstructure of the surface. It was reported that electrons and ions produced during the corona pulses can only penetrate to a limited extent into the rough surface features for pore sizes of about 1 μm, comparable to the Debye length. Furthermore, the penetration depends on the discharge polarity: in negative discharges, there is limited penetration of electrons, which locally produce reactive species by electron impact, while in positive discharges, a sheath-like region near the surface prevents electrons from penetrating into the surface features, and hence, there is less local producton of radicals.

Figure 4.19 illustrates the calculated density profiles of O, OH, and O_3 in the bulk plasma at 50 μs, i.e., halfway during the first interpulse period of a negative corona discharge, as well as the densities of O, OH, O_3, and HO_2 in the vicinity of the surface at 100 μs, i.e., at the end of the interpulse period. It is clear that the densities of the longer-lived species, which react slowly with the surface, are quite uniform over the scale of the roughness, due to diffusion, and these species are able to penetrate into all surface features over time scales of hundreds of microseconds, while more reactive radicals, such as OH, are more depleted near the surface, because they are consumed before they can diffuse deeper into the surface features. In [99] the treatment of inside surfaces of porous polypropylene microbeads, placed on an electrode, was simulated, and it was revealed that this treatment was sensitive to the view angles to the discharge and the pore size and is controlled by the relative rates of radical transport and surface reactions deep into the pores.

Furthermore, Wang et al. have also studied the propagation of an air plasma through a porous dielectric sheet, with a pore diameter of 100 μm, by means of a fluid model [100]. The model also includes radiation transport and photoionization of O_2 by VUV radiation. Figure 4.20 illustrates the electron density inside the pores, at successive times, assuming a photoionization cross section of 10^{-19} cm^2. The maximum values of the electron density are indicated under each frame. The model reveals that less than 1 ns is required for the plasma to penetrate through the porous sheet of 600 μm thickness and that the plasma propagation is controlled by a balance between retarding, due to charging of the internal surfaces of the pores, and photoionization that extends the plasma around corners [100].

Recently, within our group PLASMANT, we also developed some models to study the behavior of plasma species inside catalyst pores, and some results will be illustrated below.

(a) *Fluid Model for the Formation of Microdischarges in Catalyst Pores*

Zhang et al. developed a 2D fluid model to answer the question whether plasma can be formed inside catalyst pores with μm dimensions in a helium DBD

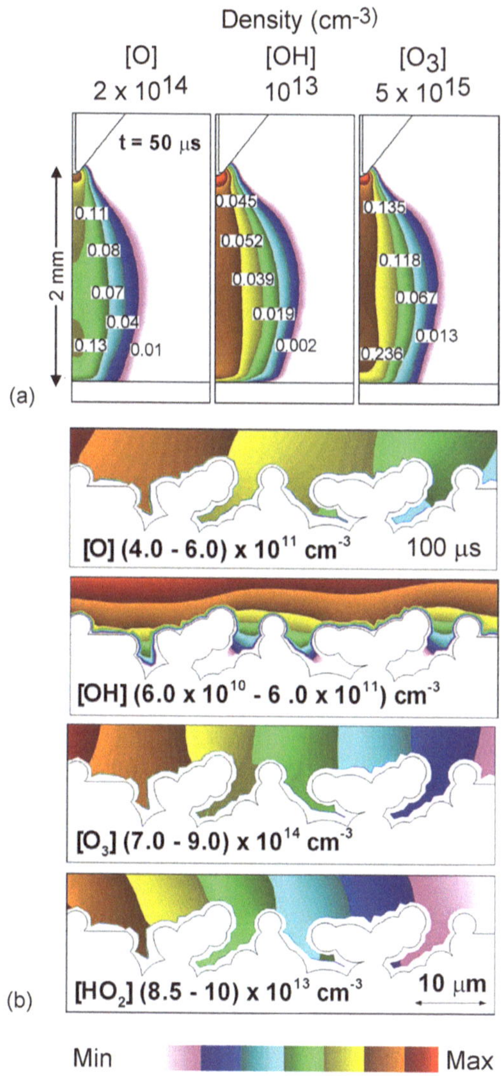

Fig. 4.19 Density profiles of O, OH, and O_3 in the bulk plasma at 50 μs, i.e., halfway during the first interpulse period of a negative corona discharge (**a**), and density profiles of O, OH, O_3, and HO_2 in the vicinity of the surface at 100 μs, i.e., at the end of the interpulse period (**b**), as calculated with a multi-scale fluid-surface kinetics model for an atmospheric pressure dielectric barrier-corona discharge in humid air. The contour labels are fractions of the maximum density noted in each figure. (Reproduced with permission from [98])

[101]. Figure 4.21 illustrates the calculation results for a pore size of 100 μm, at an applied voltage of 20 kV. The total ion density (Fig. 4.21a) increases significantly inside the pore, with a maximum value 7 times higher than in the center of the discharge. The electron density shows a slight increase near the pore, but inside the pore it drops to low values (see Fig. 4.21b). The latter can be explained because the electrons are more easily lost at the walls and also because the electric field pushes the electrons back to the bulk region.

The electric field inside the pore (Fig. 4.21c) is greatly enhanced, yielding a significant rise in electron temperature as well (Fig. 4.21d). As a result, the electron impact ionization is also greatly enhanced inside the pore, as is clear from Fig. 4.21e,

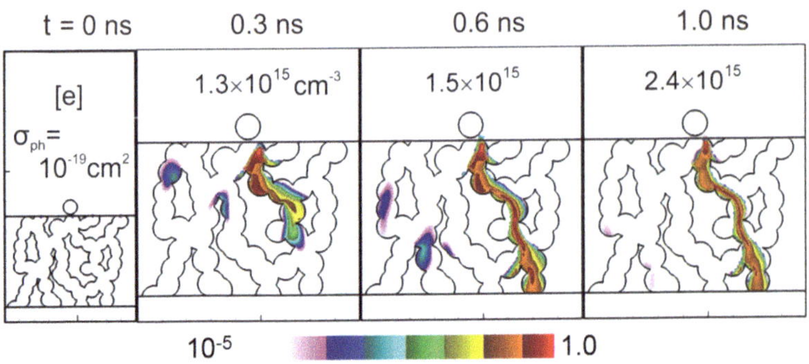

Fig. 4.20 Calculated electron density at three moments in time, illustrating plasma propagation through a porous dielectric sheet, assuming a photoionization cross section of 10^{-19} cm^2. (Reproduced with permission from [100])

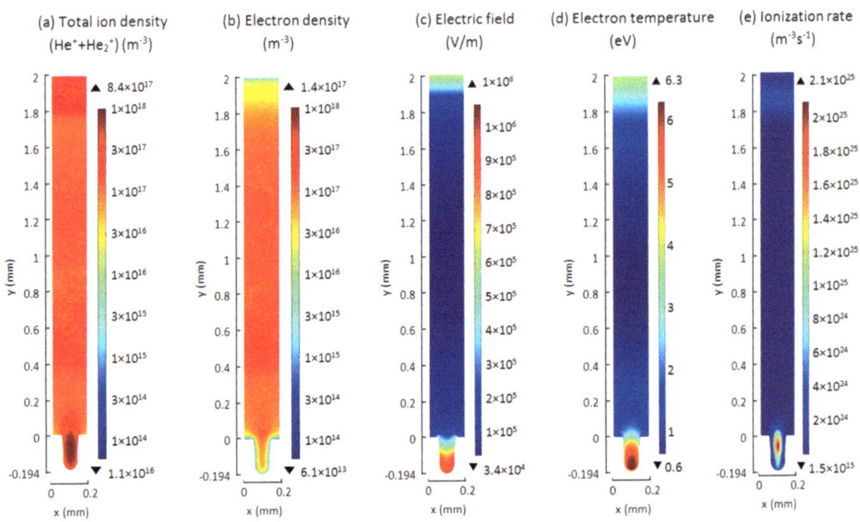

Fig. 4.21 Calculated distributions of the total ion density (**a**), electron density (**b**), electric field (**c**), electron temperature (**d**), and electron impact ionization rate (**e**), averaged over one period, for a helium discharge with applied potential of 20 kV, with a 100-μm pore. Note that in (**a**) and (**b**) the same color scale is used, to allow comparison, but the values above and below the color scale indicate the maximum and minimum densities in each case

giving rise to the formation of the ions and electrons. The same applies to electron impact excitation and the formation of excited plasma species [101]. Hence, the model predicts that for a pore size of 100 μm, the plasma species are effectively generated inside the pore, and in addition, the ions might also migrate into the pore due to the strong electric field.

When investigating the effect of various pore sizes and applied voltages, the modeling results indicate that the microdischarge formation inside the pore occurs more easily at larger pore size and applied voltage [101], which was also observed by Hensel et al. [41, 43]. At a fixed pore size of 30 μm, the presence of the pore seems to have no effect on the plasma properties for an applied voltage of only 2 kV, but above 10 kV, the ionization is significantly enhanced due to the strong electric field and high electron temperature, and the ion density shows a remarkable increase near and in the pore.

Furthermore, at a fixed applied voltage of 20 kV, the model predicts that plasma species can only be created inside catalyst pores with dimensions above 10 μm. Furthermore, while the resulting ion density inside the pores is indeed enhanced for pore sizes of 10 μm and more, the electron density only exhibits a significant increase near and inside the pore for pore sizes above 200 μm, as the electric field pushes the electrons back to the bulk region (see above). Finally, the electric potential shows a strikingly different distribution, depending on the pore size, as illustrated in Fig. 4.22. Inside a 10-μm pore, the potential reaches its maximum at the bottom of the pore, as is clear from Fig. 4.22a. For pore sizes of 20 μm and above, the potential drops gradually from the top of the pore to the bottom; see Fig. 4.22b. Finally, for a 400-μm pore (which may occur in structured catalysts), a clear potential drop in both the vertical and the horizontal direction is predicted, corresponding to the presence of a sheath near the bottom of the pore and the side walls. This is a strong evidence for the presence of a microdischarge inside the pore. The latter is as expected, because the theoretical Debye length is in the order of 40 μm for the typical conditions under study here (i.e., helium plasma with electron temperature and density of 3 eV and 10^{17} m^{-3}).

In plasma catalysis, various catalytic support materials can be used, with various dielectric constants. Zhang et al. therefore also investigated the effect of the dielectric constant of the material on the plasma behavior inside the catalyst pores, as well as in the sheath in front of the pores [102]. Figure 4.23 illustrates the calculated electron impact ionization rate, which is characteristic for the plasma generation as

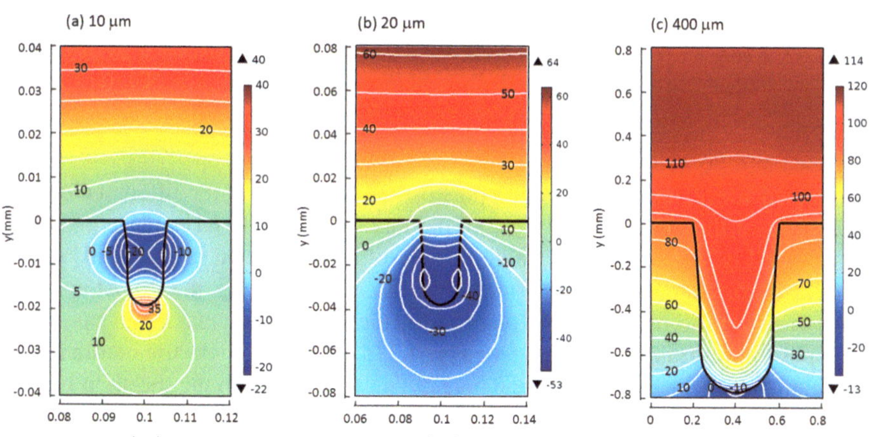

Fig. 4.22 Calculated electric potential distribution inside catalyst pores of 10 μm (**a**), 20 μm (**b**), and 400 μm (**c**), for a helium discharge with applied potential of 20 kV

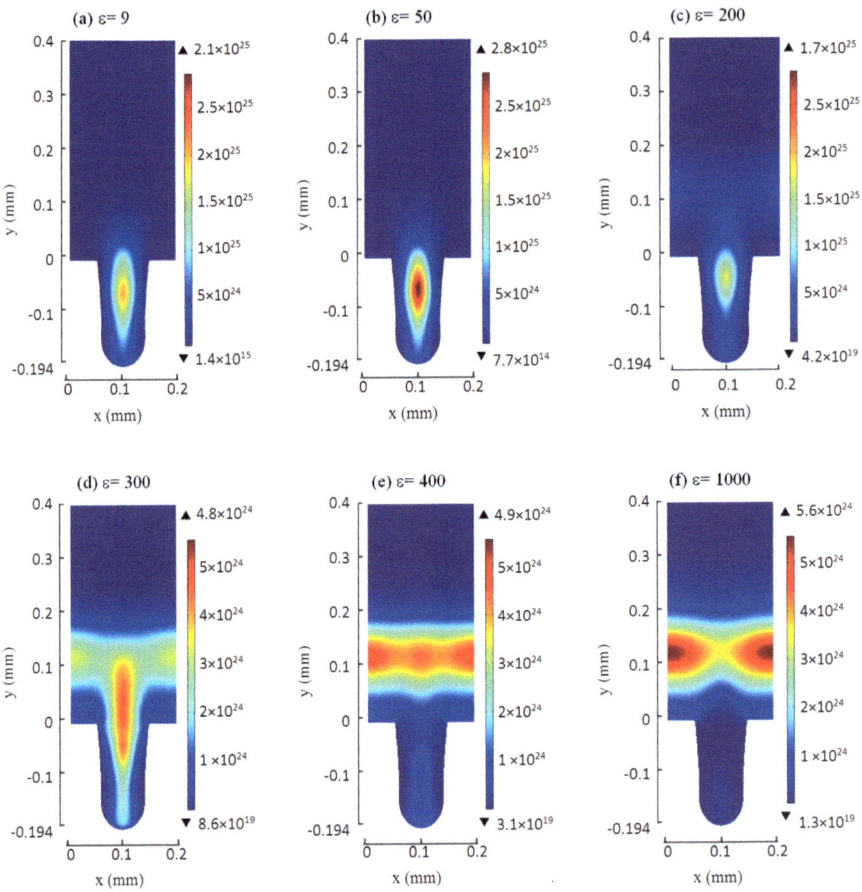

Fig. 4.23 Calculated electron impact ionization rate profiles above and inside a pore of 100 μm diameter, averaged over time in 1 AC cycle, for a helium discharge sustained at 20 kV, and for different dielectric constants: (**a**) ε = 9, (**b**) ε = 50, (**c**) ε = 200, (**d**) ε = 300, (**e**) ε = 400, (**f**) $\varepsilon = 1000$

explained above, inside and above a pore of 100 μm diameter, for different dielectric constants of the material. At $\varepsilon \leq 200$, the ionization clearly takes place inside the pore (see Fig. 4.23a–c). The absolute value of the ionization rate depends on the dielectric constant and reaches a maximum at $\varepsilon = 50$ (see Fig. 4.23b). At $\varepsilon = 300$, the ionization still partially takes place inside the pore, but the maximum has shifted to a position in front of the pore (see Fig. 4.23d). At larger dielectric constants, the ionization does not occur inside the pore anymore, but only in the sheath in front of the pore, as illustrated in Fig. 4.23e for $\varepsilon = 400$. At $\varepsilon = 1000$, the ionization is very pronounced in the sheath near the dielectric, but in front of the pore the ionization is reduced (see Fig. 4.23f). The strong ionization in the sheath in front of the dielectric can be explained because the dielectric material is more effectively polarized for larger dielectric constants, and this gives rise to a stronger local electric field in the sheath near the dielectric. The latter yields significant electron heating and thus also enhanced ionization in the sheath. Inside the pore, the electric field is slightly enhanced when ε increases from 4 to 25, but for larger dielectric constants, the polarization of the left sidewall counteracts that of the right sidewall, and as a consequence, the net electric field inside the pore is reduced. Especially for $\varepsilon = 1000$, the electric field inside the pore is very weak, resulting in a low electron temperature, and this significantly limits the plasma generation inside the pore for very high dielectric constants, as is indeed obvious from Fig. 4.23.

The model calculations reveal that the ionization enhancement inside pores smaller than 100 μm only occurs for materials with smaller dielectric constants, for the same reason as explained above (i.e., the polarization of the left sidewall counteracts that of the right sidewall). More specifically, for pore sizes of 50 μm, 30 μm, and 10 μm, only materials with dielectric constants below 200, 150, and 50, respectively, seem to yield enhanced ionization inside the pores. Note that $\varepsilon = 300$ is a typical value for $SrTiO_3$, $\varepsilon = 200$ corresponds to $CaTiO_3$, and $\varepsilon = 50$ is a typical value for TiO_2 [55].

Although some experiments are performed in discharges with different ferroelectric packed bed materials, to our knowledge, no experiments are available for the different plasma behavior inside catalyst pores with different dielectric constants. This behavior is probably different from the behavior in between the beads of packed bed discharges, due to the different dimensions. Hence, the model predictions still have to be validated by experiments, but they suggest that the most common catalyst supports, i.e., Al_2O_3 and SiO_2, with dielectric constants around $\varepsilon = 8$–11 and 4.2, respectively, should allow more easily that microdischarges can be formed inside catalyst pores, while for ferroelectric materials with dielectric constants above 300, it seems much more unlikely that plasma enhancement inside the pores occurs, even not for 100 μm pore sizes.

Finally, Zhang et al. also studied the plasma production inside catalyst pores with different pore shapes and reported that the electric field is significantly enhanced near tip-like structures [103]. A conical pore with small opening yields the strongest electric field at the opening and bottom corners of the pore, causing significant ionization throughout the pore. A cylindrical pore only yields electric field enhancement at the bottom corners, causing only slight enhancement of the ionization rate inside the pore. Finally, a conical pore with large opening yields a maximum electric field at the bottom. In conclusion, the pore shape seems to greatly affect the electric field enhancement and thus the plasma properties [103].

To summarize, this fluid model reveals that plasma species can only be created inside catalyst pores with dimensions above 10 μm, for materials with dielectric constants below 50. These pore sizes are of interest for structured catalysts, but catalytic supports typically have pores in the nm range, and the latter might thus be too small for microdischarge formation inside the pores, according to the fluid model predictions. However, the above studies were applied to a helium plasma, which yields a homogeneous discharge. In contrast, reactive gases, which are more relevant for plasma catalysis applications, exhibit streamer formation, and the latter might be characterized by much higher electron densities. Hence, it is well possible that in this case, the plasma can even be created in nm-sized catalyst pores. This is too small for fluid simulations, so we developed a PIC-MCC model for an air plasma, accounting for streamer formation, and this will be illustrated in the next section.

(b) *PIC-MCC Model for the Formation of Microdischarges in Catalyst Pores*

In [104] the formation of microdischarges inside both μm- and nm-sized catalyst pores was investigated by a 2D PIC-MCC model, for a DBD operating in dry air in filamentary mode. The calculations reveal that the streamers can penetrate into the surface features of a porous catalyst, and microdischarges can be formed inside both μm- and nm-sized pores. For the μm-sized pores, the ionization mainly occurs inside the pore, in agreement with the fluid model results presented in previous section, while for the nm-sized pores, the ionization is strongest either inside or just above the pores, depending on the pore size. The smaller pores (~ 4–10 nm) are characterized by a more pronounced surface discharge along the dielectric surface, due to the relatively large surface area, while in the larger pores (~ 100 nm or μm-sized) the ionization rate is more pronounced, due to the larger effective ionization region. This is illustrated in Fig. 4.24 for pore sizes of 4, 10, and 100 nm. In the 4-nm pore, the ionization is located above the pore, while in the 10-nm pore, it is both inside and above the pore, and in the 100-nm pore, the maximum ionization rate is clearly inside the pore. The reason why it is mainly in the upper part of the pore is because the electric field, due to space charge separation inside the pore, pushes the electrons out of the pore, inducing many ionization collisions in the upper part of the pore. It is

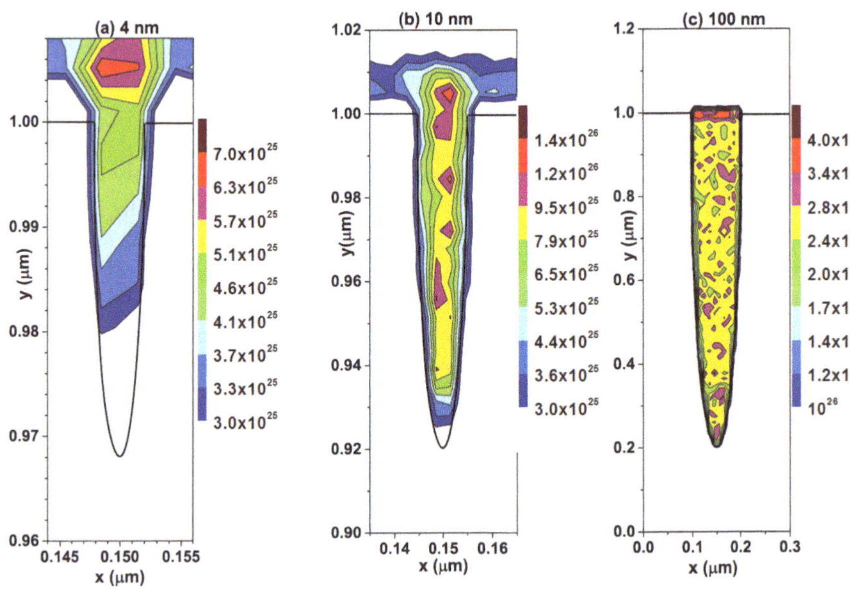

Fig. 4.24 Calculated electron impact ionization rate ($m^{-3}\ s^{-1}$) for pore sizes of 4 (**a**), 10 (**b**), and 100 nm (**c**), at a time of 1.2 ps. (Reproduced with permission from [104])

also clear from Fig. 4.24 that the maximum ionization rate drastically increases with pore size. Indeed, although the maximum local electric field decreases for increasing pore sizes between 10 and 100 nm, leading to a lower electron density, the effective ionization region increases with pore size, so there are more electrons available to generate ionization, resulting in an overall higher ionization rate [104].

Zhang and Bogaerts [105] applied a PIC-MCC model to study the mechanism of streamer propagation in catalyst pores, for various pore diameters in the nm to μm range. They concluded that the Debye length is an important criterion for plasma penetration into catalyst pores, i.e., plasma streamers can only penetrate into pores with diameter larger than the Debye length. The latter depends on the electron density and temperature in the plasma streamer, but is in the order of a few 100 nm up to 1 μm at typical DBD conditions in air. For pores in the range of ~ 50 nm, the calculations revealed that plasma can only penetrate to some extent and only at the beginning of a microdischarge, before the actual plasma streamer reaches the catalyst surface and a sheath is formed. Upon increasing applied voltage, plasma streamers can penetrate into smaller pores, due to the higher plasma density and thus shorter Debye length.

In [106] it was demonstrated with the same model that surface charging (of the dielectric surface or the catalyst pore sidewalls) plays a crucial role in the plasma distribution along the dielectric surface, as well as in the streamer propagation and discharge enhancement inside catalyst pores. The importance greatly depends on the dielectric constant of the material. At $\varepsilon_r \leq 50$, surface charging causes the plasma to spread along the dielectric surface and inside the pores, leading to deeper plasma streamer penetration, while for $\varepsilon_r > 50$ or for metallic coatings, the discharge was found to be more localized, due to very weak surface charging.

Finally, it is worth to stress again that the PIC-MCC simulations reveal that microdischarges can be formed inside nm-sized pores, while the fluid model in previous section predicted that this was only possible for μm-sized pores. The reason is that the fluid model results were obtained for a helium discharge, operating in glow mode, as explained above, while the results presented here are for an air discharge, operating in filamentary mode, which is characterized by much higher electron densities in the streamers, and in this case, the plasma can thus be created even in nm-sized catalyst pores, as long as they are larger than the Debye length at these conditions.

4.4.3 *Adsorption of Plasma Species on the Catalyst Surface, and Desorption of Newly Formed Species, that Might Affect the Plasma Chemistry*

Plasma species that adsorb on the catalyst surface can give rise to the formation of new species upon reaction at the surface. When these new species desorb from the catalyst surface, they arrive back in the plasma. Hence, when modeling the plasma chemistry for plasma catalysis application, the desorbed species need to be included in the model as well. There exist several models for describing the plasma chemistry for various environmental applications relevant for plasma catalysis, e.g., NO_x destruction [107–111], SO_2 removal [112, 113], the oxidation of carbon soot [108, 110, 114], VOC remediation [115–119], and hydrocarbon reforming and/or CO_2 conversion [120–133], but they are mostly applied to plasma without catalysis, and thus, the effect of desorbed species from a catalyst surface was typically not yet taken into account. Indeed, most often, 0D chemical kinetics models are used for this purpose, so that surface effects are typically not included at all. There exist also some (1D or 2D) fluid models for this application (e.g., [120, 125, 132, 133]), but they typically consider only simplified boundary conditions, based on sticking coefficients at the walls, thus not accounting for possible chemical reactions at a catalyst surface either.

A few papers, however, report on the plasma chemistry, accounting also for the effect of a catalyst. Kim et al. developed a simple kinetic model for the plasma catalytic decomposition of VOCs, predicting zero-order kinetics with respect to the specific energy input, in good agreement with their experimental observations [83]. Tochikubo simulated NO_x removal by plasma catalysis, based on a fluid model for a filamentary DBD, dealing with the plasma chemistry and a limited set of 17 surface reactions [81]. However, the synergy expected for plasma catalysis could not be observed, as the plasma simulations were not directly coupled to the surface reaction modeling. Moreover, the authors stated that the input data (activation barriers, rate constants) will need to be improved for better correspondence with experiments. Istadi and Amin developed an artifical neural network for a catalytic DBD reactor for dry reforming of methane, suggesting some synergism between the plasma and the CaO-MnO/CeO_2 catalyst, which affects the selectivity toward hydrocarbons with two or more C atoms [134]. Finally, Jiwu and Lei modeled the flue gas desulfurization process by a corona discharge combined with Mn^{2+} catalysis [135]. The Mn^{2+} catalyst was however in liquid phase, and thus no surface reactions were included.

There is still a lot of work to be performed in this field, but the latter can only be done when the surface reaction probabilities are known, pointing out the need of more atomic-scale simulations to obtain information on the surface reactions in case of plasma catalysis (see Sect. 4.3 above). When such reaction coefficients become available, the above plasma chemistry models could be upgraded by including the effects of a catalyst surface, through appropriate boundary conditions for the species continuity equations, so that these models, preferably in 2D, become effectively applicable to plasma catalysis.

4.5 Conclusion and Outlook: Ultimate Goal for Modeling Plasma Catalysis

The ultimate goal for modeling plasma catalysis will be to include all the above effects into one comprehensive, multi-level model. This includes describing the dynamic behavior at the atomic and molecular level, which takes place at very short time scales (order of nanoseconds), up to the slower bulk kinetic behavior, as well as mass and heat transfer, taking place at macroscopic time scales. It is clear that such a model has not yet been realized.

In other application fields, however, like for plasma etching and deposition, surface effects have been incorporated already. More specifically, the effect of surface reactions has been accounted for by updating the plasma chemistry with new species formed at the walls (e.g., [136–141]). Also the effect of surface temperature on the temperature in the plasma, which could also be relevant for plasma catalysis in case of hot spots (see Sect. 4.3.2 above), has been described already [142]. Typical codes which account for such effects are 2D hybrid models, like the Hybrid Plasma Equipment Model (HPEM) [141] and non-PDPSIM (e.g., [97–99]), both developed by Kushner and coworkers. The non-PDPSIM code has already been applied for multi-scale fluid-surface kinetics modeling of plasma treatment of rough polymer surfaces, which can be considered comparable to a porous catalyst surface (see Sect. 4.4.2 above). Furthermore, the HPEM code also allows to calculate the formation and evolution of trench profiles due to etching and the behavior of plasma species inside trenches, by means of a Monte Carlo simulation [143]. Such features would also be of great value for the simulation of plasma species inside catalyst pores.

It is thus clear that such a hybrid model would also be of great value for plasma catalysis applications. This concept is illustrated in Fig. 4.25. As plasma catalysis applications typically entail a comprehensive plasma chemistry, the most suitable type of model for this purpose is a 0D (or global) model, because of its reasonable computational cost, even when including a large number of species and chemical reactions. However, such a model does not account for geometrical (reactor) effects. Therefore, once the plasma chemistry is built up, it should be transferred to 2D or 3D fluid models, which account for geometrical effects, like in a packed bed DBD reactor, or surface effects, to allow updating the plasma chemistry by new species formed at the catalyst surface. Because such a model is computationally very expensive, the plasma chemistry will have to be based on a reduced chemistry set, which can be developed within the 0D chemical kinetics model, by comparing and benchmarking with the full chemistry model. Such a 2D or 3D model also calculates the electric field, which might affect the plasma-catalyst interactions. Furthermore, it provides the fluxes of the various plasma species arriving at the catalyst surface, which can subsequently be used to describe the behavior of plasma species inside a catalyst pore, e.g., by another fluid model or a Monte Carlo model (cf. Sect. 4.4.2 above). Ideally, the latter should also be able to account for changes in the pore shape, similar to what has been developed already for plasma etching applications (see above).

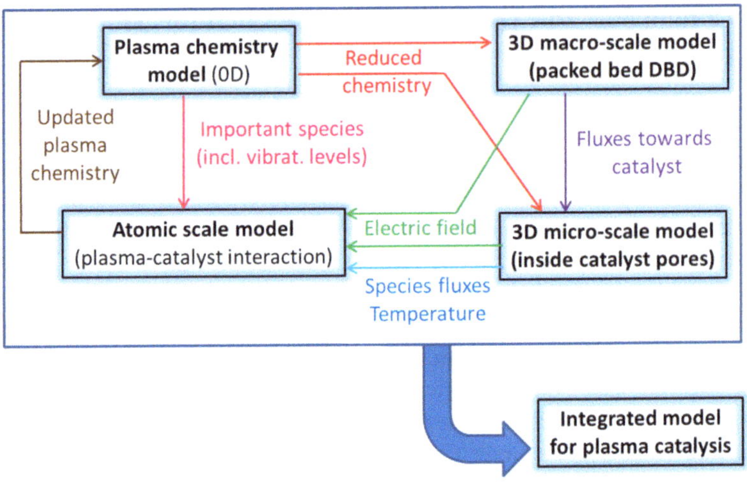

Fig. 4.25 Concept of an integrated model, which would be the ultimate goal for modeling plasma catalysis

This combined, hybrid model should furthermore be combined with atomic-scale simulations for the interaction of the plasma species with the catalyst surface. Indeed, the fluxes of the various plasma species, as well as the electric field near the catalyst surface, calculated by the plasma model, are useful input for atomic-scale simulations. Combining macro-scale plasma models and micro-scale models inside catalyst pores with atomic-scale simulations for the plasma-catalyst interactions is very challenging, due to the small time and length scales of the atomistic simulations, i.e., typically in the order of nanoseconds and nanometers, respectively [46]. Therefore, it is more realistic to use such simulations as a kind of off-line module to the plasma model, providing rate coefficients for the various surface reactions. The latter can then be used as boundary conditions for the plasma species continuity equations in the plasma model, to update the plasma chemistry. This will also allow adding new (desorbed) species from a catalyst surface to the plasma chemistry. In this way, the atomistic simulations can be integrated in a plasma model, yielding a multi-level model for plasma catalysis (see Fig. 4.25).

It is clear that developing such an integrated model will require a lot of efforts, especially because of the large number of plasma-catalyst interactions that need to be accounted for. Hence, it cannot be realized on a short time scale. Nevertheless, we should aim for such a model, which would certainly contribute to a better understanding of plasma catalysis and can help to improve this highly important and rapidly evolving application field. In the meantime, until such a comprehensive model is being developed, the individual modeling approaches, as described in Sect. 4.3 and 4.4 above, can also contribute already to a better understanding of plasma catalysis.

Acknowledgments We gratefully acknowledge K. Van Laer, Y.-R. Zhang, Y. Zhang, Q.-Z. Zhang, W. Wang, S. Huygh, and M. Shirazi (University of Antwerp) and M. Kushner (University of Michigan) for providing some of the figures used as illustrations in this chapter. The authors also acknowledge financial support from the Research Council of the University of Antwerp (TOP-BOF project), the IAP/7 (Interuniversity Attraction Poles) program "Physical Chemistry of Plasma-Surface Interactions (PSI)" by the Belgian Federal Science Policy Office (BELSPO), and the Fund for Scientific Research Flanders (FWO, grant no. G.0217.14 N).

References

1. Blin-Simiand, N., Tardivaux, P., Risacher, A., Jorand, F., & Pasquiers, S. (2005). Removal of 2-heptanone by dielectric barrier discharges - the effect of a catalyst support. *Plasma Processes and Polymers, 2*, 256–262.
2. Hong, J. P., Chu, W., Chernavskii, P. A., & Khodakov, A. Y. (2010). Cobalt species and cobalt-support interaction in glow discharge plasma-assisted Fischer-Tropsch catalysts. *Journal of Catalysis, 273*, 9–17.
3. Liu, C. J., Zou, J., Yu, K., Cheng, D., Han, Y., Zhan, J., Ratanatawante, C., & Jang, B. W. L. (2006). Plasma application for more environmentally friendly catalyst preparation. *Pure and Applied Chemistry, 78*, 1227–1238.
4. Demidyuk, V., & Whitehead, J. C. (2007). Influence of temperature on gas-phase toluene decomposition in plasma-catalytic system. *Plasma Chemistry and Plasma Processing, 27*, 85–94.
5. Shang, S., Liu, G., Chai, X., Tao, X., Li, X., Bai, M., Chu, W., Dai, X., Zhao, Y., & Yin, Y. (2009). Research on Ni/γ-Al_2O_3 catalyst for CO_2 reforming of CH_4 prepared by atmospheric pressure glow discharge plasma jet. *Catalysis Today, 148*, 268–274.
6. Tu, X., Gallon, H. J., Twigg, M. V., Gorry, P. A., & Whitehead, J. C. (2011). Dry reforming of methane over a Ni/Al_2O_3 catalyst in a coaxial dielectric barrier discharge reactor. *Journal of Physics D: Applied Physics, 44*, 274007.
7. Pylinina, A. I., & Mikhalenko, I. I. (2013). Activation of Cu-, Ag-, Au/ZrO_2 catalysts for dehydrogenation of alcohols by low-temperature oxygen and hydrogen plasma. *Theoretical and Experimental Chemistry, 49*, 65–69.
8. Guo, Y.-F., Ye, D.-Q., Chen, K.-F., He, J.-C., & Chen, W.-L. (2006). Toluene decomposition using a wire-plate dielectric barrier discharge reactor with manganese oxide catalyst in situ. *Journal of Molecular Catalysis. A, Chemical, 245*, 93–100.
9. Mahammadunnisa, S., Reddy, E. L., Ray, D., Subrahmanyam, C., & Whitehead, J. C. (2013). CO_2 reduction to syngas and carbon nanofibres by plasma-assisted in situ decomposition of water. *International Journal of Greenhouse Gas Control, 16*, 361–363.
10. Liu, C.-J., Mallison, R., & Lobban, L. (1998). Nonoxidative methane conversion to acetylene over zeolite in a low temperature plasma. *Journal of Catalysis, 179*, 326–334.
11. Wu, C. C., Wu, C. I., Sturm, J. C., & Kahn, A. (1997). Surface modification of indium tin oxide by plasma treatment: An effective method to improve the efficiency, brightness, and reliability of organic light emitting devices. *Applied Physics Letters, 70*, 1348–1350.
12. Poppe, J., Völkening, S., Schaak, A., Schütz, E., Janek, J., & Imbihl, R. (1999). Electrochemical promotion of catalytic CO oxidation on Pt/YSZ catalysts under low pressure conditions. *Physical Chemistry Chemical Physics, 1*, 5241–5249.
13. van Durme, J., Dewulf, J., Leys, C., & Van Langenhove, H. (2008). Combining non-thermal plasma with heterogeneous catalysis in waste gas treatment: A review. *Applied Catalysis B: Environmental, 78*, 324–333.

14. Liu, C. J., Wang, J. X., Yu, K. L., Eliasson, B., Xia, Q., & Xue, B. (2002). Floating double probe characteristics of non-thermal plasmas in the presence of zeolite. *Journal of Electrostatics, 54*, 149–158.
15. Kim, H. H., Ogata, A., & Futamura, S. (2006). Effect of different catalysts on the decomposition of VOCs using flow-type plasma-driven catalysis. *IEEE Transactions on Plasma Science, 34*, 984–995.
16. Löfberg, A., Essakhi, A., Paul, S., Swesi, Y., Zanota, M.-L., Meille, V., Pitault, I., Supiot, P., Mutel, B., Le Courtois, V., & Bordes-Richard, E. (2011). Use of catalytic oxidation and dehydrogenation of hydrocarbons reactions to highlight improvement of heat transfer in catalytic metallic foams. *Chemical Engineering Journal, 176-177*, 49–56.
17. Essakhi, A., Mutel, B., Supiot, P., Löfberg, A., Paul, S., Le Courtois, V., Meille, V., Pitault, I., & Bordes-Richard, E. (2011). Coating of structured catalytic reactors by plasma assisted polymerization of tetramethyldisiloxane. *Polymer Engineering and Science, 51*, 940–947.
18. Guaitella, O., Thevenet, F., Puzenat, E., Guillard, C., & Rousseau, A. (2008). C_2H_2 oxidation by plasma/TiO_2 combination: Influence of the porosity, and photocatalytic mechanisms under plasma exposure. *Applied Catalysis B: Environmental, 80*, 296–305.
19. Rousseau, A., Guaitella, O., Gatilova, L., Thevenet, F., Guillard, C., Röpcke, J., & Stancu, G. D. (2005). Photocatalyst activation in a pulsed low pressure discharge. *Applied Physics Letters, 87*, 221501.
20. Kim, H. H., Ogata, A., & Futamura, S. (2008). Oxygen partial pressure-dependent behavior of various catalysts for the total oxidation of VOCs using cycled system of adsorption and oxygen plasma. *Applied Catalysis. B, Environmental, 79*, 356–367.
21. Kim, H. H., Oh, S. M., Ogata, A., & Futamura, S. (2005). Decomposition of gas-phase benzene using plasma-driven catalyst (PDC) reactor packed with Ag/TiO_2 catalyst. *Applied Catalysis. B, Environmental, 56*, 213–220.
22. Kim, H. H., & Ogata, A. (2011). Nonthermal plasma activates catalyst: From current understanding and future prospects. *European Physical Journal Applied Physics, 55*, 13806.
23. Wallis, A. E., Whitehead, J. C., & Zhang, K. (2007). Plasma-assisted catalysis for the destruction of CFC-12 in atmospheric pressure gas streams using TiO_2. *Catalysis Letters, 113*, 29–33.
24. Mei, D., Zhu, X., He, Y., Yan, J. D., & Tu, X. (2015). Plasma-assisted conversion of CO_2 in a dielectric barrier discharge reactor: Understanding the effect of packing materials. *Plasma Sources Science and Technology, 24*, 015011.
25. Whitehead, J. C. (2016). Plasma-catalysis: The known knowns, the known unknowns and the unknown unknowns. *Journal of Physics D: Applied Physics, 49*, 243001.
26. Kang, W. S., Park, J. M., Kim, Y., & Hong, S. H. (2003). Numerical study on influences of barrier arrangements on dielectric barrier discharge characteristics. *IEEE Transactions on Plasma Science, 31*, 504–510.
27. Chang, J. S., Kostov, K. G., Urashima, K., Yamamoto, T., Okayasu, Y., Kato, T., Iwaizumi, T., & Yoshimura, K. (2000). Removal of NF_3 from semiconductor-process flue gases by tandem packed-bed plasma and adsorbent hybrid systems. *IEEE Transactions on Industry Applications, 36*, 1251–1259.
28. Takuma, T. (1991). Field behaviour at a triple junction in composite dielectric arrangements. *IEEE Transactions on Electrical Insulation, 26*, 500–509.
29. Holzer, F., Kopinke, F. D., & Roland, U. (2005). Influence of ferroelectric materials and catalysts on the performance of non-thermal plasma (NTP) for the removal of air pollutants. *Plasma Chemistry and Plasma Processing, 25*, 595–611.
30. Tu, X., Gallon, H. J., & Whitehead, J. C. (2011). Electrical and spectroscopic diagnostics of a single-stage plasma-catalysis system: Effect of packing with TiO_2. *Journal of Physics D: Applied Physics, 44*, 482003.
31. Kim, H. H., Kim, J.-H., & Ogata, A. (2009). Microscopic observation of discharge plasma on the surface of zeolites supported metal nanoparticles. *Journal of Physics D: Applied Physics, 42*, 135210.

32. Kim, H. H., Ogata, A., & Song, Y.-H. (2011). Propagation of surface streamers on the surface of HSY zeolites-supported silver nanoparticles. *IEEE Transactions on Plasma Science, 39*, 2220–2221.
33. Kim, H. H., & Ogata, A. (2012). Interaction of nonthermal plasma with catalyst for the air pollution control. *International Journal of Plasma Environmental Science and Technology, 6*, 43–48.
34. Kim, H. H., Teramoto, Y., Sano, T., Negishi, N., & Ogata, A. (2015). Effects of Si/Al ratio on the interaction of nonthermal plasma and ag/HY catalysts. *Applied Catalysis B: Environmental, 166-167*, 9–17.
35. Tu, X., Gallon, H. J., & Whitehead, J. C. (2011). Transition behavior of packed-bed dielectric barrier discharge in argon. *IEEE Transactions on Plasma Science, 39*, 2172–2173.
36. Nozaki, T., Muto, N., Kado, S., & Okazaki, K. (2004). Dissociation of vibrationally excited methane on Ni catalyst: Part 2. Process diagnostics by emission spectroscopy. *Catalysis Today, 89*, 67–74.
37. Mizuno, A. (2013). Generation of non-thermal plasma combined with catalysts and their application in environmental technology. *Catalysis Today, 211*, 2–8.
38. Malik, M. A., Minamitani, Y., & Schoenbach, K. H. (2005). Comparison of catalytic activity of aluminum oxide and silica gel for decomposition of volatile organic compounds (VOCs) in a plasmacatalytic reactor. *IEEE Transactions on Plasma Science, 33*, 50–56.
39. Roland, U., Holzer, F., & Kopinke, F.-D. (2005). Combination of non-thermal plasma and heterogeneous catalysis for oxidation of volatile organic compounds: Part 2. Ozone decomposition and deactivation of γ-Al_2O_3. *Applied Catalysis. B, Environmental, 58*, 217–226.
40. Holzer, F., Roland, U., & Kopinke, F.-D. (2002). Combination of non-thermal plasma and heterogeneous catalysis for oxidation of volatile organic compounds: Part 1. Accessibility of the intra-particle volume. *Applied Catalysis. B, Environmental, 38*, 163–181.
41. Hensel, K., Katsura, S., & Mizuno, A. (2005). DC microdischarges inside porous ceramics. *IEEE Transactions on Plasma Science, 33*, 574–575.
42. Hensel, K., Martisovits, V., Machala, Z., Janda, M., Lestinsky, M., Tardiveau, P., & Mizuno, A. (2007). Electrical and optical properties of AC microdischarges in porous ceramics. *Plasma Processes and Polymers, 4*, 682–693.
43. Hensel, K. (2009). Microdischarges in ceramic foams and honeycombs. *European Physical Journal D: Atomic, Molecular, Optical and Plasma Physics, 54*, 141–148.
44. Rousseau, A., Guaitella, O., Röpcke, J., Gatilova, L. V., & Tolmachev, Y. A. (2004). Combination of a pulsed microwave plasma with a catalyst for acetylene oxidation. *Applied Physics Letters, 85*, 2199–2201.
45. Vandenbroucke, A. M., Morent, R., De Geyter, N., & Leys, C. (2011). Non-thermal plasmas for non-catalytic and catalytic VOC abatement. *Journal of Hazardous Materials, 195*, 30–54.
46. Neyts, E. C., & Bogaerts, A. (2014). Understanding plasma catalysis through modelling and simulation - a review. *Journal of Physics D: Applied Physics, 47*, 224010.
47. Chen, H. L., Lee, H. M., Chen, S. H., Chang, M. B., Yu, S. J., & Li, S. N. (2009). Removal of volatile organic compounds by single-stage and two-stage plasma catalysis systems: A review of the performance enhancement mechanisms, current status, and suitable applications. *Environmental Science & Technology, 43*, 2216–2227.
48. Chen, H. L., Lee, H. M., Chen, S. H., Chao, Y., & Chang, M. B. (2008). Review of plasma catalysis on hydrocarbon reforming for hydrogen production - interaction, integration, and prospects. *Applied Catalysis B: Environmental, 85*, 1–9.
49. Whitehead, J. C. (2010). Plasma catalysis: A solution for environmental problems. *Pure and Applied Chemistry, 82*, 1329–1336.
50. Tu, X., Gallon, H. J., & Whitehead, J. C. (2013). Plasma-assisted reduction of a NiO/Al_2O_3 catalyst in atmospheric pressure H_2/Ar dielectric barrier discharge. *Catalysis Today, 211*, 120–125.
51. Kim, H. H. (2004). Nonthermal plasma processing for air-pollution control: A historical review, current issues, and fture prospects. *Plasma Processes and Polymers, 1*, 91–110.

52. Neyts, E. C., & Ostrikov, K. (2015). Nanoscale thermodynamic aspects of plasma catalysis. *Catalysis Today, 256*, 23–28.
53. Neyts, E. C. (2016). Plasma-surface interactions in plasma catalysis. *Plasma Chemistry and Plasma Processing, 36*, 185–212.
54. Neyts, E. C., Ostrikov, K., Sunkara, M. K., & Bogaerts, A. (2015). Plasma catalysis: Synergistic effects at the nanoscale. *Chemical Reviews, 115*, 13408–13446.
55. Kim, H. H., Teramoto, Y., Negishi, N., & Ogata, A. (2015). A multidisciplinary approach to understand the interactions of nonthermal plasma and catalyst: A review. *Catalysis Today, 256*, 13–22.
56. Kim, H. H., Teramoto, Y., Ogata, A., Takagi, H., & Nanba, T. (2016). Plasma catalysis for environmental treatment and energy applications. *Plasma Chemistry and Plasma Processing, 36*, 45–72.
57. Neyts, E. (2018). Atomistic simulations of plasma catalytic processes. *Frontiers of Chemical Science and Engineering, 12*, 145–154.
58. Somers, W., Bogaerts, A., van Duin, A. C. T., & Neyts, E. C. (2014). Interactions of plasma species on nickel catalysts: A reactive molecular dynamics study on the influence of temperature and surface structure. *Applied Catalysis B: Environmental, 154-155*, 1–8.
59. Somers, W., Bogaerts, A., van Duin, A. C. T., & Neyts, E. C. (2012). Plasma species interacting with nickel surfaces: Toward an atomic scale understanding of plasma-catalysis. *Journal of Physical Chemistry C, 116*, 20958–20965.
60. Somers, W., Bogaerts, A., van Duin, A. C. T., Huygh, S., Bal, K. M., & Neyts, E. C. (2013). Temperature influence on the reactivity of plasma species on a nickel catalyst surface: An atomic scale model. *Catalysis Today, 211*, 131–136.
61. Huygh, S., Bogaerts, A., & Neyts, E. C. (2016). How oxygen vacancies activate CO_2 dissociation on TiO_2 anatase (001). *Journal of Physical Chemistry C, 120*, 21659–21669.
62. Huygh, S., & Neyts, E. C. (2015). Adsorption of C and CH_x radicals on anatase (001) and the influence of oxygen vacancies. *Journal of Physical Chemistry C, 119*, 4908–4921.
63. Shirazi, M., Neyts, E. C., & Bogaerts, A. (2017). DFT study of Ni-catalyzed plasma dry reforming of methane. *Applied Catalysis B: Environmental, 205*, 605–614.
64. Bal, K. M., Huygh, S., Bogaerts, A., & Neyts, E. C. (2018). Effect of plasma-induced surface charging on catalytic processes: Application to CO_2 activation. *Plasma Sources Science and Technology, 27*, 024001.
65. Bal, K. M., & Neyts, E. C. (2018). Modelling molecular adsorption on charged or polarized surfaces: A critical flaw in common approaches. *Physical Chemistry Chemical Physics, 20*, 8456–8459.
66. Neyts, E. C., & Bal, K. M. (2017). Effect of electric fields on plasma catalytic hydrocarbon oxidation from atomistic simulations. *Plasma Processes and Polymers, 14*, e1600158.
67. Neyts, E. C., van Duin, A. C. T., & Bogaerts, A. (2012). Insights in the plasma-assisted growth of carbon nanotubes through atomic scale simulations: Effect of electric field. *Journal of the American Chemical Society, 134*, 1256–1260.
68. Neyts, E. C., Ostrikov, K., Han, Z. J., Kumar, S., van Duin, A. C. T., & Bogaerts, A. (2013). Defect healing and enhanced nucleation of carbon nanotubes by low-energy ion bombardment. *Physical Review Letters, 110*, 065501.
69. Somers, W., (2015). Atomic scale simulations of the interactions of plasma species on nickel catalyst surface. University of Antwerp, PhD-thesis.
70. Ni, B., Lee, C., Sun, R.-C., & Zhang, X. (2009). Microwave assisted heterogeneous catalysis: Effects of varying oxygen concentrations on the oxidative coupling of methane. *Reaction Kinetics and Catalysis Letters, 98*, 287–302.
71. Zhdanov, V. P. (1999). Simulation of surface restructuring and oscillations in CO-NO reaction on Pt(100). *The Journal of Chemical Physics, 110*, 8748–8753.
72. Kersten, H., Deutsch, H., Steffen, H., Kroesen, G. M. W., & Hippler, R. (2001). The energy balance at substrate surfaces during plasma processing. *Vacuum, 63*, 385–431.

73. Li, S., Zheng, W., Tang, Z., & Gu, F. (2012). Plasma heating and temperature difference between gas pellets in packed bed with dielectric barrier discharge under natural convection condition. *Heat Transfer Engineering, 33*, 609–617.
74. Nozaki, T., & Okazaki, K. (2013). Non-thermal plasma catalysis of methane: Principles, energy efficiency, and applications. *Catalysis Today, 211*, 29–38.
75. JiangB, G. H. (2016). Enhanced dissociative chemisorption of CO_2 via vibrational excitation. *The Journal of Chemical Physics, 144*, 091101.
76. Bal, K. M., & Neyts, E. C. (2015). Merging metadynamics into hyperdynamics: Accelerated molecular simulations reaching time scales from microseconds to seconds. *Journal of Chemical Theory and Computation, 11*, 4545–4554.
77. Guerra, V., & Marinov, D. (2016). Dynamical Monte Carlo methods for plasma-surface reactions. *Plasma Sources Science and Technology, 25*, 045001.
78. Marinov, D., Teixeira, C., & Guerra, V. (2017). Deterministic and Monte Carlo methods for simulation of plasma-surface interactions. *Plasma Processes and Polymers, 14,* 1600175.
79. Blaylock, D. W., Ogura, T., Green, W. H., & Beran, G. J. O. (2009). Computational investigation of thermochemistry and kinetics of steam methane reforming on Ni(111) under realistic conditions. *Journal of Physical Chemistry C, 113*, 4898–4908.
80. Blaylock, D. W., Zhu, Y.-A., & Green, W. H. (2011). Computational investigation of the thermochemistry and kinetics of steam methane reforming over a multi-faceted nickel catalyst. *Topics in Catalysis, 54*, 828–844.
81. Toshikubo, F. (2009). Modeling for plasma-enhanced catalytic reduction of nitrogen oxides. *Thin Solid Films, 518*, 957–961.
82. Delagrange, S., Pinard, L., & Tatibouët, J.-M. (2009). Combination of a non-thermal plasma and a catalyst for toluene removal from air: Manganese based oxide catalysts. *Applied Catalysis. B, Environmental, 68*, 92–98.
83. Kim, H. H., Ogata, A., & Futamura, S. (2005). Atmospheric plasma-driven catalysis for the low temperature decomposition of dilute aromatic compounds. *Journal of Physics D: Applied Physics, 38*, 1292–1300.
84. Van Laer, K., & Bogaerts, A. (2016). Fluid modelling of a packed bed dielectric barrier discharge plasma reactor. *Plasma Sources Science and Technology, 25*, 015002.
85. Takaki, K., Chang, J.-S., & Kostov, K. G. (2004). Atmospheric pressure of nitrogen plasmas in a ferroelectric packed bed barrier discharge reactor. Part I. Modeling. *IEEE Transactions on Dielectrics and Electrical Insulation, 11*, 481–490.
86. Russ, H., Neiger, M., & Lang, J. E. (1999). Simulation of micro discharges for the optimization of energy requirements for removal of NO_x from exhaust gases. *IEEE Transactions on Plasma Science, 27*, 38–39.
87. Babaeva, N. Y., Bhoj, A. N., & Kushner, M. J. (2006). Streamer dynamics in gases containing dust particles. *Plasma Sources Science and Technology, 15*, 591–602.
88. Kruszelnicki, J., Engeling, K. W., Foster, J. E., Xiong, Z., & Kushner, M. J. (2017). Propagation of negative electric discharges through 2-dimensional packed bed reactors. *Journal of Physics D: Applied Physics, 50*, 025203.
89. Kang, W. S., Kim, H. H., Teramoto, Y., Ogata, A., Lee, J. Y., Kim, D. W., Hur, M., & Song, Y. H. (2018). Surface streamer propagations on an alumina bead: Experimental observation and numerical modelling. *Plasma Sources Science and Technology, 27*, 015018.
90. Zhang, Y., Wang, H.-Y., Jiang, W., & Bogaerts, A. (2015). Two-dimensional particle-in cell/ Monte Carlo simulations of a packed-bed dielectric barrier discharge in air at atmospheric pressure. *New Journal of Physics, 17*, 083056.
91. Gao, M.-X., Zhang, Y., Wang, H.-Y., Guo, B., Zhang, Q.-Z., & Bogaerts, A. (2018). Mode transition of filaments in packed-bed dielectric barrier discharges. *Catalysts, 8*, 248.
92. Van Laer, K., & Bogaerts, A. (2017). Influence of gap size and dielectric constant of the packing material on the plasma behaviour in a packed bed DBD reactor: A fluid modelling study. *Plasma Processes and Polymers, 14*, e1600129.

93. Van Laer, K., & Bogaerts, A. (2017). How bead size and dielectric constant affect the plasma behaviour in a packed bed plasma reactor: A modelling study. *Plasma Sources Science & Technology, 26*, 085007.
94. Wang, W., Kim, H.-H., Van Laer, K., & Bogaerts, A. (2018). Streamer propagation in a packed bed plasma reactor for plasma catalysis applications. *Chemical Engineering Journal, 334*, 2467–2479.
95. Michielsen, I., Uytdenhouwen, Y., Pype, J., Michielsen, B., Mertens, J., Reniers, F., Meynen, V., & Bogaerts, A. (2017). CO_2 dissociation in a packed bed DBD reactor: First steps towards a better understanding of plasma catalysis. *Chemical Engineering Journal, 326*, 477–488.
96. Uytdenhouwen, Y., Van Alphen, S., Michielsen, I., Meynen, V., Cool, P., & Bogaerts, A. (2018). A packed-bed DBD micro plasma reactor for CO_2 dissociation: Does size matter? *Chemical Engineering Journal, 348*, 557–568.
97. Bhoj, A. N., & Kushner, M. J. (2006). Multi-scale simulation of functionalizationof rough polymer surfaces using atmospheric pressure plasmas. *Journal of Physics D: Applied Physics, 39*, 1594–1598.
98. Bhoj, A. N., & Kushner, M. J. (2008). Repetitively pulsed atmospheric pressure discharge treatment of rough polymer surfaces: I. Humid air discharges. *Plasma Sources Science & Technology, 17*, 035024.
99. Bhoj, A. N., & Kushner, M. J. (2008). Repetitively pulsed atmospheric pressure discharge treatment of rough polymer surfaces: II. Treatment of micro-beads in $He/NH_3/H_2O$ and $He/O_2/H_2O$ mixtures. *Plasma Sources Science & Technology, 17*, 035025.
100. Wang, X. M., Foster, J. E., & Kushner, M. J. (2011). Plasma propagation through porous dielectric sheets. *IEEE Transactions on Plasma Science, 39*, 2244–2245.
101. Zhang, Y.-R., Van Laer, K., Neyts, E. C., & Bogaerts, A. (2016). Can plasma be formed in catalyst pores? A modeling investigation. *Applied Catalysis B: Environmental, 185*, 56–67.
102. Zhang, Y.-R., Neyts, E. C., & Bogaerts, A. (2016). Influence of the material dielectric constant on plasma generation inside catalyst pores. *Journal of Physical Chemistry C, 120*, 25923–25934.
103. Zhang, Y.-R., Neyts, E. C., & Bogaerts, A. (2018). Enhancement of plasma generation in catalyst pores with different shapes. *Plasma Sources Science and Technology, 27*, 055008.
104. Zhang, Y., Wang, H.-Y., Zhang, Y.-R., & Bogaerts, A. (2017). Formation of microdischarges inside a mesoporous catalyst in dielectric barrier discharge plasmas. *Plasma Sources Science and Technology, 26*, 054002.
105. Zhang, Q.-Z., & Bogaerts, A. (2018). Propagation of a plasma streamer in catalyst pores. *Plasma Sources Science and Technology, 27*, 035009.
106. Zhang, Q.-Z., Wang, W.-Z., & Bogaerts, A. (2018). Importance of surface charging during plasma streamer propagation in catalyst pores. *Plasma Sources Science and Technology, 27*, 065009.
107. Gentille, A. C., & Kushner, M. J. (1995). Reaction chemistry and optimization of plasma remediation of N_xO_y from gas streams. *Journal of Applied Physics, 78*, 2074–2085.
108. Dorai, R., & Kushner, M. J. (2000). Consequences of propene and propane on plasma remediation of NO_x. *Journal of Applied Physics, 88*, 3739–3747.
109. Dorai, R., & Kushner, M. J. (2003). Consequences of unburned hydrocarbons on microstreamer dynamics and chemistry during plasma remediation of NO_x using dielectric barrier discharges. *Journal of Physics D: Applied Physics, 36*, 1075–1083.
110. Dorai, R., & Kushner, M. J. (2002). Repetitively pulsed plasma remediation of NO_x in soot laden exhaust using dielectric barrier discharges. *Journal of Physics D: Applied Physics, 35*, 2954–2968.
111. Teodoru, S., Kusano, Y., & Bogaerts, A. (2012). The effect of O_2 in a humid $O_2/N_2/NO_x$ gas mixture on NO_x and N_2O remediation by an atmospheric pressure dielectric barrier discharge. *Plasma Processes and Polymers, 9*, 652–689.
112. Chang, M. B., Balbach, J. H., Rood, J. J., & Kushner, M. J. (1991). Removal of SO_2 from gas streams using a dielectric barrier discharge and combined plasma photolysis. *Journal of Applied Physics, 69*, 4409–4417.

113. Chang, M. B., Kushner, M. J., & Rood, M. J. (1992). Removal of SO_2 and the simultaneous removal of SO_2 and NO from simulated flue gas streams using dielectric barrier discharge plasmas. *Plasma Chemistry and Plasma Processing, 12*, 565–580.
114. Chang, M. B., Kushner, M. J., & Rood, M. J. (1992). Gas-phase removal of NO from gas streams via dielectric barrier discharges. *Environmental Science & Technology, 26*, 777–781.
115. Storch, D. G., & Kushner, M. J. (1993). Destruction mechanisms for formaldehyde in atmospheric temperature plasmas. *Journal of Applied Physics, 73*, 51–55.
116. Evans, D., Rosocha, L. A., Anderson, G. K., Coogan, J. J., & Kushner, M. J. (1993). Plasma remediation of trichloroethylene in silent discharge plasmas. *Journal of Applied Physics, 74*, 5378–5386.
117. Aerts, R., Tu, X., De Bie, C., Whitehead, J. C., & Bogaerts, A. (2012). An investigation into the dominant reactions for ethylene destruction in non-thermal atmospheric plasmas. *Plasma Processes and Polymers, 9*, 994–1000.
118. Aerts, R., Tu, X., Van Gaens, W., Whitehead, J. C., & Bogaerts, A. (2013). Gas purification by nonthermal plasma: A case study of ethylene. *Environmental Science & Technology, 47*, 6478–6485.
119. Vandenbroucke, A. M., Aerts, R., Van Gaens, W., De Geyter, N., Leys, C., Morent, R., & Bogaerts, A. (2015). Modeling and experimental study of tricholoroethylene abatement with a negative dirrect current corona discharge. *Plasma Chemistry and Plasma Processing, 35*, 217–230.
120. De Bie, C., Martens, T., van Dijk, J., Paulussen, S., Verheyde, B., & Bogaerts, A. (2011). Dielectric barrier discharges used for the conversion of greenhouse gases: Modeling the plasma chemistry by fluid simulations. *Plasma Sources Science and Technology, 20*, 024008.
121. Yang, Y. (2003). Direct non-oxidative methane conversion by non-thermal plasma: Modeling study. *Plasma Chemistry and Plasma Processing, 23*, 327–346.
122. Pringle, K. J., Whitehead, J. C., Wilman, J. J., & Wu, J. H. (2004). The chemistry of methane remediation by a non-thermal atmospheric pressure plasma. *Plasma Chemistry and Plasma Processing, 24*, 421–434.
123. Agiral, A., Trionfetti, C., Lefferts, L., Seshan, K., & Gardeniers, J. G. E. (2008). Propane conversion at ambient temperatures C–C and C–H bond activation using cold plasma in a microreactor. *Chemical Engineering and Technology, 31*, 1116–1123.
124. Pinhao, N. R., Janeco, A., & Branco, J. B. (2011). Influence of helium on the conversion of methane and carbon dioxide in a dielectric barrier discharge. *Plasma Chemistry and Plasma Processing, 31*, 427–439.
125. De Bie, C., Verheyde, B., Martens, T., van Dijk, J., Paulussen, S., & Bogaerts, A. (2011). Fluid modelling of the conversion of methane into higher hydrocarbons in an atmospheric pressure dielectric barrier discharge. *Plasma Processes and Polymers, 8*, 1033–1058.
126. Aerts, R., Martens, T., & Bogaerts, A. (2012). Influence of vibrational states on CO_2 splitting by dielectric barrier discharges. *Journal of Physical Chemistry C, 116*, 23257–23273.
127. Snoeckx, R., Aerts, R., Tu, X., & Bogaerts, A. (2013). Plasma-based dry reforming: A computational study ranging from the nanoseconds to seconds time scale. *Journal of Physical Chemistry C, 117*, 4957–4970.
128. Snoeckx, R., Setareh, M., Aerts, R., Simon, P., Maghari, A., & Bogaerts, A. (2013). Influence of N_2 concentration in a CH_4/N_2 dielectric barrier discharge used for CH_4 conversion into H_2. *International Journal of Hydrogen Energy, 38*, 16098–16120.
129. Snoeckx, R., Zeng, Y. X., Tu, X., & Bogaerts, A. (2015). Plasma-based dry reforming: Improving the conversion and energy efficiency in a dielectric barrier discharge. *RSC Advances, 5*, 29799–29808.
130. Snoeckx, R., Heijkers, S., Van Wesenbeeck, K., Lenaerts, S., & Bogaerts, A. (2016). CO_2 conversion in a dielectric barrier discharge plasma: N_2 in the mix as a helping hand or problematic impurity? *Energy and Environmental Science, 9*, 999–1011.
131. Heijkers, S., Snoeckx, R., Kozák, T., Silva, T., Godfroid, T., Britun, N., Snyders, R., & Bogaerts, A. (2015). CO_2 conversion in a microwave plasma reactor in the presence of N_2: Elucidating the role of vibrational levels. *Journal of Physical Chemistry C, 119*, 12815–12828.

132. De Bie, C., van Dijk, J., & Bogaerts, A. (2015). The dominant pathways for the conversion of methane into oxygenates and syngas in an atmospheric pressure dielectric barrier discharge. *Journal of Physical Chemistry C, 119*, 22331–22350.
133. De Bie, C., van Dijk, J., & Bogaerts, A. (2016). CO_2 hydrogenation in a dielectric barrier discharge plasma revealed. *J. Phys. Chem, C120*, 25210–25224.
134. Istadi, A., & Amin, N. A. S. (2007). Modelling and optimization of catalytic-dielectric barrier discharge plasma reactor for methane and carbon dioxide conversion using hybrid artificial neural network - genetic algorithm technique. *Chemical Engineering Science, 62*, 6568–6581.
135. JiwuL, L. F. (2013). Modeling of corona discharge combined with Mn^{2+} catalysis for the removal of SO_2 from simulated flue gas. *Chemosphere, 91*, 1374–1379.
136. Tinck, S., Bogaerts, A., & Shamiryan, D. (2011). Simultaneous etching and deposition processes during the etching of silicon with a $Cl_2/O_2/Ar$ inductively coupled plasma. *Plasma Processes and Polymers, 8*, 490–499.
137. Tinck, S., De Schepper, P., & Bogaerts, A. (2013). Numerical investigation of SiO_2 coating deposition in wafer processing reactors with $SiCl_4/O_2/Ar$ inductively coupled plasmas. *Plasma Processes and Polymers, 10*, 714–730.
138. Tinck, S., Boullart, W., & Bogaerts, A. (2011). Modeling $Cl_2/O_2/Ar$ inductively coupled plasmas used for silicon etching: Effects of SiO_2 chamber wall coating. *Plasma Sources Science and Technology, 20*, 045012.
139. Kushner, M. J. (1987). A phenomenological model for surface deposition kinetics during plasma and sputter deposition of amorphous hydrogenated silicon. *Journal of Applied Physics, 62*, 4763–4772.
140. Zhang, D., & Kushner, M. J. (2000). Mechanisms for CF_2 radical generation and loss on surfaces in fluorocarbon plasmas. *Journal of Vacuum Science and Technology A, 18*, 2661–2668.
141. Kushner, M. J. (2009). Hybrid modelling of low temperature plasmas for fundamental investigations and equipment design. *Journal of Physics D: Applied Physics, 42*, 194013.
142. Tinck, S., Tillocher, T., Dussart, R., & Bogaerts, A. (2015). Cryogenic etching of silicon with SF_6 inductively coupled plasmas: A combined modelling and experimental study. *Journal of Physics D: Applied Physics, 48*, 155204.
143. Hoekstra, R. J., Grapperhaus, H. J., & Kushner, M. J. (1997). Integrated plasma equipment model for polysilicon etch profilesin an inductively coupled plasma reactor with subwafer and superwafer topography. *Journal of Vacuum Science and Technology A, 15*, 1913–1921.

Chapter 5
Plasma-Catalytic Removal of NO_x in Mobile and Stationary Sources

Ahmed Khacef and Patrick Da Costa

5.1 Introduction

Nitrogen oxides, which refer specifically to NO_x (NO and NO_2), are the major pollutants in the atmosphere from industries (mobile or stationary sources), which lead to acid rain and photochemical smog, and have been shown to be detrimental to human health and the environment. Automobiles and other mobile sources contribute to about 50% of the NO_x that is emitted in the atmosphere, and nitrogen monoxide (NO) is the major portion of it [1, 2].

Lean-burn engines are receiving increasing application because of their promise of improved fuel economy over stoichiometric conditions. However, they offer special challenges for meeting increasingly stringent emissions standards because of the difficulty of removing NO_x in an oxidizing environment where traditional three-way catalysts cannot work.

Consequently, several methods collectively called deNO_x processes have been developed to treat the effluents of mobile and power plants sources. Among them, selective catalytic reduction (SCR) and selective non-catalytic reduction (SNCR) are currently used to convert NO_x into nitrogen (N_2) molecules [3, 4]. Nowadays, the selective catalytic reduction of NO_x by urea (urea-SCR) and lean-NO_x trap technology (LNT) are the two most efficient NO_x reduction technologies currently available for diesel engine emission control [5–8]. Both technologies provide high NO_x reduction performance over a wide temperature range (200–400 °C) under well-controlled conditions but still have serious drawbacks [9–19].

A. Khacef (✉)
GREMI, UMR 7344 CNRS – Université d'Orléans, Orléans Cedex 02, France
e-mail: ahmed.khacef@univ-orleans.fr

P. Da Costa
Sorbonne Université, Institut Jean Le Rond d'Alembert, Saint Cyr l'école, France
e-mail: patrick.da_costa@sorbonne-universite.fr

X. Tu et al. (eds.), *Plasma Catalysis*, Springer Series on Atomic, Optical, and Plasma Physics 106, https://doi.org/10.1007/978-3-030-05189-1_5

As an alternative to catalytic processes which require high temperatures nonthermal plasma (NTP) also referred as "non-equilibrium plasma" or "cold plasma," such dielectric barrier discharges (DBDs) and corona discharges (CD) have been extensively investigated in the field of pollution control and frequently proposed in the literature for the removal of volatile organic compounds (VOCs), NO_x, and SO_2 [20–26].

In NTP, background gaseous species are chemically excited or dissociated directly by electronic impact, while the temperature of the reactants (i.e., gas temperature) remains relatively low, and thus the product distributions far from the chemical equilibrium may be obtained. In that case, the most useful deposition of energy is associated with the production of excited species (atoms, molecules) and activated species (radicals, ions) that eventually lead to the chemical conversion of pollutants. Although NTPs present attractive properties (low temperature, atmospheric pressure, compactness, low cost) and a unique way to induce gas-phase reactions by electron collisions, the formation of unwanted by-products and poor energy efficiency are serious obstacles toward their industrial implementation. To overcome these drawbacks, a more effective use of NTP is possible by exploiting its inherent synergetic potential through combination with heterogeneous catalysts as emphasized by different groups [27–29]. This innovative technique called plasma catalysis that combines the advantages of both NTP and catalysis has become a hot topic over the last decade. Catalyst can be combined with plasma in two ways: in-plasma catalysis (IPC), with the catalyst directly into the discharge zone, or post-plasma catalysis (PPC), with the catalyst downstream the discharge zone.

The conventional NTP reactors that are widely used for various environmental applications are subdivided according to the type of discharge mode (pulse, DC, AC, RF, microwave), presence of a dielectric barrier or catalyst, and geometry (cylinder, plane). It is important to note that the chemical potential of each discharge mode differs enormously from one discharge to another. Roughly speaking, the efficiency of a plasma discharge to remove pollutant from gas stream depends mainly on its ability to produce large amount of active species in the plasma.

In an attempt to overcome the inherent shortcomings of these existing technologies, plasma- assisted catalysis for lean-NO_x reduction through the improvement of the hydrocarbons (HC)- SCR technology has emerged as a promising alternative to the more mature urea-SCR and LNT technologies [30–37].

5.2 State of the Art on Catalytic NO_x Abatement Systems

Since 1970, NO_x abatement technologies were majorly based on selective catalytic reduction with ammonia (NH_3-SCR). Nowadays, two mature technologies are used at least for mobile sources: NO_x trap or NO_x storage/reduction (NSR) and NH_3-SCR [5–8]. Other promising technologies such as HC/HCR are still studied in order to use the reducing agent present in the feed for reducing the NO_x in situ.

The NO_x trap is one of the actual technologies already in used in automotive posttreatment lines. The used materials are based on noble metal dispersed on a support such as alumina or ceria zirconia containing an adsorbent material such as barium oxide. Noble metal can be deactivated or poisoned since sulfur is present in the feed. The deactivation mechanisms of NSR catalysts, especially the sulfur poisoning and thermal degradation, were already extensively reviewed. Finally, recent developments of NSR catalysts were addressed in detail, concentrating on the improvements over precious metals, NO_x storage materials, and metal oxide supports.

Despite the first-generation NSR catalysts being quite successful in NO_x emission control, new-generation NSR catalysts with low cost, high efficiency, and durability are urgently needed to meet the ever rigorous NO_x emission regulations and develop the NSR technology. Thus, the development of new materials (e.g., perovskite oxide-based materials) [38] as alternatives of precious metals is very promising to reduce the cost of NSR catalysts. Finally, a novel synthesis method and a novel structured NSR catalyst (e.g., 3-D structured catalysts) would help to increase the NSR activity, sulfur poisoning resistance, and thermal stability needed for the NSR regeneration [39].

The SCR of NO_x with NH_3 was also considered for a long time to be the most efficient technology for reducing NO_x emission in the presence of excess oxygen [40]. The V_2O_5-WO_3(MoO_3)/TiO_2 material has been commercially employed as a NH_3-SCR catalyst for a number of years. However, several serious problems with this catalyst still remain, e.g., the narrow temperature window (i.e., only applicable in 300–400 °C), the high activity of SO_2 oxidation, and the toxicity of V_2O_5 [41]. Therefore, new catalysts with environmentally benign characteristics and high SCR performance in a wide temperature range are required. Since the discovery of Cu-ZSM-5 catalyst as an efficient catalyst for NO_x removal in 1986 by Iwamoto et al. [42], copper- or iron-exchanged zeolites have been and are still widely investigated [43]. However, the major disadvantage of zeolite-based materials is their poor resistance to high temperature treatment. New zeolitic systems such as SSZ-13 and SAPO-34 are in development to give to such system high SCR performance even after hydrothermal aging [44–47].

For SCR of NO_x with HC, according to Burch et al. [5], more than 1000 catalysts were tested in laboratory, and more than 80% NO_x conversion was obtained over specific temperature ranges under lean-burn conditions. It was never possible to translate this laboratory success to real exhaust automotive systems. The authors proposed three factors in order to explain that point:

(i) Laboratory tests mostly focused on pure reducing agents such as the lower alkanes and alkenes, whereas onboard reducing agents, such as gasoline or diesel, contain hundreds of compounds, some of which may poison or inhibit the SCR reaction.
(ii) The space velocities used in laboratory micro-reactors are frequently much lower than those found in real exhausts.
(iii) The temperature range in an engine tests is much wider than the narrow window of high activity observed in typical laboratory tests.

Thus, coupled plasma catalysis technologies could be a solution to the points (i) and (iii).

5.3 Nonthermal Plasma Reactors for NO_x Treatment

Nonthermal plasmas for chemical treatment of industrial exhaust gases may be produced by a variety of electrical discharge reactors (corona discharge, surface discharge, dielectric-barrier discharge, etc.) [48–51] or by electron beam irradiation [52]. The electron beam technique, which has been first used, is based on irradiation of the gas flow by high energetic electrons being generated in ultrahigh vacuum accelerators with voltages ranging from tens of kV to some MV and transmitted into the gas flow through titanium foils. Each of the primary beam electrons generates a large number of secondary electrons in ionizing collisions, which have the right energy for efficient radical generation finally resulting in efficient pollutant removal. For some industrial and domestic applications, the pulsed corona and DBD reactors are much more suited than e-beam devices because of their high selectivity, moderate operating conditions (atmospheric pressure and room temperature), and relatively low maintenance requirements resulting in relatively low energy costs of the pollutant treatment.

5.3.1 Corona Reactors

Corona discharge is a transient luminous discharge that appears in regions of high electric field near sharp points, edges, or wires in electrically stressed gases prior to the point of electrical breakdown. The corona discharges are always nonuniform: a strong electric field, ionization, and luminosity are located in the vicinity of one electrode. The corona can be observed, for example, around high-voltage transmission lines, around lightning rods, along wire surrounded by a coaxial cylinder, or near irregularities in the form of sharp points, on the surface of a conductor at high voltage. The main physical and engineering principles of the corona discharges can be found in Loeb [53], Fridman et al. [54], and Rutgers and van Veldhuizen [55].

Various reactor configurations are used to generate corona discharges. In laboratory studies of corona discharges, the most common geometry is a pin-plane geometry (Fig. 5.1), where a needle is placed above a grounded plane. The high voltage is applied to the needle electrode. However, for industrial applications, this geometry is not sufficient, as it does not fill the whole gas volume with the discharge. The most popular geometries are the multi-pins-plane, wire- cylinder, and the saw-blade geometries [56, 57] as in electrostatic precipitators. The wire- cylinder geometry is probably used the most. It ensures a quite homogeneous distribution of the discharge that maximizes the active discharge volume and is easy to implement

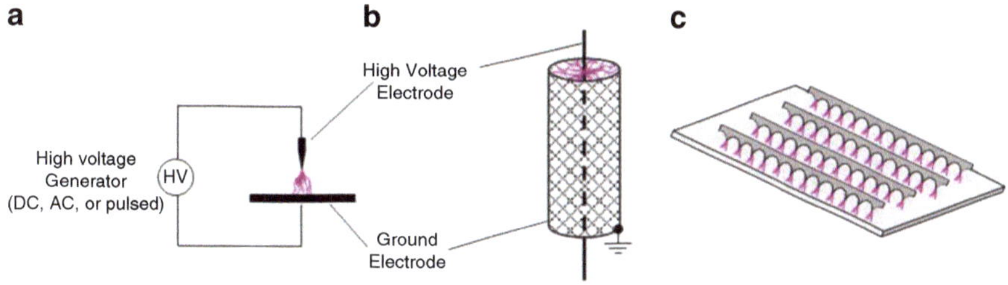

Fig. 5.1 Common corona reactor geometries: (**a**) pin-plane, (**b**) wire-cylinder, and (**c**) saw-blade

in a gas flow system. Often, multiple wire-cylinder reactors are mounted in parallel with regard to the gas flow to enable high gas throughput.

Corona discharges are separated into two different categories: continuous and pulsed. Continuous corona discharges occur at DC or low-frequency AC voltages. A recent example of work on DC-excited corona discharges is given by Eichwald et al. [58]. Application of the continuous coronas is limited by low current and power, which results in a low treatment rate of exhaust gas. Increasing the corona power could result in a large current short circuit, and a large amount of NO_x can be produced when spark breakdown occurs. For the purpose of gas cleaning, spark breakdown should be avoided.

Increasing the corona power without transition to sparks becomes possible by using voltage pulses having nanosecond or sub-microsecond duration. Since discharge gaps of several tens centimeters can be used, the pulsed corona discharge (PCD) reactor is ideal for treatment of very large gas flow rates [59–61]. The costs of the PCD reactor are low compared to a DBD reactor for the same gas flow rate. However, the development of cost-effective pulsed power supplies is challenging. Further, for good efficiency and long lifetime of the power supply, care has to be taken for good electrical matching between power supply and reactor [62].

PCD reactors have been considered for the removal of pollutants both in the gas phase [48, 60, 63] and in the liquid phase [64]. A combination of pulsed corona with catalysts can be practical for applications to achieve improved results in the treatment of automotive exhausts and for hydrogen production from heavy hydrocarbons. Another interesting technological hybrid is related to the pulsed corona coupled with water flow. Such a system can be arranged either in the form of a shower, which is called a spray corona, or with a thin water film on the walls, which is usually referred to as a wet corona.

Figure 5.2 shows an example of pulsed corona discharges developed from each pin in pure N_2 at atmospheric pressure for a multi-pins-plane corona reactor. The illuminating paths indicate the position where the streamer heads have moved. Below a certain applied voltage, the corona discharge filaments are almost invisible to the human eyes, and only at the optimum voltage, which is not sufficient to generate arc discharge, and after a long adaptation in the dark, they can just be observed. The practical advantages of the PCD are that the short duration of the pulse ensures that no transition to spark takes place; therefore, it can be used at voltages

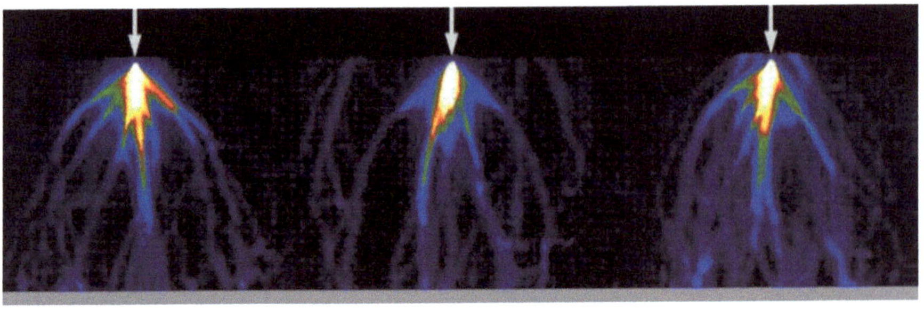

Fig. 5.2 Streamer discharges in 12 mm gap in N_2 at atmospheric pressure with a 16 kV pulse, 30 ns rise time, and half-width of about 80 ns (multi-pins-plane corona reactor) [63]

and currents higher than that at continuous corona can be used. Large voltages lead to higher electric fields and electron temperatures, which increases ionization and dissociation rates. Additionally, due to the short pulse duration, only electrons are significantly accelerated in few or tens of nanoseconds, and the gas heating can be kept to a minimum.

5.3.2 *Dielectric Barrier Discharge Reactors*

Another approach that avoids spark formation in streamer channels and instabilities in space (both effects are undesired in a volume discharge for chemical conversions) is based on the use of at least one electrically insulating barrier between the electrodes. The introduction of a dielectric barrier between the electrodes limits the direct current. Such discharges are referred to as dielectric barrier discharges or silent discharges (SD). The electrical circuit of the discharge reactor can be represented by the dielectric barrier capacity being in series with the discharge gap represented by the variable ohmic impedance of the gas discharge plasma paralleled by the capacitance of the discharge. Because of the capacitive coupling of the current to the discharge gap, DBD reactors have to be operated with AC or pulse repetitive voltages. Further reliable operation requires high-quality, low dielectric loss barrier materials having a high dielectric breakdown strength (>10 kV mm^{-1}) and high bulk resistivity (>108 Ω cm^{-1}). In case of low temperatures, polymeric barrier materials such as PTFE can be used. For exhaust gases having temperatures higher than 100 °C, rather expensive densely sintered ceramics or quartz glass barriers are required. An advantage of DBD reactors is that barriers having certain catalytic properties can be used for plasma-catalytic hybrid treatment of gases.

DBD processing is a very mature technology, first investigated by Siemens [65] in the 1850's for ozone generation. Ozone formation requires three criteria: (i) energetic electrons that are able to efficiently dissociate O_2 (4–5 eV, see Fig. 5.4), (ii) high pressure because of the three-body reaction which is responsible for ozone production, and (iii) low gas temperature because of the reduced lifetime of ozone at high gas temperature. DBDs can deliver all these requirements as the bulk gas temperature can be maintained close to room temperature (for high power, cooling the electrodes is required), while the electron temperature is typically 2–5 eV. DBDs are now routinely used in different industrial and fundamental applications such as water purification, polymer treatment, UV light generation, biological and medical treatment, pollution control, and exhaust cleaning from CO, NO_x, SO_x, and VOCs. DBDs could be generated in parallel-plane or in coaxial cylindrical reactor geometries, very common in ozonizers and other cases of gas treatment, as illustrated in Fig. 5.3. The flexibility of DBD configurations with respect to geometrical shape, operating medium, and operating parameters (input power, frequency, and gas flow) is remarkable. In many situations, discharge conditions optimized in small laboratory experiments can be scaled up to large industrial installations.

Kogelschatz, Eliasson, and their group at ABB [67–69] have greatly contributed to the fundamental understanding and industrial applications of DBDs. Typical characteristics of the DBD micro-discharges in a 1 mm gap in atmospheric air are summarized in Table 5.1.

Atmospheric pressure DBDs usually are filamentary. However, in certain gas mixtures, a quasi- homogeneous glow mode can be observed [70, 71]. Since the plasma-chemical efficiency of DBD reactors decreases with increasing discharge gap, typical discharge gap widths are in the range of a few millimeters only. When pulsed excitation with sub-microsecond rise time is applied, the discharge gap of DBD reactors can be increased from typically 1 mm to about 1 cm without loss in plasma-chemical efficiency [72]. Nevertheless, the treatment of industrial-scale gas flows requires a lot of paralleled discharge gaps in order to keep the flow resistance low.

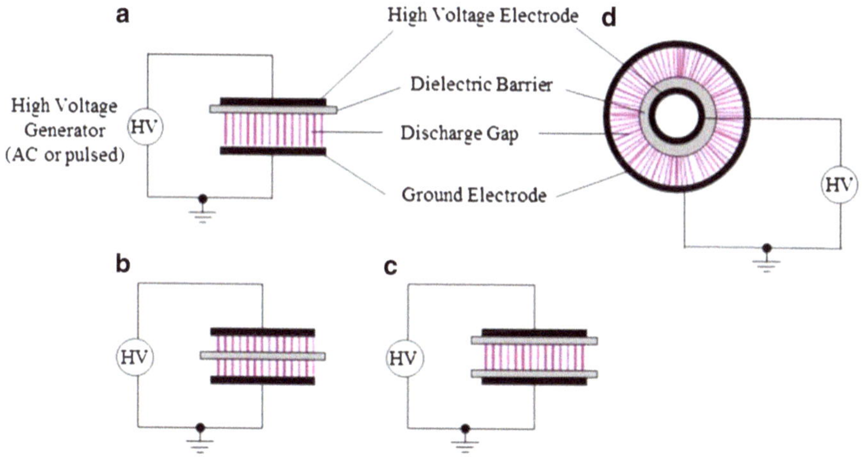

Fig. 5.3 Common dielectric barrier discharge electrode configurations: (**a–c**) planar and (**d**) cylindrical

Table 5.1 Characteristic of DBD in air at atmospheric pressure

Duration	10^{-9}–10^{-8} s
Filament radius	About 10^{-4} m
Peak current	0.1 A
Current density	10^{6}–10^{7} A m^{-2}
Total charge	10^{-10}–10^{-9} C
Electron density	10^{20}–10^{21} m^{-3}
Mean electron energy	1–10 eV
Filament temperature	Close to average gas temperature in the gap

5.4 NO_x Chemistry in Nonthermal Plasma

The energetic electrons produced in the plasma discharge mainly initiate the plasma chemistry. The elementary process of radical formation and reactions in NTPs can be broadly divided into a primary process and a secondary process based on the timescale of streamer propagation [73]. The primary process that has a typical timescale in the range of tenths of nanoseconds includes ionization, excitation, dissociation, light emission, and charge transfer. The efficiency of the primary process is highly dependent on energization methods and their parameters, such as nature of the power source (pulse, DC, or AC), voltage rise time, duration, and frequency. The secondary process is the subsequent chemical reactions involving the products of primary processes (electrons, radicals, ions, and excited molecules) to form additional radical species and reactive molecules (O_3, HO_2, and H_2O_2) by radical-neutral recombination. The secondary process is usually completed within approximately milliseconds. Oxidation is dominant for plasma processing of exhausts (with or without hydrocarbons) containing dilute concentrations of NO_x (mainly NO with concentrations ranged from few ppm to hundreds ppm) in N_2-O_2-H_2O mixtures, particularly when O_2 concentration is 5% or higher.

5.4.1 *Plasma Without Hydrocarbons*

In NTP treatment of NO-N_2-O_2-H_2O mixtures, the kinetic energy of the electrons is mainly deposited into the major gas components N_2, O_2, and H_2O. The result is the generation of N and O radicals through electron-impact dissociation:

$$e + O_2 \rightarrow e + O(^3P) + O(^3P, ^1D)$$

$$e + N_2 \rightarrow e + N(^4S) + N(^4S, ^2D)$$

where $O(^3P)$ and $O(^1D)$ are the ground-state and metastable excited-state atomic oxygen radicals, respectively, and $N(^4S)$ and $N(^2D)$ are ground-state and metastable excited-state atomic nitrogen radicals, respectively. The $N(^4S)$ is the only plasma-produced species that could effectively lead to the chemical reduction of NO [66, 74] accomplished via:

$$N(^4S) + NO \rightarrow N_2 + O$$

The O radical resulting from previous reaction can lead to the oxidation of NO to NO_2. However, the amount of O radicals is, as most, equal to the number of $N(^4S)$ produced in the plasma. Under this condition, most of the NO react with $N(^4S)$, and only a very small amount of NO_2 is produced [75].

Most experimental investigations focused on the oxidation of NO to NO_2 with different gaseous mixtures and energization systems. It was found that the oxidation of NO to NO_2 significantly depends on gaseous mixtures. It is very important to make a distinction between NO removal by chemical oxidation and NO removal by chemical reduction. Oxygen amount in the gas mixtures could greatly suppress the reduction process and enhance the oxidation process or the opposite. Yan et al. [76] show that the reduction process-induced NO removal can be negligible in the presence of O_2 with concentration of 3.6% of higher, which means that NO is completely converted into NO_2. In that case, the oxidation pathway becomes dominant for two reasons:

(a) The dissociation energy of O_2 (4.8 eV) is lower than that of N_2 (9.2 eV). For common atmospheric pressure electrical discharge plasma, the average electron kinetic energy is relatively low (3–6 eV) [77–79]. Under this condition, the rate for electron-impact dissociation of O_2 is much higher compared to that of N_2 as shown in Fig. 5.4.
(b) High electron energies are required to optimize the production of $N(^4S)$ atoms by electron-impact dissociation of N_2. Under conditions optimum for the dissociation of N_2, a large number of excited nitrogen atoms, $N(^2D)$, is produced [80]. The $N(^2D)$ species can lead to undesired reactions in the presence of O_2. Rather than reducing NO to N_2, the $N(^2D)$ species would react with O_2 to produce NO.

If oxygen concentration is less than 1%, the reduction process is dominant as showed in pulsed corona discharge under high specific input energy (SIE > 30 Wh/m) [81]. In that case, the amount of NO_2 produced could be larger than the removed NO because of $N(^2D)$-induced NO formation and NO to NO_2 conversion. The back conversion of NO_2 to NO becomes a limiting factor for the NO oxidation efficiency and energy cost. As reported by Tas et al. [82], the energy required for this process ($\geq$ 100 eV/molecule) is one order of magnitude higher than that for the oxidation process (50 eV/molecule) in the gas-phase reaction.

Additionally, the secondary radical species such as HO_2, NO_3, and O_3 are produced and consumed on longer timescales than the primary radicals. Ozone generated at low temperature through three-body reaction of $O(^3P)$ ($O(^1D)$ in humid gas) with O_2 is quickly quenched by H_2O to form hydroxyl radicals OH

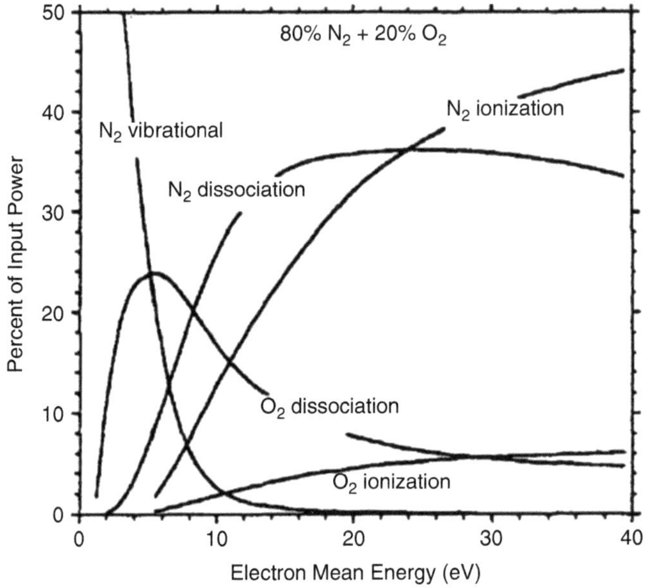

Fig. 5.4 Electrical power dissipation in a dry air discharge, showing the percent of input power consumed in the electron-impact processes leading to vibrational excitation, dissociation, and ionization of N_2 and O_2 [66]

which play an important role in the chemistry of NTP. As the oxygen and water vapor concentrations were increased, the rate of generation of OH and HO_2 radicals by reactions of atomic oxygen with H_2O molecules increases [22, 83]. These radicals (which density is highly sensitive to plasma conditions) initiate the reactions leading to chemical conversion of pollutant molecules (e.g., NO to NO_2 and NO_2 to nitric acid (HNO_3)). Figure 5.5 shows the production of oxidizing agents (O, OH, and O_3) was predicted by a self-consistent 0D-model in N_2-O_2-H_2O mixture at room temperature and atmospheric pressure [84]. O and OH maximum densities are achieved in some tens of nanoseconds, whereas the density of ozone slowly increases to reach a maximum about 1 ms after the plasma ignition.

In an oxygen-free mixture, the chemical kinetic is easier to study, but the situation is not representative of the engine or power plant exhaust gas emissions. The kinetic model in photo- triggered discharges of Fresnet et al. [85] predicts that N_2(a') singlet metastable states (which are strongly mixed by collisions and highly populated) represent an important pathway for NO removal in N_2/NO mixtures, besides NO reduction by N atoms. The drastic decrease in NO removal observed experimentally in the presence of water was attributed to de-excitation of the N_2(a') states by the H_2O molecules. For pulsed corona discharge in similar mixture, the kinetic model of Zhao et al. [86] considered reactions of NO and NO_2 with N and O atoms and with excited $N_2(A^3\Sigma)$ molecules and found that $N_2(A^3\Sigma)$ molecules are important as well, especially for the conversion of N_2O to N_2.

Fig. 5.5 O, OH, and O_3 densities as a function of time predicted by a 0D self-consistent model of photo-triggered discharge: N_2-O_2-H_2O (77.6-20-2.4) at 25 °C, 1 bar, and SIE = 100 J/L [84]

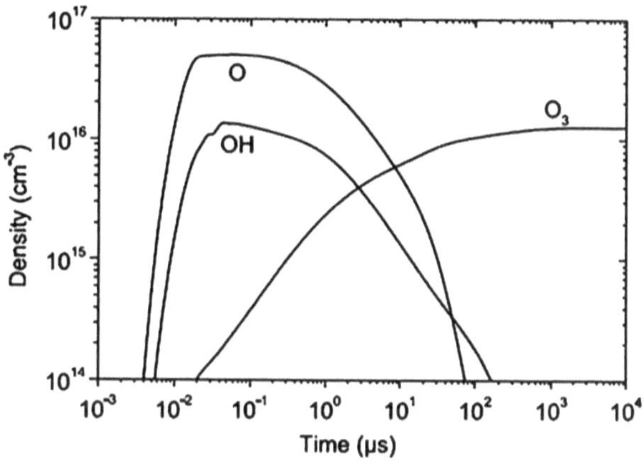

5.4.2 *Plasma with Hydrocarbons*

It is well known that, related to NTP deNO_x, promotion in the NO oxidation can be achieved with the addition of hydrocarbons. Hydrocarbon addition leads to substantial saving in the energy cost for deNO_x treatment. Among the possible hydrocarbons, ethylene (C_2H_4) [87–89], propene (C_3H_6) [1, 22, 87, 89, 90], and propane (C_3H_8) [87, 90] were selectively investigated to estimate their usefulness for the NO-NO_2 conversion. With the addition of hydrocarbons, the atomic oxygen produced in the plasma is considered to be the main initiator of the hydrocarbon chemistry, and the reactions typically produce intermediates such as hydroxy (OH) and peroxy radicals RO_2 (with R = H or C_xH_y).

The chemistry of NTP with hydrocarbons is complex and not fully understood. In the literature, there are some recommendations according to the channels and the products of the reaction. As it was remarked in the experimental work of Tsang [91], the total rate constant of the C_3H_6 + O reaction is quite accurate, and the most serious uncertainties are in the branching ratios. Usually, authors use different reaction channels and branching ratios in the modeling. Penetrante et al. [1], Shin and Yoon [87], and Park et al. [92] included the data of Tsang [91] and Wilk et al. [93] in the oxidation chemical kinetic. The effect of C_3H_6 on the conversion pathways of NO_x in pulsed corona discharge reactor was examined, and results reported by Penetrante et al. [1] show that in the very early stages of reaction, the propene is mainly consumed by O atoms following the three reaction channels:

$$
\begin{aligned}
C_3H_6 + O &\rightarrow C_2H_5 + HCO \\
&\rightarrow CH_2CO + CH_3 + H \\
&\rightarrow CH_3CHCO + H + H
\end{aligned}
$$

However, after the initial stages of the reaction chain, and following the production of hydroxyl radicals in the above reaction, the OH radical rather than the O atom becomes the main propene consuming species. The reaction products from these steps form hydrocarbon intermediates (CHO, CH_3O, CH_2OH, etc.) that react with oxygen to form HO_2 radicals. Chemical kinetic analysis shows that HO_2 radicals are the main species responsible for conversion of NO to NO_2. The main end products predicted by the mechanism of Martin et al. [94] are CO, CO_2, formaldehyde (CH_2O), and acetaldehyde (CH_3CHO), with smaller concentrations of ketene (CH_2CO) and methyl ketene (CH_3CHCO).

Dorai and Kushner [90, 95, 96] used a recommendation of Atkinson [97] to develop a mechanism in humid exhaust gas. The reaction of O with C_3H_6 they used differs from those suggested by Penetrante et al. [1] by the inclusion of the steps:

$$\begin{aligned} C_3H_6 + O \rightarrow & C_3H_6O \\ \rightarrow & CH_3CH_2CHO \\ \rightarrow & CH_2CHO + CH_3 \\ \rightarrow & C_2H_5 + HCO \end{aligned}$$

For the reaction of OH with propene, only the following step is considered:

$$C_3H_6 + OH \rightarrow C_3H_6OH$$

End products such as formaldehyde (CH_2O), acetaldehyde (CH_3CHO), methyloxirane (C_3H_6O), and propanal (CH_3CH_2CHO) were predicted as major by-products from propene-induced NO to NO_2 conversion. Experimental analysis of exhaust within similar composition made by Hoard and Panov [98] showed no presence of methyloxirane and propanal.

In the work of Filimonova et al. [89], the choice of reaction channels and their ratios was based on the experimental investigation of branching ratios in the C_3H_6 + O reaction obtained by Koda et al. [99]. Taking into account the measured yields of products and a pressure dependence of CH_2CHO formation, they used the following distribution:

$$\begin{aligned} C_3H_6 + O \rightarrow & CH_3 + CH_2CHO \\ \rightarrow & C_2H_5 + CHO \\ \rightarrow & C_2H_4 + CH_2O \end{aligned}$$

Regarding these three significant examples, an ambiguity about the choice of critical reaction channel (C_3H_6 + O) for NO_x oxidation needs to be clarified by additional experimental investigations and further refinement of the rate constants. In most cases, the comparison between experimental and simulation results is not suitable because the data is obtained in different conditions. However, some of the end products predicted by simulation results such as acetaldehyde, formaldehyde, methanol, methyl nitrate and nitrite, nitromethane, propylene oxide, and formic and nitric acids agree with experimental measurements as shown in the Figs. 5.6 and 5.7.

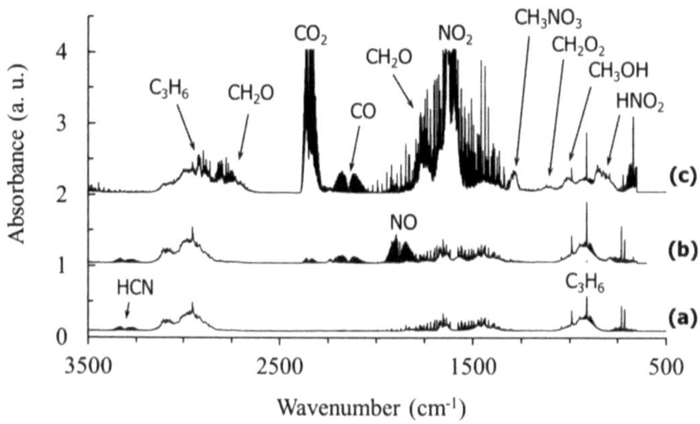

Fig. 5.6 Typical FTIR spectra produced by plasma processing (27 J/L) at room temperature of (a) C_3H_6 (500 ppm)-N_2, (b) NO (500 ppm)-C_3H_6 (500 ppm)-N_2, and (c) O_2 (10%)-NO (500 ppm)-C_3H_6 (500 ppm)-N_2 [100]

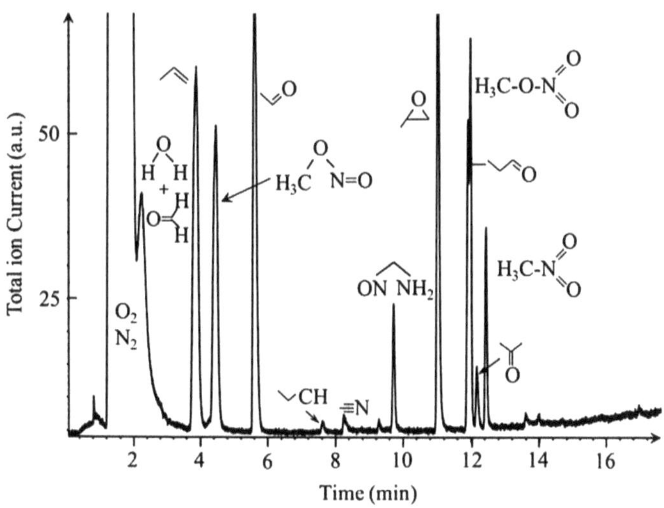

Fig. 5.7 Typical chromatogram plot of a gas phase from pulsed DBD processing O_2 (8%)-NO (300 ppm)-C_3H_6 (150 ppm)-N_2 (balance): SIE = 36 J/L, and $T = 150$ °C [101]

From these studies, the reaction products from these steps form hydrocarbon intermediates that react with oxygen to form HO_2 radicals that become the main component in the NO removal in the presence of C_3H_6 and C_2H_4. However, C_3H_6 is more beneficial compared with C_2H_4 for the cleaning process because of an additional formation of HC radicals.

For a given specific input energy, the gas temperature affects the efficiency of oxidation of NO to NO_2. Penetrante et al. [1] show that in gas mixtures without hydrocarbons, the efficiency of oxidation of NO to NO_2 by O radicals drops as the

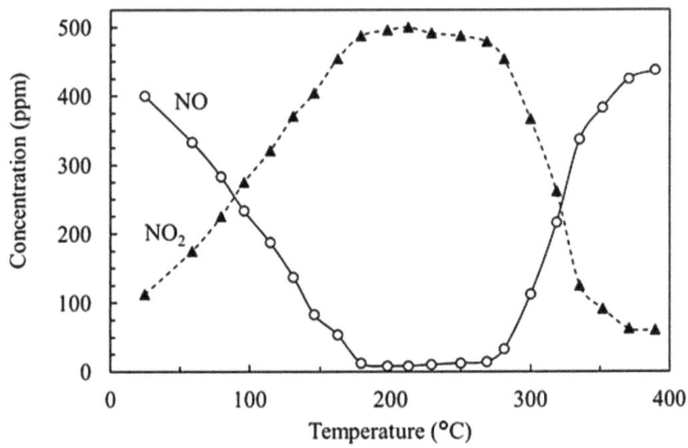

Fig. 5.8 NO and NO_2 concentrations as a function of gas temperature. Plasma processing of C_3H_6 (660 ppm)-NO (500 ppm)-O_2 (8% vol.)-N_2 at SIE = 36 J/L [101]

gas temperature increased. At high temperatures, the NO to NO_2 oxidation reaction is counteracted by the reduction reaction as demonstrated by Djéga-Mariadassou et al. [101] in mixture with propene (Fig. 5.8).

5.5 NO_x Removal by Nonthermal Plasma and Energy Cost

5.5.1 Effect of Hydrocarbons

The NO to NO_2 conversion significantly depends on gaseous mixtures. As shown in the previous section, the hydrocarbon addition leads to significantly influence the NO_x chemistry during plasma remediation leading to the promotion of the oxidation of NO to NO_2 and lower the energy cost of this oxidation. According to the calculations of Park et al. [92], the initial HC concentration more than 500 ppm (except CH_4), and O_2 content more than 10%, almost does not affect the NO oxidation to NO_2. Experimental work of Khacef et al. [22] on the NO_x remediation in the presence of propene resulted in the similar conclusion as shown in Fig. 5.9.

Figure 5.10 shows the NO and NO_x (NO + NO_2) conversion rate as a function of SIE for the cases without propene and with 500 ppm propene. In the absence of hydrocarbons, less than 55% of the initial NO is converted to NO_2 even for the highest SIE, and the NO_x removal was zero. When 500 ppm of propene is added to the gas stream, the oxidation of NO to NO_2 was enhanced due to peroxy radicals HO_2 (see previous section).

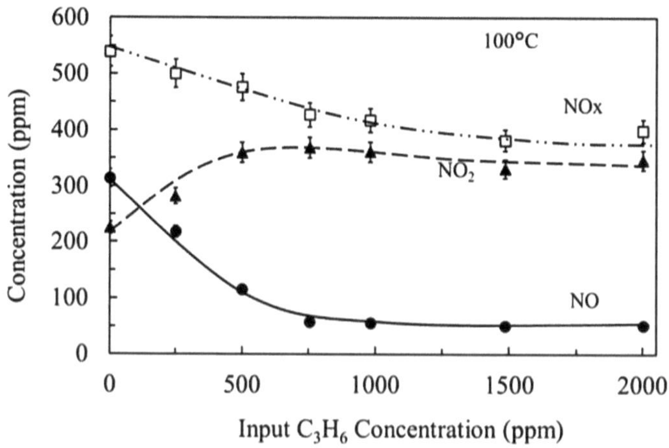

Fig. 5.9 Effect of inlet propene concentration on NO to NO_2 conversion in plasma processing of 500 ppm NO in air ($T = 100$ °C, SIE = 44 J/L) [22]

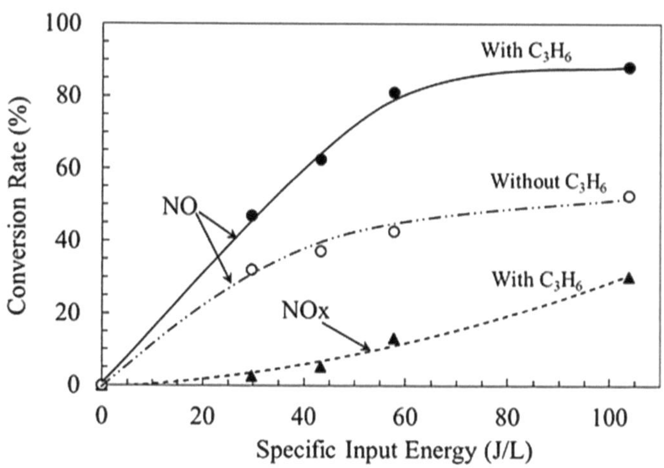

Fig. 5.10 Effect of SIE on NO and NO_x conversion without and with 500 ppm propene in NO (500 ppm)-air (195 mJ/pulse and repetition rate up to 40 Hz) [22]

5.5.2 Energy Cost

One of the relevant parameters in the cleaning of pollutant gas (NO_x, VOCs, etc.) is the energy cost for removing the unwanted molecules. To be competitive with other technologies, the nonthermal plasma technology applied for cleaning industrial or automotive exhaust gases should be the less consuming energy. The energy cost of the plasma-chemical processes is closely related to its mechanisms, and the same plasma processes in different discharge systems or under different conditions result in entirely different expenses of energy. Usually, the plasma processing literature present the fraction of NO_x removed as a function of parameters such as the specific input energy (SIE = mean input power/flow rate), the residence time of the gas into the reactor, or the applied voltage to the plasma reactor. SIE is the parameter commonly used to evaluate the NO_x conversion efficiency in the plasma. However, if this parameter is important to characterize the electrical energy consumption of the process, one should take into account the way to achieve the chosen energy value. It means that for a given plasma reactor, the plasma chemistry strongly depends on the energy transfer from the power source to plasma reactor. Many parameters, such as type of discharge (pulse, AC), high-voltage characteristics (amplitude, rise time, duration, frequency), electrode configuration, and efficiency of chemical reactions, have to be taken into account to optimize the process for a better energy efficiency.

The lowering of energy cost for NO_x conversion was obtained by many research groups for different systems. Niessen et al. [88] examine the effect of ethylene on the conversion of NO_x in a DBD. They reported an improvement of NO removal efficiency by an order of magnitude in the presence of 2000 ppm C_2H_4. Without C_2H_4 in the gas mixture, 48 eV/molecule (experiment) and 61 eV/molecule (model) are needed for the removal of one NO molecule at 90% removal rate, while adding 2000 ppm C_2H_4 leads to energetic cost of only 6 eV/molecule (experiment) and 9.6 eV/molecule (model).

In pulsed DBD experiments with gas mixture simulated diesel exhaust, Khacef et al. [22, 102] obtained a change of the energy cost for NO removal with temperature in the presence of C_3H_6. The authors demonstrate that, for a given reactor under the same gas composition and equivalent SIE, the NO removal efficiency is optimum for a low input energy per pulse and a high discharge frequency as shown in Fig. 5.11. The modeling of Filimonova and Amirov [103] on the removal of toxic impurities from air flow by a pulsed corona discharge resulted in similar conclusion. These authors confirm that it is possible to increase the removal efficiency at a constant total energy deposition by choosing the optimum pulse repetition rate and the energy deposited per discharge pulse. As example, for a SIE of about 27 J/L, NO conversion is more than 90% for 35 mJ/pulse and 200 Hz and less than 30% for 195 mJ/pulse and 30 Hz.

Whatever the pulsed discharge operations used, the main end products of the process are NO_2, CO, CO_2, CH_2O, CH_3ONO_2, CH_2O_2, HONO, and $HONO_2$. However, marked differences in final concentrations of these species were observed according the operating conditions.

Figure 5.12 shows an example of NO_x concentrations measured at the exit of a pulsed DBD reactor as a function of SIE in dry and humid (10% H_2O) gas mixture at 260 °C. With increasing energy deposition, the NO conversion and NO_2 formation efficiencies have been improved by increasing the gas temperature (up to 260 °C) and by adding 10% H_2O in the dry gas mixture. As example, the energy deposition of only 7 J/L was required to achieve maximum NO to NO_2 oxidation efficiency (up to 92%). This efficiency rate was only 61% when no water is present in the gas stream.

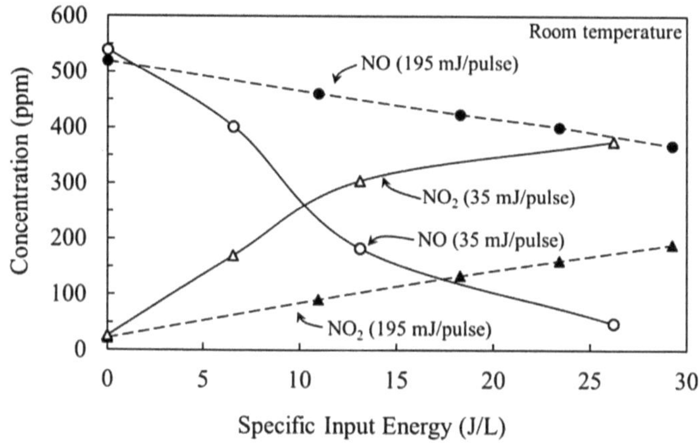

Fig. 5.11 NO and NO_2 concentrations as a function of SIE for pulse energy of 35 and 195 mJ. Gas mixture: O_2 (10%)-NO (500 ppm)-C_3H_6 (500 ppm)-N_2 (balance) [102]

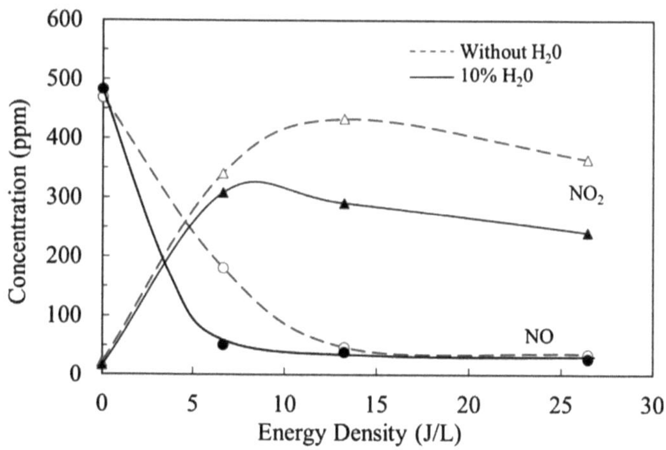

Fig. 5.12 Effect of SIE on NO_x conversion at 260 °C. Gas mixture: O_2 (10%)-NO (500 ppm)-C_3H_6 (500 ppm)-N_2 (balance) without H_2O and with 10% of H_2O [22]

The energy cost for humid plasma-induced NO removal is dramatically reduced to values scaling from ≈ 15 eV/molecule at 27 J/L down to ≈ 4 eV/molecule at 7 J/L. Bröer et al. [104] investigated synthetic gas mixture containing higher O_2 concentration (18%) with C_2H_4 instead of C_3H_6 and found that the maximum NO removal rate was 92% with an energy cost around 100 eV/molecule. In the modeling of the NO oxidation process in simulated exhaust containing O_2-H_2O-CO_2-CO-NO-H_2-C_3H_6-C_3H_8-N_2, Dorai and Kushner [90] observed that at 58 J/L, the energy cost decreases from 240 eV/molecule for a single pulse to 185 eV/molecule when the same energy was distributed over 20 pulses.

For vehicle exhaust systems, it was established that the observed chemistry in the plasma includes the conversion of NO to NO_2 as well as the partial oxidation of hydrocarbons. The presence of the unburned hydrocarbons (UHCs) in the exhaust is very important for the plasma- catalytic deNO_x process for essentially four reasons [1, 2, 22, 50, 87–89, 95].

(a) UHCs enhance the gas-phase oxidation of NO to NO_2 and lower the energy cost of this oxidation.
(b) The partial oxidation of UHCs leads to produce chemical species such as aldehydes and alcohols useful for the catalytic reduction of NO_x. For some catalysts, the partially oxygenated hydrocarbons are much more effective compared to original hydrocarbons in reducing NO_x to N_2.
(c) The UHCs minimize the formation of acid products.
(d) UHCs prevent the oxidation of SO_2 thus making the plasma-catalytic process tolerant to the sulfur content of the fuel.

5.6 Plasma-Catalytic Process for NO_x Abatement

5.6.1 Why Do We Need Plasma for NO_x Abatement in Industrial Processes

As already discussed in the literature, NO_x trap systems and assisted NO_x reduction using SCR catalysts are commonly the two industrial processes developed in order to reach the limitation in terms of NO_x emission [5–8]. An alternative solution, not industrialized yet, is the use of hydrocarbons available in the exhaust, as reductant, to reduce the NO_x. However, in order to reach low temperature NO_x abatement, no catalytic system exists nowadays; thus, the combination of plasma and catalytic system would be a solution.

5.6.2 New Processes and Industrial Technologies

Atmospheric pressure NTP hybrid exhaust aftertreatment systems in the absence of noble metals has been developed by Kuwahara et al. [105]. Two types of new environmental protection systems (a dry system and a wet system), which enable the production of ultralow CO_2, particulate matter, and NO_x emissions as well as reduced fuel consumption and low cost, are investigated for diesel engines, marine engines, and combustion boiler applications. This paper reports the principles of the dry system and some recent experimental results of laboratory tests. The NO_x reduction comprises three flow processes: (i) adsorption, (ii) heating, and (iii) cooling processes. The heating process corresponds to the regeneration process. This dry system demonstrates excellent energy efficiencies that meet Japanese national regulations regarding automobile diesel engine exhaust gas. In this study, approximately 60% of NO_x conversion in the exhaust was achieved after 35 h on stream. An energy efficiency of about 143 g $(NO_2)/^{kWh}$ was then achieved for NO_x reduction.

Recently, two types of innovative NTP environmental protection systems [106, 107] using non-noble metals and which enable the production of ultralow CO_2, particulate matter (PM), and NO_x emissions as well as reduced fuel consumption and cost were proposed. These NTP systems are realized for diesel hybrid engines, marine engines, and combustion boiler applications. In these systems, the plasma technology is combined with other environmental technologies such as adsorption or wet-type chemical scrubbers. These technologies are expected to be new global environmental protection technologies, as they will enable an avoidance of the difficulties in catalytic methods and ordinary plasma methods. Other systems using a combination of NTP process and normal EGR (exhaust gas recirculation) led to similar results [108, 109].

5.6.3 *Lab-Scale and Fundamental Studies*

5.6.3.1 SCR of NO_x by Ammonia Assisted by Plasma

Only few studies dealt with selective catalytic reduction of NO_x by ammonia assisted by plasma. Thus, the SCR of nitrogen oxides by a combination of DBD plasma and a monolithic V_2O_5-WO_3/TiO_2 catalyst using ammonia between 100 and 250 °C was proposed [30]. Figure 5.13 summarizes the results obtained in that study. The gas mixtures used were representative of diesel exhaust gases. For gas mixtures in which 95% of the nitrogen oxides were in the form of NO, the removal of NO_x below 150 °C in the absence of plasma was negligible. Using the catalytic bed in the post-discharge DBD, about 70% of the NO_x was reduced at temperatures as low as 100 °C. Due to the coexistence of NO and NO_2 on the catalyst, the selective catalytic reduction was enhanced. Similar effects were observed for the SCR in gas mixtures containing equal amounts of NO and NO_2 without plasma treatment. This study clearly shows the efficiency of the plasma catalyst system at temperatures lower than those used for the catalytic system alone.

The potential of plasma-enhanced SCR for NO_x removal from exhaust gases containing high concentration of oxygen was investigated in a combination of dielectric barrier discharges with a monolithic V_2O_5-WO_3/TiO_2 catalyst using ammonia as a reducing agent. Experiments were performed which showed the temperature dependence of NO_x removal from synthetic gas mixtures, for very low temperatures (between 100 and 250 °C). It was shown that for $T < 200$ °C, the NO reduction rate decreased sharply as long as NO_2 concentration was low compared to NO concentration. This was the case when no plasma treatment was performed prior to catalytic treatment. With plasma treatment even at temperatures

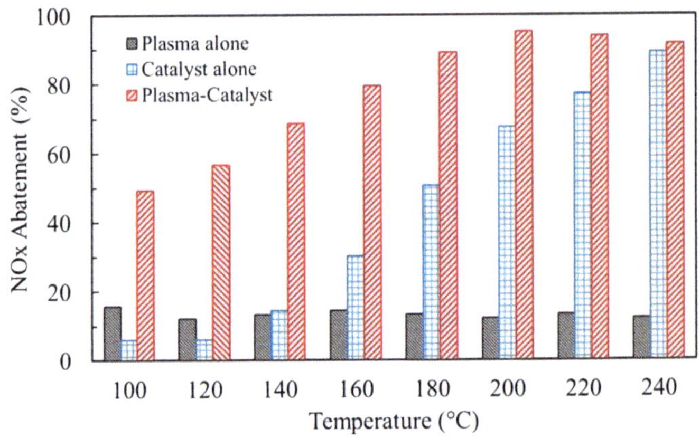

Fig. 5.13 DBD plasma-enhanced NH_3-SCR of NO_x (V_2O_5-WO_3/TiO_2 catalyst, synthetic gas mixture O_2 (13%)-H_2O (5%)-N_2 (82%)-NO_x (500 ppm)-NH_3 (500 ppm) [30]

as low as 100 °C, NO_x reduction reached values up to 70%. An explanation of these results could be given by experiments with gas mixtures containing NO_2 concentration, which were nearly equal to the NO concentration: without plasma treatment at 100 °C, the degree of catalytic NO_x reduction was 70%, also. From this result, it is concluded that on a V_2O_5-WO_3/TiO_2 catalyst at low temperatures, NO can be reduced in the presence of NO_2. Similar kinetic behavior was found in reactions of NO_x with zeolite-fixed ammonium ions [110].

In summary, one can conclude that the pre-treatment of the exhaust gas by nonthermal plasma enhances selective catalytic reduction rates at temperatures below 200 °C by oxidation of NO to NO_2, which was already proposed by McLarnon and Penetrante [75].

5.6.3.2 SCR of NO_x by Hydrocarbons Assisted by Plasma

Published literature reports on the application of the plasma-assisted HC/SCR to the lean-NO_x catalysis have focused primarily on the in situ activation by nonthermal plasma of the gas-phase reactants such as NO and/or HCs leading to NO_2 and oxygenated species [101, 111, 112]. In these studies, alumina- or zeolite-based catalysts were used coupled with NTP DBD in the wide temperature range (100–500 °C). The results clearly show that in the presence of the plasma, the catalytic activity is shifted to low temperature with high conversion of NO, and industrial application was clearly shown [110].

5.7 Role and Beneficial Effect of Plasma

5.7.1 Importance of NO_2 in the Catalytic Reduction of NO_x

Djéga-Mariadassou and co-workers showed that a three-function catalyst model for SCR NO_x with hydrocarbons could be proposed in the absence of plasma [101, 112, 113]. That model is based on experimental evidence for each function, during temperature programmed surface reactions (TPSR), and has been verified during stationary experiments. A general scheme of the model is shown in Fig. 5.14. The first function F1 leads to the oxidation of NO to NO_2. The second function F2 is the mild oxidation of hydrocarbons through organic nitrogen-containing intermediates (R-NO_x) leading to $C_xH_yO_z$ intermediate species such as aldehydes or alcohols. The third function F3 involves the NO reduction by subsequent formation of N2 assisted by the oxidation of reductants to CO_2/H_2O over transition metal cations. The previous $C_xH_yO_z$ intermediate species can achieve their own total oxidation by cleaning the adsorbed oxygen species left by NO dissociation.

It is very difficult to find the best design of the catalyst to simultaneously initiate the three functions by itself. Thus, an external device can be developed to substitute functions F1 and F2, providing the catalyst the appropriate oxygenated species, in the full range of temperature. For the vehicle exhaust systems, it was established that the plasma chemistry includes the conversion of NO to NO_2 as well as the partial oxidation of hydrocarbons. The presence of the unburned hydrocarbons in the exhaust is very important for the plasma-catalytic deNO_x process for multiple reasons. First, unburned hydrocarbons enhance the gas-phase oxidation of NO to NO_2 and lower the energy cost for this oxidation. Secondly, their partial oxidation leads to produce, in the whole range of reaction temperature, chemical species such as aldehydes, alcohols, and R-NO_x useful for the catalytic reduction of NO_x. For some catalysts, the partially oxygenated hydrocarbons are much more effective compared to original hydrocarbons in reducing NO_x to N_2. Thirdly, unburned hydrocarbons prevent the oxidation of SO_2 thus making the plasma- catalytic process tolerant to the sulfur content of the fuel. These “intermediate” species are needed for function F1 and F2 and, furthermore, for the third function itself. The NTP could substitute for the first two functions (F1, F2) of the catalysts as already proposed elsewhere [30, 101, 105–109, 112] (Fig. 5.15).

Fig. 5.14 Scheme for the three-function catalyst: concerted actions for the selective reduction of NO by hydrocarbons ($C_xH_yO_z$, partially oxidized HC)

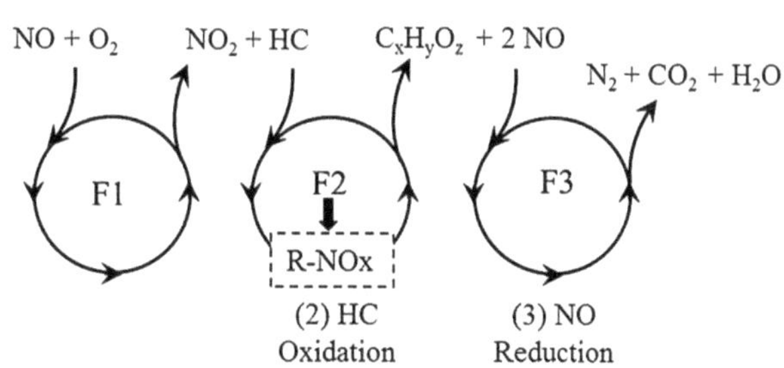

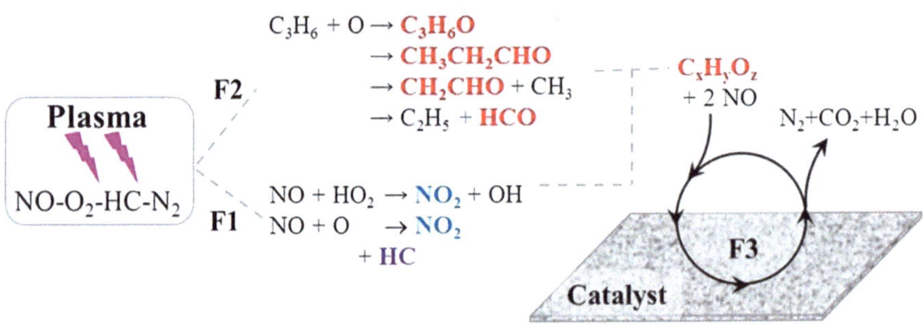

Fig. 5.15 Three-function mechanism adapted for plasma-catalytic processes

This result demonstrates what can be the role of plasma in nonthermal plasma-assisted catalytic NO_x remediation. Furthermore, plasma is able to provide both NO_2 and $C_xH_yO_z$ intermediate species at low temperature. However, the behavior of the R-NO_x species formed in the plasma still remains to be studied. The NO_x removal leads mainly to N_2 as reported previously [114].

5.7.2 *On the Effect of the Catalysts in the Coupled Process*

Miessner et al. [34] have shown that the combination of HC/HCR and nonthermal plasma enhances the overall reaction and allows an effective removal of NO_x at relatively low temperature (Fig. 5.16). The oxidative potential of an NTP in off-gases with excess oxygen results in an effective conversion of NO to NO_2 that can be converted synergistically by HC/HCR to molecular nitrogen with appropriate catalysts. The hydrocarbon added has two essential functions: first, it assists the gas-phase oxidation of NO to NO_2 by the electric discharge in excess oxygen and, secondly, it reacts with NO_2 in the hydrocarbon SCR. Besides CO_2 and CO, significant amounts of formaldehyde and acetaldehyde are formed in the plasma-initiated gas-phase reaction. These and other by-products are involved together with the remaining propene in the subsequent catalytic reaction. With a specific input energy density lower than 15 Wh/m^3, a temperature of 300 °C, and a space velocity of 20,000 h^{-1}, NO_x conversions higher than 50% are obtained. The synergistic combination of NTPs and HC/HCR has been verified under real conditions with exhaust gas from a diesel engine.

The catalyst support itself, Al_2O_3, seems to be an appropriate catalyst for the reaction [34, 114, 115], and some modifications of Al_2O_3 and ZrO_2 were found to be effective as catalysts in this reaction. The role of plasma processing on NO_x reduction over alumina and basic zeolite NaY was examined by Cho et al. [111] and compared to a conventional NH_3/SCR system, and the results are comparable at low temperature (Fig. 5.17).

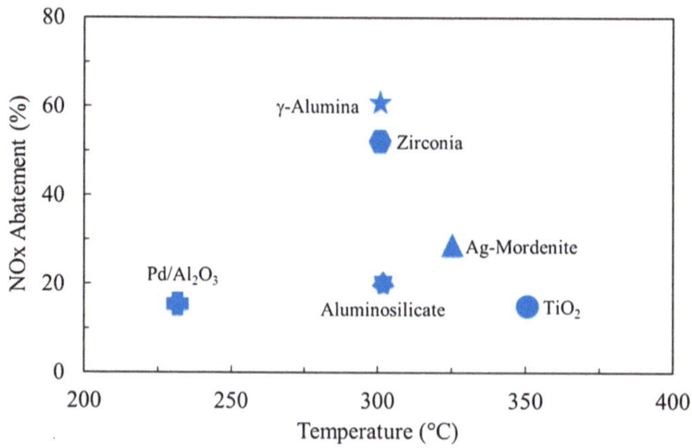

Fig. 5.16 Efficiency of various catalysts in the plasma-catalytic HC/HCR of NO_x (500 ppm NO and 1000 ppm C_3H_6 in N_2/O_2 (13%); SIE = 14 Wh/m) [34]

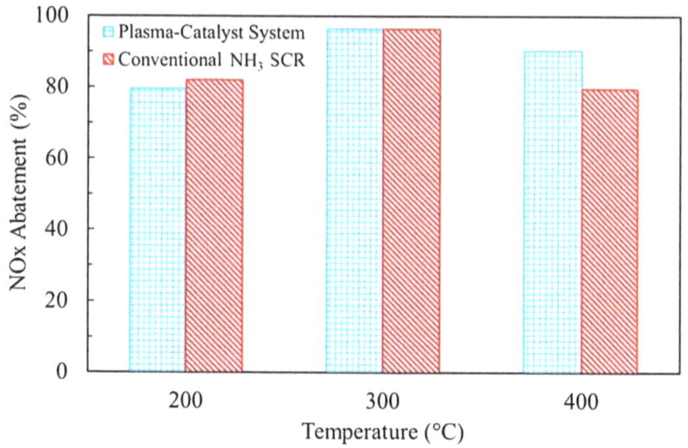

Fig. 5.17 Plasma-catalytic system and conventional NH_3/SCR in engine dyno tests as a function of temperature [111]

During the plasma treatment, NO is oxidized to NO_2, and propylene is partially oxidized to CO, CO_2, acetaldehyde, and formaldehyde. With plasma treatment, NO as the NO_x gas, and a NaY catalyst, the maximum NO_x conversion was 70% between 180 and 230 °C. The activity decreased at higher and lower temperatures. As high as 80%, NO_x removal over alumina was measured. For both catalysts, a simultaneous decrease in NO_x and aldehyde compound concentrations was observed, which suggests that aldehydes may be important components for NO_x reduction in plasma-treated exhaust, which was already proposed by Djéga-Mariadassou et al. [101, 112, 113, 116]. The authors proposed a simplified reaction mechanism to explain the enhanced deNO_x activity induced by air plasma as follows: ozone production, NO and HC oxidation by ozone, and NO_x reduction assisted by oxidized hydrocarbons to form N_2, H_2O, and CO_2. That later reaction proceeds on BaY catalyst at low temperatures and on CuCoY catalyst at high temperatures as evidenced by Schmieg et al. [115] and Lee et al. [117].

5.8 Conclusion

NTP catalytic processes are in development in order to reduce the NO_x in the exhaust gases from industrial processes such as burners for stationary sources or in the exhaust line for mobile applications. It is clear that the plasma is able to create at low temperature the needed species for NO_x abatement. Some coupled processes have been already tested at the pilot scale, but their industrialization will be possible only if the energy due to the plasma makes this process a low-cost process versus the existing ones. Another possible way to use the plasma to reduce NO_x emission is the use of this one in the combustion process to avoid the NO_x formation.

Thus, Lee et al. [118] have recently presented a new combustion technology that can reduce NO_x emissions within industrial burners to single-digit parts per million levels without employing exhaust gas recirculation or other NO_x reduction mechanisms. This new technology uses a simple modification of commercial burners, such that they are able to perform plasma- assisted staged combustion without altering the outer configuration of the commercial reference burner. Because this plasma burner acted as a low NO_x burner and was able to reduce NO_x by more than half compared to the commercial reference burner, this methodology offers important cost-effective possibilities for NO_x reduction in industrial applications at source.

Acknowledgments The authors appreciate the considerable contribution of Professor Gérald Djéga-Mariadassou and Doctor Francois Baudin as well as engineers from Renault SA and PSA Peugeot-Citröen for their fruitful discussions on deNO_x processes and engine chemistry.

References

1. Penetrante, B. M., Brusasco, R. M., Merrit, B. T., Pitz, W. J., Vogtlin, G. E., Kung, M. C., Kung, H. H., Wan, C. Z., & Voss, K. E. (1998). Plasma-assisted catalytic reduction of NO_x. SAE Tech Paper 982508.
2. Balmer, M. L., Tonkin, R., Yoon, S., Kolwaite, A., Barlow, S., Maupin, G., & Hoard, J. (1999). NO_x destruction behaviour of select materials when combined with a non-thermal plasma. SAE Tech Paper 1999-01-3640.
3. Zhang, D. S., Zhang, L., Fang, C., Gao, R. H., Qian, Y. L., Shi, L. Y., & Zhang, J. P. (2013). MnO_x-CeO_x/CNTs pyridine-thermally prepared via a novel in situ deposition strategy for selective catalytic reduction of NO with NH_3. *RSC Advances, 3*, 8811–8819.
4. Koebel, M., Madia, G., & Elsener, M. (2002). Selective catalytic reduction of NO and NO_2 at low temperatures. *Catalysis Today, 73*, 239–247.

5. Burch, R., Breen, J. P., & Meunier, F. C. (2002). A review of the selective reduction of NO_x with hydrocarbons under lean-burn conditions with non-zeolitic oxide and platinum group metal catalysts. *Applied Catalysis B: Environmental, 39*, 283–303.
6. Kaspar, J., Fornasiero, P., & Hickey, N. (2003). Automotive catalytic converters: Current status and some perspectives. *Catalysis Today, 77*, 419–449.
7. Liu, G., & Gao, P. X. (2011). A review of NO_x storage/reduction catalysts: Mechanism, materials and degradation studies. *Catalysis Science & Technology, 1*, 552–568.
8. Fu, M., Li, C., Lu Pei, Q. L., Zhang, M., Zhou, Y., Yu, M., & Fang, Y. (2014). A review on selective catalytic reduction of NO_x by supported catalysts at 100-300°C -catalysts, mechanism, kinetics. *Catalysis Science & Technology, 4*, 14–25.
9. Miller, W. R., Klein, J. T., Mueller, R., Doelling, W., & Zuerbig, J. (2000). The development of urea-SCR technology for US heavy duty trucks. SAE Tech Paper 2000-01-0190.
10. Madia, G., Koebel, M., Elsener, M., & Wokaum, A. (2002). The effect of an oxidation pre-catalyst on the NO_x reduction by ammonia SCR. *Industrial and Engineering Chemistry Research, 41*, 3512–3517.
11. Baik, J. H., Yim, S. D., Nam, I. S., Mok, Y. S., Lee, J. H., Cho, B. K., & Oh, S. H. (2004). Control of NO_x emissions from diesel engine by selective catalytic reduction (SCR) with urea. *Topics in Catalysis, 30*, 37–41.
12. Hirata, K., Masaki, N., Ueno, H., & Akagawa, H. (2005). Development of urea-SCR system for heavy-duty commercial vehicles. SAE Tech Paper 2005-01-1860.
13. Shinjoh, H., Takahashi, N., Yokoda, K., & Sugiura, N. (1998). Effect of periodic operation over Pt catalysts in simulated oxidizing exhaust gas. *Applied Catalysis B: Environmental, 15*, 189–201.
14. Hodjati, S., Semelle, F., Moral, N., Bert, C., & Rigaud, M. (2000). Impact of sulphur on the NO_x trap catalyst activity-poisoning and regeneration behaviour. SAE Tech Paper 2000-01-1874.
15. Olsson, L., Persson, H., Friedell, E., Skoglundh, M., & Andersson, B. (2001). A kinetic study of NO oxidation and NO_x storage on Pt/Al_2O_3 and $Pt/BaO/Al_2O_3$. *The Journal of Physical Chemistry. B, 105*, 6895–6906.
16. Monroe, D. R., & Li, W. (2002). Desulafation dynamics of NO_x storage catalysts. SAE Tech Paper 2002-01-2886.
17. James, D., Fourré, E., Ishii, M., & Bowker, M. (2003). Catalytic decomposition/regeneration of $Pt/Ba(NO_3)_2$ catalysts: NO_x storage and reduction. *Applied Catalysis B: Environmental, 45*, 147–159.
18. Nova, I, Castoldi, L., Lietti, L., Tronconi, E., & Forzatti, P. (2005). The Pt-Ba interaction in lean NO_x trap systems. SAE Tech Paper 2005-01-1085.
19. Ura, J. A., Goralski, C. T., Graham G. W., McCabe, R. W., & Theis, J. R. (2005). Laboratory study of lean NO_x trap desulfation strategies. SAE Tech Paper 2005-01-1114.
20. Higashi, M., Uchida, S., Suzuki, N., & Fujii, K. (1992). Soot elimination and NO_x and SO_x reduction in diesel-engine exhaust by a combination of discharge plasma and oil dynamics. *IEEE Transactions on Plasma Science, 20*, 1–12.
21. Penetrante, B. M., Brusasco, R. M., Merrit, B. T., Pitz, W. J., & Vogtlin, G. E. (1999). Feasibility of plasma aftertreatment for simultaneous control of NO_x and particulates. SAE Tech Paper 1999-01-3637.
22. Khacef, A., Cormier, J. M., & Pouvesle, J. M. (2002). NO_x remediation in oxygen-rich exhaust gas using atmospheric pressure non-thermal plasma generated by a pulsed nanosecond dielectric barrier discharge. *Journal of Physics D: Applied Physics, 35*, 1491–1498.
23. Khacef, A., & Cormier, J. M. (2006). Pulsed sub-microsecond dielectric barrier discharge treatment of simulated glass manufacturing industry flue gas: Removal of SO_2 and NO_x. *Journal of Physics D: Applied Physics, 39*, 1078–1083.
24. Kim, H. H. (2004). Non-thermal plasma processing for air-pollution control: A historical review, current issues, and future prospects. *Plasma Processes and Polymers, 1*, 91–110.

25. Park, J. Y., Jung, J. G., Kim, J. S., Rim, G. H., & Kim, K. S. (2003). Effect of non-thermal plasma reactor for CF_4 decomposition. *IEEE Transactions on Plasma Science, 31*, 1349–1354.
26. Roland, U., Holzer, F., & Kopinke, F. D. (2005). Combination of non-thermal plasma and heterogeneous catalysis for oxidation of volatile organic compounds: Part 2. Ozone decomposition and deactivation of γ-Al_2O_3. *Applied Catalysis B: Environmental, 58*, 217–226.
27. Demidiouk, V., Moon, S. I., & Chae, J. O. (2003). Toluene and butyl acetate removal from air by plasma-catalytic system. *Catalysis Communications, 4*, 51–56.
28. Baylet, A., Marécot, P., Duprez, D., Jeandel, X., Lombaert, K., & Tatibouët, J. M. (2012). Synergetic effect of plasma/catalysis hybrid system for CH_4 removal. *Applied Catalysis B: Environmental, 113*, 31–36.
29. van Veldhuizen, E. M. (Ed.). (2000). *Electrical discharge for environment purpose. Fundamentals and applications*. New York: Nova Science Publishers Inc.
30. Bröer, S., & Hammer, T. (2000). Selective catalytic reduction of nitrogen oxides by combining a non-thermal plasma and a V_2O_5-WO_3/TiO_2 catalyst. *Applied Catalysis B: Environmental, 28*, 101–111.
31. Aardahl, C. L., Habeger, C. F., Balmer, M. L., Rappe, K. G., Tran, D. N., Avila, M., Park, P. W., Koshkarian, K. A., & Chanda, A. (2000). Plasma-enhanced catalytic reduction of NO_x in simulated lean exhaust. SAE Tech Paper 2000-01-2961.
32. Panov, A. G., Tonkyn, R. G., Balmer, M. L., Peden, C. H. F., Malkin, A., & Hoard, J. W. (2001) Selective reduction of NO_x in oxygen rich environments with plasma-assisted catalysis: The role of plasma and reactive intermediates. SAE Tech Paper 2001-01-3513.
33. Yoon, S., Panov, A. G., Tonkyn, R. G., Ebeling, A. C., Barlow, E. E., & Balmer, M. L. (2002). An examination of the role of plasma treatment for lean NO_x reduction over sodium zeolite Y and gamma alumina: Part 1. Plasma assisted NO_x reduction over NaY and Al_2O_3. *Catalysis Today, 72*, 243–250.
34. Miessner, H., Francke, K. P., & Rudolph, R. (2002). Plasma-enhanced HC/HCR of NO_x in the presence of excess oxygen. *Applied Catalysis B: Environmental, 36*, 53–62.
35. Hoard, J., Schmieg, S. J., Brooks, D. J., Peden, C. H. F., Barlow, S. E., & Tonkyn, R. G. (2003, August 28). Dynamometer evaluation of plasma-catalyst for diesel NO_x reduction. Paper presented at the 9th diesel engine emissions reduction conference, Newport, Rhode Island.
36. Ravi, V., Mok, Y. S., Rajanikanth, B. S., & Kang, H. C. (2003). Temperature effect on hydrocarbon-enhanced nitric oxide conversion using a dielectric barrier discharge reactor. *Fuel Processing Technology, 81*, 187–189.
37. Cho, B. K., Lee, J. H., Crellin, C. C., & Toner, J. G. (2005, October 30–November 4) Plasma-catalyst system for diesel NO_x reduction using ethanol and E-diesel: Laboratory reactor and engine dynamometer tests. Paper presented at 2005 annual meeting of AIChE, Cincinnati, OH.
38. Kim, C. H., Qi, G., Dahberg, K., & Li, W. (2010). Strontium-doped perovskites rival platinum catalysts for treating NO_x in simulated diesel exhaust. *Science, 327*, 1624–1627.
39. Jian, D. L., Gao, P. X., Cai, W. J., Allimi, B. S., Alpay, S. P., Ding, Y., Wang, Z. L., & Brooks, C. (2009). Synthesis, characterization, and photocatalytic properties of ZnO/(La,Sr)CoO_3 composite nanorod arrays. *Journal of Materials Chemistry, 19*, 970–975.
40. Busca, G., Lietti, L., Ramis, G., & Berti, F. (1998). Chemical and mechanistic aspects of the selective catalytic reduction of NO_x by ammonia over oxide catalysts: A review. *Applied Catalysis B: Environmental, 18*, 1–36.
41. Li, J., Chang, H., Ma, L., Hao, J., & Yang, R. T. (2011). Low-temperature selective catalytic reduction of NO_x with NH_3 over metal oxide and zeolite catalysts – A review. *Catalysis Today, 175*, 147–156.
42. Iwamoto, M., Furukawa, H., Mine, Y., Uemura, F., Mikuriya, S., & Kagawa, S. (1986). Copper(II) ion-exchanged ZSM-5 zeolites as highly active catalysts for direct and continuous decomposition of nitrogen monoxide. *Journal of the Chemical Society, Chemical Communications, 16*, 1272–1273.
43. Liu, F., Shan, W., Shi, X., Zhang, C., & He, H. (2011). Research progress in vanadium-free catalysts for the selective catalytic reduction of NO with NH_3. *Chinese Journal of Catalysis, 32*, 1113–1128.

44. Bull, I., Xue, W. M., Burk, P., Boorse, R. S., Jaglowski, W. M., Koermer, G. S., Moini, A., Patchett, J. A., Dettling, J. C., & Caudle, M. T. (2009). Copper CHA zeolite catalysts. Patent US 7601662 B2.
45. Kwak, J. H., Tonkyn, R. G., Kim, D. H., Szanyi, J., & Peden, C. H. F. (2010). Excellent activity and selectivity of Cu-SSZ-13 in the selective catalytic reduction of NO_x with NH_3. *Journal of Catalysis, 275*, 187–190.
46. Ma, L., Cheng, Y., Cavataio, G., McCabe, R. W., Fu, L., & Li, J. (2013). Characterization of commercial Cu-SSZ-13 and Cu-SAPO-34 catalysts with hydrothermal treatment for NH_3-SCR of NO_x in diesel exhaust. *Chemical Engineering Journal, 225*, 323–330.
47. Jihui, W., Zhao, H., Haller, G., & Li, Y. (2017). Recent advances in the selective catalytic reduction of NO_x with NH_3 on Cu-Chabazite catalysts. *Applied Catalysis B: Environmental, 202*, 346–354.
48. Masuda, S., & Nakao, H. (1990). Control of NO_x by positive and negative pulsed corona discharges. *IEEE Transactions on Industry Applications, 26*, 374–383.
49. Dhali, S. K., & Sardja, I. (1991). Dielectric barrier discharge for processing of SO_2/NO_x. *Journal of Applied Physics, 69*, 6319–6324.
50. Penetrante, B. M., Hsiao, M. C., Bardsley, J. N., Merrit, B. T., Vogtlin, G. E., Wallman, P. H., Kuthi, A., Burkhart, C. P., & Bayless, J. R. (1996). Electron beam and pulsed corona processing of volatile organic compounds in gas streams. *Pure and Applied Chemistry, 68*, 1083–1087.
51. Evans, D., Rosocha, L. A., Anderson, G. K., Coogan, J. J., & Kushner, M. J. (1993). Plasma remediation of trichloroethylene in silent discharge plasmas. *Journal of Applied Physics, 74*, 5378–5386.
52. Frank, N. W. (1995). Introduction and historical review of electron beam processing for environmental pollution control. *Radiation Physics and Chemistry, 45*, 989–1002.
53. Loeb, L. B. (1965). *Electrical coronas, their basic physical mechanisms*. Berkeley/Los Angeles: University of California Press.
54. Fridman, A., Chirokov, A., & Gutsol, A. (2005). Non-thermal atmospheric pressure discharges. *Journal of Physics D: Applied Physics, 38*, R1–R24.
55. Rutgers, W. R., & van Veldhuizen, E. M. (2000). Fundamentals of non-thermal gas discharges at atmospheric pressure. In E. M. van Veldhuizen (Ed.), *Electrical discharge for environment purpose, fundamentals and applications* (pp. 5–19). New York: Nova Science Publishers Inc.
56. Chang, J., Lawless, P., & Yamamoto, T. (1991). Corona discharge processes. *IEEE Transactions on Plasma Science, 19*, 1152–1166.
57. Yan, K., van Heesch, E. J. M., Pemen, A. J. M., & Huijbrechts, P. A. H. J. (2001). From chemical kinetics to streamer corona reactor and voltage pulse generator. *Plasma Chemistry and Plasma Processing, 21*, 107–137.
58. Eichwald, O., Ducasse, O., Dubois, D., Abahazem, A., Merbahi, N., Benhenni, M., & Yousfi, M. (2008). Experimental analysis and modelling of positive streamer in air: Towards an estimation of O and N radical production. *Journal of Physics D: Applied Physics, 41*, 234002.
59. Chang, J. S., & Masuda, S. (1988) Mechanism of pulse corona induced plasma chemical process for removal of NO_x and SO_2 from combustion gases. IEEE Proc. Annual Meeting Industry Applications Society, pp. 1628–1635.
60. Dinelli, G., Civitano, L., & Rea, M. (1990). Industrial experiments on pulsed corona simultaneous removal of NO_x and SO_2 from flue gases. *IEEE Transactions on Industry Applications, 26*, 535–541.
61. Mok, Y. S., & Nam, I. S. (1999). Positive pulsed corona discharge process for simultaneous removal of SO_2 and NO_x from iron-ore sintering flue gas. *IEEE Transactions on Plasma Science, 27*, 1188–1196.
62. Yan, K. (2001). Corona plasma generation. PhD Thesis, Eindhoven University of Technology, The Netherlands.

63. Sivachandiran, L., & Khacef, A. (2016). In-situ and ex-situ NO oxidation assisted by sub-microsecond pulsed multi-pin-to-plane corona discharge: The effect of pin density. *RSC Advances, 6*, 29983–29995.
64. Joshi, A. A., Lock, B. R., Arce, P., & Finney, W. C. (1995). Formation of hydroxyl radicals, hydrogen peroxide and aqueous electrons by pulsed corona discharge in aqueous solution. *Journal of Hazardous Materials, 41*, 3–30.
65. Siemens, W. (1857). Ueber die elektrostatische induction und die verzoegerung des stroms in flaschendraehten. *Poggendorffs Annals of Physical Chemistry, 102*, 66–122.
66. Penetrante, B. M., Bardsley, J. N., & Hsiao, M. C. (1997). Kinetic analysis of non-thermal plasmas used for pollution control. *Japanese Journal of Applied Physics, 36*, 5007–5017.
67. Eliasson, B., & Kogelschatz, U. (1991). Nonequilibrium volume plasma chemical processing. *IEEE Transactions on Plasma Science, 19*, 1063–1077.
68. Eliasson, B., Hirth, M., & Kogelschatz, U. (1987). Ozone synthesis from oxygen in dielectric barrier discharges. *Journal of Physics D: Applied Physics, 20*, 1421–1437.
69. Eliasson, B., & Kogelschatz, U. (1991). Modeling and applications of silent discharge plasmas. *IEEE Transactions on Plasma Science, 19*, 309–323.
70. Roth, J. R., Rahel, J., Dai, X., & Sherman, D. M. (2005). The physics and phenomenology of one atmosphere uniform glow discharge plasma (OAUGDP™) reactors for surface treatment applications. *Journal of Physics D: Applied Physics, 38*, 555–567.
71. Kozlov, K. V., Brandenburg, R., Wagner, H. E., Morozov, A. M., & Michel, P. (2005). Investigation of the filamentary and diffuse mode of barrier discharges in N_2/O_2 mixtures at atmospheric pressure by cross-correlation spectroscopy. *Journal of Physics D: Applied Physics, 38*, 518–529.
72. Hammer, T. (2000). Pulsed electrical excitation of dielectric barrier discharge reactors using semiconductor power supplies. SAE Tech Paper 2000-01-2894.
73. Kim, H. H., Prieto, G., Takashima, K., Katsura, S., & Mizuno, A. (2002). Performance evaluation of discharge plasma process for gaseous pollutant removal. *Journal of Electrostatics, 55*, 25–41.
74. Penetrante, B. M., Hsiao, M. C., Merritt, B. T., & Vogtlin, G. E. (1997). Fundamental limits on NO_x reduction by plasma. SAE Tech Paper 971715.
75. McLarnon, C. R., & Penetrante, B. M. (1998). Effect of gas composition on the NO_x chemistry in a plasma. SAE Tech paper 982433.
76. Yan, K., Kanazawa, S., Ohkubo, T., & Nomoto, Y. (1999). Oxidation and reduction processes during NO_x removal with corona-induced non-thermal plasma. *Plasma Chemistry and Plasma Processing, 19*, 421–443.
77. Penetrante, B. M., Hsiao, M. C., Merritt, B. T., Vogtlin, G. E., Wallman, P. H., Neiger, M., Wolf, O., Hammer, T., & Bröer, S. (1996). Pulsed corona and dielectric-barrier discharge processing of NO in N_2. *Applied Physics Letters, 68*, 3719–3721.
78. Penetrante, B. M., Hsiao, M. C., Merritt, B. T., Vogtlin, G. E., Wallman, P. H., Kuthi, A., Burkhart, C. P., & Bayless, J. R. (1995). Electron-impact dissociation of molecular nitrogen in atmospheric-pressure non-thermal plasma reactors. *Applied Physics Letters, 67*, 3096–3098.
79. Penetrante, B. M., Hsiao, M. C., Merritt, B. T., Vogtlin, G. E., & Wallman, P. H. (1995). Comparison of electrical discharge techniques for non-thermal plasma processing of NO in N_2. *IEEE Transactions on Plasma Science, 23*, 679–687.
80. Cosby, P. C. (1993). Electron-impact dissociation of nitrogen. *The Journal of Chemical Physics, 98*, 9544–9553.
81. Tas, M. A. (1995). Plasma induced catalysis: A feasibility study and fundamentals. Ph.D. Dissertation. Eindhoven University of Technology, The Netherlands.
82. Tas, M. A., van Hardeveld, R., & van Veldhuizen, E. M. (1997). Reactions of NO in a positive streamer corona plasma. *Plasma Chemistry and Plasma Processing, 17*, 371–391.
83. Chang, M. B., Kushner, M. J., & Rood, M. J. (1992). Gas-phase removal of NO from gas streams via dielectric barrier discharges. *Environmental Science and Technology, 26*, 777–781.
84. Pasquiers, S. (2004). Removal of pollutants by plasma catalytic processes. *European Physical Journal Applied Physics, 28*, 319–324.

85. Fresnet, F., Baravian, G., Magne, L., Pasquiers, S., Postel, C., Puech, V., & Rousseau, A. (2002). Influence of water on NO removal by pulsed discharge in $N_2/H_2O/NO$ mixtures. *Plasma Sources Science and Technology, 11*, 152–160.
86. Zhao, G. B., Hu, X., Yeung, M. C., Plumb, O. A., & Radosz, M. (2004). Non-thermal plasma reactions of dilute nitrogen oxide mixtures: NO_x in nitrogen. *Industrial and Engineering Chemistry Research, 43*, 2315–2323.
87. Shin, H. H., & Yoon, W. S. (2000). Effect of hydrocarbons on the promotion of NO-NO_2 conversion in non-thermal plasma deNO_x treatment. SAE Tech Paper 2000-01-2969.
88. Niessen, W., Wolf, O., Schruft, R., & Neiger, M. (1998). The influence of ethene on the conversion of NO_x in a dielectric barrier discharge. *Journal of Physics D: Applied Physics, 31*, 542–550.
89. Filimonova, E. A., Kim, Y. H., Hong, S. H., & Song, Y. H. (2002). Multi-parametric investigation on NO_x removal from simulated diesel exhaust with hydrocarbons by pulsed corona discharge. *Journal of Physics D: Applied Physics, 35*, 2795–2807.
90. Dorai, R., & Kushner, M. J. (2001). Effect of multiple pulses on the plasma chemistry during the remediation of NO_x using dielectric barrier discharges. *Journal of Physics D: Applied Physics, 34*, 574–583.
91. Tsang, W. (1991). Chemical kinetic data base for combustion chemistry Part V. Propene. *Journal of Physical and Chemical Reference Data, 20*, 221–274.
92. Park, K. S., Kim, D. I., Lee, H. S., Chun, K. M., & Chun, B. H. (2001). Effect of various hydrocarbons on the plasma deNO_x process. SAE Tech Paper 2001-01-3515.
93. Wilk, R. D., Cernansky, N. P., Pitz, W. J., & Westbrook, C. K. (1989). Propene oxidation at low and intermediate temperatures: A detailed chemical kinetic study. *Combustion Flame, 77*, 145–170.
94. Martin, A. R., Shawcross, J. T., & Whitehead, J. C. (2004). Modelling of non-thermal plasma aftertreatment of exhaust gas streams. *Journal of Physics D: Applied Physics, 37*, 42–49.
95. Dorai, R., & Kushner, M. J. (1999). Effect of propene on the remediation of NO_x from engine exhausts. SAE Tech Paper 1999-01-3683.
96. Dorai, R., & Kushner, M. J. (2000). Consequences of propene and propane on plasma remediation of NO_x. *Journal of Applied Physics, 88*, 3739–3747.
97. Atkinson, R. (1997). Gas-phase tropospheric chemistry of volatile organic compounds: 1. Alkanes and alkenes. *Journal of Physical and Chemical Reference Data, 26*, 215–290.
98. Hoard, J. W., & Panov, A. (2001). Products and intermediates in plasma-catalyst treatment of simulated diesel exhaust. SAE Tech Paper 2001-01-3512.
99. Koda, S., Endo, Y., Tsuchiya, S., & Hirota, E. (1991). Branching ratios in atomic oxygen (^3P) reactions of terminal olefins studied by kinetic microwave absorption spectroscopy. *The Journal of Physical Chemistry, 95*, 1241–1244.
100. Khacef, A., Cormier, J. M., & Pouvesle, J. M. (2005). Non-thermal plasma NO_x remediation: From binary gas mixture to lean-burn gasoline and diesel engine exhaust. *Journal of Advanced Oxidation Technologies, 8*, 150–157.
101. Djéga-Mariadassou, G., Baudin, F., Khacef, A., & Da Costa, P. (2012). NO_x abatement by plasma catalysis. In V. I. Parvelescu, M. Magureanu, & P. Lukes (Eds.), *Plasma chemistry and catalysis in gases and liquids* (pp. 89–129). Weinheim: Wiley-VCH Verlag GmbH.
102. Khacef, A., Cormier, J. M., & Pouvesle, J. M. (2006). Energy deposition effect on the NO_x remediation in oxidative media using atmospheric non-thermal plasmas. *European Physical Journal Applied Physics, 33*, 195–198.
103. Filimonova, E. A., & Amirov, R. K. (2001). Simulation of ethylene conversion initiated by a streamer corona in an air flow. *Plasma Physics Reports, 27*, 708–714.
104. Bröer, S., Hammer, T., & Kishimoto, T. (1997, September 8–12) NO-removal in hydrocarbon containing gas mixtures induced by dielectric barrier discharges. Paper presented at the 12th international conference of gas discharges and their applications, Greifswald, Germany, pp 188–192.

105. Kuwahara, T., Yoshida, K., Kannaka, Y., Kuroki, T., & Okubo, M. (2011). Improvement of NO_x reduction efficiency in diesel emission control using non-thermal plasma combined exhaust gas recirculation process. *IEEE Transactions on Industry Applications, 47*, 2359–2366.
106. Okubo, M. (2008). Air and water pollution control technologies using atmospheric pressure low temperature plasma hybrid processes. *Journal of Plasma and Fusion Research, 84*, 121–134.
107. Okubo, M., Arita, N., Kuroki, T., & Yamamoto, T. (2008). Total diesel emission control system using ozone injection and plasma desorption. *Plasma Chemistry and Plasma Processing, 28*, 173–187.
108. Yoshioka, Y., Takahashi, T., Togashi, T., & Shoyama, T. (2007). Efficient NO removal from diesel exhaust gases by a combination of ozone injection and exhaust gas recirculation. *Journal of Advanced Oxidation Technologies, 10*, 304–310.
109. Yoshioka, Y. (2007). Recent development in plasma De-NO_x and PM (particular matter) removal technologies from diesel exhaust gases. *International Journal of Plasma Environmental Science and Technology, 1*, 110–122.
110. Richter, M., Eckelt, R., Parlitz, B., & Fricke, R. (1998). Low-temperature conversion of NO_x to N_2 by zeolite-fixed ammonium ions. *Applied Catalysis B: Environmental, 15*, 129–146.
111. Cho, B. K., Lee, J. H., Crellin, C. C., Olson, K. L., Hilden, D. L., Kim, M. K., Kim, P. S., Heo, I., Oh, S. H., & Nam, I. S. (2012). Selective catalytic reduction of NO_x by diesel fuel: Plasma-assisted HC/SCR system. *Catalysis Today, 191*, 20–24.
112. Gorce, O., Jurado, H., Thomas, C., Djéga-Mariadassou, G., Khacef, A., Cormier, J. M., Pouvesle, J. M., Blanchard, G., Calvo, S., & Lendresse, Y. (2001) Non-thermal plasma assisted catalytic NO_x remediation from a lean model exhaust. SAE Tech Paper 2001-01-3508.
113. Djéga-Mariadassou, G., Berger, M., Gorce, O., Park, J. W., Pernot, H., Potvin, C., Thomas, C., & Da Costa, P. (2007). A three-function model reaction for designing deNO_x catalysts. In P. Granger & V. I. Parvulescu (Eds.), *Past and present in deNO_x catalysis: From molecular modelling to chemical engineering* (pp. 145–173). The Netherlands: Elsevier.
114. Baudin, F. (2005). PhD Thesis, Université Pierre et Marie Curie Paris, France.
115. Schmieg, S. J., Cho, B. K., & Oh, S. H. (2004). Selective catalytic reduction of nitric oxide with acetaldehyde over NaY zeolite catalyst in lean exhaust feed. *Applied Catalysis B: Environmental, 49*, 113–125.
116. Djéga-Mariadassou, G., Fajardie, F., Tempère, J. F., Manoli, J. M., Touret, O., & Blanchard, G. (2000). A general model for both three-way and deNO_x catalysis: Dissociative or associative nitric oxide adsorption, and its assisted decomposition in the presence of a reductant: Part I. Nitric oxide decomposition assisted by CO over reduced or oxidized rhodium species supported on ceria. *Journal of Molecular Catalysis A: Chemical, 161*, 179–189.
117. Kim, M. K., Kim, P. S., Nam, I. S., Cho, B. K., & Oh, S. H. (2012). Enhanced NO_x reduction and byproduct removal by (HC+OHC)/SCR over multifunctional dual-bed monolith catalyst. *Catalysis Today, 184*, 95–106.
118. Lee, D. H., Kim, K. T., Kang, H. S., Song, Y. H., & Park, J. E. (2013). Plasma assisted combustion technology for NO_x reduction. *Environmental Science and Technology, 47*, 10964–10970.

Chapter 6
Plasma-Catalytic Removal of VOCs

Pieter Cools, Nathalie De Geyter, and Rino Morent

6.1 Introduction

6.1.1 Relevance

The aim of this chapter is to discuss the abatement of volatile organic compounds (VOCs) using nonthermal plasma (NTP) in combination with a catalyst. The first experiments, using this technique, were done in the 1950s by Dennis and Burton [1]. These, along with other pioneers like Gicquel et al., laid the foundation for plasma catalysis as it is known today [2–4]. Since the Kyoto protocol in 1997, there has been a growing public awareness and concern around environmental pollution. This has stimulated research groups around the world to further explore the possibilities of NTP catalysis as an eco-friendly technique for the abatement of dilute VOCs present in waste gas streams and indoor air.

NTP catalysis is an attractive technique for controlling air pollution for a number of reasons. It is a quick start-up process; can be operated in ambient conditions, thus avoiding the use of extensive vacuum equipment; and reacts quickly to changes in waste gas composition [5]. NTP without the assistance of a catalyst used to be the more popular technique for VOC abatement, but a review of the papers collected through Web of Science shows that since 2004, plasma catalysis has become the more popular technique, with a steady 15–20 publications a year (Fig. 6.1). In the last 2–3 years, notably more papers have been published, which seems to point toward a positive trend.

P. Cools · N. De Geyter · R. Morent (✉)
Research Unit Plasma Technology, Department of Applied Physics, Ghent University, Sint-Pietersnieuwstraat, Ghent, Belgium
e-mail: Pieter.Cools@Ugent.be; Rino.Morent@UGent.be

X. Tu et al. (eds.), *Plasma Catalysis*, Springer Series on Atomic, Optical, and Plasma Physics 106, https://doi.org/10.1007/978-3-030-05189-1_6

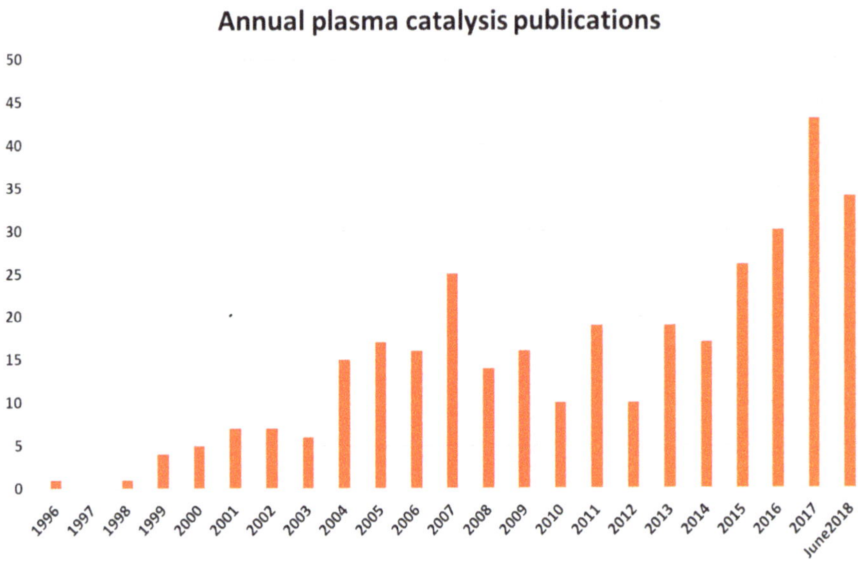

Fig. 6.1 Number of publications on plasma catalysis since 1996

6.1.2 What Is Plasma Catalysis?

The use of NTP as such for the abatement of VOCs, SO_x, NO_x, and odors was, and still is, an attractive technique, but there is a general consensus in the scientific community that there are some drawbacks inherently linked with this approach, i.e., a poor energy efficiency, a low degree of mineralization, and incomplete destruction and oxidation resulting in the production of harmful by-products such as NO_x, CO, and new VOCs.

Two approaches are available for combining NTP with a heterogeneous catalyst: in-plasma catalysis (IPC) and post-plasma catalysis (PPC). The first is a single-step process putting the catalyst in the active plasma zone, while the latter consists of a two-stage process, positioning the catalysts downstream of the active plasma region (see Fig. 6.2). The used nomenclature in literature for IPC and PPC is not unilaterally spread and might cause confusion for readers not experienced in the field. For PPC, one can find terms such as post-plasma catalysis reactor (PPCR) [6], plasma-enhanced catalysis (PEC) [7], or two-stage plasma catalysis (TSPC) [8]. Synonyms of IPC are combined plasma catalysis (CPC) [9–11], in-plasma catalysis reactor [6], plasma-driven catalysis [7], plasma and catalyst integrated technologies (PACT) [12], and single-stage plasma catalysis [8]. Plasma catalysis is generally considered to be a synergetic process, as in some cases the plasma is responsible for the (re) activation of the heterogeneous catalyst. UV, local heating, O_3, activation of lattice oxygen, adsorption/desorption, changes in work function, direct interaction of gas-phase radicals with adsorbed pollutants, and the creation of electron-hole pairs are considered to be the most prominent activation mechanisms [13].

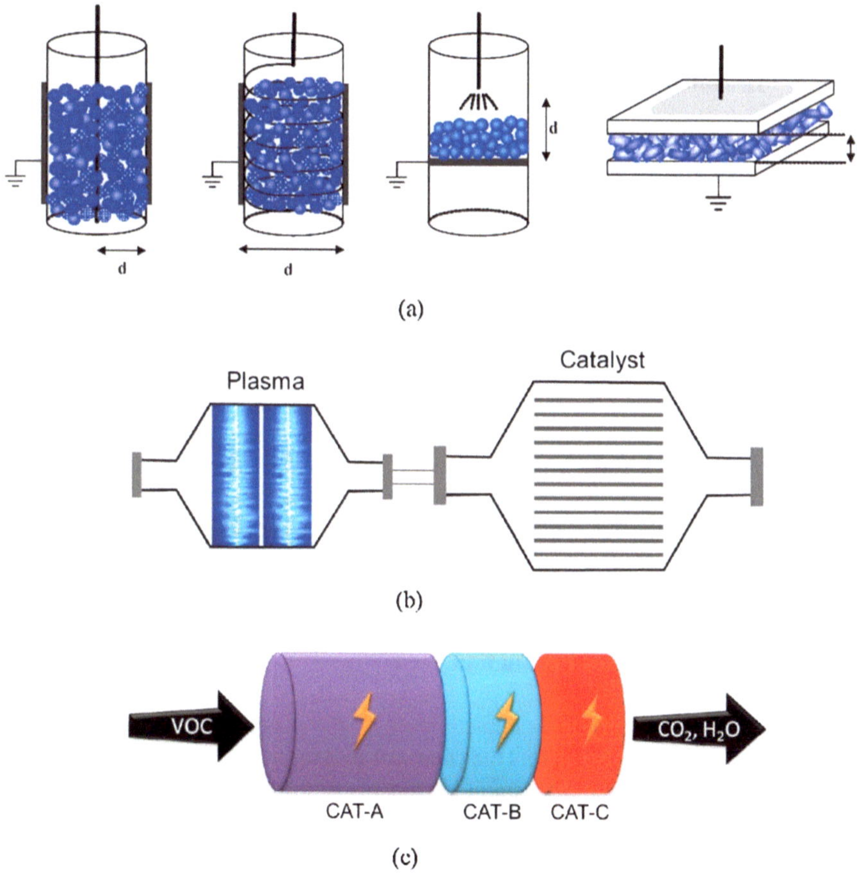

Fig. 6.2 Different configurations for plasma catalysis systems [15]. (**a**) Single-stage plasma catalysis reactor. (**b**) Two-stage plasma catalysis reactor. (**c**) Multi-stage plasma catalysis reactor

The interactions between catalyst and plasma described in the next few paragraphs can all be directly linked to one or more of the activations mechanisms mentioned above. The current experimental results and modeling data are originating from specific working conditions [14]. As such, the collection of data available might appear to be no more than scattered pieces of information. It is therefore indeed correct to assume that a continued effort is required from the plasma community to further unravel the mechanisms leading to those results and fill in the remaining blanks. However, at this stage, it is still of great value to collect and organize an overview on the general trends, dependencies, and limitations for NTP catalysis.

6.1.3 *Influence of the Catalyst on the Plasma Processes*

6.1.3.1 Discharge Mode

The introduction of a heterogeneous catalyst into the discharge zone will affect the physical properties of the discharge. A simple example with complicated consequences is the introduction of a dielectric surface in between the gap of a filamentary discharge. Whereas the discharge mode was originally characterized by bulk streamers, it now converts to a surface flashover system which will be characterized by more intense streamers running along the surface [16]. By changing the catalyst pellet size, it is possible to further distort the electrical field in the voids between the pellets, as the dielectric constant will affect the electric field generated in the voids between the pellets and therefore alter the mean electron energy [17]. By decreasing the pellet size, the number of micro-discharges will increase progressively, but the charge intensity that is transferred per micro-discharge will decrease. The addition of ferroelectric pellets into the discharge is another common practice which, under normal conditions, leads to higher average electron energies, resulting in a more oxidative discharge [18].

6.1.3.2 Reactive Species Production

Besides changing the physical characteristics of the plasma, the introduction of a heterogeneous catalyst also affects the chemical activity of the system. The oxidation of various organic compounds immobilized on (non)-porous silica and alumina catalysts was studied by Roland et al. [19]. The overall conclusion was that the several short-living species are generated in the pore volume of the porous materials when exposed to the plasma. Then again, the opposite effect was observed by other researchers, the catalyst being responsible for a reduction in ionic species [20]. This said, in this case, the reduction of ionic species did not inhibit the catalyst's efficiency in reducing the emission of CO and O_3.

6.1.4 *Influence of the Plasma on the Catalytic Processes*

6.1.4.1 Catalyst Properties

It is possible to use NTPs for catalyst preparation [21–26]. Plasma activation of the catalyst surface improves the catalytic activity and stimulates the dispersion of the catalytic components [27–29]. NTPs are furthermore capable of changing the oxidative state of the catalyst. Mn_2O_3, a commonly used catalyst, can convert into Mn_3O_4 upon prolonged exposure to a DBD plasma [29]. This lower-valent manganese oxide is known for its higher oxidation potential. For titanium dioxide, less Ti-O bonds are registered after long-term exposure to a plasma discharge [30]. Also, more uncommon metastables can be formed on the catalyst surface. Al-O-O* is the best known example of a long-term metastable found within the pores of an aluminum oxide catalyst up to 2 weeks after the IPC run [30]. Finally, plasma exposure is also known to actively change the specific surface area and/or structure of the exposed catalyst [28, 29, 31].

6.1.4.2 Thermal Activation

The heat generated by the plasma during IPC will result in a higher catalyst surface temperature but is as such insufficient to cause thermal activation of the catalyst. However, due to intense micro-discharges running between sharp edges and the pellets, local hot spots can occur in packed bed reactors [32], at those localized points, catalyst temperatures can reach sufficiently high levels to enhance the catalytic breakdown of VOCs [33].

6.1.4.3 Adsorption

The role adsorption processes play during IPC cannot be neglected and might considerably alter the reaction mechanisms. Depending on the adsorption capacity of the catalyst, pollutant retention time can be increased (good adsorption capacity) or reduced (low adsorption capacity). The use of a more porous catalyst normally results in a prolonged residence time of both the VOC and the generated plasma species [34]. If the VOC residence within the plasma is prolonged, the probability of collision with an active species within the plasma increases, enhancing the overall VOC removal efficiency. Besides the adsorption of pollutants, other compounds such as water and ozone might compete for the active sites on the catalyst surface. Humidity is an essential component for both IPC and PPC processes and often causes a reduction in VOC conversion due to a decrease in reaction probability between the catalyst surface and the pollutant [35]. The adsorption of ozone on certain catalysts results in its dissociation into highly reactive atomic oxygen, a process that is considered favorable if the plasma-generated ozone in itself is not able to break down the VOC [36].

Besides sequential adsorption/activation, the cyclic operation of VOC adsorption and NTP-assisted regeneration has received considerable interest in the last few years, as first proposed by Ogata et al. [37, 38]. As discussed in the review paper by Sultana et al., the use of a cyclic adsorption/activation scheme can lead to a more optimized energy consumption and a better cost effectiveness [39]. Furthermore, under certain conditions, it is found to lead to a higher CO_2 selectivity and a reduced production of ozone. Several papers will be discussed further on in the chapter that makes use of cyclic adsorption, which can be recognized by their improved energy density [40–44].

6.1.4.4 Plasma-Mediated Activation of Photocatalysts

Photocatalysis is a process where upon VOC adsorption, the semiconducting catalyst is exposed to a UV source. The UV radiation then generates electron-hole pairs within the semiconducting material, inducing oxidation of the adsorbed pollutant through valence band holes. In a final step, the newly generated oxidized end-products are then resorbed within the gas stream. TiO_2 is by far the most popular photocatalyst, as it is most suitable for the processing of a wide range of VOCs. When combined with an NTP, the conversion efficiency of TiO_2 substantially increases, and a higher selectivity toward CO_2 can be achieved. Although it was initially believed that the UV light generated by the NTP was responsible for the photocatalytic effect, later research has shown that the wavelength shorter than 388 nm is needed to form up an electron-hole pair, as the bandgap for anatase phase TiO_2 is 3.2 eV. Those few excited nitrogen states that do emit radiation beneath the maximum wavelength are not sufficient to explain the synergetic effect experimentally observed by several research groups. Sano et al. performed a series of experiments where the walls of their reactor were coated with a TiO_2 layer to analyze the changes in acetylene conversion efficiency [45]. The UV light emitted by the plasma was adsorbed by the photocatalyst, but the intensity of the radiation was considered too low to initiate photoactivation. Similar experiments performed by other research groups confirmed these experimental findings [46–50]. Lee et al., among others, ran a series of experiments, using plasma-assisted photocatalysis for the abatement of benzene [48]. Besides the addition of O_2, the used dilution gas was changed from N_2 to Ar, resulting in considerably higher abatement efficiencies. These data strongly indicate that the contribution of the plasma generated UV radiation is negligible, as the emission range for excited Ar lies somewhere between 400 and 850 nm. Even though, there are some groups that assigned some of the photocatalytic activation to the UV light alone or in combination with the (re) activation of the catalyst emitted by the plasma [51]. The exact mechanisms driving the plasma-assisted photocatalysis are not known to date. Different theories on how the TiO_2 band gap can be bridged could be formulated, but unfortunately, there is still insufficient information available on the relative importance that ions, electrons, metastables, surface recombination effects, and charging play within the complex mix that forms up NTP.

6.1.5 Different Types of Catalysts

Both for IPC and PPC systems, the heterogeneous catalyst can be introduced in a number of ways. The most common practice is the introduction of pellets (packed bed reactor), but also the use of foams and honeycomb monoliths has been studied [18, 20, 28, 29, 35, 52–61]. Besides filling up the reactor system, it is also possible to use a single layer of catalyst material, coat the reactor walls, or use the electrodes themselves as catalyst material [12, 51, 62–68].

Many different catalysts have been used in IPC and PPC systems for the abatements of VOCs. Porous adsorbents placed within the plasma were the first materials tested, as it was the common belief that the increase in retention time for the pollutants would give cause to a higher interaction probability between the active plasma components and said pollutants [37, 69]. Common examples of such materials were Al_2O_3, molecular sieves, and zeolites [19, 37, 38, 70–75]. Metals such as silver, palladium, nickel, copper, manganese, rhodium, cobalt, platinum, and manganese were incorporated into the porous catalyst support structure to provide for the catalytic activity [11, 13, 31, 33, 36, 72, 76–91].

In more recent years, TiO_2 has been combined with a number of different supports and adsorbents and enriched with metals and metal oxides [13, 36, 49, 50, 73, 92–96]. Common examples of supports include beads, (glass) fibers, activated carbon, silica gel pellets, and nickel foam [47, 48, 56, 63, 97–102].

6.2 Critical Process Parameters

Before giving an in-depth discussion on the abatement of the most common VOCs by plasma catalysis, a number of critical process parameters will be listed that define the initial condition of the gas waste stream fed into the plasma catalysis system. For each parameter, the impact on the abatement efficiency will be reviewed and compared if possible.

6.2.1 Humidity Level

Humidity is an essential parameter within the plasma catalytic process. From an industrial point of view, it is highly relevant, as waste streams usually contain fluctuating concentrations of water vapor. The influence of the vapor on the abatement efficiency is highly dependent on the initial VOC concentration, the type of applied discharge, and the type of VOC targeted.

In the plasma chemical process, water plays a prominent role as it decomposes into the highly active OH and H radicals according the to the following equations:

$$H_2O + e^- \rightarrow OH^\bullet + H^\bullet + e^- \tag{6.1}$$

$$H_2O + N_2\left(A^3\Sigma^+{}_u\right) \rightarrow N_2 + OH^\bullet + H^\bullet \tag{6.2}$$

$$H_2O + O\left(^1D\right) \rightarrow 2\,OH^\bullet \tag{6.3}$$

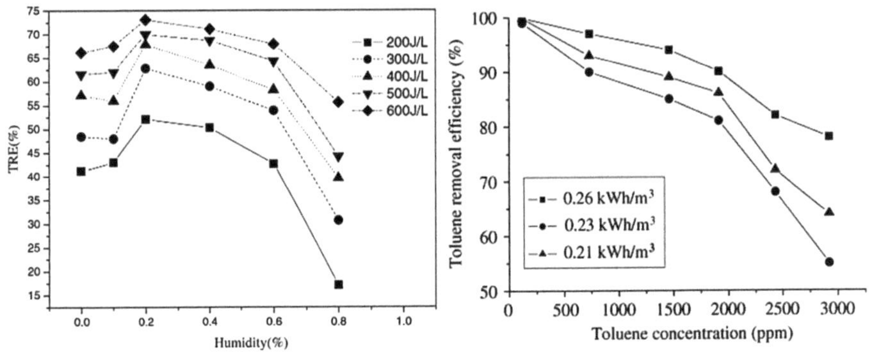

Fig. 6.3 Negative correlation between toluene removal efficiency (TRE) as a function of relative humidity (left) and initial VOC concentration (right) [35, 136]

The oxidative capacity of the OH radical is considerably higher compared to other oxidative species such as atomic oxygen or peroxyl radicals.

The influence of humidity has been mainly investigated on (packed-bed) DBD systems. For such systems, high humidity is known to negatively affect the total charge transfer within a micro-discharge, ultimately lowering the total volume of the active plasma region [103]. For corona discharges, similar findings have been reported [104]. The addition of water vapor to the waste stream reduces the currents observed for a given voltage, which can be assigned to the higher probability of the plasma attachment process [35]. This in turn leads to a lowered OH radical generation. The electronegativity of the water molecules also negatively affects the VOC decomposition, as it lowers the electron density and quenches other reactive chemical species formed within the active plasma region, as shown in Fig. 6.3, left. For a negative corona discharge, Ge et al. found that the optimal relative humidity was between 40 and 60% at room temperature [105].

Independent of the selected VOC, a negative correlation has been found between the presence of water vapor and the characteristics of the plasma discharge. Not all is negative though, as the addition of water does improve the generation of OH radicals for certain VOCs, which is a process directly in competition with the abovementioned observations [106]. Depending on the target VOC, these competing processes either can result in a suppression [102, 107–121] and an enhancement [100, 111, 122–130] or have no effect at all on the abatement efficiency [75, 111, 114, 131–137]. There are several papers available describing the possibility of having an optimal relative humidity for the abatement of a target VOC. For both toluene and TCE, this lies around 20% [35, 104, 138]. Additionally, water is known to compete for the consumption of atomic oxygen, thus lowering the production of O_3 during plasma catalysis [139]. Finally, several papers also describe the positive influence of higher relative humidity on the selectivity toward CO_2 generation at the expense of CO [99, 112, 133, 140].

For systems set up in a post-plasma catalysis configuration, catalytic ozonation will be of minor importance, as the formation of O_3 will be prevented by the water vapor present. Furthermore, the water vapor directly competes for the active adsorption sited of the catalyst with the O_3 molecules and the target VOC, thus lowering the probability of direct VOC/catalyst interaction [36, 57, 104, 141]. This kept in mind, it is of the uttermost importance to select catalysts with those chemical compositions and morphologies that are not greatly influenced by the adsorption of water. Besides competing for the active sites, increased concentrations of water vapor can effectively poison the catalyst, thus progressively lowering the catalyst performance [57, 141].

6.2.2 *Initial VOC Concentration*

The study on the variation of the initial VOC concentration is of great importance, as the VOC concentration in waste streams generated during industrial processes can strongly vary. A higher initial VOC concentration usually means a decrease in the electron density per VOC molecule as well as a reduction of the other active plasma species per VOC molecule. A general consensus is found in literature that a too high initial pollutant concentration is detrimental for the abatement efficiency in plasma-catalytic systems (see Fig. 6.3, right). In literature, several papers can be found that make the correlation between an increase in the characteristic energy (energy needed to abate 63% of the initially fed VOC concentration) and the energy yield as a progressive function of the pollutant concentration [114, 136, 142–147]. For certain halogenated hydrocarbons, the starting concentration only marginally influences the decomposition efficiency [132, 148–152]. This alternative behavior can be explained by a secondary decomposition step, the result of fragment ion and radical interactions, species that were produced in the primary decomposition step [152, 153]. Others claim that for this class of molecules, the primary decomposition by the active plasma species forms the rate-determining step, thus resulting in analog abatement efficiencies regardless of the used starting concentration [148].

6.2.3 *Temperature*

For most plasma-catalytic processes, an increase in temperature usually results in more efficient pollutant decomposition. This trend can be attributed to a higher reaction rate between atomic oxygen/ hydroxyl radicals and the pollutant exposed to them, as the degradation process is usually considered to be endothermic [58, 130, 154–161]. For those processes where electron impact is the predominant form of pollutant decomposition (e.g., CCl_4), rather than radical interactions, no temperature dependency is observed, because the electron density is only marginally dependent of temperature [162, 163]. Still, an increase in temperature can lead to an improved abatement efficiency, as the overall energy density will increase, leading to an improved destruction of the VOC [164].

A second plausible explanation for the improved abatement at elevated temperatures is the increase of the reduced electric field (E/n), being the ratio of the electric field (E) and the gas density (n). The reduced electric field is an essential factor in determining the overall electron energy of the discharge. At a constant pressure, an increase in gas temperature usually leads to a reduction of the gas density; the generated discharge will therefore be generated at an improved reduced electrical field [130, 136]. For post-plasma catalysis systems, the ozone produced by the plasma discharge can be dissociated in the gas phase as described in Eq. 6.4.

$$O_3 + O_2 \rightarrow O + O_2 + O_2 \tag{6.4}$$

At higher temperatures, the reaction rate for the O_3 will profoundly accelerate (e.g., from 100 °C to 300 °C, a five times increase is found). Even though, the lifetime of the atomic oxygen is too short-lived to interact with the adsorbed pollutants at the catalyst surface. Besides a net increase in ozone decomposition, the reaction rate between adsorbed oxygen and pollutant will also be significantly increased. As these two effects are in direct competition, it forms the most plausible explanation for the different influences of temperature found in literature: with a higher temperature setting, the VOC abatement can increase [83, 85, 161, 165], decrease [58], or be independent of temperature variation [72].

6.2.4 Oxygen Content

The addition of oxygen to the waste gas stream influences the character of the generated discharge and takes a crucial role in the plasma chemistry. Small increments of oxygen concentration have a pronounced impact on the production of atomic oxygen, one of the key components for efficient VOC abatement. The addition of oxygen can also negatively impact the process, as its electronegative character can lead to electron attachment reactions, thus limiting the overall electron density and altering the electron energy distribution functions [119, 166]. Furthermore, both oxygen and atomic oxygen are capable of reacting with nitrogen and its excited states, thus competing directly for reactive species otherwise available for VOC abatement [167–169].

Taking the abovementioned characteristics of oxygen into account, it comes as no surprise that it is possible to optimize the oxygen content for the abatement of target VOCs. For most pollutants, the ideal oxygen concentration range lies between 1% and 5%. However, for most industrial processes, it is cumbersome to keep the oxygen level constant, as the waste stream composition is highly dependent on the industrial environment, more often than not consisting of ambient air.

For IPC, the direct interaction between VOCs adsorbed on the surface of the catalyst and the generated oxygen radicals leads to a more pronounced VOC abatement [170]. Furthermore, it is known from literature that the NO_x production is inhibited upon the addition of moderate amounts of oxygen [171].

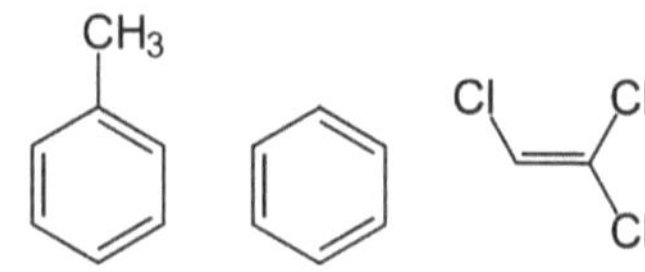

Fig. 6.4 From left to right: structural representation of toluene, benzene, and TCE

6.2.5 *Gas Flow Rate*

In literature, the used gas flow rates typically vary between 0.1 and 10 L/min. Increasing the flow rate results in a reduced VOC residence time within the hybrid catalytic system. This means that the reaction probability between the target VOC and the plasma-generated active species will be reduced, resulting in a lowered VOC abatement efficiency. For the VOC adsorption probability on the catalyst surface, the same argumentation can be followed. With the industrial process in mind, some research groups have tested multistage nonthermal plasma reactors, trying to improve the VOC residence time at higher gas flow rates, although this rarely has a positive impact on the energy efficiency of the VOC abatement process [50, 165, 172–175].

6.3 VOC Abatement

In the following chapter paragraphs, a more detailed overview will be given on the most frequently studied VOCs, i.e., toluene, benzene, and trichloroethylene (TCE) (see Fig. 6.4).

6.3.1 *Toluene*

Toluene, a by-product of the production of gasoline, is a commonly used industrial solvent found in paint thinners and glues and as a precursor for the synthesis of benzene. Compared to other chemicals typically studied in the field of plasma catalysis, it is not a carcinogenic, but exposure to low levels of the compound can lead to confusion, memory loss, hearing loss, nausea, and loss of color vision.

Li et al. used a DC corona discharge in combination with a titanium dioxide photocatalyst [176]. The placement of the catalyst between the needle and mesh electrodes resulted in a more intense plasma discharge caused by a higher streamer repetition rate. This system configuration performed at its best giving an energy efficiency of 7.2 g/kWh and was able to reach 76% abatement. Without the presence of the photocatalyst, both the abatement efficiency (44%) and energy efficiency (3.2 g/kWh) are significantly lower. The increase of both parameters upon addition of the catalyst is attributed to the simultaneous oxidation of toluene in the waste stream and adsorbed on the catalyst surface and the activation of the titanium dioxide by the active plasma species. By operating the system in an intermittent regime, Li et al. claim that the efficiency can be further improved, as the catalyst surface is able to regenerate through surface desorption during the discharge.

A macro-porous γ–Al_2O_3 catalyst packing within a DBD system was compared to a packing of non-adsorbing glass beads by Song et al., and they observed a higher absorbance of toluene at 100 °C [70]. The production of ozone and nitric acid by-products was reduced as well. A paper published on a similar topic by Malik et al. showed that a higher overall surface area of the alumina packing allowed for a reduction of ozone production without influencing the overall abatement efficiency [177].

In 2007, Guo et al. performed a systematic study on IPC, using different catalysts [178]. Among those tested, a MnO_x/Al_2O_3/Ni foam used in an IPC configuration came out as being the most effective [29]. Compared to a plasma alone system, the manganese oxide catalyst significantly enhances the energy yield. Based on an OH-rad detection method developed by the group, they were able to conclude that the improved toluene abatement could be primarily contributed to the efficient reaction process between the OH^* and the toluene adsorbed on the activated catalyst as well as with the other activated species present on the catalyst. An increased humidity resulted in the coverage of the catalysts' active sites, lowering the probability of reaction between catalyst and pollutant [179]. These experimental findings were confirmed by both Van Dormer et al. and Huang et al., as they both showed that water adsorbs onto the catalyst clustering into mono- and/or multilayer systems that protect the active sites from the waste stream [56, 57, 180, 181]. For the toluene to interact with an active site, it now has to diffuse through the watery layer. In the 2011 paper by Huang et al., a wire-plate DBD IPC system is filled with the aforementioned manganese oxide catalyst [182]. A systematic investigation was performed on the effect of humidity on the overall abatement efficiency of toluene, the generated CO_2 selectivity, carbon balance, and O_3 concentration. The addition of progressive amounts of water to the waste stream leads to an increasing reduction in O_3 formation, as the high-energetic electrons are quenched by the H_2O. Furthermore, there is active competition between the adsorption of water and O_3 on the catalyst's active sites. Optimized CO_2 selectivity and carbon balance were reached for relative humidity between 25 and 75%. For a NiO-enriched alumina catalyst, Wu et al. derived the desorption activation energies of H_2O and toluene and found that the adsorption of toluene is considerably weaker compared to H_2O [183].

In 2016, Qin et al. published two papers on the use of Ag/Mn type of catalyst deposited on an alumina support and compared the abatement efficiency and CO_x selectivity to the use of Ag or Mn as such [42, 184]. Both parameters were found to be substantially enhanced for the combined catalyst, which, according to the authors, could be attributed to the combination of a better toluene adsorption on Ag^+, while the metallic silver was held responsible for assisting the manganese oxide in decomposing the adsorbed VOC. Furthermore, it was shown that the impregnation sequence of either catalyst plays an important role on the type of by-products formed.

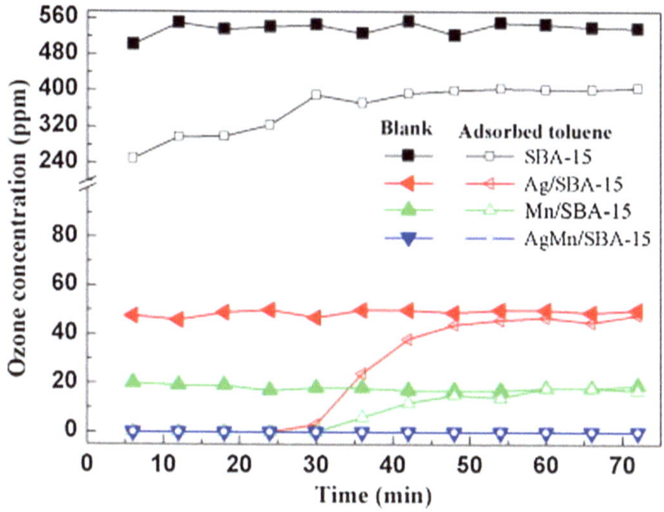

Fig. 6.5 Ozone concentration for different catalyst compositions [186]

Xu et al. performed similar studies but focused on the use of SBA-15 as support instead of the alumina [185, 186]. They also link the improvement in abatement efficiency and carbon balance to the π-complexation between the silver ions and the toluene double bonds (Fig. 6.5). Besides the variation in catalyst composition, also the feeding method was changed from a continuous mode to an on/off system, resulting in an alternated adsorption/decomposition mode, which lead to a more efficient abatement of the target VOC. FT-IR and GC-MS analysis showed that the generation of other carbonaceous intermediates forms the basis for this altered behavior.

For PPC systems, similar conclusions were found: when the relative humidity is increased, the abatement efficiency of the PPC system decreases [36]. For a palladium-enriched alumina oxide, the decomposition efficiency was found to be more than 90% for intermediate relative humidity. For extremely dry or humid air, the conversion efficiency drops to 37%. The lowered abatement efficiency is mostly contributed to changes in Van der Waals interactions.

The use of $MnPO_4$, Mn–SAPO–11, and Mn–APO–5 as catalysts for a system in a PPC configuration at 400 °C was studied by Magureanu et al. [72]. Without the application of a plasma discharge, no catalytic activity was noted for the system, which stands in sharp contrast to the hybrid system used, where a pronounced synergetic effect was observed. Although not tested, the authors expect that converting to an IPC system would further increase the abatement efficiency, as the active species generated in the discharge would, under normal conditions, lead to an increased oxidation efficiency of the catalyst surface.

Huang et al. also studied a PPC system, using a TiO_2/Al_2O_3/Ni foam as catalyst [102]. Different active species were followed during the runs, and the catalytic formation of O_3 was found to be the principal process for toluene abatement, activated oxygen being the predominant species interacting with the toluene, as described in paper [187]. Besides CO_2, formic acid, benzaldehyde, benzene, acetic acid, and 2-methylamylene were found within the waste stream leaving the catalytic system.

Others also studied the use of manganese oxide and cobalt oxide-coated metal fibers (MF), incorporated as the inner electrode, for the abatement of toluene (100 ppm) [67]. On bare MF, a 50% CO_2 selectivity was noted at an energy density of 235 J/L. Upon the addition of the catalyst, the selectivity increased to 80%, and a near to 100% decomposition efficiency could be reached. After a 3-hour run, the reactor was scanned for polymeric deposits, but those were not found, indicating that the electrodes do not lose any activity during toluene abatement.

The literature on toluene abatement is extensive, and not all papers found were included in the above discussion. The use of other catalysts (combinations) and supports such as Fe_2O_3, MnO_x, CoO_x, Cu, Ti, $BaTiO_3$, TiO_2, AgO, Ag, Pt, Ni, V, cordierit, zeolites, and honeycombs was also mentioned in literature. Abatement efficiencies were situated between 45 and 95%, depending on the plasma configuration (PPC vs IPC) and energy density used (2.5–1000 J/L) [20, 40, 41, 58, 60, 67, 68, 75, 137, 188–200].

6.3.2 *Benzene*

Benzene, one of the most common derivates from crude oil, is produced in mass both for its use as an industrial solvent as well as the starting material for a broad spectrum of more complex organic molecules. Furthermore, thanks to its high octane number, it is an important compound for gasoline. Despite its widespread use, benzene is a notorious carcinogenic directly linked to leukemia and bone marrow failure, and already since 1948 the safe work concentration has been set to 0.

Several NTP reactor configurations and catalyst compositions for the abatement of benzene have been tested by Kim et al. The ferroelectric material used in a barium titanate packed-bed reactor was swapped for titanium dioxide pellets as such or in combination with silver or platinum [201]. Temperature for the oven encapsulating the NTP system was set at 100 °C. The catalytic activity for the different catalyst combinations was found to be $Ag/TiO_2 > TiO_2 > Pt/TiO_2$. CO_2 selectivity was also found to be highest for the Ag containing catalyst (+15%) compared to the packed-bed barium titanate reactor. Analysis of the by-products revealed only the presence of CO and CO_2, which was further confirmed by the carbon balance. Varying the catalyst concentration and/or the pollutant residence time had no discernable effects besides a slight reduction in N_2O formation, whereas the adjustment of the energy density had a more pronounced effect on the abatement efficiency [189, 202]. For a sufficiently low energy density, small amounts of formic acid were found as by-product.

In a third study, the effect of silver concentration on the benzene oxidation efficiency was analyzed [203]. An increase of the metal additive did not influence the pollutant breakdown efficiency as such, but it is a crucial factor in the oxidation of intermediates adsorbing on the TiO_2 surface, as indicated by the CO_2 selectivity and carbon balance. Yet another study focused on unraveling the activation mechanisms for the Ag/TiO_2 catalyst [49]. As mentioned earlier in the section on photocatalysis, the effect on the abatement efficiency of the UV light generated by the plasma is negligible. Also variations in temperature and dilution gasses (N_2, Ar) did not give cause to any notable changes. The authors therefore concluded that the in situ decomposition of O_3 by the Ag-enriched photocatalyst at higher energy density is the main factor influencing the oxidation of benzene. Kinetic studies showed a zero-order relation to benzene concentration, further validating these conclusions. Long-duration testing revealed that the catalyst can withstand deactivation for runs up to 150 h [189]. In a final study, other types of catalysts (γ-Al_2O_3, zeolites, TiO_2) were tested in combination with a cycled system of adsorption and oxygen plasma, similar as what was done by Song et al. and Ogata et al. [13, 37, 70]. The progressive oxidation of benzene was studied for a systematic variation of the oxygen partial pressure between 0 and 80%. As to be expected, a higher oxygen partial pressure greatly enhances the abatement efficiency as well as the CO_2 selectivity, regardless of the applied catalyst. To prevent the formation of NO_x during the cycled plasma catalysis, regeneration has to occur with pure oxygen. A likely reaction mechanism, as proposed by Kim et al., suggests that the decomposition of such target VOCs primarily occurs through the surface interactions with the catalysts (see Fig. 6.6) [204].

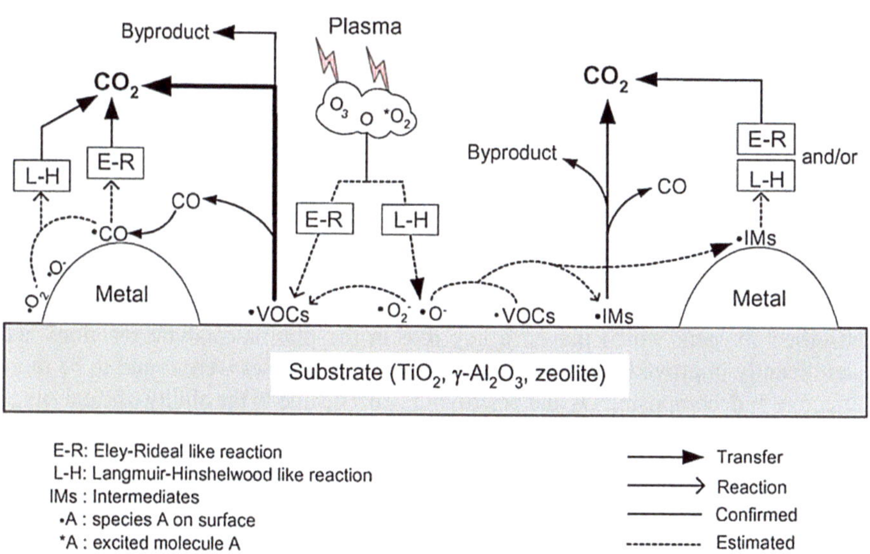

Fig. 6.6 Plausible mechanism for IPC for VOCs on various catalysts [13]

Ogata et al. used a hybrid packed-bed reactor filled with a mixture of aluminum oxide and barium titanate and compared the benzene decomposition efficiency to a barium titanate reactor as such and a barium titanate packed-bed reactor with the aluminum oxide position downstream of the IPC reactor [37]. For all the important process parameters (energy efficiency, N_2O suppression, CO_2 selectivity), the hybrid system gave the best results. The adsorption of benzene on the aluminum oxide combined with subsequent surface decomposition and gas phase reaction is believed to be, according to the authors, the primordially mechanism for the stimulated destruction efficiency of benzene. When applied in a cyclic process of adsorption and plasma activation, a further reduction in energy consumption could be obtained. In a follow-up study, the barium titanate was replaced with different metal supports including Co, Cu, Ag, and Ni, giving cause to a minor increase in CO_x selectivity and N_2O suppression [71]. Another study from the same group describes the use of a zeolite combined with barium titanate in an IPC configuration for the abatement of diluted waste streams containing benzene [38]. An improved decomposition efficiency and CO_x selectivity were reported and linked to the higher adsorption capacity of the zeolite. Surprisingly, the decomposition efficiency of the benzene molecules adsorbed on the outer surface of the zeolite was higher compared to those adsorbing inside of the porous material. In 2003, Ogata et al. continued expanding the study by analyzing the influences of barium titanate pellet size and the effects of changing mixing ratios between the titanate and the catalyst, adsorbent, or zeolite [81]. With the exception of pellet sizes larger than 2 mm, which led to sparking, the plasma energy was not influenced by pellet size. Overall, it was found that a larger pellet size leads to a higher energy density near the contact points of the pellets. These results point out the need for hybrid techniques to optimize the catalytic process.

In a more recent paper, Fan et al. performed a similar study, using Ag/HZSM-5, a metal supported zeolite as the catalytic component for a cycled IPC system [205]. Results showed excellent oxidation of the benzene at moderate discharge powers and at a low energy consumption (3.7×10^{-3} kWh/m^3). Long-term experiments showed a negligible deactivation of the catalyst.

Jiang et al. studied the use of a Ag/Ce oxide catalyst in different reactor configurations, as shown in Fig. 6.7 [206]. Surprisingly, a PPC configuration showed a higher abatement efficiency and improved CO_2 selectivity of benzene compared to an IPC configuration. The combination of Ag and Ce species with certain proportion facilitated the surface lattice of the catalyst and increased the formation of surface adsorbed oxygen, which played a key role in the plasma-catalytic reactions and significantly improved benzene degradation. PPC processes were found to be more effective at decomposing O_3 and destroying benzene due to the ability of catalysts to adequately decompose NTP generated ozone, especially when the catalysts were packed downstream.

The use of photocatalysis for benzene abatement is also a popular route of investigation [48]. Glass beads were coated with variations of γ-Al_2O_3, examining the effects of pore volume, surface area, and pore diameter. The use of highly porous alumina had a pronounced positive effect on the benzene decomposition and the degree of mineralization.

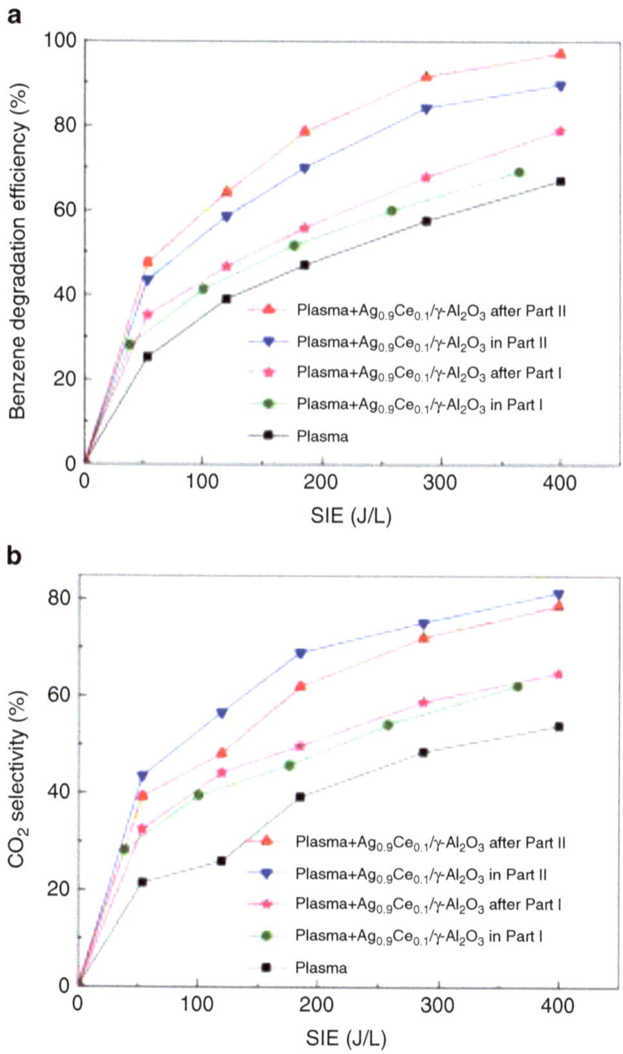

Fig. 6.7 Effect of plasma catalysis configurations on (**a**) benzene degradation efficiency and (**b**) CO_2 selectivity as a function of *SIE*. Condition: $Ag_{0.9}Ce_{0.1}/\gamma\text{-}Al_2O_3$ was used as the catalyst; gas relative humidity was 0% [206]

Others used different sheet type catalysts (TiO_2, V_2O_5/TiO_2, and Pt/TiO_2) deposited on the DBD layer of the electrodes. The efficiency in benzene abatement was as followed: $TiO_2 < Pt/TiO_2 < V_2O_5/TiO_2$ [207]. Both an improvement in mineralization degree and a strong reduction in N_2O production were noted and this for all catalysts used. The mechanism driving the benzene decomposition can, according to the authors, be attributed to the combination of UV in combination with high energy electrons, both generated by the plasma discharge.

Futamura et al. filled a DBD system with a combination of manganese dioxide, titanium dioxide, and a titanium dioxide-silica gel [63]. The manganese dioxide, responsible for the formation of active oxygen out of adsorbed O_3, would be the main driving force for the oxidative destruction of benzene. Hu et al. combined the use of TiO_2 and MnO_x catalysts on a zeolite support in an IPC configuration [208]. Compared to the use of either the titanium oxide or the manganese oxide, a higher abatement efficiency of the target VOC as well as an improved CO_x selectivity could be achieved. The authors attribute this synergetic effect to the charge transfer between Ti^{4+} and Mn^{4+}, which resulted in an effective separation of photogenerated electrons and holes. This process contributed to the production of hydroxyl radicals, an essential component for the oxidative decomposition of benzene. Zhu et al. examined the influence of humidity on the abatement of benzene, using an IPC system filled with titanium dioxide-coated Raschig rings [139]. As was to be expected, an increased humidity reduced the decomposition efficiency. The influence of temperature and catalyst positioning (PPC vs IPC) were studied by Harling et al., using an γ-Al_2O_3 support combined with silver [165]. In comparison with thermal catalysis, a significant increase in NO_x concentration could be detected, and a progressive increase of the greenhouse gas was found at higher temperatures. For both the IPC and PPC configurations, a more complete oxidation of benzene was found at lower temperatures.

Other types of catalysts used in an IPC or PPC system configuration include, but not limited to, Kr/I_2, silica gel, and TiO_2 sol-gels in combination with Pt, Ni, Pd, and ferrierite. Most of which were used in combination with a support structure such as barium titanate and titanium dioxide of aluminum oxide [13, 16, 40, 110, 122, 172, 201, 209–214].

6.3.3 *Trichloroethylene*

TCE is a commonly used solvent in industry but is considered highly controversial, as it is a known carcinogenic and nonhuman health hazard. Therefore, its abatement is one of the most studied processes in IPC and PPC.

Oda et al. studied the use of manganese oxide in a PPC configuration [96]. The effect of direct processing of the contaminated waste stream was compared to indirect processing, where plasma-treated clean air was mixed with the waste stream. Results show that manganese oxide is sufficiently effective to improve the abatement efficiency in both cases, as it is a known ozone dissociator, generating in both cases a sufficiently high concentration of oxygen radicals to efficiently break down the TCE molecular structure. Also, Han et al. studied the synergetic effect of PPC and MnO_2 on the direct and indirect processing of TCE-containing waste streams [215]. The main component generated during the direct processing of the waste stream is dichloroacetylchloride (DCAC), a result from the collisions between plasma species and O_2. Compared to the indirect process, the degradation efficiency is higher due to the oxidation of the rest-TCE into tricholoracetaldehyde by atomic oxygen generated from the dissociation of ozone at the MnO_2 catalyst surface. When increasing the energy density from 120 J/L to 400 J/L, the CO_x yield was further increased from 35% to 98%.

Oda et al. also studied IPC, using the photocatalytic TiO_2 [216]. The effect of initial TCE concentration, sintering temperature for TiO_2, and pellet size on the TCE abatement performance were investigated. When the sintering temperature for TiO_2 was changed from 1100 °C to 400 °C, the breakdown voltage needed to generate the NTP was significantly lowered. The authors suggested that the electric field was disturbed by the non-uniformity of the disk-like pellets, resulting in a more concentrated electric field at the contact points of the pellets. This gave cause to the formation of contact point discharges, which decreased the breakdown voltage and thus improved the overall energy efficiency of the system. A decrease of the pellet size affected the gas flow distribution, resulting in an incomplete filling of the active area with plasma.

Magureanu et al. used a DBD system which consisted of an inner electrode fabricated from sintered MF and coated with different transition metal oxides [64]. CO_2 and CO selectivity were studied for a range of energy densities and they found that the MF as such reached a CO_2 selectivity of 25%. Though, when combined with a manganese oxide coating, selectivity increased to 60%. Unfortunately, the use of manganese oxide in combination with the MF did not decrease the selectivity toward CO. The explanation as to why this change in selectivity could be observed was again assigned to the production of atomic oxygen from ozone adsorbed on MnO_2. The active oxygen is considered to be responsible for the enhanced oxidation of TCE, thus leading to a higher selectivity toward CO_2 [51, 64, 68]. X-Ray photoelectron spectroscopy (XPS) analysis of the catalyst surface after running the system showed that both Fe and Mn did not change their oxidation state. In terms of elemental composition, an enrichment of Fe was found on the catalyst surface, indication of an active reorganization of the Mn on the surface during the process run. Besides the enrichment of Fe, also small traces of Cl were detected, by-products of the TCE abatement, that bonded onto the catalyst surface.

A study performed within the same research group focused on the use of SBA-15 embedded with gold nanoparticles for PPC [217]. Gold concentration was varied systematically, and the effect on CO_2 selectivity was analyzed. Surprisingly, the lowest concentration of gold (0.5 wt%) resulted in the best selectivity. Similar to the manganese oxide, the SBA-Au system is able to generate oxygen radicals out of dissociated ozone, which in turn effectively decomposes the TCE. The authors suspected that the isolated Au^+ acts as the active catalytic site.

Subrahmanyam et al. looked for the existence of activated nitrogen species within their DBD discharge, using UV-VIS spectroscopy [51]. Measurements taken in the 250–500 nm wavelength region showed the emission of excited nitrogen, within the desired UV range of their used photocatalyst. Therefore, the authors suggest that the increased activity of the photocatalyst can be attributed to the presence of the UV light generated by the plasma as well as the (re)activation of the catalyst by the various active species present within the discharge zone.

At Ghent University, Morent et al. reached a more complete oxidation at a reduced energy cost using an IPC system filled with cylindrical TiO_2 pellets, which they assigned to the efficient adsorption of TCE onto the catalyst surface, thus considerably increasing the residence time within the discharge zone [52]. From the same group, Vandenbroucke et al. performed several studies using a PPC system driven by a DC glow discharge combined with either a Pd/γ–Al_2O_3 or a Pd/$LaMnO_3$ catalyst located in a downstream oven [82, 218]. At 100 °C, the PPC system could be operated in a synergetic regime, resulting in an increase in abatement efficiency of 12–22% more TCE compared to multiplication of the theoretically expected decomposition ratios [153]. Dinh et al. used the same PPC configuration but combined it with a manganese oxide catalyst. Improved CO_x yields were detected for the abatement of TCE compared to the plasma alone system [219]. This result was mainly attributed to the capacity of the catalyst to convert the ozone generated by the plasma into atomic oxygen, thus resulting in the complete oxidation of the target VOC. In a similar study, MnO_x was combined with Ce, resulting in a lowered poisoning of the catalyst by the chlorinated by-products compared to the other catalysts used for this setup [220].

Vandenbroucke et al. have also defined a what they call a "synergy factor" in their work on plasma–catalytic decomposition of TCE with a PPC reactor [221]. This synergy factor equals to the ratio of the degree of TCE destruction by plasma catalysis to the sum of the degree of TCE destruction by plasma alone and the degree of TCE destruction by catalyst alone. A synergistic system is a system with a synergy factor greater than one. In their work, this factor ranges from 0.78 to 4.78 by varying the operating parameters indicating that synergy is only observed under specific conditions. According to Whitehead, this measure deserves adoption as it is an unambiguous indication of synergy [222].

Besides the use of catalysts discussed here, a number of other catalysts have been combined with NTP (DBD and DC) for the abatement of TCE. These include, but are not limited to, cobalt oxide, vanadium oxide, tungsten oxide, palladium, Cu–ZSM–5, and $LaMnO_{3+\delta}$ [68, 82, 92, 96, 127, 153, 221, 223]. For most, the destruction efficiency of TCE is above 95%, but the energy efficiency varies between 50 and 670 J/L. For a more elaborate review on the abatement of TCE, using IPC and PPC in combination with the previously mentioned catalyst, the reader is referred to the review by Vandenbroucke et al. [224].

6.3.4 Other VOCs

Besides toluene, benzene, and TCE, a number of other VOCs have been studied as models for IPC and PPC systems. Although they are not the focus of this chapter, they are listed in Table 6.1, thus giving the reader a complete overview on the available literature for plasma catalysis.

Table 6.1 Published papers on removal of other VOCs with plasma catalysis

VOC	Reference
Methane	[16, 79, 125, 225–229]
Ethylene	[230, 231]
Propane	[70, 83, 87]
Propene	[83]
Acetylene	[34, 99, 101, 232, 233]
Cyclohexane	[234]
Styrene	[84, 170, 189, 235–237]
Xylene	[43, 95, 189, 238–244]
Methanol	[245, 246]
Ethanol	[247]
Acetaldehyde	[45, 248–254]
Buteraldehyde	[255]
Formaldehyde	[44, 123, 171, 256–258]
Diethylether	[259]
Acetone	[18, 94, 97, 131, 247, 260–264]
Dichloromethane	[31, 77, 265–267]
Tetrachloromethane	[80, 164, 267]
Chlorobenzene	[268]

6.4 Conclusion and Outlook

In this chapter, a thorough overview has been given on the abatement of some of the most common VOCs through plasma catalysis. As mentioned in the introduction, there still remain a lot of unknowns for the exact mechanisms driving the VOC decomposition. Yet, depending on the target VOC, certain trends were found for the influence of relative humidity, temperature and initial concentration on the decomposition efficiency, and CO_x selectivity. Also, for some catalysts or catalyst combinations, the principal decomposition mechanisms could be identified, (mostly) independent of the target VOC. Still, more fundamental research is required to identify all essential interactions between the mixture of active plasma species, the (adsorbed) VOC, and the catalyst and their relative importance. For the near future, a shift toward the abatement of VOC mixtures will become of greater importance if plasma-assisted catalysis wants to become a valid alternative for the current industrial solutions.

References

1. Devins, J. C., & Burton, M. (1954). Formation of Hydrazine in electric discharge decomposition of Ammonia1, 2. *Journal of the American Chemical Society, 76*, 2618–2626.
2. Gicquel, A., Cavadias, S., & Amouroux, J. (1986). Heterogeneous catalysis in low-pressure plasmas. *Journal of Physics D: Applied Physics, 19*, 2013–2029.
3. Venugopalan, M., & Vepřek, S. (1983). Kinetics and catalysis in plasma chemistry. In *Plasma chemistry IV* (pp. 1–58). New York: Springer-Verlag.
4. Badyal, J. (1996). Catalysis and plasma chemistry at solid surfaces. *Topics in Catalysis, 3*, 255–264.
5. Vandenbroucke, A. M., Morent, R., De Geyter, N., & Leys, C. (2011). Non-thermal plasmas for non-catalytic and catalytic VOC abatement. *Journal of Hazardous Materials, 195*, 30–54.
6. Holzer, F., Roland, U., & Kopinke, F.-D. (2002). Combination of non-thermal plasma and heterogeneous catalysis for oxidation of volatile organic compounds: Part 1. Accessibility of the intra-particle volume. *Applied Catalysis B: Environmental, 38*, 163–181.
7. Kim, H. H. (2004). Nonthermal plasma processing for air-pollution control: A historical review, current issues, and future prospects. *Plasma Processes and Polymers, 1*, 91–110.
8. Chen, H. L., Lee, H. M., Chen, S. H., Chang, M. B., Yu, S. J., & Li, S. N. (2009). Removal of volatile organic compounds by single-stage and two-stage plasma catalysis systems: A review of the performance enhancement mechanisms, current status, and suitable applications. *Environmental Science & Technology, 43*, 2216–2227.
9. Yu, S. J., & Chang, M. B. (2001). Oxidative conversion of PFC via plasma processing with dielectric barrier discharges. *Plasma Chemistry and Plasma Processing, 21*, 311–327.
10. Chang, M. B., & Huang, C. P. (1999). Oxidative conversion of methane via plasma processing with dielectric barrier discharges. *Journal of Advanced Oxidation Technologies, 4*, 333–338.
11. Chang, M. B., & Lee, H. M. (2004). Abatement of perfluorocarbons with combined plasma catalysis in atmospheric-pressure environment. *Catalysis Today, 89*, 109–115.
12. Chen, X., Rozak, J., Lin, J.-C., Suib, S. L., Hayashi, Y., & Matsumoto, H. (2001). Oxidative decomposition of chlorinated hydrocarbons by glow discharge in PACT (plasma and catalyst integrated technologies) reactors. *Applied Catalysis A: General, 219*, 25–31.
13. Kim, H.-H., Ogata, A., & Futamura, S. (2008). Oxygen partial pressure-dependent behavior of various catalysts for the total oxidation of VOCs using cycled system of adsorption and oxygen plasma. *Applied Catalysis B: Environmental, 79*, 356–367.
14. Neyts, E., & Bogaerts, A. (2014). Understanding plasma catalysis through modelling and simulation—A review. *Journal of Physics D: Applied Physics, 47*, 224010.
15. Kim, H.-H., Teramoto, Y., Negishi, N., & Ogata, A. (2015). A multidisciplinary approach to understand the interactions of nonthermal plasma and catalyst: A review. *Catalysis Today, 256, Part 1*, 13–22.
16. Malik, M. A., Minamitani, Y., & Schoenbach, K. H. (2005). Comparison of catalytic activity of aluminum oxide and silica gel for decomposition of volatile organic compounds (VOCs) in a plasmacatalytic reactor. *IEEE Transactions on Plasma Science, 33*, 50–56.
17. Chen, H. L., Lee, H. M., Chen, S. H., & Chang, M. B. (2008). Review of packed-bed plasma reactor for ozone generation and air pollution control. *Industrial & Engineering Chemistry Research, 47*, 2122–2130.
18. Holzer, F., Kopinke, F., & Roland, U. (2005). Influence of ferroelectric materials and catalysts on the performance of non-thermal plasma (NTP) for the removal of air pollutants. *Plasma Chemistry and Plasma Processing, 25*, 595–611.
19. Roland, U., Holzer, F., & Kopinke, F.-D. (2005). Combination of non-thermal plasma and heterogeneous catalysis for oxidation of volatile organic compounds: Part 2. Ozone decomposition and deactivation of γ-Al_2O_3. *Applied Catalysis B: Environmental, 58*, 217–226.
20. Chae, J. O., Demidiouk, V., Yeulash, M., Choi, I. C., & Jung, T. G. (2004). Experimental study for indoor air control by plasma-catalyst hybrid system. *IEEE Transactions on Plasma Science, 32*, 493–497.

21. Zhu, Y.-R., Li, Z.-H., Zhou, Y.-H., Lv, J., & Wang, H.-T. (2005). Plasma treatment of Ni and Pt catalysts for partial oxidation of methane. *Reaction Kinetics and Catalysis Letters, 87*, 33–41.
22. Liu, C.-J., Zou, J., Yu, K., Cheng, D., Han, Y., Zhan, J., Ratanatawanate, C., & Jang, B. W.-L. (2006). Plasma application for more environmentally friendly catalyst preparation. *Pure and Applied Chemistry, 78*, 1227–1238.
23. Li, Z.-H., Tian, S.-X., Wang, H.-T., & Tian, H.-B. (2004). Plasma treatment of Ni catalyst via a corona discharge. *Journal of Molecular Catalysis A: Chemical, 211*, 149–153.
24. Ratanatawanate, C., Macias, M., & Jang, B. W.-L. (2005). Promotion effect of the nonthermal RF plasma treatment on Ni/Al_2O_3 for benzene hydrogenation. *Industrial & Engineering Chemistry Research, 44*, 9868–9874.
25. Zhu, X., Huo, P.-P., Zhang, Y.-P., & Liu, C.-J. (2006). Characterization of argon glow discharge plasma reduced Pt/Al_2O_3 catalyst. *Industrial & Engineering Chemistry Research, 45*, 8604–8609.
26. Wang, J.-G., Liu, C.-J., Zhang, Y.-P., Yu, K.-L., Zhu, X.-L., & He, F. (2004). Partial oxidation of methane to syngas over glow discharge plasma treated Ni–Fe/Al_2O_3 catalyst. *Catalysis Today, 89*, 183–191.
27. Zhang, Y.-P., Ma, P.-S., Zhu, X., Liu, C.-J., & Shen, Y. (2004). A novel plasma-treated Pt/NaZSM-5 catalyst for NO reduction by methane. *Catalysis Communications, 5*, 35–39.
28. Guo, Y.-F., Ye, D.-Q., Chen, K.-F., He, J.-C., & Chen, W.-L. (2006). Toluene decomposition using a wire-plate dielectric barrier discharge reactor with manganese oxide catalyst in situ. *Journal of Molecular Catalysis A: Chemical, 245*, 93–100.
29. Guo, Y.-F., Ye, D.-Q., Chen, K.-F., & He, J.-C. (2007). Toluene removal by a DBD-type plasma combined with metal oxides catalysts supported by nickel foam. *Catalysis Today, 126*, 328–337.
30. Pribytkov, A., Baeva, G., Telegina, N., Tarasov, A., Stakheev, A. Y., Tel'nov, A., & Golubeva, V. (2006). Effect of electron irradiation on the catalytic properties of supported Pd catalysts. *Kinetics and Catalysis, 47*, 765–769.
31. Wallis, A. E., Whitehead, J. C., & Zhang, K. (2007). The removal of dichloromethane from atmospheric pressure nitrogen gas streams using plasma-assisted catalysis. *Applied Catalysis B: Environmental, 74*, 111–116.
32. Lu, B., Zhang, X., Yu, X., Feng, T., & Yao, S. (2006). Catalytic oxidation of benzene using DBD corona discharges. *Journal of Hazardous Materials, 137*, 633–637.
33. Kim, H.-H., Ogata, A., & Futamura, S. (2006). Effect of different catalysts on the decomposition of VOCs using flow-type plasma-driven catalysis. *IEEE Transactions on Plasma Science, 34*, 984–995.
34. Rousseau, A., Guaitella, O., Röpcke, J., Gatilova, L., & Tolmachev, Y. (2004). Combination of a pulsed microwave plasma with a catalyst for acetylene oxidation. *Applied Physics Letters, 85*, 2199–2201.
35. Guo, Y. F., Ye, D. Q., Tian, Y. F., & Chen, K. F. (2006). Humidity effect on toluene decomposition in a wire-plate dielectric barrier discharge reactor. *Plasma Chemistry and Plasma Processing, 26*, 237–249.
36. Van Durme, J., Dewulf, J., Demeestere, K., Leys, C., & Van Langenhove, H. (2009). Post-plasma catalytic technology for the removal of toluene from indoor air: Effect of humidity. *Applied Catalysis B: Environmental, 87*, 78–83.
37. Ogata, A., Yamanouchi, K., Mizuno, K., Kushiyama, S., & Yamamoto, T. (1999). Oxidation of dilute benzene in an alumina hybrid plasma reactor at atmospheric pressure. *Plasma Chemistry and Plasma Processing, 19*, 383–394.
38. Ogata, A., Ito, D., Mizuno, K., Kushiyama, S., & Yamamoto, T. (2001). Removal of dilute benzene using a zeolite-hybrid plasma reactor. *IEEE Transactions on Industry Applications, 37*, 959–964.

39. Sultana, S., Vandenbroucke, A. M., Leys, C., De Geyter, N., & Morent, R. (2015). Abatement of VOCs with alternate adsorption and plasma-assisted regeneration: A review. *Catalysts, 5*, 718–746.
40. Dang, X. Q., Qin, C. H., Huang, J. Y., Teng, J. J., & Huang, X. M. (2016). Adsorbed benzene/toluene oxidation using plasma driven catalysis with gas circulation: Elimination of the byproducts. *Journal of Industrial and Engineering Chemistry, 37*, 366–371.
41. Huang, J. Y., Dang, X. Q., Qin, C. H., Shu, Y., Wang, H. C., & Zhang, F. (2016). Toluene decomposition using adsorption combined with plasma-driven catalysis with gas circulation. *Environmental Progress & Sustainable Energy, 35*, 386–394.
42. Qin, C. H., Dang, X. Q., Huang, J. Y., Teng, J. J., & Huang, X. M. (2016). Plasma-catalytic oxidation of adsorbed toluene on Ag-Mn/gamma-Al_2O_3: Comparison of gas flow-through and gas circulation treatment. *Chemical Engineering Journal, 299*, 85–92.
43. Kuroki, T., Hirai, K., Kawabata, R., Okubo, M., & Yamamoto, T. (2010). Decomposition of adsorbed xylene on adsorbents using nonthermal plasma with gas circulation. *IEEE Transactions on Industry Applications, 46*, 672–679.
44. Zhao, D.-Z., Li, X.-S., Shi, C., Fan, H.-Y., & Zhu, A.-M. (2011). Low-concentration formaldehyde removal from air using a cycled storage–discharge (CSD) plasma catalytic process. *Chemical Engineering Science, 66*, 3922–3929.
45. Sano, T., Negishi, N., Sakai, E., & Matsuzawa, S. (2006). Contributions of photocatalytic/catalytic activities of TiO_2 and γ-Al_2O_3 in nonthermal plasma on oxidation of acetaldehyde and CO. *Journal of Molecular Catalysis A: Chemical, 245*, 235–241.
46. Kang, M., Ko, Y.-R., Jeon, M.-K., Lee, S.-C., Choung, S.-J., Park, J.-Y., Kim, S., & Choi, S.-H. (2005). Characterization of Bi/TiO_2 nanometer sized particle synthesized by solvothermal method and CH_3CHO decomposition in a plasma-photocatalytic system. *Journal of Photochemistry and Photobiology A: Chemistry, 173*, 128–136.
47. Kang, M., Kim, B.-J., Cho, S. M., Chung, C.-H., Kim, B.-W., Han, G. Y., & Yoon, K. J. (2002). Decomposition of toluene using an atmospheric pressure plasma/TiO_2 catalytic system. *Journal of Molecular Catalysis A: Chemical, 180*, 125–132.
48. Lee, B.-Y., Park, S.-H., Lee, S.-C., Kang, M., & Choung, S.-J. (2004). Decomposition of benzene by using a discharge plasma–photocatalyst hybrid system. *Catalysis Today, 93*, 769–776.
49. Kim, H.-H., Oh, S.-M., Ogata, A., & Futamura, S. (2005). Decomposition of gas-phase benzene using plasma-driven catalyst (PDC) reactor packed with Ag/TiO_2 catalyst. *Applied Catalysis B: Environmental, 56*, 213–220.
50. Chavadej, S., Saktrakool, K., Rangsunvigit, P., Lobban, L. L., & Sreethawong, T. (2007). Oxidation of ethylene by a multistage corona discharge system in the absence and presence of Pt/TiO_2. *Chemical Engineering Journal, 132*, 345–353.
51. Subrahmanyam, C., Magureanu, M., Laub, D., Renken, A., & Kiwi-Minsker, L. (2007). Nonthermal plasma abatement of trichloroethylene enhanced by photocatalysis. *The Journal of Physical Chemistry C, 111*, 4315–4318.
52. Morent, R., Dewulf, J., Steenhaut, N., Leys, C., & Van Langenhove, H. (2006). Hybrid plasma-catalyst system for the removal of trichloroethylene in air. *Journal of Advanced Oxidation Technologies, 9*, 53–58.
53. Leys, C., Neirynck, D., Morent, R., & Temmerman, E. (2006). DC-excited cold atmospheric pressure plasmas. *Czechoslovak Journal of Physics, 56*, B896–B902.
54. Urashima, K., Kostov, K. G., Chang, J.-S., Okayasa, Y., Iwaizumi, T., Yoshimura, K., & Kato, T. (2001). Removal of C_2F_6 from a semiconductor process flue gas by a ferroelectric packed-bed barrier discharge reactor with an adsorber. *IEEE Transactions on Industry Applications, 37*, 1456–1463.
55. Takaki, K., Urashima, K., & Chang, J.-S. (2004). Ferro-electric pellet shape effect on C_2F_6 removal by a packed-bed-type nonthermal plasma reactor. *IEEE Transactions on Plasma Science, 32*, 2175–2183.

56. Huang, H., Ye, D., & Guan, X. (2008). The simultaneous catalytic removal of VOCs and O_3 in a post-plasma. *Catalysis Today, 139*, 43–48.
57. Huang, H., & Ye, D. (2009). Combination of photocatalysis downstream the non-thermal plasma reactor for oxidation of gas-phase toluene. *Journal of Hazardous Materials, 171*, 535–541.
58. Hayashi, K., Yasui, H., Tanaka, M., Futamura, S., Kurita, S., & Aoyagi, K. (2009). Temperature dependence of toluene decomposition behavior in the discharge–catalyst hybrid reactor. *IEEE Transactions on Industry Applications, 45*, 1553–1558.
59. Krawczyk, K., Ulejczyk, B., Song, H., Lamenta, A., Paluch, B., & Schmidt-Szałowski, K. (2009). Plasma-catalytic reactor for decomposition of chlorinated hydrocarbons. *Plasma Chemistry and Plasma Processing, 29*, 27–41.
60. Demidiouk, V., & Chae, J. O. (2005). Decomposition of volatile organic compounds in plasma-catalytic system. *IEEE Transactions on Plasma Science, 33*, 157–161.
61. Blin-Simiand, N., Tardiveau, P., Risacher, A., Jorand, F., & Pasquiers, S. (2005). Removal of 2-Heptanone by dielectric barrier discharges–the effect of a catalyst support. *Plasma Processes and Polymers, 2*, 256–262.
62. Hensel, K., Katsura, S., & Mizuno, A. (2005). DC microdischarges inside porous ceramics. *IEEE Transactions on Plasma Science, 33*, 574–575.
63. Futamura, S., Einaga, H., Kabashima, H., & Hwan, L. Y. (2004). Synergistic effect of silent discharge plasma and catalysts on benzene decomposition. *Catalysis Today, 89*, 89–95.
64. Magureanu, M., Mandache, N., Parvulescu, V., Subrahmanyam, C., Renken, A., & Kiwi-Minsker, L. (2007). Improved performance of non-thermal plasma reactor during decomposition of trichloroethylene: Optimization of the reactor geometry and introduction of catalytic electrode. *Applied Catalysis B: Environmental, 74*, 270–277.
65. Subrahmanyam, C., Renken, A., & Kiwi-Minsker, L. (2007). Novel catalytic dielectric barrier discharge reactor for gas-phase abatement of isopropanol. *Plasma Chemistry and Plasma Processing, 27*, 13–22.
66. Kirkpatrick, M. J., Finney, W. C., & Locke, B. R. (2004). Plasma–catalyst interactions in the treatment of volatile organic compounds and NO_x with pulsed corona discharge and reticulated vitreous carbon Pt/Rh-coated electrodes. *Catalysis Today, 89*, 117–126.
67. Subrahmanyam, C., Renken, A., & Kiwi-Minsker, L. (2010). Catalytic non-thermal plasma reactor for abatement of toluene. *Chemical Engineering Journal, 160*, 677–682.
68. Subrahmanyam, C., Renken, A., & Kiwi-Minsker, L. (2007). Novel catalytic non-thermal plasma reactor for the abatement of VOCs. *Chemical Engineering Journal, 134*, 78–83.
69. Roland, U., Holzer, F., & Kopinke, F.-D. (2002). Improved oxidation of air pollutants in a non-thermal plasma. *Catalysis Today, 73*, 315–323.
70. Song, Y.-H., Kim, S.-J., Choi, K.-I., & Yamamoto, T. (2002). Effects of adsorption and temperature on a nonthermal plasma process for removing VOCs. *Journal of Electrostatics, 55*, 189–201.
71. Ogata, A., Yamanouchi, K., Mizuno, K., Kushiyama, S., & Yamamoto, T. (1999). Decomposition of benzene using alumina-hybrid and catalyst-hybrid plasma reactors. *IEEE Transactions on Industry Applications, 35*, 1289–1295.
72. Magureanu, M., Mandache, N. B., Eloy, P., Gaigneaux, E. M., & Parvulescu, V. I. (2005). Plasma-assisted catalysis for volatile organic compounds abatement. *Applied Catalysis B: Environmental, 61*, 12–20.
73. Oda, T., & Yamaji, K. (2003). Dilute trichloroethylene decomposition in air by using non-thermal plasma-catalyst effect. *Journal of Advanced Oxidation Technologies, 6*, 93–99.
74. Oh, S.-M., Kim, H.-H., Einaga, H., Ogata, A., Futamura, S., & Park, D.-W. (2006). Zeolite-combined plasma reactor for decomposition of toluene. *Thin Solid Films, 506*, 418–422.
75. Oh, S.-M., Kim, H.-H., Ogata, A., Einaga, H., Futamura, S., & Park, D.-W. (2005). Effect of zeolite in surface discharge plasma on the decomposition of toluene. *Catalysis Letters, 99*, 101–104.

76. Oda, T., Takahashi, T., & Kohzuma, S. (1998). Decomposition of dilute trichloroethylene by using non-thermal plasma processing-frequency and catalyst effect. In *Industry applications conference, 1998. Thirty-Third IAS annual meeting. The 1998 IEEE*. IEEE.
77. Wallis, A. E., Whitehead, J. C., & Zhang, K. (2007). The removal of dichloromethane from atmospheric pressure air streams using plasma-assisted catalysis. *Applied Catalysis B: Environmental, 72*, 282–288.
78. Grossmannova, H., Neirynck, D., & Leys, C. (2006). Atmospheric discharge combined with Cu-Mn/Al_2O_3 catalyst unit for the removal of toluene. *Czechoslovak Journal of Physics, 56*, B1156–B1161.
79. Da Costa, P., Marques, R., & Da Costa, S. (2008). Plasma catalytic oxidation of methane on alumina-supported noble metal catalysts. *Applied Catalysis B: Environmental, 84*, 214–222.
80. Yamamoto, T., Mizuno, K., Tamori, I., Ogata, A., Nifuku, M., Michalska, M., & Prieto, G. (1996). Catalysis-assisted plasma technology for carbon tetrachloride destruction. *IEEE Transactions on Industry Applications, 32*, 100–105.
81. Ogata, A., Einaga, H., Kabashima, H., Futamura, S., Kushiyama, S., & Kim, H.-H. (2003). Effective combination of nonthermal plasma and catalysts for decomposition of benzene in air. *Applied Catalysis B: Environmental, 46*, 87–95.
82. Vandenbroucke, A., Morent, R., De Geyter, N., Nguyen Dinh, M., Giraudon, J., Lamonier, J., & Leys, C. (2010). Plasma-catalytic decomposition of TCE. *International Journal of Plasma Environmental Science and Technology, 4*, 135–138.
83. Blackbeard, T., Demidyuk, V., Hill, S. L., & Whitehead, J. C. (2009). The effect of temperature on the plasma-catalytic destruction of propane and propene: A comparison with thermal catalysis. *Plasma Chemistry and Plasma Processing, 29*, 411–419.
84. Chang, C.-L., Bai, H., & Lu, S.-J. (2005). Destruction of styrene in an air stream by packed dielectric barrier discharge reactors. *Plasma Chemistry and Plasma Processing, 25*, 641–657.
85. Demidyuk, V., & Whitehead, J. C. (2007). Influence of temperature on gas-phase toluene decomposition in plasma-catalytic system. *Plasma Chemistry and Plasma Processing, 27*, 85–94.
86. Chang, C.-L., & Lin, T.-S. (2005). Elimination of carbon monoxide in the gas streams by dielectric barrier discharge systems with Mn catalyst. *Plasma Chemistry and Plasma Processing, 25*, 387–401.
87. Jarrige, J., & Vervisch, P. (2009). Plasma-enhanced catalysis of propane and isopropyl alcohol at ambient temperature on a MnO_2-based catalyst. *Applied Catalysis B: Environmental, 90*, 74–82.
88. Zhu, T., Li, J., Liang, W., & Jin, Y. (2009). Synergistic effect of catalyst for oxidation removal of toluene. *Journal of Hazardous Materials, 165*, 1258–1260.
89. Demidiouk, V., Moon, S.-I., Chae, J.-O., & Lee, D.-Y. (2003). Application of a plasma-catalytic system for decomposition of volatile organic compounds. *Journal of the Korean Physical Society, 42*, S966–S970.
90. Chen, H. L., Lee, H.-M., Cheng, L. C., Chang, M. B., Yu, S. J., & Li, S.-N. (2008). Influence of nonthermal plasma reactor type on and abatements. *IEEE Transactions on Plasma Science, 36*, 509–515.
91. Fan, X., Zhu, T., Wang, M., & Li, X. (2009). Removal of low-concentration BTX in air using a combined plasma catalysis system. *Chemosphere, 75*, 1301–1306.
92. Oda, T., Yamaji, K., & Takahashi, T. (2004). Decomposition of dilute trichloroethylene by nonthermal plasma processing-gas flow rate, catalyst, and ozone effect. *IEEE Transactions on Industry Applications, 40*, 430–436.
93. Kim, H.-H., Kobara, H., Ogata, A., & Futamura, S. (2005). Comparative assessment of different nonthermal plasma reactors on energy efficiency and aerosol formation from the decomposition of gas-phase benzene. *IEEE Transactions on Industry Applications, 41*, 206–214.

94. Besov, A. S., & Vorontsov, A. V. (2007). Acceleration of acetone destruction process under synergistic action of photocatalytic oxidation and barrier discharge. *Plasma Chemistry and Plasma Processing, 27*, 624–634.
95. Hakoda, T., Matsumoto, K., Mizuno, A., & Hirota, K. (2009). Role of metals loaded on a TiO_2 surface in the oxidation of xylene in air using an electron beam irradiation/catalytic process. *Applied Catalysis A: General, 357*, 244–249.
96. Oda, T., Takahahshi, T., & Yamaji, K. (2002). Nonthermal plasma processing for dilute VOCs decomposition. *IEEE Transactions on Industry Applications, 38*, 873–878.
97. Chang, C.-L., & Lin, T.-S. (2005). Decomposition of toluene and acetone in packed dielectric barrier discharge reactors. *Plasma Chemistry and Plasma Processing, 25*, 227–243.
98. Sun, R.-B., Xi, Z.-G., Chao, F.-H., Zhang, W., Zhang, H.-S., & Yang, D.-F. (2007). Decomposition of low-concentration gas-phase toluene using plasma-driven photocatalyst reactor. *Atmospheric Environment, 41*, 6853–6859.
99. Thevenet, F., Guaitella, O., Puzenat, E., Guillard, C., & Rousseau, A. (2008). Influence of water vapour on plasma/photocatalytic oxidation efficiency of acetylene. *Applied Catalysis B: Environmental, 84*, 813–820.
100. Kogoma, M., Miki, Y., Tanaka, K., & Takahashi, K. (2006). Highly efficient VOC decomposition using a complex system (OH radical, ozone-UV, and TiO_2). *Plasma Processes and Polymers, 3*, 727–733.
101. Guaitella, O., Thevenet, F., Puzenat, E., Guillard, C., & Rousseau, A. (2008). C_2H_2 oxidation by plasma/TiO_2 combination: Influence of the porosity, and photocatalytic mechanisms under plasma exposure. *Applied Catalysis B: Environmental, 80*, 296–305.
102. Huang, H., Ye, D., & Leung, D. Y. (2010). Removal of toluene using UV-irradiated and nonthermal plasma–driven photocatalyst system. *Journal of Environmental Engineering, 136*, 1231–1236.
103. Falkenstein, Z., & Coogan, J. J. (1997). Microdischarge behaviour in the silent discharge of nitrogen-oxygen and water-air mixtures. *Journal of Physics D: Applied Physics, 30*, 817–825.
104. Van Durme, J., Dewulf, J., Sysmans, W., Leys, C., & Van Langenhove, H. (2007). Abatement and degradation pathways of toluene in indoor air by positive corona discharge. *Chemosphere, 68*, 1821–1829.
105. Ge, H., Zhang, L., Yan, L., Mi, D., & Zhu, Y. (2011). Parameter optimization of excited OH radical in multi-needle to plate negative DC corona discharge. *Journal of Electrostatics, 69*, 529–532.
106. Sugasawa, M., Terasawa, T., & Futamura, S. (2010). Additive effect of water on the decomposition of VOCs in nonthermal plasma. *IEEE Transactions on Industry Applications, 46*, 1692–1698.
107. Ogata, A., Shintani, N., Yamanouchi, K., Mizuno, K., Kushiyama, S., & Yamamoto, T. (2000). Effect of water vapor on benzene decomposition using a nonthermal-discharge plasma reactor. *Plasma Chemistry and Plasma Processing, 20*, 453–467.
108. Cal, M. P., & Schluep, M. (2001). Destruction of benzene with non-thermal plasma in dielectric barrier discharge reactors. *Environmental Progress, 20*, 151–156.
109. Futamura, S., Zhang, A., Einaga, H., & Kabashima, H. (2002). Involvement of catalyst materials in nonthermal plasma chemical processing of hazardous air pollutants. *Catalysis Today, 72*, 259–265.
110. Xu, N., Fu, W., He, C., Cao, L., Liu, X., Zhao, J., & Pan, H. (2014). Benzene removal using non-thermal plasma with CuO/AC catalyst: Reaction condition optimization and decomposition mechanism. *Plasma Chemistry and Plasma Processing, 34*, 1387–1402.
111. Fan, X., Zhu, T., Wan, Y., & Yan, X. (2010). Effects of humidity on the plasma-catalytic removal of low-concentration BTX in air. *Journal of Hazardous Materials, 180*, 616–621.
112. Futamura, S., Zhang, A., & Yamamoto, T. (1997). The dependence of nonthermal plasma behavior of VOCs on their chemical structures. *Journal of Electrostatics, 42*, 51–62.
113. Krawczyk, K., & Ulejczyk, B. (2004). Influence of water vapor on CCl_4 and $CHCl_3$ conversion in gliding discharge. *Plasma Chemistry and Plasma Processing, 24*, 155–167.

114. Futamura, S., & Sugasawa, M. (2008). Additive effect on energy efficiency and byproduct distribution in VOC decomposition with nonthermal plasma. *IEEE Transactions on Industry Applications, 44*, 40–45.
115. Abdelaziz, A. A., Seto, T., Abdel-Salam, M., & Otani, Y. (2013). Influence of nitrogen excited species on the destruction of naphthalene in nitrogen and air using surface dielectric barrier discharge. *Journal of Hazardous Materials, 246*, 26–33.
116. Zhang, X., Zhu, J., Li, X., Liu, Z., Ren, X., & Yan, K. (2011). Characteristics of styrene removal with an AC/DC streamer corona plasma system. *IEEE Transactions on Plasma Science, 39*, 1482–1488.
117. Zhang, X., Feng, W., Yu, Z., Li, S., Zhu, J., & Yan, K. (2013). Comparison of styrene removal in air by positive and negative DC corona discharges. *International journal of Environmental Science and Technology, 10*, 1377–1382.
118. Futamura, S., Einaga, H., & Zhang, A. (2001). Comparison of reactor performance in the nonthermal plasma chemical processing of hazardous air pollutants. *IEEE Transactions on Industry Applications, 37*, 978–985.
119. Tonkyn, R., Barlow, S., & Orlando, T. (1996). Destruction of carbon tetrachloride in a dielectric barrier/packed-bed corona reactor. *Journal of Applied Physics, 80*, 4877–4886.
120. Korzekwa, R., Grothaus, M., Hutcherson, R., Roush, R., & Brown, R. (1998). Destruction of hazardous air pollutants using a fast rise time pulsed corona reactor. *Review of Scientific Instruments, 69*, 1886–1892.
121. Agnihotri, S., Cal, M. P., & Prien, J. (2004). Destruction of 1, 1, 1-trichloroethane using dielectric barrier discharge nonthermal plasma. *Journal of Environmental Engineering, 130*, 349–355.
122. Karuppiah, J., Reddy, E. L., Reddy, P. M., Ramaraju, B., Karvembu, R., & Subrahmanyam, C. (2012). Abatement of mixture of volatile organic compounds (VOCs) in a catalytic non-thermal plasma reactor. *Journal of Hazardous Materials, 237–238*, 283–289.
123. Wan, Y., Fan, X., & Zhu, T. (2011). Removal of low-concentration formaldehyde in air by DC corona discharge plasma. *Chemical Engineering Journal, 171*, 314–319.
124. Mfopara, A., Kirkpatrick, M. J., & Odic, E. (2009). Dilute methane treatment by atmospheric pressure dielectric barrier discharge: Effects of water vapor. *Plasma Chemistry and Plasma Processing, 29*, 91–102.
125. Baylet, A., Marécot, P., Duprez, D., Jeandel, X., Lombaert, K., & Tatibouët, J. (2012). Synergetic effect of plasma/catalysis hybrid system for CH_4 removal. *Applied Catalysis B: Environmental, 113*, 31–36.
126. Karuppiah, J., Karvembu, R., & Subrahmanyam, C. (2012). The catalytic effect of MnO_x and CoO_x on the decomposition of nitrobenzene in a non-thermal plasma reactor. *Chemical Engineering Journal, 180*, 39–45.
127. Dinh, M. N., Giraudon, J.-M., Lamonier, J.-F., Vandenbroucke, A., De Geyter, N., Leys, C., & Morent, R. (2014). Plasma-catalysis of low TCE concentration in air using $LaMnO_{3+\delta}$ as catalyst. *Applied Catalysis B: Environmental, 147*, 904–911.
128. Lu, S. Y., Sun, X. M., Li, X. D., Yan, J. H., & Du, C. M. (2012). Decomposition of toluene in a rotating glidarc discharge reactor. *IEEE Transactions on Plasma Science, 40*, 2151–2156.
129. Trushkin, A., Grushin, M., Kochetov, I., Trushkin, N., & Akishev, Y. S. (2013). Decomposition of toluene in a steady-state atmospheric-pressure glow discharge. *Plasma Physics Reports, 39*, 167–182.
130. Lee, H. M., & Chang, M. B. (2003). Abatement of gas-phase p-xylene via dielectric barrier discharges. *Plasma Chemistry and Plasma Processing, 23*, 541–558.
131. Zhu, Y.-P., Liu, Y.-L., Renb, T.-Z., & Yuan, Z.-Y. (2014). Plasma-catalytic removal of a low concentration of acetone in humid conditions. *RSC Advances, 4*, 37796–37805.
132. Kuroki, T., Oishi, T., Yamamoto, T., & Okubo, M. (2013). Bromomethane decomposition using a pulsed dielectric barrier discharge. *IEEE Transactions on Industry Applications, 49*, 293–297.

133. Assadi, A. A., Bouzaza, A., Vallet, C., & Wolbert, D. (2014). Use of DBD plasma, photocatalysis, and combined DBD plasma/photocatalysis in a continuous annular reactor for isovaleraldehyde elimination–synergetic effect and byproducts identification. *Chemical Engineering Journal, 254*, 124–132.
134. Aubry, O., & Cormier, J.-M. (2009). Improvement of the diluted propane efficiency treatment using a non-thermal plasma. *Plasma Chemistry and Plasma Processing, 29*, 13–25.
135. Yamashita, R., Takahashi, T., & Oda, T. (1996). Humidify effect on non-thermal plasma processing for VOCs decomposition. In *Industry applications conference, 1996. Thirty-First IAS annual meeting, IAS'96., conference record of the 1996 IEEE.* IEEE.
136. Du, C. M., Yan, J. H., & Cheron, B. (2007). Decomposition of toluene in a gliding arc discharge plasma reactor. *Plasma Sources Science and Technology, 16*, 791–797.
137. Wang, J. T., Cao, X., Zhang, R. X., Gong, T., Hou, H. Q., Chen, S. P., & Zhang, R. N. (2016). Effect of water vapor on toluene removal in catalysis-DBD plasma reactors. *Plasma Science and Technology, 18*, 370–375.
138. Nakagawa, Y., Fujisawa, H., Ono, R., & Oda, T. (2010). Dilute trichloroethylene decomposition by using high pressure non-thermal plasma: humidity effects. In *Industry Applications Society Annual Meeting (IAS), 2010 IEEE.* IEEE.
139. Zhu, T., Li, J., Jin, Y.-q., Liang, Y., & Ma, G. (2008). Decomposition of benzene by non-thermal plasma processing: Photocatalyst and ozone effect. *International Journal of Environmental Science & Technology, 5*, 375–384.
140. Ogata, A., Ito, D., Mizuno, K., Kushiyama, S., Gal, A., & Yamamoto, T. (2002). Effect of coexisting components on aromatic decomposition in a packed-bed plasma reactor. *Applied Catalysis A: General, 236*, 9–15.
141. Einaga, H., Ibusuki, T., & Futamura, S. (2001). Performance evaluation of a hybrid system comprising silent discharge plasma and manganese oxide catalysts for benzene decomposition. *IEEE Transactions on Industry Applications, 37*, 1476–1482.
142. Byeon, J. H., Park, J. H., Jo, Y. S., Yoon, K. Y., & Hwang, J. (2010). Removal of gaseous toluene and submicron aerosol particles using a dielectric barrier discharge reactor. *Journal of Hazardous Materials, 175*, 417–422.
143. Delagrange, S., Pinard, L., & Tatibouet, J.-M. (2006). Combination of a non-thermal plasma and a catalyst for toluene removal from air: Manganese based oxide catalysts. *Applied Catalysis B: Environmental, 68*, 92–98.
144. Blin-Simiand, N., Jorand, F., Magne, L., Pasquiers, S., Postel, C., & Vacher, J.-R. (2008). Plasma reactivity and plasma-surface interactions during treatment of toluene by a dielectric barrier discharge. *Plasma Chemistry and Plasma Processing, 28*, 429–466.
145. Chiper, A. S., Blin-Simiand, N., Heninger, M., Mestdagh, H., Boissel, P., Jorand, F., Lemaire, J., Leprovost, J., Pasquiers, S., & Popa, G. (2009). Detailed characterization of 2-heptanone conversion by dielectric barrier discharge in N_2 and N_2/O_2 mixtures. *The Journal of Physical Chemistry A, 114*, 397–407.
146. Rosocha, L. A., & Korzekwa, R. A. (1999). Advanced oxidation and reduction processes in the gas phase using non-thermal plasmas. *Journal of Advanced Oxidation Technologies, 4*, 247–264.
147. Vertriest, R., Morent, R., Dewulf, J., Leys, C., & Van Langenhove, H. (2003). Multi-pin-to-plate atmospheric glow discharge for the removal of volatile organic compounds in waste air. *Plasma Sources Science and Technology, 12*, 412–416.
148. Oh, J. H., Mok, Y. S., Lee, S. B., & Chang, M. S. (2009). Destruction of HCFC-22 and distribution of byproducts in a nonthermal plasma reactor packed with dielectric pellets. *Journal of the Korean Physical Society, 54*, 1539–1546.
149. Gandhi, M. S., & Mok, Y. (2012). Decomposition of trifluoromethane in a dielectric barrier discharge non-thermal plasma reactor. *Journal of Environmental Sciences, 24*, 1234–1239.
150. Mok, Y., Demidyuk, V., & Whitehead, J. (2008). Decomposition of hydrofluorocarbons in a dielectric-packed plasma reactor. *The Journal of Physical Chemistry A, 112*, 6586–6591.

151. Kang, H.-C. (2002). Decomposition of chlorofluorocarbon by non-thermal plasma. *Journal of Industrial and Engineering Chemistry, 8*, 488–492.
152. Futamura, S., & Yamamoto, T. (1997). Byproduct identification and mechanism determination in plasma chemical decomposition of trichloroethylene. *IEEE Transactions on Industry Applications, 33*, 447–453.
153. Vandenbroucke, A., Aerts, R., Van Gaens, W., De Geyter, N., Leys, C., Morent, R., & Bogaerts, A. (2015). Modeling and experimental study of trichloroethylene abatement with a negative direct current corona discharge. *Plasma Chemistry and Plasma Processing, 35*, 217–230.
154. Hsiao, M., Merritt, B., Penetrante, B., Vogtlin, G., & Wallman, P. (1995). Plasma-assisted decomposition of methanol and trichloroethylene in atmospheric pressure air streams by electrical discharge processing. *Journal of Applied Physics, 78*, 3451–3456.
155. Blin-Simiand, N., Pasquiers, S., Jorand, F., Postel, C., & Vacher, J. R. (2009). Removal of formaldehyde in nitrogen and in dry air by a DBD: Importance of temperature and role of nitrogen metastable states. *Journal of Physics D: Applied Physics, 42* 122003.
156. Storch, D. G., & Kushner, M. J. (1993). Destruction mechanisms for formaldehyde in atmospheric pressure low temperature plasmas. *Journal of Applied Physics, 73*, 51–55.
157. Hill, S. L., Kim, H.-H., Futamura, S., & Whitehead, J. C. (2008). The destruction of atmospheric pressure propane and propene using a surface discharge plasma reactor. *The Journal of Physical Chemistry A, 112*, 3953–3958.
158. Demidyuk, V., Hill, S. L., & Whitehead, J. C. (2008). Enhancement of the destruction of propane in a low-temperature plasma by the addition of unsaturated hydrocarbons: Experiment and modeling. *The Journal of Physical Chemistry A, 112*, 7862–7867.
159. Li, J., Bai, S.-P., Shi, X.-C., Han, S.-L., Zhu, X.-M., Chen, W.-C., & Pu, Y.-K. (2008). Effects of temperature on benzene oxidation in dielectric barrier discharges. *Plasma Chemistry and Plasma Processing, 28*, 39–48.
160. Penetrante, B., Hsiao, M., Bardsley, J., Merritt, B., Vogtlin, G., Kuthi, A., Burkhart, C., & Bayless, J. (1997). Decomposition of methylene chloride by electron beam and pulsed corona processing. *Physics Letters A, 235*, 76–82.
161. Kim, D.-H., Mok, Y., & Lee, S. (2011). Effect of temperature on the decomposition of trifluoromethane in a dielectric barrier discharge reactor. *Thin Solid Films, 519*, 6960–6963.
162. Harling, A. M., Kim, H.-H., Futamura, S., & Whitehead, J. C. (2007). Temperature dependence of plasma catalysis using a nonthermal, atmospheric pressure packed bed; the destruction of benzene and toluene. *The Journal of Physical Chemistry C, 111*, 5090–5095.
163. Penetrante, B., Hsiao, M., Bardsley, J., Merritt, B., Vogtlin, G., Wallman, P., Kuthi, A., Bukhart, C., & Bayless, J. (1996). Electron beam and pulsed corona processing of volatile organic compounds in gas streams. *Pure and Applied Chemistry, 68*, 1083–1087.
164. Huang, L., Nakajyo, K., Hari, T., Ozawa, S., & Matsuda, H. (2001). Decomposition of carbon tetrachloride by a pulsed corona reactor incorporated with in situ absorption. *Industrial & Engineering Chemistry Research, 40*, 5481–5486.
165. Harling, A. M., Demidyuk, V., Fischer, S. J., & Whitehead, J. C. (2008). Plasma-catalysis destruction of aromatics for environmental clean-up: Effect of temperature and configuration. *Applied Catalysis B: Environmental, 82*, 180–189.
166. Lee, H. M., & Chang, M. B. (2001). Gas-phase removal of acetaldehyde via packed-bed dielectric barrier discharge reactor. *Plasma Chemistry and Plasma Processing, 21*, 329–343.
167. Ogata, A., Mizuno, K., Kushiyama, S., & Yamamoto, T. (1998). Methane decomposition in a barium titanate packed-bed nonthermal plasma reactor. *Plasma Chemistry and Plasma Processing, 18*, 363–373.
168. Mok, Y. S., Lee, S. B., Oh, J. H., Ra, K. S., & Sung, B. H. (2008). Abatement of Trichloromethane by using nonthermal plasma reactors. *Plasma Chemistry and Plasma Processing, 28*, 663–676.

169. Pringle, K. J., Whitehead, J. C., Wilman, J. J., & Wu, J. (2004). The chemistry of methane remediation by a non-thermal atmospheric pressure plasma. *Plasma Chemistry and Plasma Processing, 24*, 421–434.
170. Zhang, H., Li, K., Sun, T., Jia, J., Lou, Z., & Feng, L. (2014). Removal of styrene using dielectric barrier discharge plasmas combined with sol–gel prepared TiO_2 coated γ-Al_2O_3. *Chemical Engineering Journal, 241*, 92–102.
171. Ding, H. X., Zhu, A. M., Lu, F. G., Xu, Y., Zhang, J., & Yang, X. F. (2006). Low-temperature plasma-catalytic oxidation of formaldehyde in atmospheric pressure gas streams. *Journal of Physics D: Applied Physics, 39*, 3603–3608.
172. Chavadej, S., Kiatubolpaiboon, W., Rangsunvigit, P., & Sreethawong, T. (2007). A combined multistage corona discharge and catalytic system for gaseous benzene removal. *Journal of Molecular Catalysis A: Chemical, 263*, 128–136.
173. Shi, Y., Wang, X., Li, W., Tan, T.-e., & Ruan, J.-j. (2006). Evaluation of multiple corona reactor modes and the application in odor removal. *Plasma Chemistry and Plasma Processing, 26*, 187–196.
174. Harling, A. M., Glover, D. J., Whitehead, J. C., & Zhang, K. (2008). Industrial scale destruction of environmental pollutants using a novel plasma reactor. *Industrial & Engineering Chemistry Research, 47*, 5856–5860.
175. Hubner, M., Guaitella, O., Rousseau, A., & Ropcke, J. (2013). A spectroscopic study of ethylene destruction and by-product generation using a three-stage atmospheric packed-bed plasma reactor. *Journal of Applied Physics, 114*, 033301.
176. Li, D., Yakushiji, D., Kanazawa, S., Ohkubo, T., & Nomoto, Y. (2002). Decomposition of toluene by streamer corona discharge with catalyst. *Journal of Electrostatics, 55*, 311–319.
177. Malik, M. A., & Xuanzhen, J. (2000). Catalyst assisted destruction of trichloro ethylene and toluene in corona discharges. *Journal of Environmental Sciences, 12*, 7–11.
178. Guo, Y., Liao, X., He, J., Ou, W., & Ye, D. (2010). Effect of manganese oxide catalyst on the dielectric barrier discharge decomposition of toluene. *Catalysis Today, 153*, 176–183.
179. Liao, X.-B., Guo, Y.-F., He, J.-H., Ou, W.-J., & Ye, D.-Q. (2010). Hydroxyl radicals formation in dielectric barrier discharge during decomposition of toluene. *Plasma Chemistry and Plasma Processing, 30*, 841–853.
180. Van Durme, J., Dewulf, J., Sysmans, W., Leys, C., & Van Langenhove, H. (2007). Efficient toluene abatement in indoor air by a plasma catalytic hybrid system. *Applied Catalysis B: Environmental, 74*, 161–169.
181. Huang, H. B., Ye, D. Q., & Leung, D. Y. C. (2011). Abatement of toluene in the plasma-driven catalysis: Mechanism and reaction kinetics. *IEEE Transactions on Plasma Science, 39*, 877–882.
182. Huang, H., Ye, D., & Leung, D. Y. (2011). Plasma-driven catalysis process for toluene abatement: Effect of water vapor. *IEEE Transactions on Plasma Science, 39*, 576–580.
183. Wu, J., Huang, Y., Xia, Q., & Li, Z. (2013). Decomposition of toluene in a plasma catalysis system with NiO, MnO_2, CeO_2, Fe_2O_3, and CuO catalysts. *Plasma Chemistry and Plasma Processing, 33*, 1073–1082.
184. Qin, C. H., Huang, X. M., Dang, X. Q., Huang, J. Y., Teng, J. J., & Kang, Z. L. (2016). Toluene removal by sequential adsorption-plasma catalytic process: Effects of Ag and Mn impregnation sequence on Ag-Mn/gamma-Al_2O_3. *Chemosphere, 162*, 125–130.
185. Xu, W. C., Wang, N., Chen, Y. D., Chen, J. D., Xu, X. X., Yu, L., Chen, L. M., Wu, J. L., Fu, M. L., Zhu, A. M., & Ye, D. Q. (2016). In situ FT-IR study and evaluation of toluene abatement in different plasma catalytic systems over metal oxides loaded gamma-Al_2O_3. *Catalysis Communications, 84*, 61–66.
186. Xu, X. X., Wang, P. T., Xu, W. C., Wu, J. L., Chen, L. M., Fu, M. L., & Ye, D. Q. (2016). Plasma-catalysis of metal loaded SBA-15 for toluene removal: Comparison of continuously introduced and adsorption-discharge plasma system. *Chemical Engineering Journal, 283*, 276–284.
187. Huang, H., Ye, D., Leung, D. Y., Feng, F., & Guan, X. (2011). Byproducts and pathways of toluene destruction via plasma-catalysis. *Journal of Molecular Catalysis A: Chemical, 336*, 87–93.

188. Ban, J.-Y., Son, Y.-H., Kang, M., & Choung, S.-J. (2006). Highly concentrated toluene decomposition on the dielectric barrier discharge (DBD) plasma–photocatalytic hybrid system with Mn-Ti-incorporated mesoporous silicate photocatalyst (Mn-Ti-MPS). *Applied Surface Science, 253*, 535–542.
189. Kim, H. H., Ogata, A., & Futamura, S. (2005). Atmospheric plasma-driven catalysis for the low temperature decomposition of dilute aromatic compounds. *Journal of Physics D: Applied Physics, 38*, 1292–1300.
190. Liang, W.-J., Ma, L., Liu, H., & Li, J. (2013). Toluene degradation by non-thermal plasma combined with a ferroelectric catalyst. *Chemosphere, 92*, 1390–1395.
191. Zhu, T., Wan, Y., Li, H., Chen, S., & Fang, Y. (2011). VOCs decomposition via modified ferroelectric packed bed dielectric barrier discharge plasma. *IEEE Transactions on Plasma Science, 39*, 1695–1700.
192. Huang, R., Lu, M. J., Wang, P. T., Chen, Y. D., Wu, J. L., Fu, M. L., Chen, L. M., & Ye, D. Q. (2015). Enhancement of the non-thermal plasma-catalytic system with different zeolites for toluene removal. *RSC Advances, 5*, 72113–72120.
193. Lu, M. J., Huang, R., Wu, J. L., Fu, M. L., Chen, L. M., & Ye, D. Q. (2015). On the performance and mechanisms of toluene removal by FeO_x/SBA-15-assisted non-thermal plasma at atmospheric pressure and room temperature. *Catalysis Today, 242*, 274–286.
194. Teramoto, Y., Kosuge, K., Sugasawa, M., Kim, H. H., Ogata, A., & Negishi, N. (2015). Zirconium/cerium oxide solid solutions with addition of SiO_2 as ozone-assisted catalysts for toluene oxidation. *Catalysis Communications, 61*, 112–116.
195. Wang, W. Z., Wang, H. L., Zhu, T. L., & Fan, X. (2015). Removal of gas phase low-concentration toluene over Mn, Ag and Ce modified HZSM-5 catalysts by periodical operation of adsorption and non-thermal plasma regeneration. *Journal of Hazardous Materials, 292*, 70–78.
196. Ye, Z. L., Han, Q. Y., Jiang, Y. F., Zhang, S. D., Shu, L., & Zhang, R. X. (2015). Synergetic effect between plasma and UV for toluene conversion in integrated combined plasma photolysis reactor with KrCl/KrBr/XeCl/Xe-2 Excilamp. *Canadian Journal of Chemical Engineering, 93*, 1168–1175.
197. Zabihi, M., Khorasheh, F., & Shayegan, J. (2015). Supported copper and cobalt oxides on activated carbon for simultaneous oxidation of toluene and cyclohexane in air. *RSC Advances, 5*, 5107–5122.
198. Chen, J., Xie, Z. M., Tang, J. H., Zhou, J., Lu, X. T., & Zhao, H. T. (2016). Oxidation of toluene by dielectric barrier discharge with photo-catalytic electrode. *Chemical Engineering Journal, 284*, 166–173.
199. Giroir-Fendler, A., Alves-Fortunato, M., Richard, M., Wang, C., Diaz, J. A., Gil, S., Zhang, C. H., Can, F., Bion, N., & Guo, Y. L. (2016). Synthesis of oxide supported $LaMnO_3$ perovskites to enhance yields in toluene combustion. *Applied Catalysis B: Environmental, 180*, 29–37.
200. Rodrigues, A., Tatibouet, J. M., & Fourre, E. (2016). Operando DRIFT spectroscopy characterization of intermediate species on catalysts surface in VOC removal from air by non-thermal plasma assisted catalysis. *Plasma Chemistry and Plasma Processing, 36*, 901–915.
201. Kim, H.-H., Lee, Y.-H., Ogata, A., & Futamura, S. (2003). Plasma-driven catalyst processing packed with photocatalyst for gas-phase benzene decomposition. *Catalysis Communications, 4*, 347–351.
202. Kim, H., Lee, Y., Ogata, A., & Futamura, S. (2003). Decomposition of gas-phase benzene using hybrid systems of non-thermal plasma and catalysts. In *Plasma science, 2003. ICOPS 2003. IEEE conference record-abstracts. The 30th international conference on. 2003*. IEEE.
203. Kim, H.-H., Oh, S.-M., Ogata, A., & Futamura, S. (2004). Decomposition of benzene using Ag/TiO_2 packed plasma-driven catalyst reactor: Influence of electrode configuration and Ag-loading amount. *Catalysis Letters, 96*, 189–194.
204. Higashi, M., Uchida, S., Suzuki, N., & Fujii, K.-I. (1992). Soot elimination and NO_x and SO_x reduction in diesel-engine exhaust by a combination of discharge plasma and oil dynamics. *IEEE Transactions on Plasma Science, 20*, 1–12.

205. Fan, H. Y., Shi, C., Li, X. S., Zhao, D. Z., Xu, Y., & Zhu, A. M. (2009). High-efficiency plasma catalytic removal of dilute benzene from air. *Journal of Physics D: Applied Physics, 42*, 225105.
206. Jiang, N., Hu, J., Li, J., Shang, K. F., Lu, N., & Wu, Y. (2016). Plasma-catalytic degradation of benzene over Ag-Ce bimetallic oxide catalysts using hybrid surface/packed-bed discharge plasmas. *Applied Catalysis B: Environmental, 184*, 355–363.
207. Park, D.-W., Yoon, S.-H., Kim, G.-J., & Sekiguchi, H. (2002). The effect of catalyst on the decomposition of dilute benzene using dielectric barrier discharge. *Journal of Industrial and Engineering Chemistry, 8*, 393–398.
208. Hu, J., Jiang, N., Li, J., Shang, K. F., Lu, N., & Wu, Y. (2016). Degradation of benzene by bipolar pulsed series surface/packed-bed discharge reactor over MnO_2-TiO_2/zeolite catalyst. *Chemical Engineering Journal, 293*, 216–224.
209. Ye, Z. L., Shen, Y., Xi, R. Z., & Hou, H. Q. (2007). Destruction of benzene in an air stream by the outer combined plasma photolysis method. *Journal of Physics D: Applied Physics, 41*, 025201.
210. Ge, H., Hu, D. X., Li, X. G., Tian, Y., Chen, Z. B., & Zhu, Y. M. (2015). Removal of low-concentration benzene in indoor air with plasma-MnO_2 catalysis system. *Journal of Electrostatics, 76*, 216–221.
211. He, C., Cao, L., Liu, X., Fu, W., & Zhao, J. (2015). Catalytic behavior and synergistic effect of nonthermal plasma and CuO/AC catalyst for benzene destruction. *International journal of Environmental Science and Technology, 12*, 3531–3540.
212. Liu, Y., Li, X. S., Liu, J. L., Wu, J. L., Ye, D. Q., & Zhu, A. M. (2016). Cycled storage-discharge (CSD) plasma catalytic removal of benzene over AgMn/HZSM-5 using air as discharge gas. *Catalysis Science & Technology, 6*, 3788–3796.
213. Ma, T. P., Jiang, H. D., Liu, J. Q., & Zhong, F. C. (2016). Decomposition of benzene using a pulse-modulated DBD plasma. *Plasma Chemistry and Plasma Processing, 36*, 1533–1543.
214. Pangilinan, C. D. C., Kurniawan, W., Salim, C., & Hinode, H. (2016). Effect of Ag/TiO_2 catalyst preparation on gas-phase benzene decomposition using non-thermal plasma driven catalysis under oxygen plasma. *Reaction Kinetics, Mechanisms and Catalysis, 117*, 103–118.
215. Han, S., Oda, T., & Ono, R. (2005). Improvement of the energy efficiency in the decomposition of dilute trichloroethylene by the barrier discharge plasma process. *IEEE Transactions on Industry Applications, 41*, 1343–1349.
216. Oda, T., Takahashi, T., & Yamaji, K. (2004). TCE decomposition by the nonthermal plasma process concerning ozone effect. *IEEE Transactions on Industry Applications, 40*, 1249–1256.
217. Magureanu, M., Mandache, N. B., Hu, J., Richards, R., Florea, M., & Parvulescu, V. I. (2007). Plasma-assisted catalysis total oxidation of trichloroethylene over gold nano-particles embedded in SBA-15 catalysts. *Applied Catalysis B: Environmental, 76*, 275–281.
218. Vandenbroucke, A. M., Dinh, M. T. N., Nuns, N., Giraudon, J. M., De Geyter, N., Leys, C., Lamonier, J. F., & Morent, R. (2016). Combination of non-thermal plasma and Pd/$LaMnO_3$ for dilute trichloroethylene abatement. *Chemical Engineering Journal, 283*, 668–675.
219. Dinh, M. T. N., Giraudon, T. M., Vandenbroucke, A. M., Morent, R., De Geyter, N., & Lamonier, J. F. (2016). Manganese oxide octahedral molecular sieve K-OMS-2 as catalyst in post plasma-catalysis for trichloroethylene degradation in humid air. *Journal of Hazardous Materials, 314*, 88–94.
220. Dinh, M. T. N., Giraudon, J. M., Vandenbroucke, A. M., Morent, R., De Geyter, N., & Lamonier, J. F. (2015). Post plasma-catalysis for total oxidation of trichloroethylene over Ce-Mn based oxides synthesized by a modified "redox-precipitation route". *Applied Catalysis B: Environmental, 172*, 65–72.
221. Vandenbroucke, A., Mora, M., Jiménez-Sanchidrián, C., Romero-Salguero, F., De Geyter, N., Leys, C., & Morent, R. (2014). TCE abatement with a plasma-catalytic combined system using MnO_2 as catalyst. *Applied Catalysis B: Environmental, 156*, 94–100.
222. Whitehead, J. C. (2016). Plasma–catalysis: The known knowns, the known unknowns and the unknown unknowns. *Journal of Physics D: Applied Physics, 49*, 243001.

223. Oda, T., Yamaji, K., & Takahashi, T. (2001). Decomposition of dilute trichloroethylene by nonthermal plasma processing-catalyst and ozone effect. In *Industry applications conference, 2001. Thirty-Sixth IAS annual meeting. Conference record of the 2001 IEEE.*
224. Vandenbroucke, A., Morent, R., De Geyter, N., & Leys, C. (2011). Decomposition of trichloroethylene with plasma-catalysis: A review. *Journal of Advanced Oxidation Technologies, 14*, 165–173.
225. Lee, H., Lee, D. H., Song, Y. H., Choi, W. C., Park, Y. K., & Kim, D. H. (2015). Synergistic effect of non-thermal plasma-catalysis hybrid system on methane complete oxidation over Pd-based catalysts. *Chemical Engineering Journal, 259*, 761–770.
226. Marques, R., Da Costa, S., & Da Costa, P. (2008). Plasma-assisted catalytic oxidation of methane: On the influence of plasma energy deposition and feed composition. *Applied Catalysis B: Environmental, 82*, 50–57.
227. Abd Allah, Z., & Whitehead, J. C. (2015). Plasma-catalytic dry reforming of methane in an atmospheric pressure AC gliding arc discharge. *Catalysis Today, 256*, 76–79.
228. Huu, T. P., Gil, S., Da Costa, P., Giroir-Fendler, A., & Khacef, A. (2015). Plasma-catalytic hybrid reactor: Application to methane removal. *Catalysis Today, 257*, 86–92.
229. Lee, H., Lim, T. H., & Kim, D. H. (2015). Complementary effect of plasma-catalysis hybrid system on methane complete oxidation over non-PGM catalysts. *Catalysis Communications, 69*, 223–227.
230. Trinh, Q. H., Lee, S. B., & Mok, Y. S. (2015). Removal of ethylene from air stream by adsorption and plasma-catalytic oxidation using silver-based bimetallic catalysts supported on zeolite. *Journal of Hazardous Materials, 285*, 525–534.
231. Trinh, Q. H., & Mok, Y. S. (2015). Effect of the adsorbent/catalyst preparation method and plasma reactor configuration on the removal of dilute ethylene from air stream. *Catalysis Today, 256*, 170–177.
232. Hoard, J., Wallington, T. J., Bretz, R. L., Malkin, A., Dorai, R., & Kushner, M. J. (2003). Importance of O (^{3}P) atoms and OH radicals in hydrocarbon oxidation during the nonthermal plasma treatment of diesel exhaust inferred using relative-rate methods. *International Journal of Chemical Kinetics, 35*, 231–238.
233. Rousseau, A., Guaitella, O., Gatilova, L., Thevenet, F., Guillard, C., Ropcke, J., & Stancu, G. (2005). Photocatalyst activation in a pulsed low pressure discharge. *Applied Physics Letters, 87*, 221501–221900.
234. Gatica, J. M., Garcia-Cabeza, A. L., Yeste, M. P., Marin-Barrios, R., Gonzalez-Leal, J. M., Blanco, G., Cifredo, G. A., Guerra, F. M., & Vidal, H. (2016). Carbon integral honeycomb monoliths as support of copper catalysts in the Kharasch-Sosnovsky oxidation of cyclohexene. *Chemical Engineering Journal, 290*, 174–184.
235. Kim, J., Han, B., Kim, Y., Lee, J.-H., Park, C.-R., Kim, J.-C., Kim, J.-C., & Kim, K.-J. (2004). Removal of VOCs by hybrid electron beam reactor with catalyst bed. *Radiation Physics and Chemistry, 71*, 429–432.
236. Kim, K.-J., Kim, J.-C., Kim, J., & Sunwoo, Y. (2005). Development of hybrid technology using E-beam and catalyst for aromatic VOCs control. *Radiation Physics and Chemistry, 73*, 85–90.
237. Zhang, H. B., Li, K., Sun, T. H., Jia, J. P., Lou, Z. Y., Yao, S. A., & Wang, G. (2015). The combination effect of dielectric barrier discharge (DBD) and TiO_2 catalytic process on styrene removal and the analysis of the by-products and intermediates. *Research on Chemical Intermediates, 41*, 175–189.
238. Ye, Z., Wang, C., Shao, Z., Ye, Q., He, Y., & Shi, Y. (2012). A novel dielectric barrier discharge reactor with photocatalytic electrode based on sintered metal fibers for abatement of xylene. *Journal of Hazardous Materials, 241*, 216–223.
239. Hakoda, T., Matsumoto, K., Shimada, A., Narita, T., Kojima, T., & Hirota, K. (2008). Application of ozone decomposition catalysts to electron-beam irradiated xylene/air mixtures for enhancing carbon dioxide production. *Radiation Physics and Chemistry, 77*, 585–590.
240. Hakoda, T., Matsumoto, K., Mizuno, A., Narita, T., Kojima, T., & Hirota, K. (2008). Oxidation process of xylene in air using under electron beam irradiation. *IEEE Transactions on Industry Applications, 44*, 1950–1956.

241. Hakoda, T., Matsumoto, K., Mizuno, A., & Hirota, K. (2008). Oxidation of xylene and its irradiation byproducts using an electron-beam irradiating a γ-Al_2O_3 bed. *Journal of Physics D: Applied Physics, 41*, 155202.
242. Francke, K.-P., Miessner, H., & Rudolph, R. (2000). Cleaning of air streams from organic pollutants by plasma–catalytic oxidation. *Plasma Chemistry and Plasma Processing, 20*, 393–403.
243. Wei, B. L., Chen, Y. P., Ye, M. J., Shao, Z. H., He, Y., & Shi, Y. (2015). Enhanced degradation of gaseous xylene using surface acidized TiO_2 catalyst with non-thermal plasmas. *Plasma Chemistry and Plasma Processing, 35*, 173–186.
244. Wang, L., Zhang, C. B., He, H., Liu, F. D., & Wang, C. X. (2016). Effect of doping metals on OMS-2/gamma-Al_2O_3 catalysts for plasma-catalytic removal of o-xylene. *Journal of Physical Chemistry C, 120*, 6136–6144.
245. Norsic, C., Tatibouët, J.-M., Batiot-Dupeyrat, C., & Fourré, E. (2016). Non thermal plasma assisted catalysis of methanol oxidation on Mn, Ce and Cu oxides supported on γ-Al_2O_3. *Chemical Engineering Journal, 304*, 563–572.
246. Zhu, X. B., Liu, S. Y., Cai, Y. X., Gao, X., Zhou, J. S., Zheng, C. H., & Tu, X. (2016). Post-plasma catalytic removal of methanol over Mn-Ce catalysts in an atmospheric dielectric barrier discharge. *Applied Catalysis B: Environmental, 183*, 124–132.
247. Lyulyukin, M. N., Besov, A. S., & Vorontsov, A. V. (2016). Acetone and ethanol vapor oxidation via negative atmospheric corona discharge over titania-based catalysts. *Applied Catalysis B: Environmental, 183*, 18–27.
248. Li, Y., Fan, Z., Shi, J., Liu, Z., Zhou, J., & Shangguan, W. (2014). Removal of volatile organic compounds (VOCs) at room temperature using dielectric barrier discharge and plasma-catalysis. *Plasma Chemistry and Plasma Processing, 34*, 801–810.
249. Jia, Z. X., Vega-Gonzalez, A., Ben Amar, M., Hassouni, K., Tieng, S. T., Touchard, S., Kanaev, A., & Duten, X. (2013). Acetaldehyde removal using a diphasic process coupling a silver-based nano-structured catalyst and a plasma at atmospheric pressure. *Catalysis Today, 208*, 82–89.
250. Ohshima, T., Kondo, T., Kitajima, N., & Sato, M. (2010). Adsorption and plasma decomposition of gaseous acetaldehyde on fibrous activated carbon. *IEEE Transactions on Industry Applications, 46*, 23–28.
251. Mizuno, A., Kisanuki, Y., Noguchi, M., Katsura, S., Lee, S. H., Hong, Y. K., Shin, S. Y., & Kang, J. H. (1999). Indoor air cleaning using a pulsed discharge plasma. *IEEE Transactions on Industry Applications, 35*, 1284–1288.
252. Klett, C., Duren, X., Tieng, S., Touchard, S., Jestin, P., Hassouni, K., & Vega-Gonzalez, A. (2014). Acetaldehyde removal using an atmospheric non-thermal plasma combined with a packed bed: Role of the adsorption process. *Journal of Hazardous Materials, 279*, 356–364.
253. Li, Y. Z., Fan, Z. Y., Shi, J. W., Liu, Z. Y., Zhou, J. W., & Shangguan, W. F. (2015). Modified manganese oxide octahedral molecular sieves M '-OMS-2 (M ' = Co,Ce,Cu) as catalysts in post plasma-catalysis for acetaldehyde degradation. *Catalysis Today, 256*, 178–185.
254. Thevenet, F., Olivier, L., Batault, F., Sivachandiran, L., & Locoge, N. (2015). Acetaldehyde adsorption on TiO_2: Influence of NO_2 preliminary adsorption. *Chemical Engineering Journal, 281*, 126–133.
255. Gharib-Abou Ghaida, S., Assadi, A. A., Costa, G., Bouzaza, A., & Wolbert, D. (2016). Association of surface dielectric barrier discharge and photocatalysis in continuous reactor at pilot scale: Butyraldehyde oxidation, by-products identification and ozone valorization. *Chemical Engineering Journal, 292*, 276–283.
256. Liang, W.-J., Li, J., Li, J.-X., Zhu, T., & Jin, Y.-Q. (2010). Formaldehyde removal from gas streams by means of $NaNO_2$ dielectric barrier discharge plasma. *Journal of Hazardous Materials, 175*, 1090–1095.
257. Ding, H. X., Zhu, A. M., Yang, X. F., Li, C. H., & Xu, Y. (2005). Removal of formaldehyde from gas streams via packed-bed dielectric barrier discharge plasmas. *Journal of Physics D: Applied Physics, 38*, 4160–4167.

258. Zhu, X. B., Gao, X., Qin, R., Zeng, Y. X., Qu, R. Y., Zheng, C. H., & Tu, X. (2015). Plasma-catalytic removal of formaldehyde over Cu-Ce catalysts in a dielectric barrier discharge reactor. *Applied Catalysis B: Environmental, 170*, 293–300.
259. Trinh, Q. H., & Mok, Y. S. (2015). Non-thermal plasma combined with cordierite-supported Mn and Fe based catalysts for the decomposition of Diethylether. *Catalysts, 5*, 800–814.
260. Trinh, H. Q., & Mok, Y. S. (2014). Plasma-catalytic oxidation of acetone in annular porous monolithic ceramic-supported catalysts. *Chemical Engineering Journal, 251*, 199–206.
261. Barakat, C., Gravejat, P., Guaitella, O., Thevenet, F., & Rousseau, A. (2014). Oxidation of isopropanol and acetone adsorbed on TiO_2 under plasma generated ozone flow: Gas phase and adsorbed species monitoring. *Applied Catalysis B: Environmental, 147*, 302–313.
262. Trinh, Q. H., Gandhi, M. S., & Mok, Y. S. (2015). Adsorption and plasma-catalytic oxidation of acetone over zeolite-supported silver catalyst. *Japanese Journal of Applied Physics, 54*,1S 01AG04.
263. Zhu, X. B., Gao, X., Yu, X. N., Zheng, C. H., & Tu, X. (2015). Catalyst screening for acetone removal in a single-stage plasma-catalysis system. *Catalysis Today, 256*, 108–114.
264. Zhu, X. B., Tu, X., Mei, D. H., Zheng, C. H., Zhou, J. S., Gao, X., Luo, Z. Y., Ni, M. J., & Cen, K. F. (2016). Investigation of hybrid plasma-catalytic removal of acetone over CuO/gamma-Al_2O_3 catalysts using response surface method. *Chemosphere, 155*, 9–17.
265. Harling, A. M., Wallis, A. E., & Whitehead, J. C. (2007). The effect of temperature on the removal of DCM using non-thermal, atmospheric-pressure plasma-assisted catalysis. *Plasma Processes and Polymers, 4*, 463–470.
266. Ogata, A., Saito, K., Kim, H.-H., Sugasawa, M., Aritani, H., & Einaga, H. (2010). Performance of an ozone decomposition catalyst in hybrid plasma reactors for volatile organic compound removal. *Plasma Chemistry and Plasma Processing, 30*, 33–42.
267. Abedi, K., Ghorbani-Shahna, F., Bahrami, A., Jaleh, B., & Yarahmadi, R. (2015). Effect of TiO_2-ZnO/GAC on by-product distribution of CVOCs decomposition in a NTP-assisted catalysis system. *Polish Journal of Chemical Technology, 17*, 32–40.
268. Zhu, R. Y., Mao, Y. B., Jiang, L. Y., & Chen, J. M. (2015). Performance of chlorobenzene removal in a nonthermal plasma catalysis reactor and evaluation of its byproducts. *Chemical Engineering Journal, 279*, 463–471.

Chapter 7
Plasma-Catalytic Decomposition of Ammonia for Hydrogen Energy

Yanhui Yi, Li Wang, and Hongchen Guo

7.1 Introduction

7.1.1 Hydrogen Energy

The energy and environmental crises have become a common concern in our human society. The use of hydrogen as a kind of sustainable and environmental-friendly energy can not only reduce human dependence on fossil fuels but also reduce greenhouse gas emissions [1]. Due to the oil crisis of the last century, countries around the world are committed to the development of the hydrogen economy. In 1993, Japan put forward the "sunshine plan" and proposed that the key to solve the problem of clean energy is by promoting the use of hydrogen energy. At the beginning of the twenty-first century, the United States issued the "blueprint report" on the development of hydrogen energy. In 2003, the European Union published the "future prospect report" of hydrogen energy and fuel cells. Between 2011 and 2015, China also increased the investment for the research and development (R & D) of hydrogen energy and fuel cells in the "12th Five-Year science and technology plan." Clearly, hydrogen energy has been considered as a "future fuel"; its development and utilization has become one of the world's development trends for the future [2].

So far, the direct application of hydrogen on fuel cell vehicles has been limited by hydrogen storage technologies, which refer to hydrogen storage vessels and

---

Y. Yi · H. Guo (✉)
State Key Laboratory of Fine Chemicals, School of Chemical Engineering, Dalian University of Technology, Dalian, Liaoning, China

L. Wang
State Key Laboratory of Fine Chemicals, School of Chemical Engineering, Dalian University of Technology, Dalian, Liaoning, China

College of Environmental Sciences and Engineering, Dalian Maritime University, Dalian, Liaoning, China

Department of Electrical Engineering and Electronics, University of Liverpool, Liverpool, UK

X. Tu et al. (eds.), *Plasma Catalysis*, Springer Series on Atomic, Optical, and Plasma Physics 106, https://doi.org/10.1007/978-3-030-05189-1_7

hydrogen storage materials. In 2015, the US department of energy proposed requirements for hydrogen storage technology [3]:

(i) High hydrogen content (≧9 wt.% or ≧81 g/L)
(ii) Low cost (≦67 \$/kg_{H2})
(iii) Low hydrogen loss (≦0.12%/day)
(iv) Feasible transportation temperature (−40 to 80 °C) and transportation pressure (≦10 MPa)

Nevertheless, the current status is that:

(i) The gas-state hydrogen storage vessels not only need high pressure (20–70 MPa), but the hydrogen content is also too low (≦1 wt.%);
(ii) The liquid-state hydrogen storage vessels have high costs in order to liquefy hydrogen and inevitably lose a lot of hydrogen (2–3%/day) because of the low boiling point of hydrogen (−252 °C).
(iii) The dehydrogenation temperature and recycling stability of hydrogen storage materials also need to be improved to meet the requirements of practical application [4].

Recently, Pandith and Dahari reviewed the research progresses on hydrogen storage materials [5, 6].

On-site hydrogen production is another alternative for application of hydrogen energy on fuel cell vehicles. Currently, the feed stocks for on-site hydrogen production mainly include hydrocarbons [7, 8], alcohol [9, 10], biomass [11, 12], water [13, 14], and ammonia [15, 16]. However, the hydrogen production from hydrocarbons, alcohol, and biomass inevitably produces CO_2 and CO, which not only causes the deactivation of the platinum electrode in the fuel cell but also causes the greenhouse effect. Water electrolysis is a clean and pollution-free hydrogen production method, but it can not be used in fuel cell vehicles. The reason is that the H_2/O_2 fuel cell reaction is the reverse reaction of H_2O electrolysis, which means no surplus energy can be used to drive the car.

The hydrogen production by ammonia decomposition, however, does not contain any greenhouse gases or pollutants. In addition, NH_3 features a low liquefied pressure, high hydrogen content, low cost and wide availability, meaning it is an excellent hydrogen carrier. Thus, on-site NH_3 decomposition to supply hydrogen for fuel cell vehicles has drawn more and more attention. Recently, Enrique G. B. and Torrente-Murciano L. reviewed the research progress of NH_3 decomposition for hydrogen energy [17, 18].

7.1.2 Thermodynamics of NH_3 Decomposition

NH_3 decomposition reaction (7.1) is endothermic, and its standard molar enthalpy change (ΔH) is about 54.4 kJ/mol. The H_2/O_2 fuel cell reaction (7.2) is exothermic, and its ΔH is about −285.8 kJ/mol. One mole of NH_3 can produce 1.5 moles of H_2, that is, consuming 54.4 kJ energy (NH_3 decomposition) can produce 428.7 kJ energy

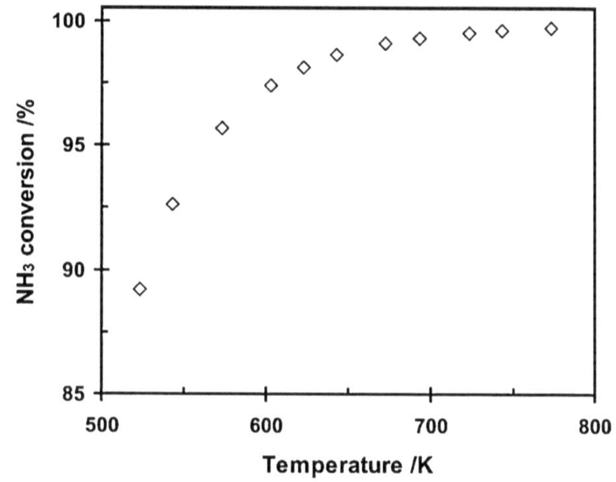

Fig. 7.1 Equilibrium conversion of NH_3 in NH_3 decomposition reaction as a function of temperature at atmospheric pressure [19]

(fuel cell reaction), which means that under ideal conditions, 87.3% of energy produced can be used to drive the vehicle. However, in the actual operation process, the energy efficiency of reactions 7.1 and 7.2 is always lower than 100%. Currently, the energy conversion efficiency of H_2/O_2 proton exchange membrane fuel cell (PEMFC) (reaction 7.2) is about 40–60%, which means that 1.5 mole H_2 in PEMFC could produce 171.5–257.7 kJ surplus energy. In addition to improving the energy efficiency of PEMFC, efforts should also be focused on reducing the energy consumption of NH_3 decomposition, because the less energy consumed in NH_3 decomposition, the more energy can be extracted to drive the car.

$$NH_3(g) \rightarrow 0.5\,N_2(g) + 1.5\,H_2(g) \quad 54.4\ \text{kJ/mol} \tag{7.1}$$

$$H_2(g) + 0.5O_2(g) \rightarrow H_2O(l) \quad -285.8\ \text{kJ/mol} \tag{7.2}$$

Figure 7.1 shows the function of NH_3 equilibrium conversion vs temperature at atmospheric pressure [19]. At the temperature of 573, 673 and 773 K, the equilibrium conversion of NH_3 is about 95.8%, 99.1% and 99.9%, respectively. However, the optimal NH_3 conversion in recent studies is much lower than the equilibrium conversion, especially when cheap metal catalysts were used. Therefore, there is still plenty of room to study the NH_3 decomposition reaction to increase NH_3 conversion at low temperatures as much as possible, especially at high space velocity.

7.1.3 *Catalysts for NH_3 Decomposition*

7.1.3.1 Active Components

The active component is the most important factor in heterogeneous catalysis. The investigated catalysts for NH_3 decomposition reaction mainly focused on noble-metal Ru catalysts [19–25], cheap-metal Fe catalysts [20, 26], Ni catalysts [20, 27–

32], Co catalysts [33–37], as well as transition metal carbides and nitrides catalysts [20, 38–42]. Among these catalysts, Ru is the most active metal component for thermal catalytic NH_3 decomposition.

1. Ruthenium

Ru is the most active component for NH_3 decomposition, and the B5 type Ru atom is the real active site of Ru-based catalysts [43, 44]. The catalytic activities are related to the grain size and morphology of the Ru particle. Zheng's group found that the supported Ru cluster catalysts with an average particle size range of 1.9–4.6 nm exhibited high ammonia decomposition activity, and the best particle size is 2.2 nm [45]. Garcia's group also obtained the same conclusion [44]. The activity of B5 site is also closely related to grain shape, and hemispherical Ru grain with grain size in the range of 1.8–3 nm exhibited the highest ammonia decomposition activity; while the flat and elongated Ru grain with a grain size of about 7 nm also exhibited high ammonia decomposition activity [43]. Although the Ru based catalyst in ammonia decomposition showed a high catalytic activity, it is not suitable for large-scale application because of its scarcity and high cost.

2. Iron

Fe is a cost-effective catalyst, but the NH_3 decomposition activity of the Fe-based catalyst was two orders of magnitude lower than that of the Ru-based catalyst. Nevertheless, it has drawn many people's attentions since Fe is abundant and cheap. The highest NH_3 decomposition activity of Fe-based catalysts was reported by Lu's group [46]. They used the carbon material CMK-5 as a support and encapsulated Fe_2O_3 nanoparticles with grain size of about 6 nm in the channels of CMK-5, which can prevent migration and sintering of active components at high temperature. Using this Fe_2O_3@CMK-5 catalyst, NH_3 can be completely decomposed with 60,000 ml/(g·h) space velocity at 973 K; NH_3 can also be completely decomposed with 7500 ml/(g·h) space velocity at 873 K. In the NH_3 decomposition process, Fe can be easily transformed into iron nitrides. Kowalczyk and co-workers found that both the metallic iron phase and iron nitrides phase can catalyze the decomposition of NH_3 [47]. While Pelka and co-workers found that the catalytic activity of metallic iron phase is much higher than the nano iron nitrides site in NH_3 decomposition, and the corresponding apparent activation energy was 68 and 143 kJ/mol, respectively [48, 49].

3. Nickel

Ni is a high-efficiency active component for ammonia decomposition. Ganley and co-workers, using Al_2O_3 as the support, found that the sequence of catalytic activity of NH_3 decomposition is Ru > Ni > Rh > Co > Ir > Fe [50]. Yin and co-workers, using CNTs as the support, found that the sequence of catalytic activity of NH_3 decomposition is Ru > Rh > Ni > Pd $\approx$ Pt > Fe [19]. Liu and co-workers found that 89% NH_3 conversion could be achieved over a Ni/SBA-15 catalyst at 823 K with a space velocity of 30,000 ml/(g·h) [51]. Xu and co-workers found that 98.3% NH_3 conversion could be achieved over a CeO_2-modified Ni/Al_2O_3 catalyst at 823 K with a space velocity of 30,000 ml/(g·h) [52]. Since Ni is much more abundant and cheaper than Ru, and its catalytic activity is relatively higher than some other transition metals, Ni-based catalysts provide a feasibility for NH_3 decomposition in large-scale application.

4. Cobalt

Co is also an efficient active component in NH_3 decomposition, and its activity is higher than Fe catalysts [33, 34, 53]. Arabczyk and co-workers found that the activation energies of NH_3 synthesis reaction over Co and Fe catalysts are 268 and 180 kJ/mol, respectively. However, the activation energy of NH_3 decomposition reaction over Co and Fe catalysts are 111 and 138 kJ/mol, respectively [33]. In addition, some reports indicate that when Co is used as the second active component it can regulate the electronic properties of the bimetallic catalyst, which can weaken the strong interactions between metal atoms and N, resulting in acceleration of N atom recombination to improve the ammonia decomposition activity [54].

5. Transition Metal Carbides and Nitrides

In 1973, Levy found that tungsten carbide catalysts have similar catalytic behaviors to a noble metal Pt catalyst [55], since transition metal nitrides or carbides have similar electronic structures to noble metals [56]. This finding extended the application of transition metal nitrides and carbides in heterogeneous catalysis. So far, carbide and nitride catalysts used in NH_3 decomposition include tungsten carbide [38, 57], molybdenum nitride [58, 59] and iron nitride [48, 49, 60]. Among the investigated transition metal carbides and nitrides, Mo_2N exhibited the highest activity in thermal catalytic NH_3 decomposition. In addition, introduction of Co into molybdenum nitride (Co_xMo_yN/gamma-Al_2O_3) can further improve the catalytic activity. The reason is that Co reduces the adsorption energy of N atoms on Mo active site [61].

7.1.3.2 Supports

Supports can influence the catalytic properties of active component metals via regulating the size, morphology and electronic properties of the metal particles. The reported supports of catalysts for NH_3 decomposition mainly include carbon materials (activated carbon, carbon nanotubes, fullerene and carbon fiber) [62–64], zeolites (MCM-41 and SBA-15) [27], and metal oxides (MgO, Al_2O_3, SiO_2, ZrO_2 and TiO_2) [65–67]. The specific surface area, acidity, alkalinity and electron transfer capability are the important properties of supports, which have a great influence on NH_3 decomposition. Generally, acidic supports have an inhibitory effect on NH_3 decomposition; however, alkaline supports have a lower activation energy for ammonia decomposition; a large specific area of support favors the high dispersion of metals. Therefore, supports with large specific surface areas, strong alkalinity, and high electron transfer capability are beneficial to the NH_3 decomposition reaction. With regard to Ru catalysts, carbon materials, especially some carbon materials with high thermal stability and strong electrical conductivity, are the best supports.

Regarding other transition metal catalysts, however, alkaline metal oxide and silicon materials exhibited better performance than some other support materials.

7.1.3.3 Promoters

Generally, promoters have a dramatical influence on heterogeneous catalysis. As literature reported, alkali metals, alkaline earth metals and rare earth metals usually play the role of catalyst promoter to accelerate the NH_3 decomposition reaction. The functions of promoters mainly include:

1. Adjusting the electronic structure of the main active components as an electron promoter. Yin et al. reported that promoter modification has dramatically improved the catalytic activity of the Ru/CNTs catalysts, and the catalytic activities sequence follow K-Ru > Na-Ru > Li-Ru > Ce-Ru > La-Ru > Ca-Ru > Ru [66]. Rarog et al. also found that the catalytic activity sequence of the modified Ru/CNTs catalysts follow Cs-Ru > K-Ru > Ba-Ru > Ca-Ru [68]. These studies clearly indicate that, with regard to Ru catalysts, the electronegativity of the promoters is a key factor; the lower the electronegativity, the better the catalytic performance. The reason is that the promoters with low electronegativity can transfer electrons to Ru easier, which can weaken the bond energy of Ru–N, resulting in promoting the desorption recombination of the surface adsorbed N atoms and eventually improving NH_3 decomposition activity.
2. Improving the dispersion of the metal and inhibiting the sintering of nano metal particles (Ru, Ni, etc.) at high temperature as a structure promoter [52, 66, 69].
3. Adjusting the interaction between the active metal component and the support. When the Ni/Al_2O_3 catalyst was used in NH_3 decomposition, strong interactions between Ni and Al lead to the formation of nickel aluminum spinel, which strongly inhibited the reduction of Ni, and thus resulted in decrease of NH_3 conversion [52]. While CeO_2 modification has weakend the interaction between Ni and support Al_2O_3, which improved the reducibility of NiO, leading to enhancement of NH_3 conversion [52]. CeO_2 modification has also improved the catalytic performance of the Ni/SBA-15 catalyst in NH_3 decomposition [69].

Some experimental results of NH_3 decomposition by heterogeneous catalysis are shown in Table 7.1. It can be seen that Ru catalysts exhibit excellent activity in NH_3 decomposition, even at relatively low temperature and high space velocity. However, the other transition metal catalysts can only show high NH_3 conversion at high temperatures.

7.1.4 Mechanism of Catalytic NH_3 Decomposition

The methanism of NH_3 decomposition is more complicated than that of NH_3 synthesis. It is widely believed that NH_3 decomposition on the catalysts surface is mainly composed of a series of step-by-step dehydrogenation processes, as shown in

Table 7.1 Studies of hydrogen production from ammonia decomposition by thermal catalysis

Catalysts	NH_3 conversion (%)	H_2 productivity (mmol/g·min)	Reaction conditions	Reference
$LiNH_2$-Ru/MgO	100	68.3	547 °C, 60,000 ml/(g·h)	[21]
Ru-Ba$(NH_2)_2$	–	8.07	400 °C, 60,000 ml/(g·h)	[25]
K-Ru/MgO-CNTs	99	66.3	450 °C, 0.1 g, 100 ml/min NH_3, 60,000 ml/(g·h)	[65]
Microcapsular-like Ru@SiO_2	100	33.5	550 °C, 0.1 g, 50 ml/min NH_3, 30,000 ml/(g·h)	[70]
Fe_2O_3/CMK-5	100	66.9	700 °C, 0.025 g, 25 ml/min NH_3, 60,000 ml/(g·h)	[46]
Cs-Fe@micro-SiO_2	100	33.5	650 °C, 0.1 g, 50 ml/min NH_3, 30,000 ml/(g·h)	[71]
CeO_2-Ni/Al_2O_3	98	32.9	550 °C, 0.1 g, 50 ml/min NH_3, 30,000 ml/(g·h)	[52]
Co/MgO-La_2O_3	91	–	550 °C, 6000 ml/(g·h)	[37]
5% $CoFe_5$ on CNTs	48	19.3	600 °C, 0.05 g, 30 ml/min NH_3, 36,000 ml/(g·h)	[72]
WC	100	–	500 °C, 0.05 g, NH_3:He =1:54, 55 ml/min total gas flow	[38]
MoN_x/SBA-15	100	–	650 °C, 15,800 ml/(g·h)	[39]
MnN-Li_2NH	–	37.5	500 °C, 60,000 ml/(g·h)	[40]

reactions 7.3, 7.4, 7.5, and 7.6, where (g) represents a gas-phase species and (ad) represents an adsorbed surface species. Then, the adsorbed H and N atoms may combine into adsorbed H_2 and N_2 molecule, respectively (reactions 7.7 and 7.8). The adsorbed H_2 and N_2 molecule eventually desorbed from catalyst surface (reactions 7.9 and 7.10).

$$NH_3(g) \rightarrow NH_3(ad) \tag{7.3}$$

$$NH_3(ad) \rightarrow NH_2(ad) + H\ (ad) \tag{7.4}$$

$$NH_2(ad) \rightarrow NH(ad) + H(ad) \tag{7.5}$$

$$NH(ad) \rightarrow N(ad) + H(ad) \tag{7.6}$$

$$2N(ad) \rightarrow N_2(ad) \tag{7.7}$$

$$2H(ad) \rightarrow H_2(ad) \tag{7.8}$$

$$N_2(ad) \rightarrow N_2(g) \tag{7.9}$$

$$H_2(ad) \rightarrow H_2(g) \tag{7.10}$$

Table 7.2 Dissociation energy of N–H bond

Species	Bond	Dissociation energy		
		kcal/mol	kJ/mol	eV
NH_3	H-NH_2	107.6	450.2	4.7
NH_2	H-NH	93.0	389.4	4.0
NH	H-N	78.4	328.0	3.4

With regard to the mechanism of catalytic NH_3 decomposition, the rate controlling step is considered to be either the breaking of the first N–H bond of NH_3 molecule (reaction 7.4) or the recombinative desorption of adsorbed N atom (reactions 7.7 and 7.9). As shown in Table 7.2, the dissociation energy for the N–H bond of NH_3, NH_2 and NH species reduces successively, which means that the breaking of the first N–H bond of NH_3 molecule requires the highest energy. If the dissociation energy of the first N–H bond of NH_3 molecule (E_{dis}) is higher than the energy of N_{ad} atoms recombinative desorption (E_{des}), i.e., $E_{dis} > E_{des}$, the breaking of the first N–H bond of the NH_3 molecule (reaction 7.4) is the rate controlling step. On the contrary, in the case of $E_{des} > E_{dis}$, the recombinative desorption of adsorbed N atom (reactions 7.7 and 7.9) is the rate controlling step. Many researches have demonstrated that the breaking of the first N–H bond of NH_3 molecule (reaction 7.4) is the rate controlling step over Ru, Ir, Pd, Pt and Cu Catalysts [50, 73, 74]; however, on the surface of Fe, Co, and Ni catalysts, the recombinative desorption of adsorbed N atom (reactions 7.7 and 7.9) is the rate controlling step [50, 75, 76].

The core issue of NH_3 decomposition is how to accelerate the rate controlling step, which is mainly dependent on the strength of the M–N bond (M represents the active site of the metal catalysts). (1) A weak M–N bond is beneficial to the desorption of N atoms, but it is difficult to dissociate the N–H bond of an NH_3 molecule; (2) a strong M–N bond is in favor of dissociating N–H bond, but strongly adsorbed N atoms are difficult to desorb from the surface of the catalyst; and (3) moderate M–N bond can not only benefit the dissociation of N–H bond but also favor the desorption of the strong adsorbed N atoms. Therefore, the ideal catalyst for NH_3 decomposition should have a moderate M–N bond.

Vlachos and co-workers, using first principles calculation and modeling, studied the NH_3 conversion versus the nitrogen binding energies with different metal catalysts (Fig. 7.2). A volcano curve was established, and Ru with a moderate Ru–N bond energy of ca. 134 kcal/mol was found to have the highest activity, which is consistent with some experimental results [77]; when the metal–nitrogen bonding energy is too high or too low, it is unfavorable to NH_3 decomposition This finding reveals that the binding energy of metal–nitrogen bond is a good reaction activity descriptor for catalytic NH_3 decomposition. With regards to the catalysts Fe, Co and Ni, the recombinative desorption of the strong adsorbed N atoms is the bottleneck of NH_3 decomposition, and the key issue for scholars has become how to find a way to solve this problem. Therefore, in addition to the innovation of catalyst preparation process, it is necessary to seek some new methods to help the catalyst to achieve efficient NH_3 decomposition. The following will introduce the plasma-catalytic NH_3 decomposition in a gliding arc discharge (GAD) reactor and dielectric barrier discharge (DBD) reactor, respectively.

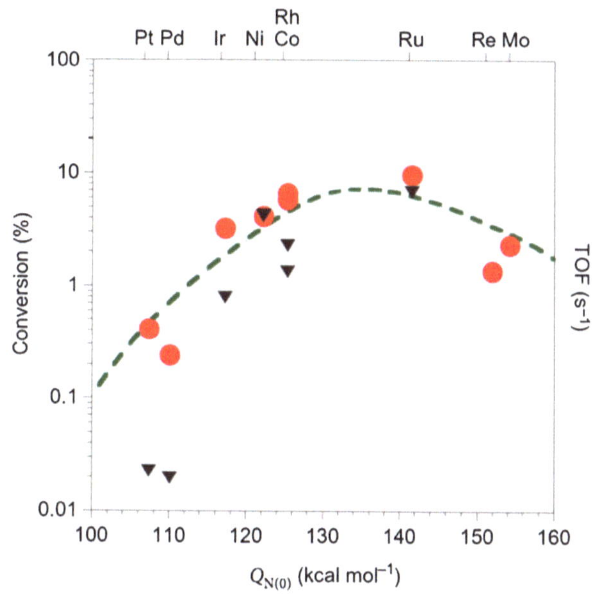

Fig. 7.2 Ammonia decomposition volcano curve. Ammonia conversion calculated from microkinetic modeling (circles, left axis) at 850 K for various transition-metal catalysts and experimental supported catalyst turnover frequencies (TOF, triangles, right axis) at 850 K plotted against the nitrogen binding energies $Q_{N(0)}$ [77]

7.2 NH_3 Decomposition in a Gliding Arc Discharge Reactor

Zhao and co-workers investigated NH_3 decomposition in gliding arc discharge reactors [78–80]. A schematic of the experimental apparatus used for ammonia decomposition in GAD reactor is shown in Fig. 7.3. It consists of an ammonia gas supply system (ammonia tank and mass flow controller), a reaction system (reactor and power supply) and an analytical system (on-line gas chromatograph with thermal conductivity detector for N_2 and NH_3 detection). The discharge voltages, discharge currents and input powers were measured by the Tektronix DPO-3012 Digital phosphor Oscilloscope (Tektronix, America) with a power module. The reactors temperature was taken by thermocouples at two sites: one was fixed on the external surface of the reactor shell facing the electrode gap, and the other was bound to the lower end of the ground electrode. The discharge image (the size of arc filament) was obtained with a digital camera at exposure time 1/25 s (Sony H10, Japan). The active species of the ammonia plasma in the electrode gap were detected by the optical emission spectrometry under the conditions of 0.5 s exposure and 10 μm slit width (SP2758 Princeton Instruments, America). In order to calculate the ammonia conversion, the on-line mass spectrometer was used in advance to qualitatively analyze the products of ammonia decomposition in the reactor. It was confirmed that the products of ammonia decomposition in this type of plasma reactor contained only H_2 and N_2, and no hydrazine was found.

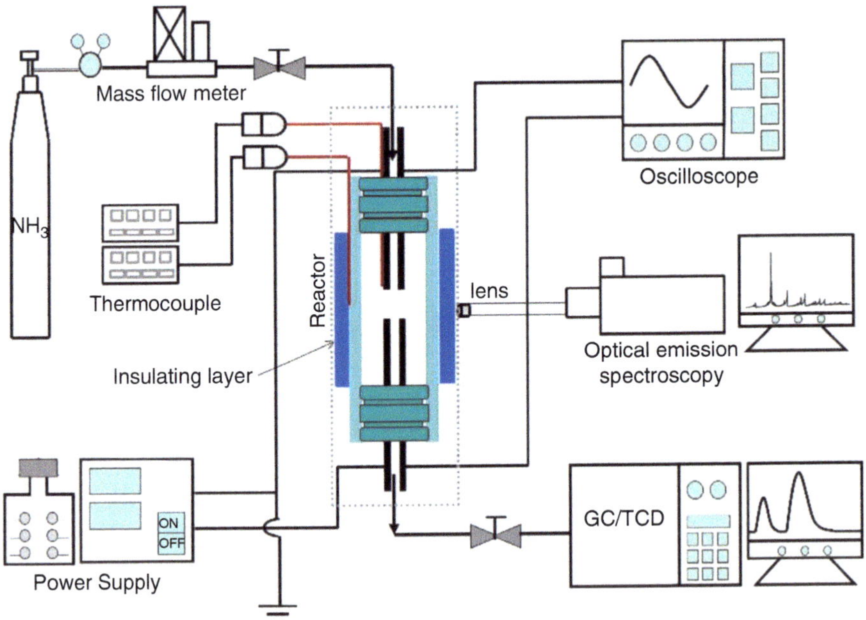

Fig. 7.3 The schematic illustration of the experimental apparatus and process for ammonia decomposition [80]

7.2.1 *Influence of Reactor Configurations*

The reactor configuration has a dramatic effect on NH_3 decomposition [78]. Gliding arc discharge (GAD) can be triggered by alternating current plasma power in flat-flat, point-flat, tube-flat and tube-tube plasma reactors, as shown in Fig. 7.4. The flat-flat, point-flat, and tube-flat plasma reactors have the same high voltage electrode, i.e., stainless steel circular plate (3 mm thickness and 50 mm diameter); however, they have different ground electrodes. In the flat-flat, point-flat, and tube-flat plasma reactors, the ground electrodes consist of a stainless steel circular plate (3 mm thickness and 50 mm diameter), stainless steel bar (3 mm diameter) with sophisticated structure, and a stainless steel tube (3 mm diameter). In the tube-tube reactor, both the high voltage electrode and ground electrode are stainless steel tubes (3 mm diameter).

In the flat-flat, point-flat and tube-flat plasma reactors, a dielectric barrier discharge (DBD) can be generated when the high voltage electrode is covered by an intact insulate dielectric. Otherwise, when the high voltage electrode was naked or was covered by an insulate dielectric with a hole in the center, the GAD would be generated. However, in the tube-tube reactor, only GAD can be generated. As shown in Fig. 7.5, the DBD plasma (labeled by intact dielectric barrier) exhibited the lowest NH_3 conversion, no matter what kind of the reactors were used or how much power was inputted (20, 30, 40 and 50 W). In other words, the GAD plasma showed much

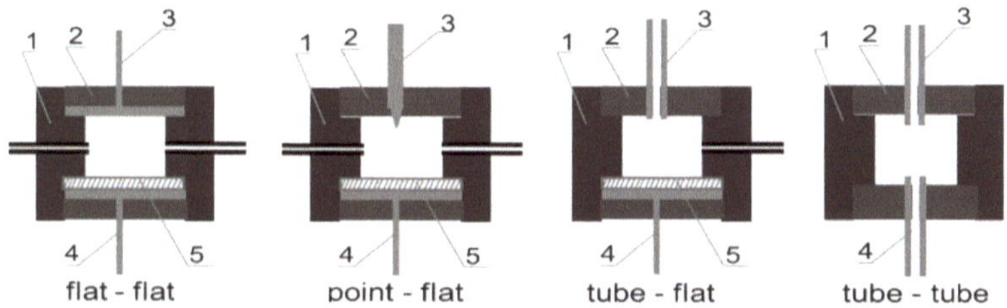

Fig. 7.4 Scheme of configuration of the plasma reactor for NH_3 decomposition [78]. (1) reactor shell; (2) fixing devices; (3) ground electrode; (4) high-voltage electrode; (5) dielectric

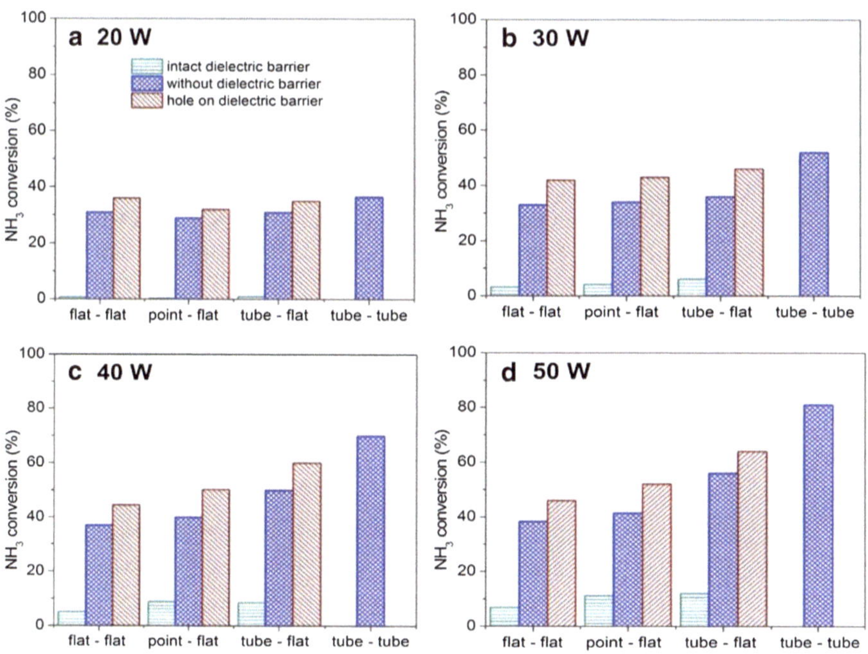

Fig. 7.5 The influence of reactor configuration and discharge mode on NH_3 conversion [78]

higher NH_3 conversion than the DBD plasma. In addition, when the high voltage electrode was covered by a dielectric with a hole (labeled by hole on dielectric barrier), higher NH_3 conversion was obtained than that of naked high voltage electrode (labeled by without dielectric barrier). That is, in the flat-flat, point-flat and tube-flat plasma reactors, the NH_3 conversion followed the sequence of hole on dielectric barrier > without dielectric barrier > intact dielectric barrier. However, at the condition of GAD plasma, the tube-tube reactor has the highest NH_3 conversion in the investigated four reactors, and the NH_3 conversion followed the sequence of tube-tube > tube-flat > point-flat > flat-flat.

7.2.2 Mechanism of NH_3 Decomposition in a GAD Reactor

In the mode of DBD plasma, the optical emission spectroscopy (OES) has been collected. The OES of NH_3 DBD plasma was dominated by the band of the electronic excited NH_3^* molecule, i.e., 563.5–567 nm (Fig. 7.6). There were also some bands of N_2^* (300–500 nm) and NH^* (336.0 nm), and a line of H^* (656.3 nm), but their intensities were extremely weak. The attribution of these bands and the line are shown in Table 7.3. However, in the mode of GAD plasma, the OES (Fig. 7.7) was dominated by the line of H^* (656.3 nm), as well as the bands of NH^* (336.0 nm) and NH_2^* (661.9 nm) species. There were also some bands of N_2^* (300–500 nm), N_2^+ (391.4 and 427.8 nm) and H_2^* (400–700 nm) with weak intensity. The attribution of these bands and line are shown in Table 7.4. The most important information is that the band of NH_3^* species did not appear in the OES of NH_3 GAD plasma (Fig. 7.7), but it was collected as the main peak in the OES of NH_3 DBD plasma (Fig. 7.6), indicating different mechanism of NH_3 decomposition in the modes of DBD and GAD plasma.

The NH_3 conversion of DBD plasma was much lower than that of the GAD plasma. The above mentioned OES data indicates that, in the NH_3 DBD plasma, the decomposition of NH_3 may be mainly achieved by the NH_3^* species, and the most

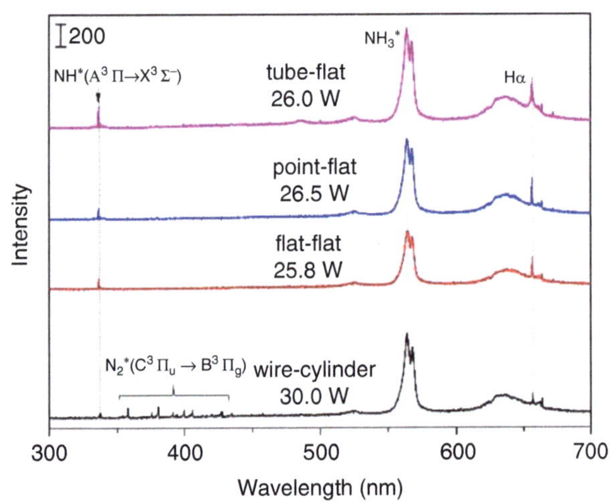

Fig. 7.6 The OES of DBD NH_3 plasma

Table 7.3 Active species in DBD NH_3 plasma [81–84]

Species	NH	N_2^*	NH_3^*	H_a
λ, nm	336	337max	563.5 and 567 continuum	656.3
Electronic transition	$A^3\Pi \rightarrow X^3S$-	$C^3\Pi_u \rightarrow B^3\Pi_g$	Schuster's bands	$2p^2p^0{}_{3/2}$–$3d^2D_{3/2}$

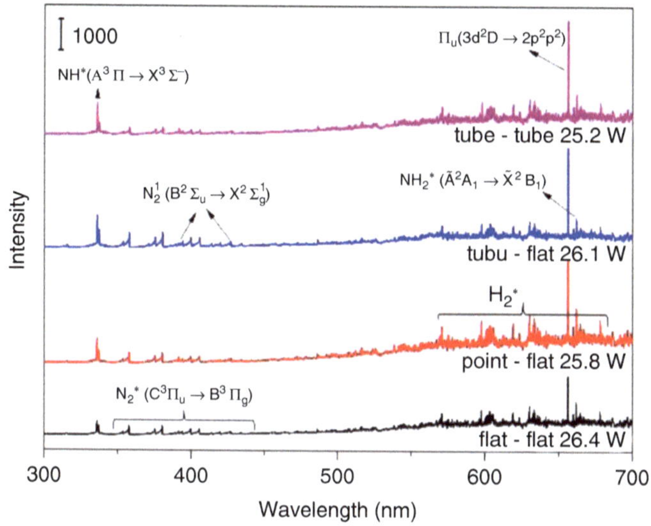

Fig. 7.7 The OES of GAD NH_3 plasma

Table 7.4 Active species in GAD NH_3 plasma [83–86]

Species	λ (nm)	Electronic transition	Species	λ (nm)	Electronic transition
NH	336, 337	$A^3\Pi \rightarrow X^3S^-$	N_2^*	300–500	$C^3\Pi_u \rightarrow B^3\Pi_g$
H_α	656.3	$2p^2p^0{}_{3/2}$–$3d^2D_{3/2}$	H_2^*	400–700	–
N_2^*	391.4, 427.8	$B^2\Sigma_u \rightarrow X^2\Sigma_g{}^1$	NH_2	661.9	$\tilde{A}^2A_1 \rightarrow X^2B_1$

possible NH_3 decomposition mechanism in DBD plasma is shown as the reactions 7.11, 7.12, and 7.13.

$$e + NH_3 \rightarrow NH_3{}^* + e \tag{7.11}$$

$$e + NH_3{}^* \rightarrow NH_2 + H + e \tag{7.12}$$

$$e + NH_2 \rightarrow NH + H + e \tag{7.13}$$

However, in the NH_3 GAD plasma, the decomposition of NH_3 may be mainly achieved by the NH_2 and NH species, and the most possible NH_3 decomposition mechanism in GAD plasma is shown as the reactions 7.14, 7.15, and 7.16.

$$e + NH_3 \rightarrow NH_2 + H + e \tag{7.14}$$

$$e + NH_3 \rightarrow NH + H_2 + e \tag{7.15}$$

$$e + NH_3 \rightarrow NH + H + H + e \tag{7.16}$$

Generally, GAD plasma has a much higher electron energy and electron density than the DBD plasma. In the NH_3 DBD plasma, NH_3 was mainly activated into

NH_3^* species, then the NH_3^* species was further activated by using an electron to dehydrogenate and generate NH_2 and NH species. In the NH_3 GAD plasma, however, NH_3 was directly activated and transformed into NH_2 and NH species by high energy electrons, which could be the reason why GAD plasma exhibited much higher efficiency in NH_3 decomposition than the DBD plasma. As literature reported, there were many reactions that produce H_2 and N_2 from NH_2 and NH species, but the fastest reaction is the self-disproportionation reaction (7.17) [82]. So, the H_2 and N_2, in both DBD and GAD NH_3 plasma may be mostly generated from NH species through reaction 7.17.

$$NH + NH \rightarrow N_2 + H_2 \tag{7.17}$$

7.2.3 *Synergy of the Plasma and Electrode Catalysts in the Tube-Tube Reactor*

GAD plasma in the tube-tube reactor was the most efficient path for NH_3 conversion. In the tube-tube reactor, electrode materials were investigated, and it has been found that some metal electrodes also played the role of a catalyst for NH_3 decomposition [79]. As shown in Fig. 7.8, when Ni and stainless steel (SS) electrodes were used in the tube-tube reactor, there were clear induction periods, during which NH_3 conversion increased remarkably with time and remained constant after 100 min and 240 min, respectively. When Cu electrodes were used, however, a constant NH_3 conversion appeared after the beginning of the discharge. To compare the performance of different electrodes in NH_3 decomposition, data were taken after the induction period. As shown in Fig. 7.9, the ammonia conversion increased with the input power no matter which electrodes were used. At the same input power, however, the ammonia conversion varied with electrode materials and followed the sequence of Ni > SS > Cu.

The catalytic function of the Ni and SS electrodes was investigated by Zhao and co-workers [79]. The external surfaces of both high-voltage electrode and ground electrode were defunctionalized by tight quartz annular tubes. As shown in Fig. 7.10, the defunctionalization of the external surface of Cu electrode had no influence on the NH_3 conversion, indicating that the surface of Cu electrodes was inert. In other words, the NH_3 conversion was contributed to only by gas-phase plasma reactions when the Cu electrode was used. In contrast, the defunctionalization of the external surface of Ni and SS electrodes caused remarkable decreases in NH_3 conversion, which implied that the surfaces of Ni and SS electrodes had obvious catalytic activity for NH_3 decomposition. Taking the Cu electrodes as a reference, the contributions to the NH_3 conversion by the catalytic decomposition of Ni and SS electrodes were estimated to be 51.7% and 49.8%, respectively.

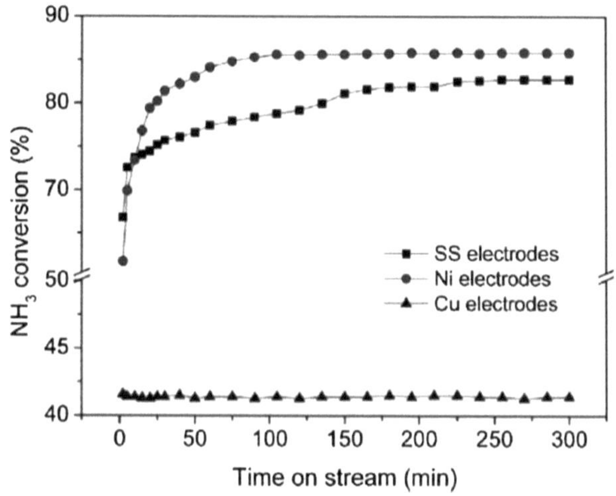

Fig. 7.8 Time dependence of ammonia conversion on different electrodes [79]

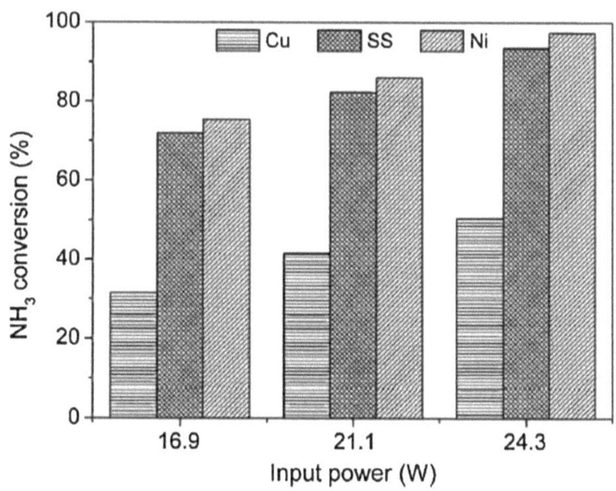

Fig. 7.9 Comparison of the ammonia conversion obtained with different electrodes. (Data was taken after ammonia conversion induction) [79]

The electrode catalysis of the high-voltage electrode and the ground electrode were investigated, using the tube-tube reactors with Cu–Ni and Cu–SS electrodes, respectively. As shown in Fig. 7.11, when Cu was used as a high-voltage electrode and a ground electrode, respectively, and Ni and SS were used as the counter electrode accordingly, the NH_3 conversion decreased significantly in all cases when compared with those double Ni and SS electrodes. Comparatively, the use of Cu as high-voltage electrode caused a bigger drop in NH_3 conversion. This means that in the cases of double Ni and SS electrodes, the high-voltage electrodes had a higher catalytic activity than the ground electrode. However, it seemed that the relative catalytic activity of the high-voltage electrode varied with electrode material. For example, the catalytic activity ratio of the high-voltage electrode to the ground electrode was estimated to be about 3:2 for Ni electrodes and 2:1 for SS electrodes. The temperature difference between the high-voltage electrode and ground electrode

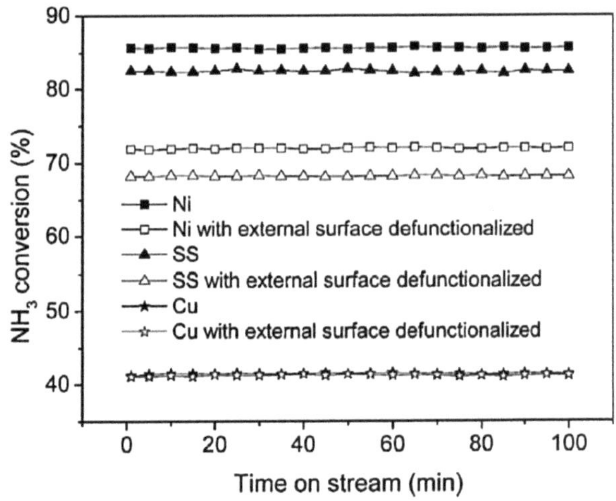

Fig. 7.10 Influence of metal electrode external surface defunctionalization with quartz annular tubes on NH_3 conversion. (Data was taken after NH_3 conversion induction) [79]

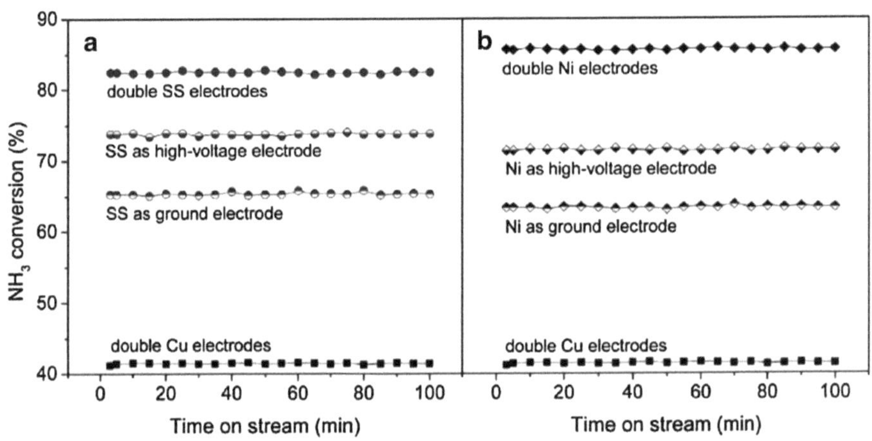

Fig. 7.11 Influence of substituting Ni and SS electrode with a Cu electrode on the NH_3 conversion. (Data was taken after ammonia conversion induction) [79]

in the SS electrode reactor is shown in Fig. 7.12. It clearly shows that the temperature of the high-voltage electrode was much higher than that of the ground electrode. A similar phenomenon also appeared with Ni and Cu electrode reactors. Thus, the different catalytic activity of the high voltage electrode and ground electrode could be mainly caused by the different temperatures.

The essence of the induction period for NH_3 decomposition in the case of both SS and Ni electrodes was investigated. The XRD patterns of the SS electrode samples with different discharge time during the induction period are shown in Fig. 7.13. It can be seen that the fresh SS electrode had the typical structure of austenitic stainless steel which was featured by the existence of (Fe, C) phase. During the induction period, the intensities of the austenitic stainless steel characteristic peaks decreased

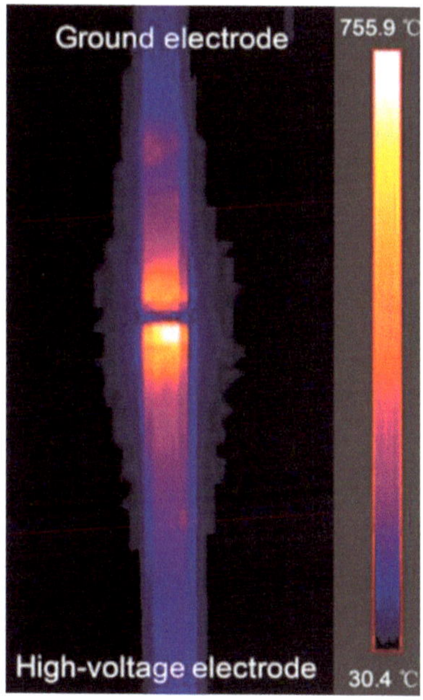

Fig. 7.12 The temperature of the SS electrodes in the tube-tube reactor detected using a thermal infrared imager [79]

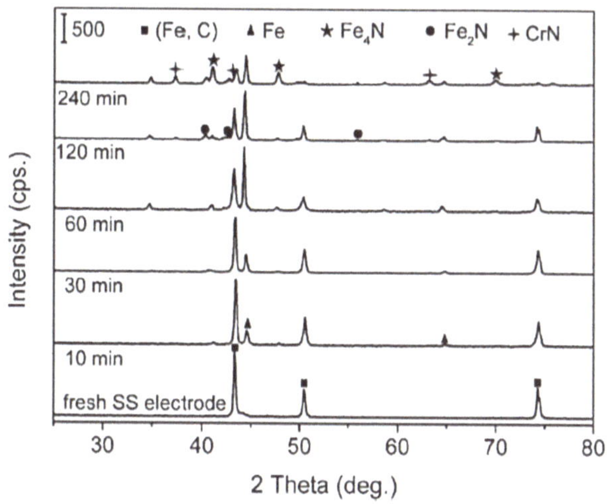

Fig. 7.13 XRD patterns of SS electrodes sampled at different discharge times during the induction period [79]

with time. Meanwhile, new phases of Fe, Fe_4N and Fe_2N appeared. Specifically, the Fe phase appeared first of all, and its peak intensities notably increased from 10 min to 120 min and decreased by 240 min. Fe_4N and Fe_2N appeared after 30 min and 120 min, respectively, and their peak intensities increased thereafter in the induction period. As before mentioned, Fe and its nitrides had catalytic activity for ammonia decomposition in traditional heterogeneous reactors [46–49, 60]. Therefore, it is clear that the induction period of the SS electrode was actually a complex process featured by the reduction of the austenitic stainless steel (Fe, C) to Fe phase first, and

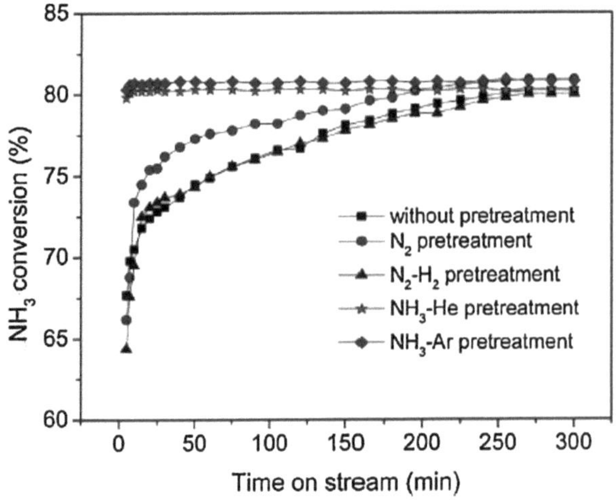

Fig. 7.14 Effect of different discharge pretreatment on the induction period of SS electrodes in the decomposition of ammonia [79]. (N_2 pretreatment (20 ml/min, 300 min); N_2-H_2 pretreatment (N_2/H_2 = 1:3, 80 ml/min, 300 min); NH_3-He pretreatment (NH_3/He = 1:1, 80 ml/min, 40 min); NH_3-Ar pretreatment (NH_3/Ar = 1:1, 80 ml/min, 40 min))

then the Fe phase was partially nitrided to Fe_4N and Fe_2N. The increase of the ammonia conversion during the induction period could be attributed to the catalytic activity of Fe and its nitride on the surface of electrode.

Four kinds of discharge pretreatments have been employed for the SS electrodes, as shown by Fig. 7.14. It can be seen that the SS electrodes pretreated in nitrogen or in the mixture of nitrogen and hydrogen needed a similar induction period compared to the fresh SS electrodes when they were used for NH_3 decomposition in the tube-tube reactor. In contrast, the SS electrodes pretreated in the mixtures of NH_3-Ar and NH_3-He for 40 min could entirely eliminate the induction period in the decomposition of ammonia. It can be inferred that it was not nitrogen but ammonia that made the SS electrodes nitrided during the induction period.

The XRD patterns of the Ni electrode samples with varied discharge times during the induction period are shown in Fig. 7.15. However, no obvious diffraction peaks related to Ni nitride were found, although the same induction period as the SS electrode was seen during ammonia decomposition. Using the on-line mass spectroscopy, it has been found that the used Ni and SS electrodes gave an obvious nitrogen desorption peak when they were treated by He GAD (Fig. 7.16). This experiment confirmed the existence of a small amount of nitrides on the surface of Ni electrodes, but it may not be identified by XRD due to small amount or small grains. In order to demonstrate this idea, the following experiments were carried out. First, the SS and Ni electrodes were treated in NH_3 GAD for 300 min and 120 min, respectively. Second, the pretreated electrodes were then used for NH_3 decomposition directly. Third, the pretreated electrodes were re-treated by He GAD for 60 min to desorb N. Fourth, the re-treated electrodes were subjected to NH_3 decomposition. Figure 7.17 shows that the Ni electrodes and the SS electrodes exhibited similar NH_3 conversion curves in each case. That is, after pretreated by NH_3 GAD, the induction period has been eliminated; however, the induction period reappeared when the He-retreated Ni and SS electrodes were used. These comparative experiments

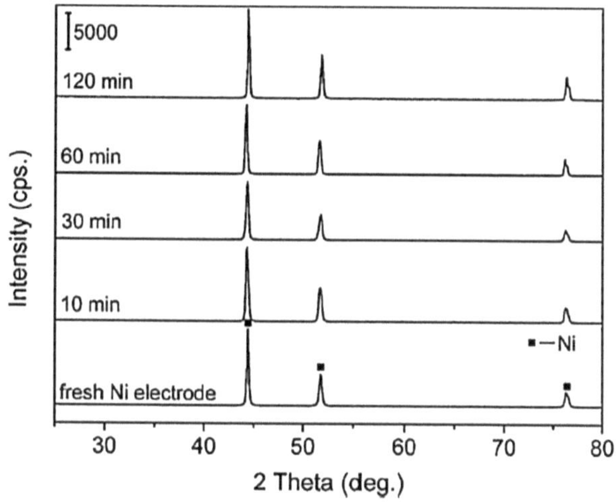

Fig. 7.15 XRD patterns of Ni electrodes sampled at different discharge times during the induction period [79]

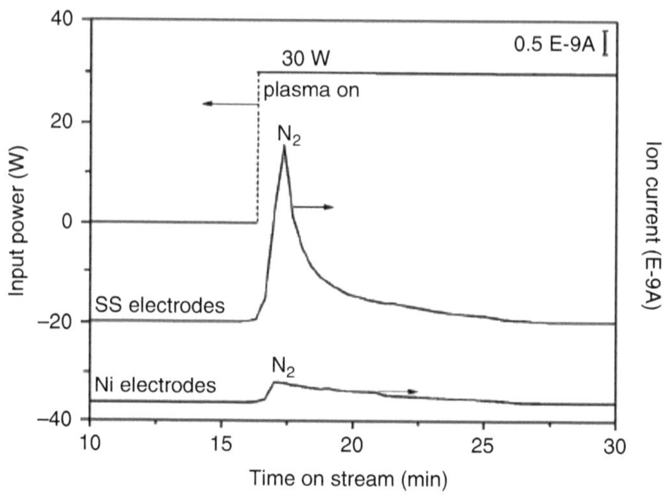

Fig. 7.16 On-line mass spectroscopy monitor of the He discharge desorption treatment of the Ni and SS electrodes that were pretreated by ammonia discharge [79]

assured that the nitridation should be related to the induction period of the Ni electrodes. Thus, the Ni nitride on the surface of the Ni electrode may have very high NH_3 decomposition activity in view of its relatively small amount compared with the nitrides on the surface of SS electrodes.

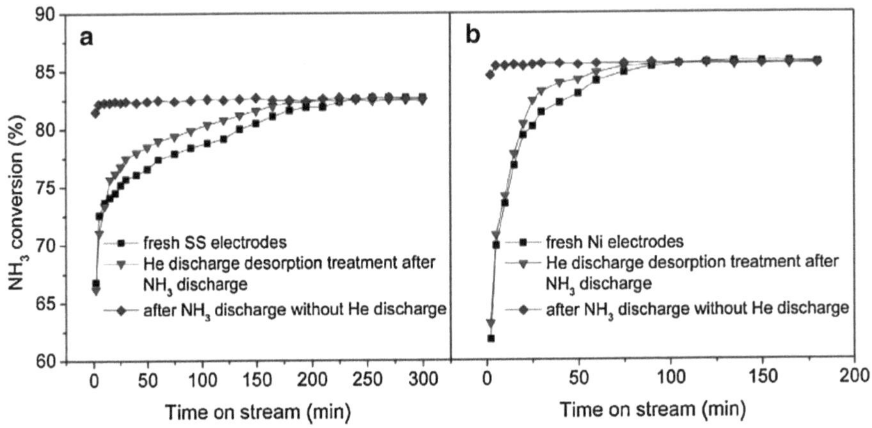

Fig. 7.17 Effects of ammonia-discharge and He-discharge treatments on the inductions of the Ni and SS electrodes [79]

7.2.4 *Influence of Process Parameters in Tube-Tube Reactor*

7.2.4.1 The Parameters of the Tube-Tube Reactor

In the tube-tube reactor, the electrode gap and electrode tube diameter had remarkable influence on NH_3 conversion [80]. It can be seen from Fig. 7.18, when the electrode tube diameter was 2 mm, the ammonia conversion increased with the decrease of electrode gap from 10 to 1 mm. However, when the electrode tube diameter was 3 mm, the ammonia conversion increased with the increase of electrode gap from 1 to 4 mm, and then decreased with the further increase of the electrode gap from 4 to 10 mm. When the electrode tube diameter was 6 mm, the ammonia conversion increased with the increase of electrode gap from 1 to 10 mm.

As mentioned in Sect. 7.2.3, the NH_3 conversion in the tube-tube reactor was contributed to by both the gas-phase plasma decomposition reaction and the electrode-surface catalytic decomposition reaction. When SS electrodes were used, the contributions of the two aspects to the total NH_3 conversion were about a 50/50 percentage under the conditions of 25 W input power, 40 ml/min ammonia flow rate, 4 mm electrode gap, and 3 mm electrode tube diameter [79]. However, the surface of Cu electrodes did not have any catalytic activity for NH_3 decomposition. Therefore, Cu electrodes have been used to gain insight into the influence of the electrode gap and electrode tube diameter on NH_3 conversion.

The discharge image and the temperature of the tube electrode were attained. As shown in Fig. 7.19, the discharge was featured by a bright arc filament between the two electrodes. The area of the bright arc filament was considered as the effective plasma volume. This means that the gas-phase plasma decomposition of NH_3 mainly occurred in the effective plasma volume where energetic electrons existed. The effective plasma volume would therefore be larger when the arc filament got longer

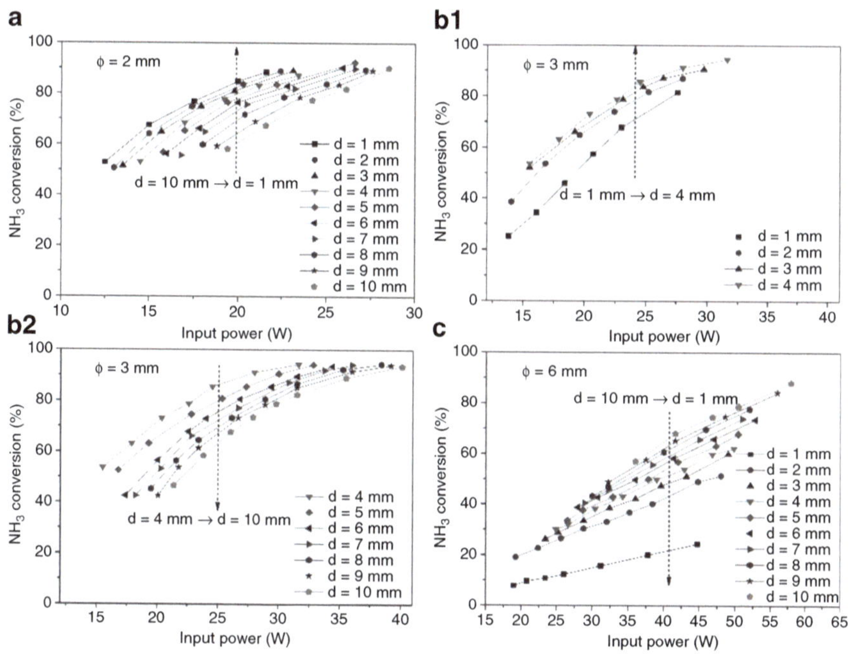

Fig. 7.18 Effect of electrode gap (d) and electrode diameter (Φ) on NH_3 conversion (double SS electrodes, 40 ml/min, 5 kHz, reactor diameter 8 mm) [80]

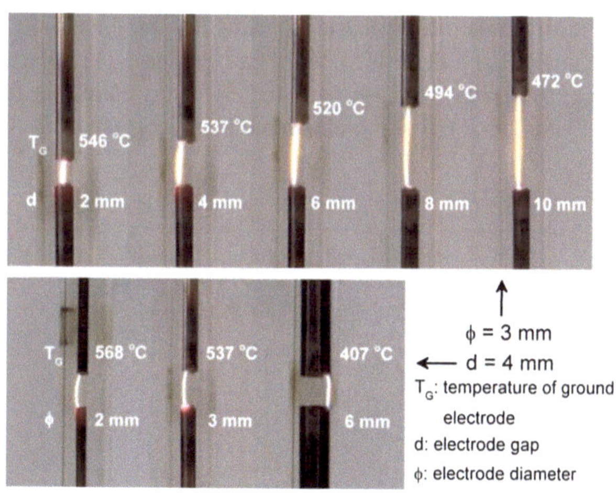

Fig. 7.19 The influence of electrode gap (d) and electrode diameter (Φ) on the effective plasma volume and electrode temperature (double Cu electrodes 40 ml/min, 22 W, 5 kHz, reactor diameter 8 mm) [80]

due to the increase of electrode gap. Thus, the gas-phase plasma decomposition of NH_3 would be enhanced by the increase of the electrode gap. When the electrode gap was fixed, however, there was no obvious change in the size of the bright arc filament. That is, the effective plasma volume kept stable during the increase of the electrode tube diameter. Nevertheless, more and more NH_3 molecules would bypass the effective area of plasma with increasing the electrode tube diameter. This means that the interaction between NH_3 molecules and the energetic electrons would be less effective when a larger electrode tube was used. Figure 7.19 also indicates that, when the electrode tube diameter was fixed, the temperature of the ground electrode decreased with the increase of electrode gap from 1 to 10 mm. Similarly, when the electrode gap was fixed, the temperature of the ground electrode decreased with the increase of electrode tube diameter from 3 to 6 mm. These results indicate that the influences of the electrode gap and the electrode tube diameter on NH_3 conversion were actually the combined effects of both the gas-phase plasma decomposition reaction and the electrode-surface catalytic decomposition reaction. This is because the increase of the effective plasma volume would result in the enhancement of the gas-phase plasma ammonia decomposition reaction, while the decrease of the electrode temperature (the temperature of both ground and high-voltage electrodes would increase or decrease at the same time) would lead to the weakening of the surface catalytic ammonia decomposition reaction.

To verify the relationship between the effective plasma volume and the gas-phase plasma NH_3 decomposition reaction, parallel experiments were carried out by using inert Cu tube electrodes (only gas-phase plasma decomposition of ammonia existed in Cu–electrode reactor [79]) instead of SS electrodes in the same reactor, and the electrode gap was regulated. Figure 7.20 shows that, in both 2 and 3 mm electrode tube diameters, the gas-phase plasma NH_3 decomposition reaction was indeed strengthened by the increase of the electrode gap.

It is interesting to note that, in the case of using a 2 mm electrode tube, the NH_3 conversion monotonously increased with the decrease of electrode gap from 10 to 1 mm; while in the case of using 6 mm electrode tube, the NH_3 conversion monotonously decreased with the decrease of electrode gap from 10 to 1 mm (Fig. 7.18). In the case of using a 3 mm electrode tube diameter, however, the NH_3 conversion first increased and then decreased with the decrease of electrode gap

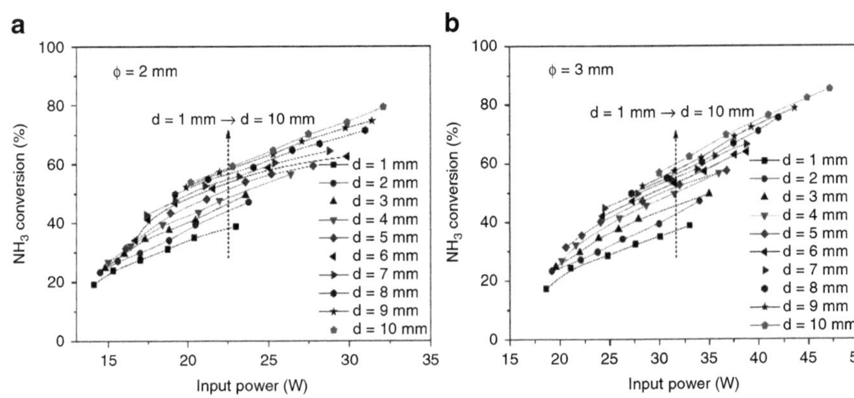

Fig. 7.20 Effect of electrode gap (*d*) and electrode diameter (Φ) on NH_3 conversion (double Cu electrodes, 40 ml/min, 5 kHz, reactor diameter 8 mm) [80]

from 10 to 1 mm and showed the maximum value for a 4 mm electrode gap (Fig. 7.18). This phenomenon is a reflection of the varying contributions of the gas-phase plasma decomposition reaction and the electrode-surface catalytic decomposition reaction. In the case of using a 2 mm electrode tube, the NH_3 decomposition was dominated by the electrode-surface catalysis. As a result, the NH_3 conversion decreased with the increase of electrode gap owing to the decrease of the electrode temperature (Fig. 7.18). In the case of using 3 mm electrode tube, the NH_3 decomposition was determined equally by the gas-phase plasma decomposition reaction and the electrode-surface catalytic decomposition reaction [79]. Therefore, the NH_3 conversion increased with a decrease of the electrode gap from 10 to 4 mm, and the NH_3 conversion decreased with the further decrease of electrode gap from 4 to 1 mm. In the case of using 6 mm electrode tube, the NH_3 decomposition was dominated by the gas-phase plasma decomposition due to the low electrode temperature as a whole. Consequently, the NH_3 conversion decreased with the decrease of electrode gap owing to the decrease of the effective plasma volume.

7.2.4.2 The Discharge Frequency

The discharge frequency is directly related to the validity of gliding arc discharge. As shown in Fig. 7.21, when the discharge frequency decreased from 17 to 5 kHz, the duration of the discharge stage (the period between the breakdown voltage of gas and the voltage when the discharge current disappeared) gradually increased. This means that the discharge time increased with the decrease of discharge frequency. When the discharge was carried out at 17 kHz, the discharge time accounted for about 55% in one cycle of applied voltage. By contrast, when the discharge was carried out at 5 kHz, the discharge time increased to about 83%. As shown in Fig. 7.22, the discharge frequency is also directly related to the effective plasma volume and the electrode temperature, which increase with the decrease of discharge frequency. Consequently, both the gas-phase plasma decomposition and the electrode-surface catalytic decomposition of ammonia were enhanced. The NH_3 conversion increased with the decrease of discharge frequency (Fig. 7.23). When the discharge frequency decreased from 17 to 5 kHz under the conditions of 40 ml/min NH_3 flow rate, 25 W input power, 8 mm reactor diameter, 3 mm electrode diameter, and 4 mm electrode gap, the NH_3 to H_2 energy efficiency increased from 3.8 to 5.7 mol/kWh. In addition, it has been found that thermal insulation has a dramatic influence on NH_3 decomposition. Under the conditions of 40 ml/min NH_3 flow rate, 20 W input power, 3 mm electrode diameter, and 4 mm electrode gap, the energy efficiency increased from 5.6 to 7.9 mol/kWh when the 8 mm diameter reactor was insulated [80].

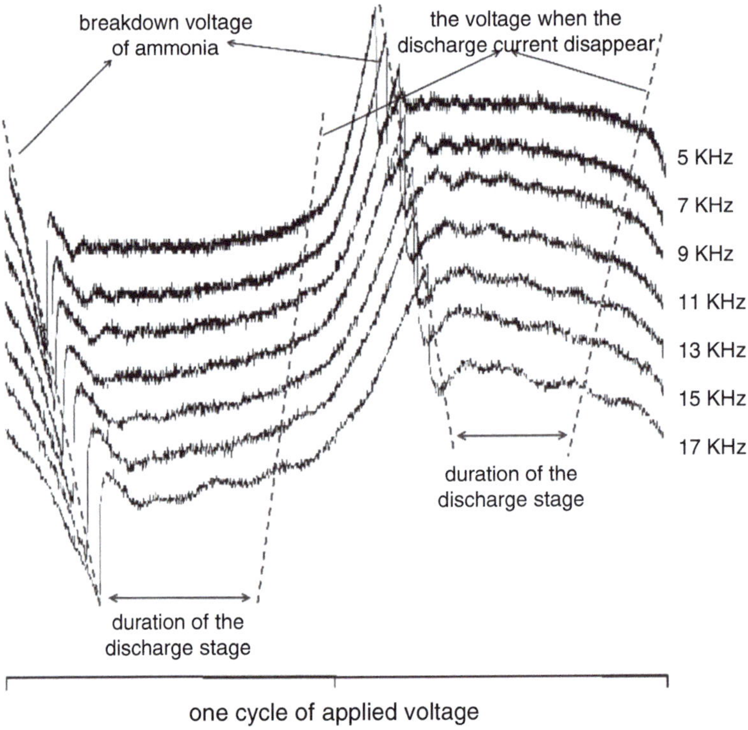

Fig. 7.21 The effect of discharge frequency on the duration of the discharge stage [80]

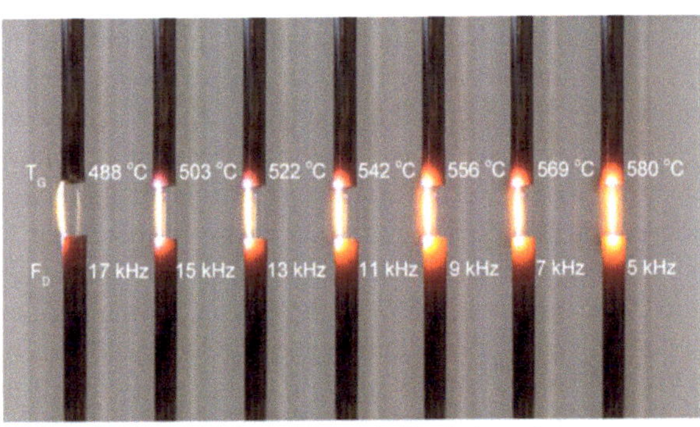

T_G: temperature of ground electrode; F_D: discharge frequency
electrode diameter 3 mm; electrode gap 4 mm

Fig. 7.22 The influence of discharge frequency on the effective plasma volume and electrode temperature (40 ml/min, 25 W, reactor diameter 8 mm) [80]

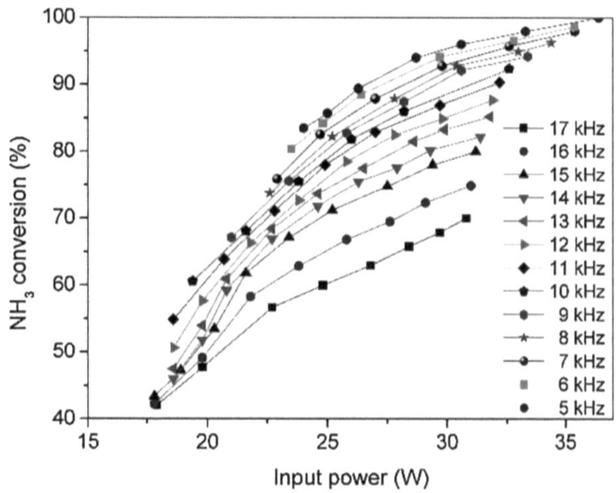

Fig. 7.23 The influence of discharge frequency on NH_3 conversion (40 ml/min, reactor diameter 8 mm, electrode diameter 3 mm, electrode gap 4 mm) [80]

7.3 NH_3 Decomposition in a Dielectric Barrier Discharge Reactor

Wang and co-workers studied the plasma-catalytic NH_3 decomposition in DBD reactors [87–89]. The scheme of the experimental setup for NH_3 decomposition through combination of DBD plasma and catalysis is shown in Fig. 7.24. NH_3 decomposition was performed in a DBD reactor at atmospheric pressure, with a catalyst bed in the discharge zone. The reactor had two coaxial bare-metal electrodes. The shell was a quartz tube (10 mm o.d. × 8 mm i.d.), which served as the dielectric barrier. The outer ground electrode was an Al foil (0.1 mm thick) wrapped tightly around the surface of the quartz tube. The inner high-voltage electrode was a stainless-steel rod (2 mm o.d.; surface composition, Fe 68.5 wt%, Cr 19.9 wt%, Ni 8.1 wt%, Mn 2.0 wt%, Si 0.5 wt%, Cu 0.7 wt%, Ag 0.1 wt%, Al 0.1 wt%, and C 0.1 wt%), and it was installed along the axis of the quartz tube. An alternating current (AC) supply was used, and catalyst was embedded in the plasma zone. Before discharge, an NH_3 feed (99.999% purity) was flushed through the reactor for 30 min to remove the air, and then the DBD plasma was generated by switching on the AC power supply. NH_3 was decomposed to N_2 and H_2 by the DBD plasma and the catalyst. The products were analyzed using an on-line gas chromatograph equipped with a thermal conductivity detector for N_2 and NH_3 detection. The input power of the plasma reactor was measured using a digital oscilloscope (Tektronix DPO 3012 digital oscilloscope equipped with a Tektronix P6015A high-voltage probe and a Pearson 6585 current probe, USA).

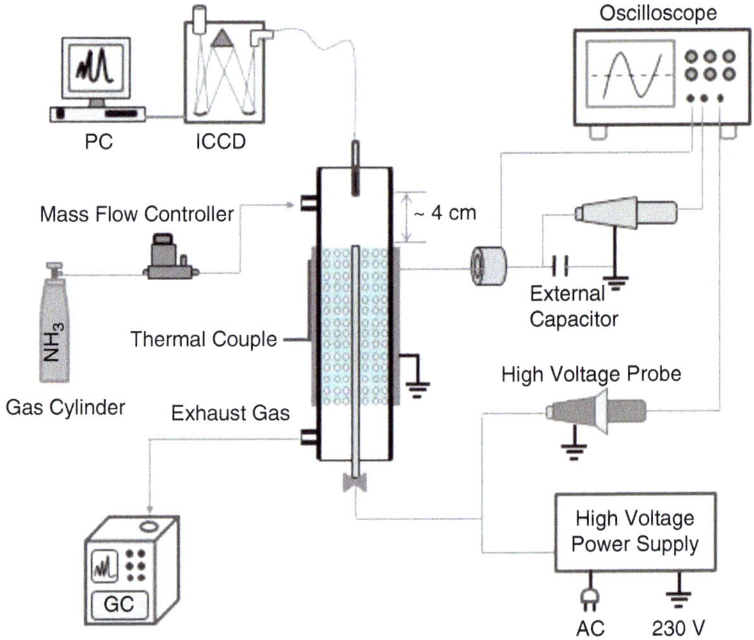

Fig. 7.24 Scheme of the experimental setup used for H_2 generation from NH_3 decomposition through combination of DBD plasma and catalysis [89]

7.3.1 Synergy of the DBD Plasma and Bulk Fe Catalyst

7.3.1.1 The Bulk Fe Catalyst

Wang and co-workers studied NH_3 decomposition using a DBD plasma and a commercial bulk Fe catalyst which is used for industrial NH_3 synthesis [87]. As shown in Table 7.5, the main component of the bulk Fe catalyst is FeO, and it also has many promoters. During NH_3 decomposition, the catalyst was placed in the discharge zone of the DBD reactor.

7.3.1.2 NH_3 Decomposition Results

When the fresh bulk Fe catalyst was used, the initial NH_3 conversion was as low as that obtained with DBD plasma (~10%) along, and NH_3 conversion then increased as time progressed, and it reached a maximum after about 6 h, as shown in Fig. 7.25. Figure 7.26 shows that, when the Fe catalyst was used after undergoing a 6-h plasma-catalytic NH_3 decomposition (Fe-based catalyst) reaction, a dramatically enhanced NH_3 conversion by plasma-driven catalysis was observed around 400 °C. At 410 °C, the plasma-driven catalysis, i.e., the combination of DBD plasma and Fe-based catalyst, gave a NH_3 conversion higher than 99.9%, while the Fe-based

Table 7.5 Analysis of fresh commercial Fe-based catalyst by X-ray fluorescence (XRF)

Component	Content (%)	Component	Content (%)
FeO	94.4	Al_2O_3	2.19
K_2O	0.166	V_2O_5	0.835
SiO_2	0.477	TiO_2	0.446
SO_3	0.235	MgO	0.968
MnO	0.0605	Lu_2O_3	0.0376
Cl	0.0762	CuO	0.0282
NiO	0.0205	MoO_3	0.0191
ZnO	0.0117		

X-ray fluorescence: SRS-3400, Germany, 40 kW, 60 kV, 150 mA, 75 mm, ±0.01 °C

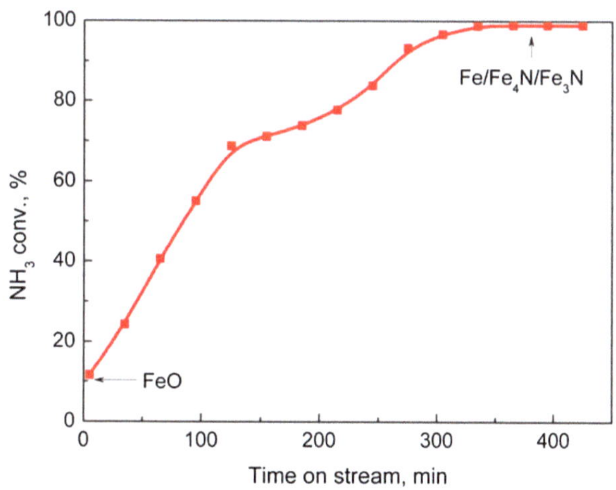

Fig. 7.25 The role of the catalyst in the synergy obtained in ammonia decomposition by plasma-driven catalysis (NH_3 40 ml/min, 2 g FeO, the reactor temperature increased to 500 °C within about 5 min and then kept constant with input power 26 W, discharge frequency 12 kHz) [87]

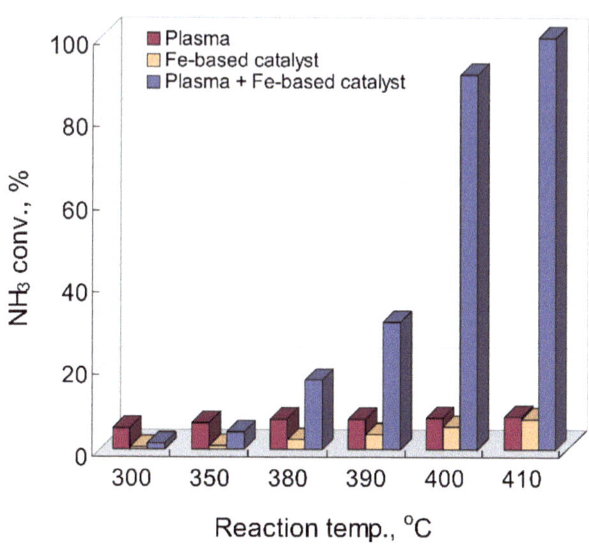

Fig. 7.26 Influence of reaction temperature on the synergy between DBD plasma and an Fe-based catalyst for ammonia decomposition [87]

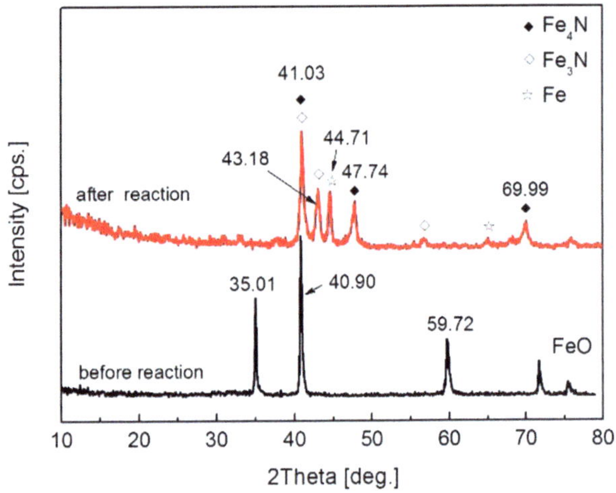

Fig. 7.27 X-ray diffraction patterns of FeO after NH_3 decomposition in plasma-driven catalysis mode [87]. (X-ray diffraction: D/MAX-2400, Japan, 12 kW, 0.01, Cu, $\lambda = 0.1541$ nm)

catalyst alone and the DBD plasma alone only gave NH_3 conversions of 7.4% and 7.8%, respectively. The plasma-driven catalysis decreased the complete decomposition temperature by 140 °C compared to the Fe-based catalyst alone and increased the energy efficiency and rates of hydrogen formation more than ten times. The above results demonstrate a synergy between the bulk Fe-based catalyst and the DBD plasma in NH_3 decomposition.

NH_3 conversion gradually increased from 10% to 99% (Fig. 7.25) when a fresh commercial bulk Fe catalyst was placed in the plasma reactor at a fixed input power (26 W). Figure 7.27 seemingly indicates that FeO was transformed into Fe/Fe_3N/Fe_4N. The FeO, however, might be transformed to metallic Fe together with H_2O during on-site NH_3 decomposition reaction process. The nitride could be formed during the cooling process. The XRD analysis of the catalyst after the reaction was done after turning off the plasma and keeping the catalyst in NH_3 flow until the catalyst was totally cooled to room temperature. When the DBD plasma was turned off, part of the metallic Fe inevitably adsorbed N to form Fe_3N/Fe_4N in NH_3 flow. The activity was obviously influenced by the structure of the catalyst; such a phenomenon reflects the crucial role of catalyst in the synergy of plasma-driven catalysis. Therefore, it can be inferred that the majority of NH_3 molecules were decomposed on the surface of catalysts.

7.3.2 *Synergy of the DBD Plasma and Supported Metal Catalysts*

7.3.2.1 Characterization of the Supports and Supported Catalysts

The supported Fe, Co, Ni and Cu catalysts were prepared using an incipient wet impregnation method [88]. Briefly, the metal (Fe, Co, Ni and Cu) nitrate (30 wt% metal in the final catalyst) was dissolved in deionized water. The supports (TS-1 um zeolite, TS-1 nm zeolite, NaZSM-5 nm zeolite, HZSM-5 nm zeolite, TiO_2, γ-Al_2O_3, fumed SiO_2 and SiO_2-ball) were calcined at 400 °C for 5 h to remove impurities such as H_2O before impregnation, and then the pretreated supports were added to the metal nitrate solution. The mixture was kept at room temperature aging for 3 h, vacuum freeze-dried at −50 °C overnight, and dried in air at 120 °C for 5 h. Dried samples were calcined in a He DBD plasma at 540 °C for 3 h.

1. Physicochemical properties of supports

The data in Table 7.6 shows that the supports have different specific surface areas (S_g), average pore sizes and electrical properties. The electrical capacity (C_d) values of the supports are in the range of 27.77–36.91 pF, and the corresponding relative dielectric constants (ε_d) are in the range of 19.99–26.58.

2. Chemical states of supported metals

The XRD patterns (Fig. 7.28) show that the supported Fe, Co, Ni, and Cu all underwent similar chemical state changes during reduction in the H_2 plasma and plasma-catalytic NH_3 decomposition, regardless of the supports used. For instance, over fumed SiO_2, the as-prepared Fe, Co, Ni, and Cu existed as oxides (Fe_2O_3, Co_3O_4, NiO, and CuO), which were transformed to the metallic states (Fe, Co, Ni, and Cu) after reduction in H_2 plasma. However, significant differences were observed when the metallic catalysts were used for plasma-catalytic NH_3 decomposition, i.e., plasma-catalytic NH_3 decomposition. The Ni and Cu catalysts remained in their metallic states, whereas the Co catalyst was partly transformed to the nitride, and the Fe catalyst was completely nitrided to form Fe_3N and Fe_4N.

3. Metal dispersion of catalysts

TEM images of the metals supported on fumed SiO_2 and Co supported on various supports are shown in Figs. 7.29 and 7.30, respectively. The average sizes of the Fe, Co, Ni, and Cu particles on the fumed SiO_2 support were about 5, 3, 3–8, and 5 nm

Table 7.6 Physicochemical properties of supports used in NH_3 decomposition through combination of supported catalysts and DBD plasma

Supports	Phase state	Specific surface area (S_g, m/g)	Average pore size (nm)	Electric capacity (C_d, pF)	Relative dielectric constant (ε_d)
Fumed SiO_2	Amorphous	297.79	–	27.77	19.99
SiO_2-ball	Amorphous	194.79	7.97	28.56	20.56
TS-1 nm zeolite	Crystal	397.37	0.55	28.72	20.68
γ-Al_2O_3	Crystal	111.66	6.97	29.20	21.02
TS-1um zeolite	Crystal	365.57	0.54	29.87	21.51
HZSM-5 zeolite	Crystal	265.09	0.55	31.75	22.86
NaZSM-5 zeolite	Crystal	273.08	0.54	33.33	24.00
TiO_2	Crystal (anatase)	14.22	–	36.91	26.58

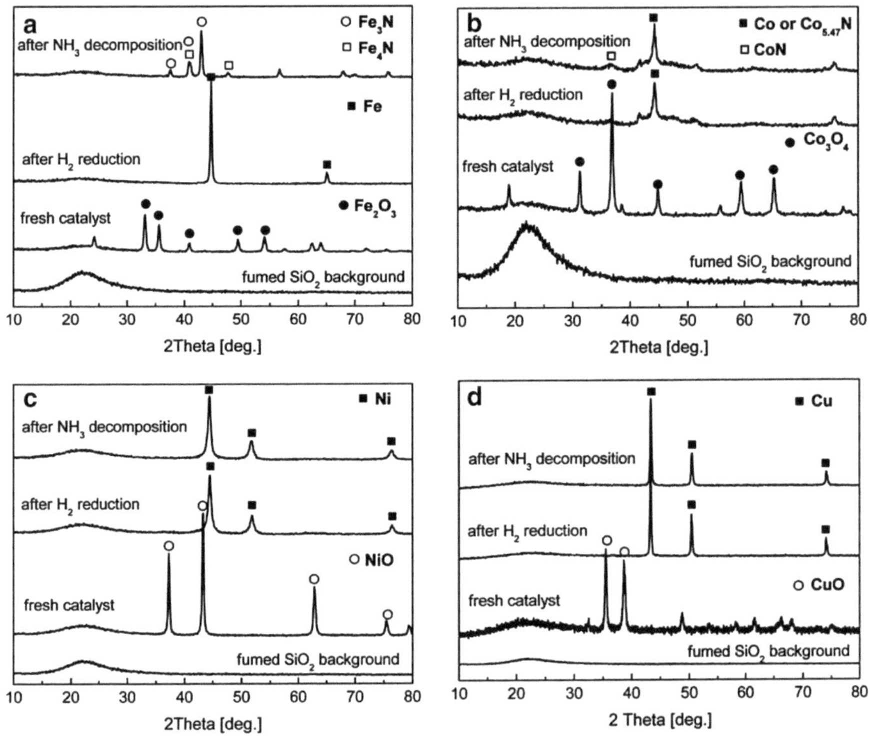

Fig. 7.28 XRD patterns of various metal catalysts supported on fumed SiO_2: (**a**) Fe, (**b**) Co, (**c**) Ni, and (**d**) Cu. (Fe_2O_3 ICCD: 33-664, Fe ICCD: 6-696, Fe_3N ICCD: 1-1236, Fe_4N ICCD: 6-627, Co_3O_4 ICCD: 42-1467, Co ICCD: 15-806, NiO ICCD: 22-1189, Ni ICCD: 4-850, CuO ICCD: 41-254, Cu ICCD: 4-836) [88]

(Fig. 7.29), respectively, indicating high dispersion of the metals. Except NaZSM-5 zeolite, highly dispersed Co particles were formed on the supports, i.e., fumed SiO_2, SiO_2-ball, TiO_2, γ-Al_2O_3, TS-1 nm zeolite, TS-1 um zeolite, and HZSM-5 nm zeolite (Fig. 7.30). The average Co particle size was less than 5 nm (mostly around 2–3 nm). In the case of NaZSM-5 zeolite, the average size of Co particles was found to be about 50 nm.

4. Metal–support interactions

H_2-TPR was used to investigate the metal–support interactions. The reduction of Co_3O_4 on the supports mainly occurred in the temperature range of 300–550 °C (Fig. 7.31). The reduction temperatures increased in the order fumed SiO_2 < SiO_2-ball < γ-Al_2O_3 < TS-1 nm zeolite < HZSM-5 nm zeolite < TS-1 um zeolite < NaZSM-5 nm zeolite < TiO_2, indicating that the strength of interaction between Co particles and these supports increased in the same order. It was worth noting that the profiles of the Co_3O_4 reductions generally show two reduction peaks. The lower-temperature peak could be attributed to the reduction of Co_3O_4 to CoO,

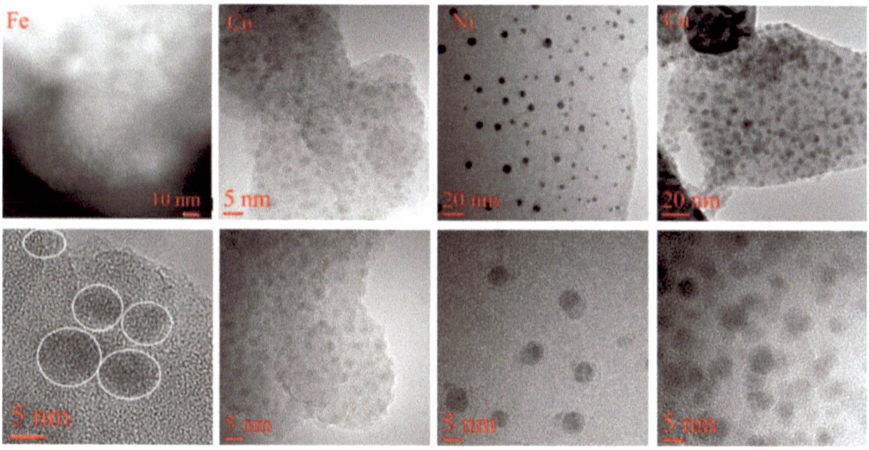

Fig. 7.29 TEM images of Fe, Co, Ni, and Cu catalysts supported on fumed SiO_2 (reduced in H_2 plasma) [88]

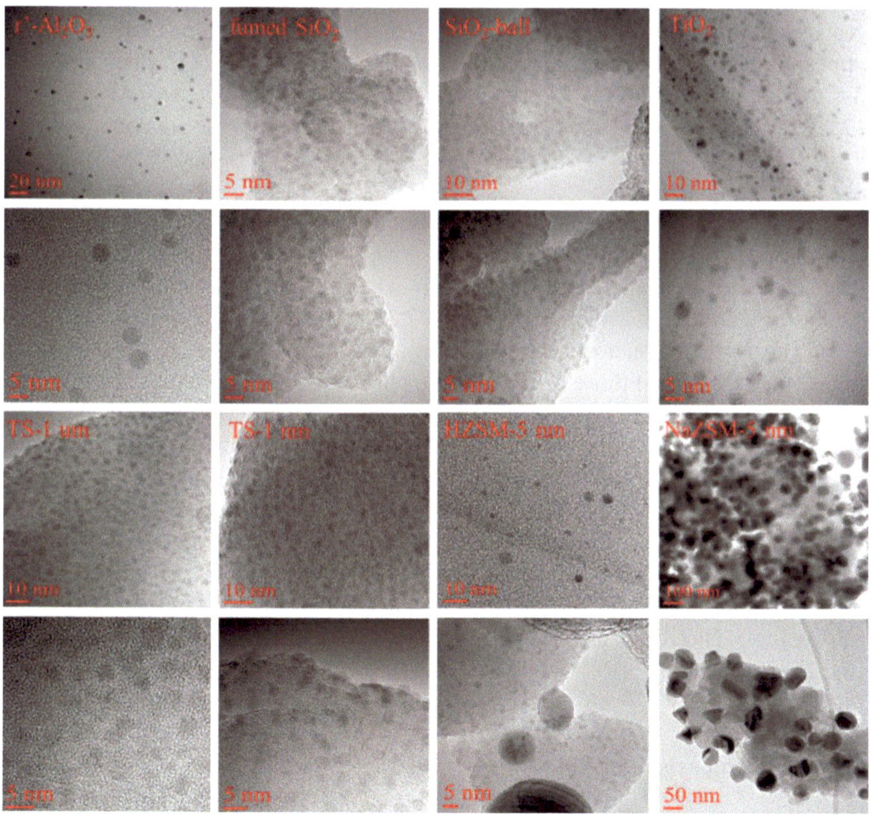

Fig. 7.30 TEM images of Co on various supports (reduced in H_2 plasma) [88]

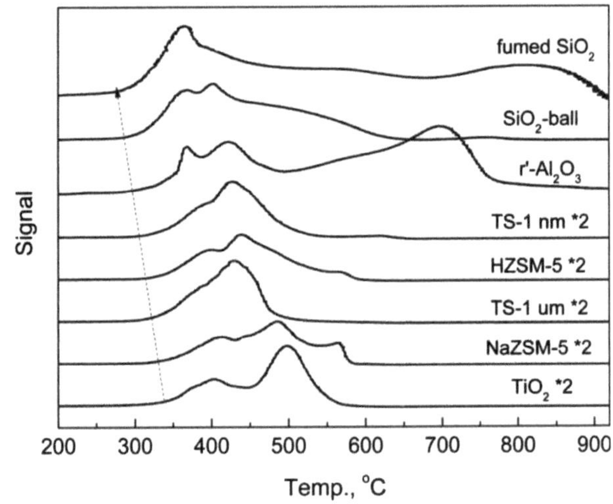

Fig. 7.31 TPR profiles of Co catalysts on various supports [88]

and the higher-temperature peak could be assigned to the reduction of CoO to metallic Co [90–92]. In the case of the γ-Al_2O_3 support, a third reduction peak appeared in a much higher temperature region, i.e., 600–750 °C, which could be assigned to reduction of Co^{x+} ions incorporated into the crystal lattice of the support [92, 93]. Such Co^{x+} ions have the strongest interactions with the support.

7.3.2.2 The NH_3 Decomposition Results

The NH_3 conversion increased significantly when a plasma was combined with the catalyst, regardless of which metals (Fig. 7.32) or supports (Fig. 7.33) were used [88]. The NH_3 conversion achieved using "plasma + catalyst" mode was much higher than the sum of those obtained using plasma or a catalyst alone. For instance, over the Co/fumed SiO_2 catalyst, the "plasma + catalyst" mode gave an NH_3 conversion of 99.2%, whereas the sum of the NH_3 conversions using only plasma or catalysis was only 26.0%. These results show a clear synergy between plasma and the catalysts. The synergy strongly depended on the types of metals and supports. It can be seen that the synergistic capability of the catalysts decreased in the order of Co > Ni > Fe > Cu, when fumed SiO_2 was used as the support, and in the order of fumed $SiO_2 \approx \gamma$-$Al_2O_3 \approx SiO_2$-ball > TS-1 nm zeolite > HZSM-5 nm zeolite > TS-1 um zeolite > NaZSM-5 nm zeolite > TiO_2 when Co metal was used as the active metal component.

7.3.2.3 Comparison of Metals

Generally, the metal dispersion has an important effect on the catalytic activity of a supported metal catalyst. However, Fig. 7.29 shows that the four metals (Fe, Co, Ni

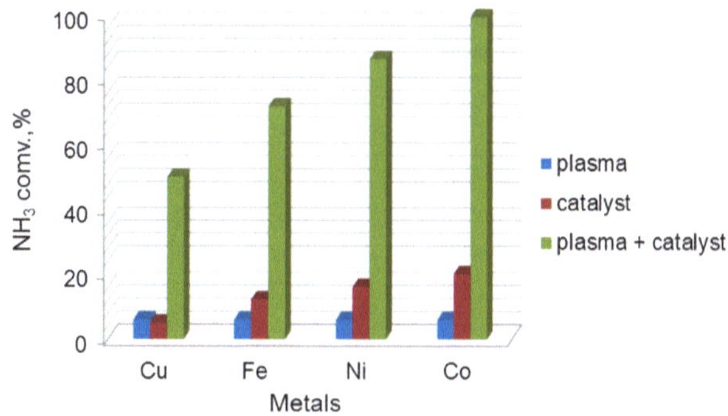

Fig. 7.32 Influence of metals on NH_3 conversion in "plasma + catalyst," "plasma" and "catalyst" modes [88]

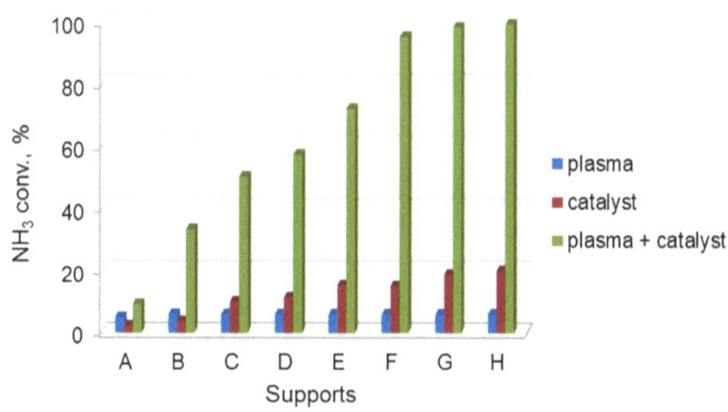

Fig. 7.33 Influence of supports on NH_3 conversion in "plasma + catalyst," "plasma" and "catalyst" modes: A TiO_2, B NaZSM-5 nm zeolite, C TS-1 um zeolite, D HZSM-5 nm zeolite, E TS-1 nm zeolite, F γ-Al_2O_3, G SiO_2-ball, H fumed SiO_2 [88]

and Cu) have similar particle size and similar loading. Figure 7.28 shows that the Fe and Co catalysts were mainly transformed into nitrides after NH_3 decomposition reaction, while the Ni catalyst was probably transformed into surface nitrides. Compared with their metallic counterparts, the nitrided Fe, Co, and Ni catalysts all showed lower catalytic activities, as Pelka and co-workers reported [94]. Wang and co-workers demonstrated that the different activities in plasma-catalytic NH_3 decomposition are closely related with the essential characteristic of the four elements (Fe, Co, Ni and Cu), i.e., the metal–N bond strength [88]. The relationship between the synergetic capability of the four supported catalysts and the metal–N strength will be discussed in Sect. 7.4.3.

7.3.2.4 Comparison of Supports

As mentioned in Sect. 7.3.2.2 (Fig. 7.33), the supports had a significant effect on the plasma−catalyst synergy in NH_3 decomposition. The effect of the support on the synergy was investigated using a series of supported Co catalysts with similar Co loadings and similar distribution of Co particle sizes (Fig. 7.30). A comparison of the reaction results of NH_3 decomposition and characterization results of the catalysts showed that the specific surface area, average pore size, phase state (crystalline or amorphous), and acidity of the support (Table 7.6) had no obvious effects on the

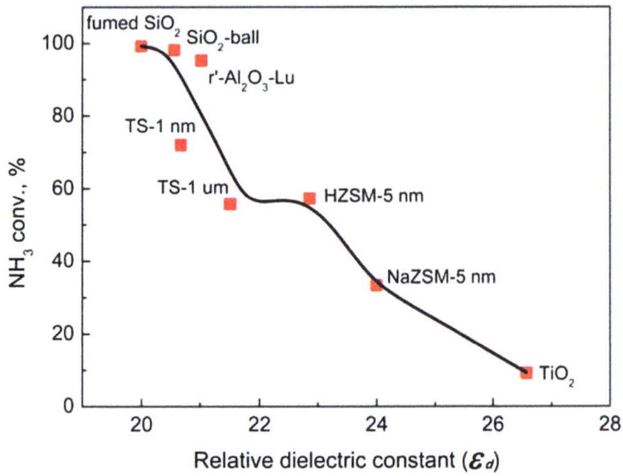

Fig. 7.34 NH_3 conversions on Co catalysts on various supports as a function of relative dielectric constants of supports in plasma + catalyst mode. (NH_3 feed 40 mL/min, temperature 450 °C, supported catalyst 0.88 g, discharge gap 3 mm, discharge frequency 12 kHz) [88]

synergy. The effect of support on the synergy was closely related to the metal–support interactions and the electrical properties of the support.

Figure 7.31 shows that the order of the strengths of the interactions of Co with the supports was fumed SiO_2 < SiO_2-ball < γ-Al_2O_3 < TS-1 nm zeolite < HZSM-5 nm zeolite < TS-1 um zeolite < NaZSM-5 nm zeolite < TiO_2, which was consistent with the order of NH_3 conversion in plasma-catalytic NH_3 decomposition (Fig. 7.33). Clearly, weak interactions between Co and supports favored NH_3 decomposition [88].

Figure 7.34 shows that the relative dielectric constant (ε_d) of the supports also correlated well with NH_3 conversion of plasma-catalytic NH_3 decomposition, i.e., the decomposition decreases significantly with increasing ε_d. In other words, a support with a small ε_d facilitated plasma-catalytic NH_3 decomposition. To understand the influence of the relative dielectric constant of the support on the plasma-catalytic NH_3 decomposition reaction, the NH_3 decomposition was performed in the mode of "plasma + support" (Fig. 7.35). It can be seen that NH_3 conversion still decreased significantly with increasing ε_d of the support. Taking NH_3 conversion in "plasma" mode as a reference (NH_3 conversion was 6%, shown by a dashed line), it can be found that fumed SiO_2, γ-Al_2O_3, and SiO_2-ball have weak but positive effects on NH_3 conversion, whereas other supports such as HZSM-5 zeolite, NaZSM-5 zeolite, and TiO_2 have negative effects on NH_3 conversion. The catalyst support usually plays an important role in heterogeneous catalysis, and the catalytic performances are generally related to the specific surface area, porosity, acidity and support–metal strong interactions. However, the electrical properties of supports have been hardly mentioned in heterogeneous catalysis. In the area of plasma catalysis, a few researchers have attempted to promote reactions by filling the plasma zone with high-dielectric-constant materials [95, 96]. The similarity in the sequences of ε_d–NH_3 conversion relationships in Figs. 7.34 and 7.35 shows that the effect of ε_d on the NH_3 conversion could be attributed to the influence of the support on the

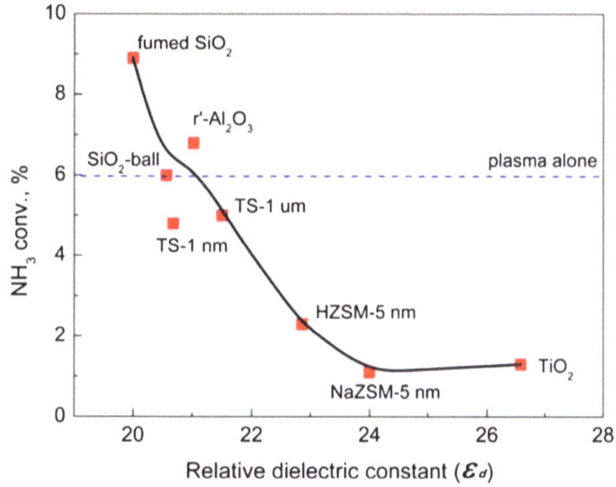

Fig. 7.35 NH_3 conversions on different supports as a function of relative dielectric constants of supports in plasma + bare support mode. (NH_3 feed 40 mL/min, temperature 450 °C, bare support 0.88 g, discharge gap 3 mm, discharge frequency 12 kHz) [88]

discharge. In the plasma zone, the net electric field decreases because of accumulation of electrons on the solid surface, especially for supports with high ε_d.

7.4 Fundamental of the Synergetic Effect Between DBD Plasma and Cheap Metals

7.4.1 Plasma Enhances the NH_3 Adsorption Step on Metals

Wang and co-workers studied the fundamentals of the synergetic effects between DBD plasma and cheap metals on NH_3 decomposition and found that plasma enhanced the NH_3 adsorption step on metals [89].

Firstly, the NH_3 plasma was diagnosed using OES. In the absence of Fe-based catalyst, the OES of NH_3 plasma exhibited two electronically excited state species (Fig. 7.36). The continuum band from 564.3 to 568.1 nm (Schuster's emission bands) is attributed to electronically excited state NH_3 (NH_3^*) [81]. The line at 337 nm ($C^3\Pi_u \rightarrow B^3\Pi_g$) is attributed to electronically excited state N_2 (N_2^*) [97]. In addition, three radical species, i.e., $NH_2\cdot$ (390–830 nm) [81], $NH\cdot$ ($A^3\Pi \rightarrow X^3S^-$, 336 nm) [98] and H ($2p^2p^0{}_{3/2} - 3d^2D_{3/2}$, 656.3 nm), were detected in NH_3 plasma as well. Among these species, NH_3^* was estimated to be the most abundant species since the signal intensity of NH_3^* species was the most intensive one. More importantly, the increase of the signal intensity of NH_3^* species as input power indicated that the amount of NH_3^* species correspondingly increased with the input power. In the presence of the Fe-based catalyst (Fig. 7.37), the OES of NH_3 plasma also showed the existence of two electronically excited state species (NH_3^* and N_2^*) and three radical species ($NH_2\cdot$, $NH\cdot$ and $H\cdot$); while the signal intensity of OES was much lower than that without the Fe-based catalyst, which might be caused by the

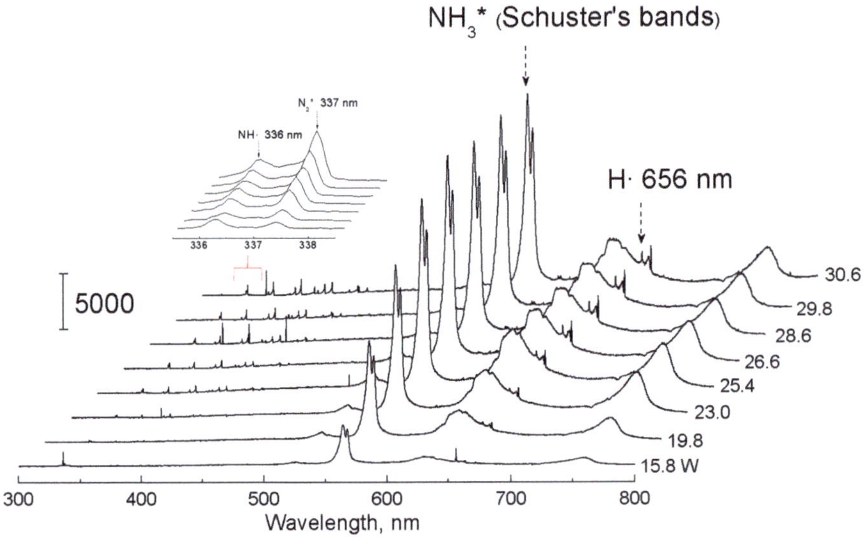

Fig. 7.36 OES of NH_3 plasma as a function of input power without a catalyst (1 atm, NH_3 40 ml/min, discharge length 4 cm, Fe-based catalyst 6 g, exposure time of OES 5 sec) [89]

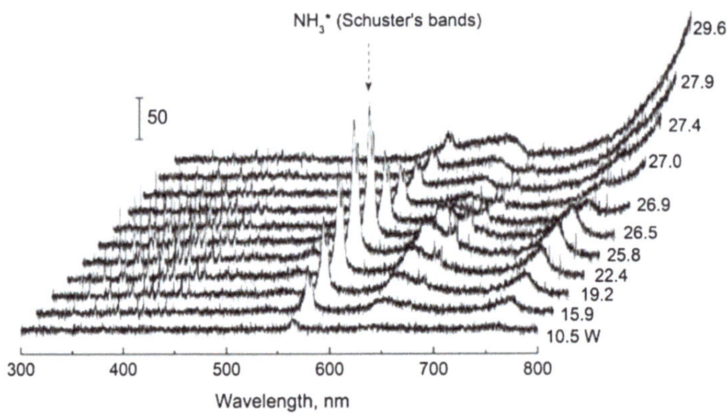

Fig. 7.37 OES of NH_3 plasma as a function of input power in the presence of an Fe-based catalyst (1 atm, NH_3 40 ml/min, discharge length 4 cm, Fe-based catalyst 6 g, exposure time of OES 5 sec) [89]

shielding and absorption of the solid catalyst. More strangely, in the presence of an Fe-based catalyst, the signal intensity of the NH_3^* species first increased significantly with input power, but then decreased suddenly when the input power was higher than 26 W. This abnormal phenomenon indicates that packing an Fe-based catalyst in the plasma zone has effectively changed the composition of NH_3 plasma under the conditions of high input power.

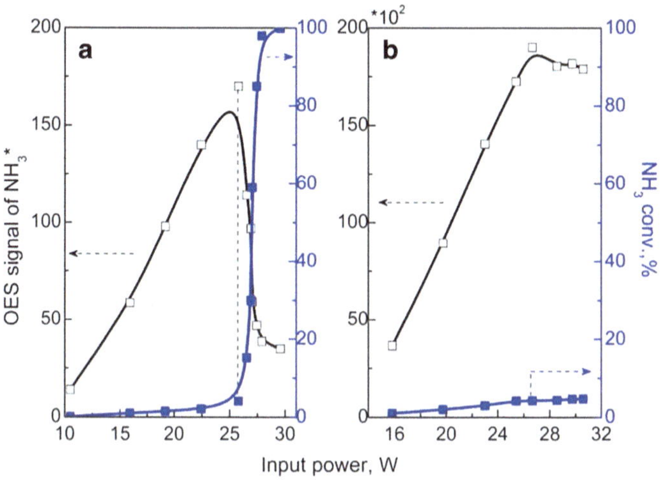

Fig. 7.38 Relationship between NH_3^* OES signal and NH_3 conversion as a function of input power in the presence of (**a**) and absence of (**b**) Fe-based catalyst (1 atm, NH_3 40 ml/min, discharge length 4 cm, Fe-based catalyst 6 g, exposure time of OES 5 sec) [89]

Then, the NH_3^* OES signal was correlated with the NH_3 conversion. As shown in Fig. 7.38a, in the presence of an Fe-based catalyst, the OES signal intensity of NH_3^* increased steadily and reached its maximum value at about 26 W input power. In this stage, the conversion of NH_3 increased very slowly and stayed at a low level. Interestingly, when the input power reached more than 26 W, the OES signal intensity of NH_3^* decreased dramatically with the increase of input power, and the NH_3 conversion soared up abruptly at the same time. As the conversion of NH_3 reached its maximum value near 30 W, the OES signal intensity of NH_3^* fell down to its lowest value. This phenomenon suggested that, in plasma-catalytic NH_3 decomposition, the abrupt increase of NH_3 conversion might originate from the contribution of NH_3^*. As shown in Fig. 7.38b, in the absence of the Fe-based catalyst, the OES signal intensity of NH_3^* first increased steadily with input power and then reached its maximum value at about 26 W. Meanwhile, the conversion of NH_3 also increases slowly at a low level. However, in contrast to the case involving the presence of the Fe-based catalyst, a sharp decrease of the OES signal intensity of NH_3* and quick increase of the NH_3 conversion was not found at higher input power. Instead, both the OES signal intensity of NH_3^* and the conversion of NH_3 remained almost constant. This distinct phenomenon indicates that, in the absence of Fe-based catalyst, the NH_3* species had not been effectively consumed at high input power. Consequently, the conversion of NH_3 was still very low even at high input power.

The relationships of the OES signal intensity of NH_3^* and the conversion of NH_3 in the presence and absence of Fe-based catalyst provide an idea that the NH_3^* species might have been efficiently consumed by the Fe-based catalyst and that the NH_3^* species as an active form of reactant showed the superiority in the adsorption activation step, as illustrated by Fig. 7.39. Wang and co-workers also designed some experiments to demonstrate this hypothesis. In addition, they found that, compared

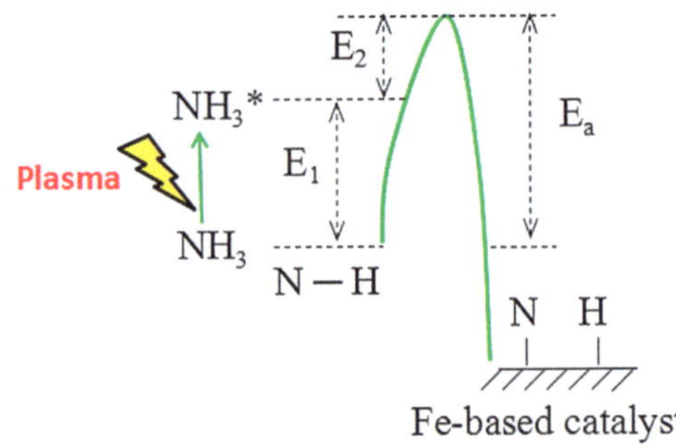

Fig. 7.39 Schematic diagram of adsorption activation of NH_3^* and ground-state NH_3 on an Fe-based catalyst (E_1, internal energy of NH_3^*; E_a, adsorption activation energy of NH_3; E_2, adsorption activation energy of NH_3^*) [89]

to the ground state NH_3, NH_3^* species could obviously increase the adsorption capacity and adsorption strength on the Fe-based catalyst. The NH_3 conversion was increased by nearly 40% due to the contribution of NH_3^* species under the condition of eliminating the influence of desorption step. Readers can find the experimental details in the reference [89].

7.4.2 Plasma Accelerates the Recombinative Desorption of Surface N Atoms

As mentioned in Sect. 7.1.2, the recombinative desorption of adsorbed N atoms is usually the rate controlling step when non-noble metal catalysts are used. Wang and co-workers demonstrated that plasma can accelerate the recombinative desorption of surface adsorbed N atoms [87].

Wang and co-workers studied the role of the DBD plasma in the synergy by monitoring the desorption process of N_{ad} during ammonia decomposition in the absence or presence of DBD plasma (Fig. 7.40). The metallic Fe (pretreated in pure H_2 flow) showed a very high capacity for producing H_2 even without DBD plasma, indicating a high activity for ammonia decomposition, as shown in Fig. 7.40a. However, only a minor amount of N_2 was detected and the H_2 generation rate quickly decreased. In the presence of DBD plasma, however, a considerable amount of N_2 was detected together with H_2, and the H_2 generation rate remained stable with time, as shown in Fig. 7.40b. These results show that metallic Fe is an active phase for NH_3 decomposition but that the metallic Fe phase is easily poisoned by N_{ad} atoms by forming less active iron nitrates. More importantly, it gives direct proof that plasma exhibits a powerful capacity to accelerate N_2 desorption from the catalyst surface.

How does the plasma accelerate N_2 desorption from the surface of the catalyst? In-situ diagnosis with optical emission spectroscopy showed that active species such as NH_3^* and NH· were present in the DBD plasma (Fig. 7.36). Then, a $^{15}NH_3$ isotope tracing experiment was designed by Wang and co-workers to uncover the desorption process accelerated by plasma, as shown in Fig. 7.41.

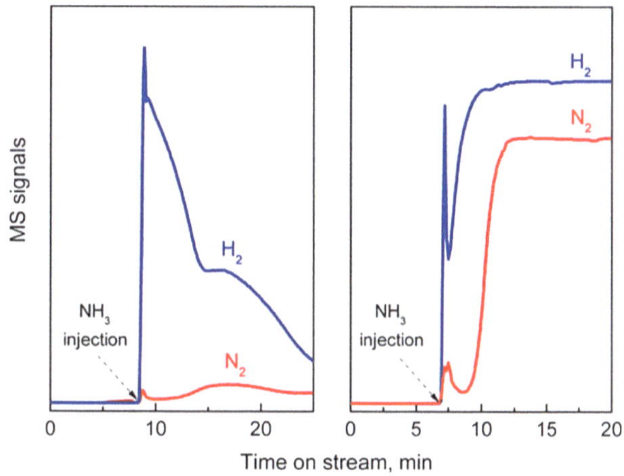

Fig. 7.40 Accelerating effect of DBD plasma on N_{ad} removal from the metallic Fe catalyst in ammonia decomposition [87]. (**a**) Without DBD. (**b**) With DBD

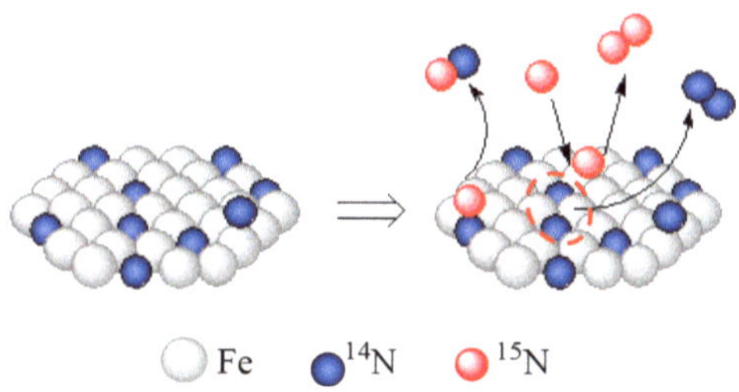

Fig. 7.41 Scheme of the desorption of adsorbed N atoms

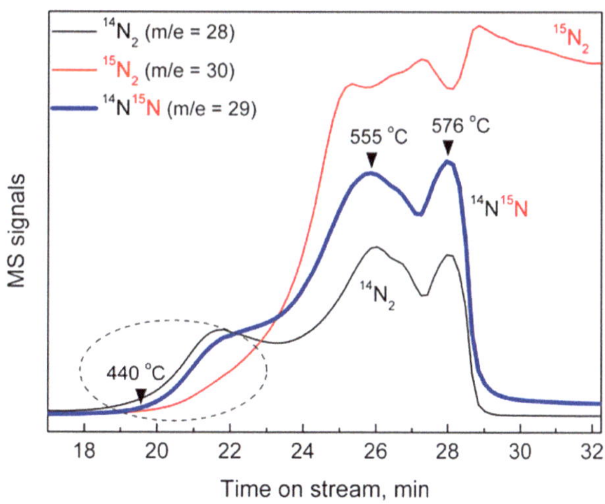

Fig. 7.42 Desorption process of ^{14}N atoms from $Fe_2{}^{14}N$ was studied by $^{15}NH_3$ isotope tracing without DBD plasma [87]

A $Fe_2{}^{14}N$ catalyst was thermally prepared in an $^{14}NH_3$ flow, and it was used as a $^{14}N_{ad}$ covered sample. Then, the $^{15}NH_3$ decomposition reaction was conducted on the $Fe_2{}^{14}N$ catalyst in the absence or presence of DBD plasma, during which the

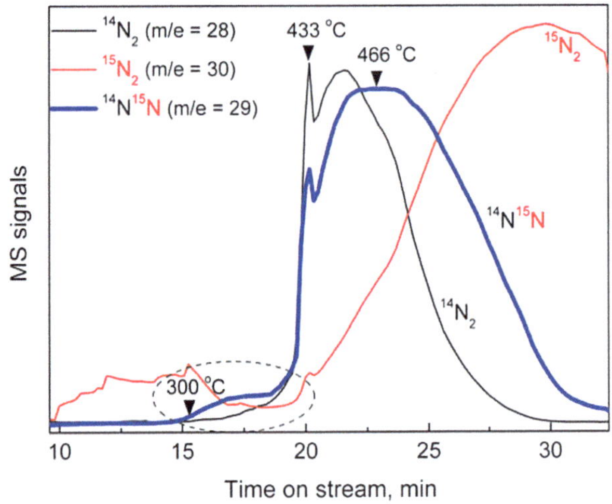

Fig. 7.43 Desorption process of ^{14}N atoms from $Fe_2{}^{14}N$ was studied by $^{15}NH_3$ isotope tracing with DBD plasma [87]

exhaust gas was detected by MS, as shown in Figs. 7.42 and 7.43, respectively [87]. In the absence of DBD plasma, $^{14}N_2$ and then $^{14}N^{15}N$ appeared successively in the initial desorption stage (Fig. 7.42), indicating that the main way to desorb N_2 was the self-recombinative desorption of $^{14}N_{ad}$. Only after part of the $^{14}N_{ad}$ desorbed from the catalyst surface, the $^{15}NH_3$ in the gas phase could adsorb on the catalyst and then produce $^{14}N^{15}N$ and $^{15}N_2$. In the presence of the DBD plasma (Fig. 7.43), however, the initial desorption product was $^{14}N^{15}N$ but not $^{14}N_2$, and the $^{14}N^{15}N$ could mostly be produced through the interaction of $^{14}N_{ad}$ and gas-phase ^{15}N active species ($^{15}NH_3{}^*$, $^{15}NH\cdot$) by an Eley-Rideal mechanism. In addition, comparison of the desorption spectra of $^{15}N_2$ in the presence or absence of DBD plasma in the later desorption stage showed that the main desorption of $^{15}N_2$ appeared after the desorption of $^{14}N^{15}N$ and $^{14}N_2$ in the presence of DBD plasma (Fig. 7.43), but appeared close together with $^{14}N^{15}N$ and $^{14}N_2$ in the absence of DBD plasma (Fig. 7.42). These results indicate that most ^{15}N species in the gas-phase were directly consumed by reaction with the $^{14}N_{ad}$ to form $^{14}N^{15}N$ (Eley-Rideal mechanism) in the presence of plasma. The results also support the idea that the reaction of the active species ($NH_3{}^*$, $NH\cdot$) with surface active N_{ad} may be the main approach for the desorption of N_{ad} accelerated by the plasma technique. Zhang and co-workers also reported an interaction of $NH\cdot$ and $NH_2\cdot$ radicals with non-noble metal catalyst surfaces [99].

7.4.3 *Dependence of Synergetic Effect on Metal–N Bond Strength*

As shown in Sects. 7.3.2.2 and 7.3.2.3, when fumed SiO_2 was used as the support, the synergistic capability of the catalysts decreased in the order of Co > Ni > Fe > Cu,

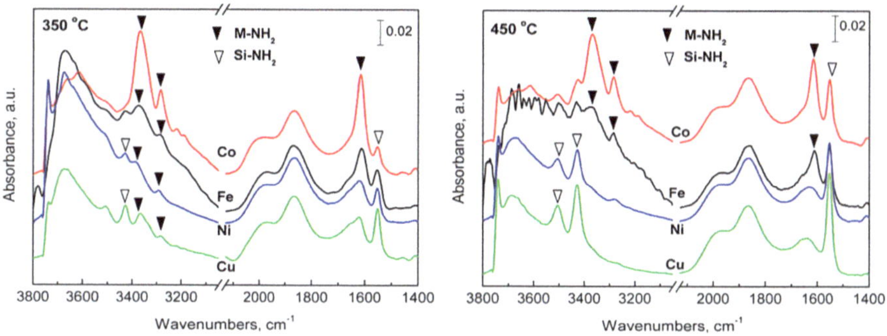

Fig. 7.44 IR spectra for NH_3 adsorption on Fe, Co, Ni, and Cu catalysts supported on fumed SiO_2. (Samples were treated by in situ H_2 reduction at 500 °C, and NH_3 adsorption was conducted at 350 and 450 °C, respectively, followed by outgassing at 200 °C) [88]

which may be closely related to the metal–N bond strength. Wang and co-workers studied the metal–N bond strength of Co, Ni, Fe and Cu, using adsorption and desorption experiments [88].

The adsorption of NH_3 on the Fe, Co, Ni, and Cu catalysts was studied using FTIR. There was a significant number of adsorbed NH_2 ($NH_{2,ad}$) species on the catalyst surfaces, but no NH_{ad} species were detected (Fig. 7.44). The attribution of these species in FTIR can be found in references [100, 101]. The amount of $NH_{2,ad}$ varied depending on the metals and the temperature. When the adsorption temperature was increased from 350 to 450 °C, the $NH_{2,ad}$ species on the Cu catalyst disappeared, but the amount of such species on the Fe catalyst was almost unchanged. These results indicate that dissociative adsorption of NH_3 occurred on these catalysts, and different metals had different bonding abilities toward $NH_{2,ad}$ species. In other words, the $Cu-NH_{2,ad}$ bond was the weakest one, while the $Fe-NH_{2,ad}$ bond was the strongest one.

During NH_3 decomposition on the supported Fe, Co and Ni catalysts in the absence of DBD plasma, an intense H_2 signal was detected almost immediately after the injection of NH_3 (Fig. 7.45). However, only a weak N_2 signal was observed for these catalysts. The H_2 signal attenuated sharply with time and then leveled off. These results show that H_2 desorption is easy, but N_2 desorption from these metals is difficult in the absence of a plasma. The majority of N atoms were adsorbed by Fe, Co, and Ni catalysts through forming strong M−N bonds. The H_2 and N_2 signals for the Cu catalyst were much weaker than those for the Fe, Co, and Ni catalysts, which could be caused by poor activity of the Cu catalyst in NH_3 decomposition.

The relative M−N bond strengths (M = Fe, Co, Ni, and Cu) were investigated using TPD. Figure 7.46 shows that considerable amounts of N_2 were released from the Fe, Co, and Ni catalysts, and it may be the reason why only small amounts of N_2 were detected by MS and the H_2 signal intensities decreased with time during NH_3 decomposition over the catalysts in the absence of a plasma (Fig. 7.45). As shown in Fig. 7.46, the temperature needed for N_2 desorption from the catalysts' surface

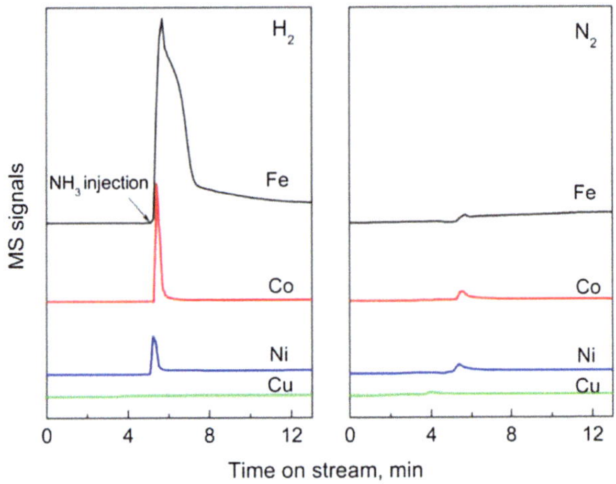

Fig. 7.45 On-line MS analysis of the initial stage of NH_3 decomposition over Fe, Co, Ni, and Cu catalysts supported on fumed SiO_2 without plasma. (0.95 g of metal catalyst was prepared in a H_2 flow for 3 h at 500 °C and cooled to 450 °C in a He flow to remove H_2; the flow was switched to 40 mL/min NH_3 for the reaction at 450 °C) [88]

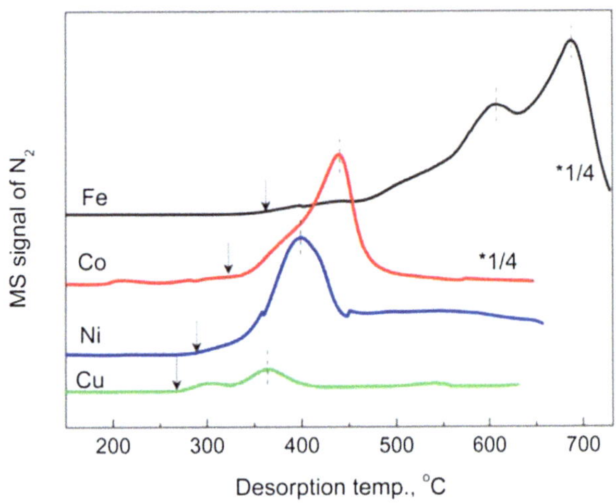

Fig. 7.46 TPD profiles of supported Fe, Co, Ni, and Cu catalysts obtained from the same experiment as in Fig. 7.45. (After the reaction, these catalysts were cooled in situ to room temperature in an NH_3 flow) [88]

increased significantly from Cu to Ni, Co, and Fe, indicating that the strength of the M–N bond increased from Cu to Ni, Co and Fe, i.e., the Fe–N bond was the strongest, and the Cu–N bond was the weakest, which is consistent with the results of FTIR (Fig. 7.44).

The M–N bond strength is one of the key factors in NH_3 decomposition over cheap metal catalysts. Metals with moderate M–N bond strengths (Co–N) showed higher activities (Figs. 7.32 and 7.46). If the M–N bond was too weak, dissociative adsorption of NH_3 on the metal is difficult, and desorption of the $NH_{2,ad}$ intermediate species takes place easier than further dehydrogenation, resulting in low catalytic activity. This was the case for the supported Cu catalysts (Fig. 7.44). Boisen and co-workers reported that the activity of Cu in catalytic NH_3 decomposition is lower than those of other metal catalysts [102]. However, if the M–N bond was too strong, dissociative adsorption of NH_3 on the metal occurred easily, but desorption of the

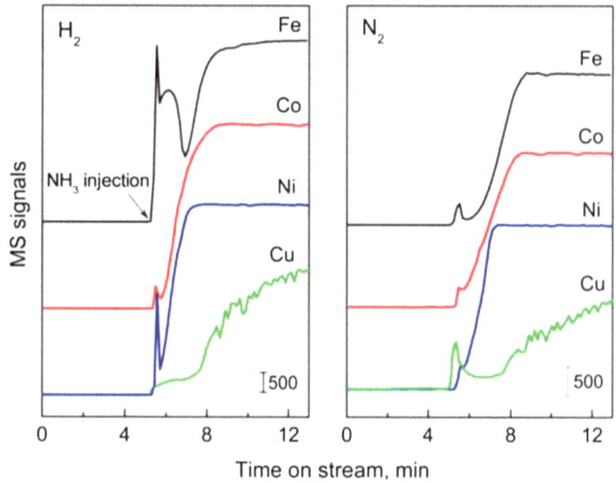

Fig. 7.47 On-line MS analysis of the initial stage of NH_3 decomposition over Fe, Co, Ni, and Cu catalysts in plasma + catalyst mode. (Fumed SiO_2 was used as the support; 0.95 g of metal catalyst were prepared in H_2 flow for 3 h at 500 °C, followed by cooling to 250 °C in He flow to remove H_2 and then switching to 40 mL/min NH_3 flow for reaction under AC supply power about 28 W, and the final temperature remained around 450 °C) [88]

bonded N atoms from the catalyst was difficult, and therefore highly active metal sites were difficult to recover. This was the case for the Fe catalysts (Figs. 7.45 and 7.46). The reason for the different synergies of Fe, Co, Ni, and Cu catalysts with plasma, as shown in Fig. 7.32, can be mainly attributed to their different M–N bond strengths. It should be pointed out that although Co has a moderate M–N bond strength, NH_3 decomposition over the Co catalyst alone gave very low conversion (~20%), as shown in Fig. 7.32. The recombinative desorption of the bonded N atoms is still the rate-limiting step. However, when NH_3 decomposition was performed in "plasma + Co catalyst" mode, NH_3 conversion reached almost 100%.

Comprehensively, the role of the plasma in the synergy is to enhance the NH_3 adsorption step on metals and accelerate the recombinative desorption of bonded N atoms. Figure 7.47 gives more direct proof to demonstrate the role of the plasma in the plasma–catalyst synergy for NH_3 decomposition. Clearly, this acceleration effect of the plasma works not only in the case of Co catalysts, but also for Fe and Ni catalysts.

7.5 Perspectives of Ammonia as a Carrier for Hydrogen Energy Circulation

Ammonia is a non-carbonyl hydrogen source, and NH_3 as a source of hydrogen does not emit greenhouse gases. The hydrogen for synthesis of NH_3, however, is mostly from carbonaceous fossil raw materials. That is, in conventional NH_3 synthesis a large amount of greenhouse gas is emitted. Therefore, the synthesis process of ammonia in industry when based on fossil raw materials is not a green route. As everyone knows, water is an inexhaustible source of hydrogen. The only by-product of hydrogen production from water is oxygen, and the hydrogen in the fuel cell will

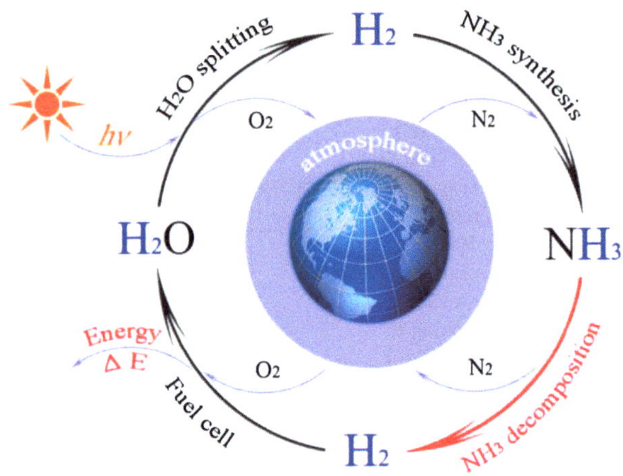

Fig. 7.48 Potential route of a hydrogen energy circle using NH_3 as a carrier

consume the same amount of oxygen. As a result, water is undoubtedly the greenest source of hydrogen. In the near future, humans may be able to adopt the following ways to use hydrogen, as shown in Fig. 7.48.

(i) Continuous hydrogen production using H_2O as the initial hydrogen source and using solar energy, wind energy, tidal energy, geothermal energy and some other green and sustainable energy as the initial energy. Hydrogen production sites can be selected in areas where water is abundant and green and sustainable energy can be obtained easily, such as the coast. Decomposition of water to hydrogen can be achieved using photocatalysis, electrolysis, etc.
(ii) In-site synthesis of NH_3 using hydrogen produced from H_2O and nitrogen obtained by air separation. The current industrial technology and process for the synthesis of NH_3 from hydrogen and nitrogen can be used. In order to avoid the emission of greenhouse gases, all the energy consumption in NH_3 synthesis and air separation should fully use solar and wind energy, tidal energy, water and other green and sustainable energy.
(iii) NH_3 liquefaction, storage and decomposition for hydrogen production. NH_3 can be liquefied at room temperature under 0.8 MPa pressure. Currently, the storage and transportation of liquid NH_3 is also a mature technology. NH_3 decomposition can be carried out using conventional heterogeneous catalytic method or the plasma catalytic method.
(iv) H_2/O_2 fuel cell reaction. PEMFC, an industry-proven process, produces a power to drive vehicles.

In general, this hydrogen cycle route includes two groups of reversible reactions (water decomposition and synthesis, ammonia synthesis and decomposition), circulation of N_2 and O_2 in the atmosphere and circulation of hydrogen on earth. This proposal aims to provide a potential route to use hydrogen energy without involving greenhouse gas emission, which is always the purpose of developing hydrogen energy.

Acknowledgment We acknowledge financial support from the China Postdoctoral Science Foundation [2016T90217 and 2015M580220], the National Natural Science Foundation of China [20473016, 20673018, 21503032], the Fundamental Research Funds for the Central Universities [DUT18JC42], and the China Petroleum Science and Technology Innovation Fund Research Project [2018D-5007-0501]. We also acknowledge Dr. Yue Zhao, Prof. Weimin Gong and Prof. Jialiang Zhang for their contribution on the research of NH_3 decomposition.

References

1. Mansilla, C., Avril, S., Imbach, J., & Le Duigou, A. (2012). CO_2-free hydrogen as a substitute to fossil fuels: What are the targets? Prospective assessment of the hydrogen market attractiveness. *International Journal of Hydrogen Energy, 37*, 9451–9458.
2. Mazloomi, K., & Gomes, C. (2012). Hydrogen as an energy carrier: Prospects and challenges. *Renewable & Sustainable Energy Reviews, 16*, 3024–3033.
3. Satyapal, S., Petrovic, J., Read, C., Thomas, G., & Ordaz, G. (2007). The US department of energy's national hydrogen storage project: Progress towards meeting hydrogen-powered vehicle requirements. *Catalysis Today, 120*, 246–256.
4. Berg, A. W. C., & Arean, C. O. (2008). Materials for hydrogen storage: Current research trends and perspectives. *Chemical Communications, 6*, 668–681.
5. Niaz, S., Manzoor, T., & Pandith, A. H. (2015). Hydrogen storage: Materials, methods and perspectives. *Renewable and Sustainable Energy Reviews, 50*, 457–469.
6. Rusman, N. A. A., & Dahari, M. (2016). A review on the current progress of metal hydrides material for solid-state hydrogen storage applications. *International Journal of Hydrogen Energy, 41*, 12108–12126.
7. Barelli, L., Bidini, G., Gallorini, F., & Servili, S. (2008). Hydrogen production through sorption-enhanced steam methane reforming and membrane technology: A review. *Energy, 33*, 554–570.
8. Abbas, H. F., & Daud, W. M. A. W. (2010). Hydrogen production by methane decomposition: A review. *International Journal of Hydrogen Energy, 35*, 1160–1190.
9. Zhang, F. B., Li, M., Yang, L., Ye, S., & Huang, L. (2014). Ni-Ma-Mn-Fe-O catalyst derived from layered double hydroxide for hydrogen production by auto-thermal reforming of ethanol. *Catalysis Communications, 43*, 6–10.
10. Murdoch, M., Waterhouse, G. I. N., Nadeem, M. A., Metson, J. B., Keane, M. A., Howe, R. F., Llora & Idriss, H. (2011). The effect of gold loading and particle size on photocatalytic hydrogen production from ethanol over Au/TiO_2 nanoparticles. *Nature Chemistry, 3*, 489–492.
11. Navarro, R. M., Pena, M. A., & Fierro, J. L. G. (2007). Hydrogen production reactions from carbon feedstocks: Fossil fuels and biomass. *Chemical Reviews, 107*, 3952–3991.
12. Song, T., Wu, J. H., Shen, L. H., & Xiao, J. (2012). Experimental investigation on hydrogen production from biomass gasification in interconnected fluidized beds. *Biomass and Bioenergy, 36*, 258–267.
13. Yeh, T. F., Syu, J. M., & Cheng, C. (2010). Graphite oxide as a photocatalyst for hydrogen production from water. *Advanced Functional Materials, 20*, 2255–2262.
14. Maeda, K., Teramura, K., Lu, D. L., Takata, T., Saito, N., Inoue, Y., & Domen, K. (2006). Photocatalyst releasing hydrogen from water – Enhancing catalytic performance holds promise for hydrogen production by water splitting in sunlight. *Nature, 440*, 295–295.
15. George Thomas and George Parks (2006) Potential roles of ammonia in a hydrogen economy: A study of issues related to the use ammonia for on-board vehicular hydrogen storage. Washington: U.S. Department of Energy

16. Li, Y. X., Yao, L. H., Song, Y. Y., Liu, S., Zhao, J., Ji, W., & Au, C. T. (2010). Core-shell structured mcrocapsular-like Ru@SiO_2 reactor for efficient generation of CO_x-free hydrogen through ammonia decomposition. *Chemical Communications, 46*, 5298–5300.
17. Enrique, G. B., Sabino, A., & Laura, R. (2014). Toward practical application of H_2 generation from ammonia decomposition guided by rational catalyst design. *Catalysis Reviews: Science and Engineering, 56*, 220–237.
18. Bell, T. E., & Torrente-Murciano, L. (2016). H_2 production via ammonia decomposition using non-noble metal catalysts: A review. *Top Catalysis, 59*, 1438–1457.
19. Yin, S. F., Zhang, Q. H., Xu, B. Q., Zhu, W. X., Ng, C. F., & Au, C. T. (2004). Investigation on the catalysis of CO_x-free hydrogen generation from ammonia. *Journal of Catalysis, 224*, 384–396.
20. Schuth, F., Palkovits, R., Schlogl, R., & Su, D. S. (2012). Ammonia as a possible element in an energy infrastructure: Catalysts for ammonia decomposition. *Energy & Environmental Science, 5*, 6278–6289.
21. Guo, J. P., Chen, Z., Wu, A., Chang, F., Wang, P., Hu, D., Wu, G., Xiong, Z., Yu, P. & Chen, P. (2015). Electronic promoter or reacting species? The role of $LiNH_2$ on Ru in catalyzing NH_3 decomposition. *Chemical Communications, 51*, 15161–15164.
22. Bajus, S., Agel, F., Kusche, M., Bhriain, N. N., & Wasserscheid, P. (2016). Alkali hydroxide-modified Ru/γ-Al_2O_3 catalysts for ammonia decomposition. *Applied Catalysis A: General, 510*, 189–195.
23. Furukawa, S., Tsuchiya, A., Kojima, Y., Endo, M., & Komatsu, T. (2016). Raney-type Ru-La catalysts prepared from a Ru-La-Al ternary alloy: Enhanced activity in ammonia decomposition. *Chemistry Letter, 45*, 158–160.
24. Takahashi, A., & Fujitani, T. (2016). Kinetic analysis of decomposition of ammonia over nickel and ruthenium catalysts. *Journal of Chemical Engineering of Japan, 49*, 22–28.
25. Yu, P., Guo, J., Liu, L., Wang, P., Chang, F., Wang, H., Ju, X., & Chen, P. (2016). Effects of alkaline earth metal amides on Ru in catalytic ammonia decomposition. *The Journal of Physical Chemistry C, 120*, 2822–2828.
26. Duan, X., Qian, G., Zhou, X. G., Sui, Z., Chen, D., & Yuan, W. (2011). Tuning the size and shape of Fe nanoparticles on carbon nanofibers for catalytic ammonia decomposition. *Applied Catalysis B: Environmental, 101*, 189–196.
27. Li, X. K., Ji, W. J., Zhao, J., Wang, S. J., & Au, C. T. (2005). Ammonia decomposition over Ru and Ni catalysts supported on fumed SiO_2, MCM-41, and SBA-15. *Journal of Catalysis, 236*, 181–189.
28. Zhang, L. F., Li, M., Ren, T. Z., Liu, X. Y., & Yuan, Z. Y. (2015). Ce-modified Ni nanoparticles encapsulated in SiO_2 for CO_x-free hydrogen production via ammonia decomposition. *International Journal of Hydrogen Energy, 40*, 2648–2656.
29. Guo, W., & Vlachos, D. G. (2015). Patched bimetallic surfaces are active catalysts for ammonia decomposition. *Nature Communications, 6*, 8619.
30. Meng, T., Xu, Q. Q., Li, Y. T., Chang, J. L., Ren, T. Z., & Yuan, Z. Y. (2015). Nickle nanoparticles highly dispersed on reduced graphene oxide for ammonia decomposition to hydrogen. *Journal of Industrial and Engineering Chemistry, 32*, 373–379.
31. Inokawa, H., Ichikawa, T., & Miyaoka, H. (2015). Catalysis of nickel nanoparticles with high thermal stability for ammonia decomposition. *Applied Catalysis A: General, 491*, 184–188.
32. Okura, K., Okanishi, T., Muroyama, H., Matsui, T., & Eguchi, K. (2015). Promotion effect of rare-earth elements on the catalytic decomposition of ammonia over Ni/Al_2O_3 catalyst. *Applied Catalysis A: General, 505*, 77–85.
33. Bielun, Z. L., Pelka, R., & Arabczyk, W. (2009). Study of the kinetics of ammonia synthesis and decomposition on iron and cobalt catalysts. *Catalysis Letters, 129*, 119–123.
34. Zhang, J., Comotti, M., Schuth, F., Schlögl, R., & Su, D. S. (2007). Commercial Fe- or Co-containing carbon nanotubes as catalysts for NH_3 decomposition. *Chemical Communication, 2007*, 1916–1918.

35. Gu, Y. Q., Fu, X. P., Du, P. P., Gu, D., Jin, Z., Huang, Y. Y., Si, rui., Zheng, L. Q., Song, Q. S., Jia, C. J., & Weidenthaler, C. (2015). In situ X-ray diffraction study of Co-Al nanocomposites as catalysts for ammonia decomposition. *The Journal of Physical Chemistry C, 119*, 17102–17110.
36. Podila, S., Alhamed, Y. A., AlZahrani, A. A., & Petrov, L. A. (2015). Hydrogen production by ammonia decomposition using Co catalyst supported on Mg mixed oxide systems. *International Journal of Hydrogen Energy, 40*, 15411–15422.
37. Podila, S., Driss, H., Zaman, S. F., Alhamed, Y. A., AlZahrani, A. A., Daous, M. A., & Petrov, L. A. (2016). Hydrogen generation by ammonia decomposition using Co/MgO-La_2O_3 catalyst: Influence of support calcination atmosphere. *Journal of Molecular Catalysis A: Chemical, 414*, 130–139.
38. Cui, X. Z., Li, H., Guo, L. M., He, D., Chen, H., & Shi, J. (2008). Synthesis of mesoporous tungsten carbide by an impregnation-compaction route, and its NH_3 decomposition catalytic activity. *Dalton Transactions, 2008*, 6435–6440.
39. Liu, H. C., Wang, H., Shen, J. H., Sun, Y., & Liu, Z. (2006). Influence of preparation conditions on the catalytic performance of MoN_x/SBA-15 for ammonia decomposition. *Journal of Natural Gas Chemistry, 15*, 178–180.
40. Guo, J., Chang, F., Wang, P., Hu, D., Yu, P., Wu, G., Xiong, Z., & Chen, P. (2015). Highly active MnN-Li_2NH composite catalyst for producing CO_x-free hydrogen. *ACS Catalysis, 5*, 2708–2713.
41. Guo, J., Wang, P., Wu, G., Wu, A., Hu, D., Xiong, Z., Wang, J. J., Yu, P., Chang, F., Chen, Z., & Chen, P. (2015). Lithium imide synergy with 3d transition-metal nitrides leading to unprecedented catalytic activities for ammonia decomposition. *Angewandte Chemie International Edition, 54*, 2950–2954.
42. Esrafili, M. D., & Nurazar, R. (2015). Catalytic decomposition of ammonia over silicon-carbide nanotube: A DFT study. *Structural Chemistry, 26*, 799–807.
43. Karim, A. M., Prasad, V., Mpourmpakis, G., Lonergan, W. W., Frenkel, A. I., Chen, J. G., & Vlachos, D. G. (2009). Correlating particle size and shape of supported Ru/γ-Al_2O_3 catalysts with NH_3 decomposition activity. *Journal of the American Chemical Society, 131*, 12230–12239.
44. Garcia-Garcia, F. R., Ruiz, A. G., & Ramos, I. R. (2009). Role of B5-type sites in Ru catalysts used for the NH_3 decomposition reaction. *Topics in Catalysis, 52*, 758–764.
45. Zheng, W. Q., Zhang, J., Xu, H. Y., & Li, W. (2007). NH_3 decomposition kinetics on supported Ru clusters: Morphology and particle size effect. *Catalysis Letters, 119*, 311–318.
46. Lu, A. H., Nitz, J. J., Comotti, M., Weidenthaler, C., Schlichte, K., Lehmann, C. W., Terasaki, O., & Schuth, F. (2010). Spatially and size selective synthesis of Fe-based nanoparticles on ordered mesoporous supports as highly active and stable catalysts for ammonia decomposition. *Journal of the American Chemical Society, 132*, 14152–14162.
47. Kowalczyk, Z., Sentek, J., Jodzis, S., Muhler, M., &. Hinrichsen, O. (1997). Effect of potassium on the kinetics of ammonia synthesis and decomposition over fused Iron catalyst at atmospheric pressure. *Journal of Catalysis, 169*, 407–414.
48. Pelka, R., Moszynska, I., & Arabczyk, W. (2009). Catalytic ammonia decomposition over Fe/Fe_4N. *Catalysis Letters, 128*, 72–76.
49. Arabczyk, W., & Pelka, R. (2009). Studies of the kinetics of two parallel reactions: Ammonia decomposition and nitriding of iron catalyst. *The Journal of Physical Chemistry A, 113*, 411–416.
50. Ganley, J. C., Thomas, F. S., Seebauer, E. G., & Masel, R. I. (2004). A priori catalytic activity correlations: The difficult case of hydrogen production from ammonia. *Catalysis Letters, 96*, 117–122.
51. Liu, H. C., Wang, H., Shen, J. H., Sun, Y., & Liu, Z. (2008). Preparation, characterization and activities of the nano-sized Ni/SBA-15 catalyst for producing CO_x-free hydrogen from ammonia. *Applied Catalysis A: General, 337*, 138–147.

52. Zheng, W. Q., Zhang, J., Ge, Q. J., Xu, H., & Li, W. (2008). Effects of CeO_2 addition on Ni/Al_2O_3 catalysts for the reaction of ammonia decomposition to hydrogen. *Applied Catalysis B: Environmental, 80*, 98–105.
53. Soensen, R. Z., Nielsen, L. J. E., Jensen, S., Hansen, O., Johannessen, T., Quaade, U., & Christensen, C. H. (2005). Catalytic ammonia decomposition: Miniaturized production of CO_x-free hydrogen for fuel cells. *Catalysis Communications, 6*, 229–232.
54. Hansgen, D. A., Thomanek, L. M., Chen, J. G., & Vlachos, D. G. (2011). Experimental and theoretical studies of ammonia decomposition activity on Fe-Pt, Co-Pt, and Cu-Pt bimetallic surfaces. *Journal of Chemical Physics, 134*, 184701–184707.
55. Levy, R. B., & Boudart, M. (1973). Platinum-like behavior of tungsten carbide in surface catalysis. *Science, 181*, 547–549.
56. Siegel, I. E. (1979). D-bandwidth contraction upon metalloid formation: A resolution of the bonding controversy in transition metal carbides, nitrides and borides. *Semiconductors and Insulators, 5*, 47–60.
57. Pansare, S. S., Torres, W., & Goodwin, J. G. (2007). Ammonia decomposition on tungsten carbide. *Catalysis Communications, 8*, 649–654.
58. Liu, H. C., Wang, H., Shen, J. H., Sun, Y., & Liu, Z. (2008). Preparation and evaluation of ammonia decomposition catalysts by high-throughput technique. *Reaction Kinetics and Catalysis Letters, 93*, 11–17.
59. Lee, H. J., Choi, J. G., Colling, C. W., Mudholkar, M. S., & Thompson, L. T. (1994). Temperature-programmed desorption and decomposition of NH_3 over molybdenum nitride film. *Applied Surface Science, 78*, 299–307.
60. Pelka, R., & Arabczyk, W. (2009). Studies of the kinetics of reaction between iron catalysts and ammonia-nitriding of nanocrystalline iron with parallel catalytic ammonia decomposition. *Topics in Catalysis, 52*, 1506–1516.
61. Lu, C. S., Li, X. N., Zhu, Y. F. (2004). Ammonia decomposition over bimetallic nitrides supported onγ-Al_2O_3. *Chinese Chemical Letters, 15*, 105–108.
62. Zheng, W. Q., Zhang, J., Zhu, B., Blume, R., Zhang, Y., Schlichte, K., & Su, D. S. (2010). Structure-function correlations for Ru/CNT in the catalytic decomposition of ammonia. *ChemSusChem, 3*, 226–230.
63. Garcia-Garcia, F. R., Alvarez-Rodriguez, J., Rodriguez-Ramos, I., & Guerrero-Ruiz, A. (2010). The use of carbon nanotubes with and without nitrogen doping as support for ruthenium catalysts in the ammonia decomposition reaction. *Carbon, 48*, 267–276.
64. Li, L., Zhu, Z. H., Lu, G. Q., Yan, Z. F., & Qiao, S. Z (2007). Catalytic ammonia decomposition over CMK-3 supported Ru catalysts: Effects of surface treatments of supports. *Carbon, 45*, 11–20.
65. Yin, S. F., Xu, B. Q., Wang, S. J., Ng, C. F., & Au, C. T. (2004). Magnesia–carbon nanotubes (MgO–CNTs) nanocomposite: Novel support of Ru catalyst for the generation of CO_x-free hydrogen from ammonia. *Catalysis Letters, 96*, 113–116.
66. Yin, S. F., Xu, B. Q., Zhou, X. P., & Au, C. T. (2004). A mini-review on ammonia decomposition catalysts for on-site generation of hydrogen for fuel cell applications. *Applied Catalysis A: General, 277*, 1–9.
67. Zhang, J., Xu, H. Y., Jin, X. L., Ge, Q., & Li, W. (2005). Characterizations and activities of the nano-sized Ni/Al_2O_3 and Ni/La–Al_2O_3 catalysts for NH_3 decomposition. *Applied Catalysis A: General, 290*, 87–96.
68. Rarog-Pilecka, W., Szmigiel, D., Kowalczyk, Z., Jodzis, S., & Zielinski, J. (2003). Ammonia decomposition over the carbon-based ruthenium catalyst promoted with barium or cesium. *Journal of Catalysis, 218*, 465–469.
69. Liu, H. C., Wang, H., Shen, J. H., Sun, Y., & Liu, Z. (2008). Promotion effect of cerium and lanthanum oxides on Ni/SBA-15 catalyst for ammonia decomposition. *Catalysis Today, 131*, 444–449.

70. Li, Y. X., Yao, L. H., Song, Y. Y., Liu, S., Zhao, J., Ji, W., & Au, C. T. (2010). Core-shell structured microcapsular-like Ru@SiO_2 reactor for efficient generation of CO_x-free hydrogen through ammonia decomposition. *Chemical Communication, 46*, 5298–5300.
71. Li, Y. X., Yao, L. H., Liu, S. Q., Zhao, J., Ji, W., & Au, C. T. (2011). Cs-modified iron nanoparticles encapsulated in microporous and mesoporous SiO_2 for CO_x-free H_2 production via ammonia decomposition. *Catalysis Today, 160*, 79–86.
72. Zhang, J., Muller, J. O., Zheng, W. Q., Wang, D., Su, D., & Schlögl, R. (2008). Individual Fe-Co alloy nanoparticles on carbon nanotubes: Structural and catalytic properties. *Nano Letters, 8*, 2738–2743.
73. Stolbov, S., & Rahman, T. S. (2005). First-principles study of some factors controlling the rate of ammonia decomposition on Ni and Pd surfaces. *Journal of Chemical Physics, 123*, 204716–204720.
74. Huang, W. Y., Lai, W. Z., & Xie, D. Q. (2008). First-principles study of decomposition of NH_3 on Ir(100). *Surface Science, 602*, 1288–1294.
75. Duan, X. Z., Qian, G., Fan, C., Zhu, Y., Zhou, X., Chen, D., & Yuan, W. (2012). First-principles calculations of ammonia decomposition on Ni(110) surface. *Surface Science, 606*, 549–553.
76. Baiker, A., & Monti, D. (1983). Interaction of ammonia with metallic copper, nickel and cobalt catalysts studied by temperature programmed desorption. *Berichte der Bunsengesellschaft für Physikalische Chemie, 87*, 602–605.
77. Hansgen, D. A., Vlachos, D. G., & Chen, J. G. (2010). Using first principles to predict bimetallic catalysts for the ammonia decomposition reaction. *Nature Chemistry, 2*, 484–489.
78. Zhao, Y., Wang, L., Zhang, J. L., & Guo, H. C. (2014). Influence of non-thermal plasma discharge mode and reactor structure on ammonia decomposition to hydrogen. *Acta Physico-Chimica Sinica, 30*, 738–744.
79. Zhao, Y., Wang, L., Zhang, J. L., & Guo H. C. (2013). Decomposition of ammonia by atmospheric pressure AC discharge: Catalytic effect of the electrodes. *Catalysis Today, 211*, 72–77.
80. Zhao, Y., Wang, L., Zhang, J. L., & Guo, H. C. (2014). Enhancing the ammonia to hydrogen (ATH) energy efficiency of alternating current arc discharge. *International Journal of Hydrogen Energy, 39*, 7655–7663.
81. Watson, J. K. G., & Majewski, W. A. (1986). Assignment of the schuster band of ammonia. *Journal of Molecular Spectroscopy, 115*, 82–87.
82. Agostino, R., Cramarossa, F., Benedictis, S. D., & Ferraro, G. (1981). Kinetic and spectroscopic analysis of NH_3 decomposition under R.F. plasma at moderate pressure. *Plasma Chemistry and Plasma Processing, 1*, 19–35.
83. Cicala, G., De Tommaso, E., Raino, A. C., Lebedev, Y. A., & Shakhatov, V. A. (2009). Study of positive column of glow discharge in nitrogen by optical emission spectroscopy and numerical simulation. *Plasma Sources Science and Technology, 18*, 025032.
84. Nicholas, J. E., Spiers, A. I., & Martin, N. A. (1986). Kinetics and mechanism in the decomposition of NH_3 in a radio-frequency pulse discharge. *Plasma Chemistry and Plasma Processing, 6*, 39–51.
85. Granier, A., Vervloet, M., Aumaille, K., & Vallée, C. (2003). Optical emission spectra of TEOS and HMDSO derived plasmas used for thin film deposition. *Plasma Sources Science & Technology, 12*, 89–96.
86. Clay, K. J., Speakman, S. P., Amaratunga, G. A. J., & Silva, S. R. P. (1996). Characterization of a-C:H:N deposition from CH_4/N_2 rf plasmas using optical emission spectroscopy. *Journal of Applied Physics, 79*, 7227–7233.
87. Wang, L., Zhao, Y., Liu, C. Y., Gong, W., & Guo, H. (2013). Plasma driven ammonia decomposition on a Fe-catalyst: Eliminating surface nitrogen poisoning. *Chemical Communications, 49*, 3787–3789.

88. Wang, L., Yi, Y. H., Zhao, Y., Zhang, R., Zhang, J., & Guo, H. (2015). NH_3 decomposition for H_2 generation: Effects of cheap metals and supports on plasma-catalyst synergy. *ACS Catalysis, 5*, 4167–4174.
89. Wang, L., Yi, Y. H., Guo, Y. J., Zhao, Y., Zhang, J., & Guo, H. (2017). Synergy of DBD plasma and Fe-based catalyst in NH_3 decomposition: Plasma enhancing adsorption step. *Plasma Processes and Polymers, 14*, e1600111.
90. Jongsomjit, B., Panpranot, J., & Goodwin, J. G. (2001). Co-support compound formation in alumina-supported cobalt catalysts. *Journal of Catalysis, 204*, 98–109.
91. Tang, C. W., Wang, C. B., & Chien, S. H. (2008). Characterization of cobalt oxides studied by FT-IR, Raman, TPR and TG-MS. *Thermochim Acta, 473*, 68–73.
92. Fischer, N., Minnermann, M., Baeumer, M., Van Steen, E., & Claeys, M. (2012). Metal support interactions in Co_3O_4/Al_2O_3 catalysts prepared from w/o microemulsions. *Catalysis Letter, 142*, 830–837.
93. Arnoldy, P., & Moulijn, J. A. (1985). Temperature-programmed reduction of CoO/Al_2O_3 catalysts. *Journal of Catalysis, 93*, 38–54.
94. Pelka, R., Kiełbasa, K., & Arabczyk, W. (2014). Catalytic ammonia decomposition during nanocrystalline iron nitriding at 475 °C with NH_3/H_2 mixtures of different nitriding potentials. *The Journal of Physical Chemistry C, 118*, 6178–6185.
95. Rico, V. J., Hueso, J. L., Cotrino, J., Gallardo, V., Sarmiento, B., Brey, J. J., & González-Elipe, A. R. (2009). Hybrid catalytic-DBD plasma reactor for the production of hydrogen and preferential CO oxidation (CO-PROX) at reduced temperatures. *Chemical Communications, 2009*, 6192–6194.
96. Pan, K. L., Chung, W. C., & Chang, M. B. (2014). Dry reforming of CH_4 with CO_2 to generate syngas by combined plasma catalysis. *IEEE Transactions on Plasma Sciences, 42*, 3809–3818.
97. Qayyum, A., Zeb, S., Naveed, M. A., Ghauri, S. A., Zakaullah, M., & Waheed, A. (2005). Diagnostics of nitrogen plasma by trace rare-gas–optical emission spectroscopy. *Journal of Applied Physics, 98*, 103303.
98. Rimmer, W. B. (1923). The spectrum of ammonia. *Royal Society of London Proceedings, 103*, 696–705.
99. Cheng, H. K., Huang, Y. Q., Wang, A. Q., Wang, X. D., & Zhang, T. (2009). Preparation of cobalt nitride from Co-Al hydrotalcite and its application in hydrazine decomposition. *Top Catalysis, 52*, 1535–1540.
100. Wang, L., Xin, Q., Zhao, Y., Zhang, G., Dong, J., Gong, W., & Guo, H. (2012). In situ FT-IR studies on catalytic nature of iron nitride: Identification of the N active site. *ChemCatChem, 4*, 624–627.
101. Ramis, G., Yi, L., & Busca, G. (1995). Adsorption, activation and oxidation of ammonia over SCR catalysts. *Journal of Catalysis, 157*, 523–535.
102. Boisen, A., Dahl, S., Norskov, J. K., & Christensen, C. H. (2005). Why the optimal ammonia synthesis catalyst is not the optimal ammonia decomposition catalyst. *Journal of Catalysis, 230*, 309–312.

Chapter 8
Plasma-Catalytic Conversion of Methane

Tomohiro Nozaki, Seigo Kameshima, Zunrong Sheng, Keishiro Tamura, and Takumi Yamazaki

8.1 Introduction

Methane is one of the cleanest and most abundant primary energy, carbon, and hydrogen sources among the available hydrocarbon resources. The growing demand for shale gas accelerates the need for methane as a promising future energy source. In addition to its direct use at thermal power plants, steam methane reforming (SMR) has been used for various chemical industries as well as being a hydrogen source for ammonia synthesis and fuel cell applications. SMR has a long and proven history and its related technology is well developed. Nevertheless, further improvement of energy and material conversion efficiencies is demanded. Moreover, an urgent demand for CO_2 free hydrogen inevitably requires innovative new technologies for the sustainable use of methane.

More recently, the power-to-gas concept has been highlighted where water electrolysis driven by renewable electricity is utilized. Renewable hydrogen is further combined with the catalytic conversion of CO_2 to synthetic natural gas (SNG) [1]. Renewable H_2 and SNG can be stored and distributed through the existing gas grid for efficient use of renewable energy with reduced CO_2 emissions. Based on this similar concept, we investigated CH_4 reforming using nonthermal plasma-enhanced catalytic reaction [2–4]. Analogous to electrolysis of water splitting, renewable electricity is converted into chemical energy via nonthermal plasma-assisted endothermic reactions. Syngas is then upgraded to H_2-rich gas or SNG for power-to-gas applications. Alternatively, liquid fuels such as gasoline and methanol are synthesized through well-established C1 chemistry using plasma-generated synthetic gas. Carbon-containing liquid fuels are particularly important because their energy density is 10–100 times greater than that of secondary batteries, meaning the transport and storage capability of renewable energy are greatly

T. Nozaki (✉) · S. Kameshima · Z. Sheng · K. Tamura · T. Yamazaki
Department of Mechanical Engineering, Tokyo Institute of Technology, Tokyo, Japan
e-mail: nozaki.t.ab@m.titech.ac.jp

X. Tu et al. (eds.), *Plasma Catalysis*, Springer Series on Atomic, Optical, and Plasma Physics 106, https://doi.org/10.1007/978-3-030-05189-1_8

improved, while CO_2 capture and utilization are also strengthened simultaneously. Currently, electrochemical reactions are primarily studied for renewable-to-chemical energy conversions. Nonthermal plasma can provide additional energy and material conversion pathways, contributing to an extended carbon recycling network and material flexibility [4–10]. Over the next decade, a sufficient amount of renewable electricity will be available through photovoltaics and wind turbines, where energy conversion efficiency and power distribution technology are rapidly growing worldwide. Nonthermal plasma features compactness, a flexible reactor design, low-temperature operation, and load followability: which are compatible with renewable energy sources and the various chemical reactors developed so far. Renewable energy-driven nonthermal plasma catalysis provides a viable solution for efficient energy and material conversion, storage, and transportation with reduced CO_2 emissions.

This chapter focuses particularly on dielectric barrier discharge (DBD) and catalyst combination for methane conversion to synthesis gas (H_2, CO). First, the plasma catalysis of methane is reviewed in terms of the endothermic and exothermic nature of the reaction systems where the role of nonthermal plasma is essentially different. After that, the dry reforming of methane ($CH_4 + CO_2 \rightarrow 2CO + 2H_2$) in DBD/catalyst hybrid reaction is highlighted based on our recent study. The basics of heterogeneous reaction, known as a Langmuir-Hinshelwood mechanism, are overviewed for the better understanding of nonthermal plasma and surface interaction as well as plasma-enhanced heterogeneous reactions, which is known as *plasma catalysis*. Pulsed reaction spectrometry is introduced as a powerful diagnostic approach of heterogeneous reaction kinetics under the influence of nonthermal plasma. Moreover, the nonthermal plasma-induced synergistic effect and energy efficiency are discussed in order to move toward a deeper understanding of plasma catalysis in methane reforming. Finally, concluding remarks and the future outlook are presented.

8.2 Plasma Catalysis for CH_4 Reforming

In general, CH_4 conversion indicates syngas (H_2 and CO) production via steam reforming (8.R1), dry reforming (8.R2), partial oxidation (8.R3), and the combination of 8.R1, 8.R2 and 8.R3 known as autothermal reforming. Syngas is converted into various synthetic chemicals by well-established C1 chemistry [11]. In particular, Fischer-Tropsch hydrocarbon fuels have attracted keen attention in view of their high energy storability and transportability in the form of liquid at ambient conditions:

$$CH_4 + H_2O \rightarrow 3H_2 + CO \quad \Delta H^{\circ} = 206\ \mathrm{kJ/mol} \quad \Delta G^{\circ} = 142\ \mathrm{kJ/mol} \quad (8.R1)$$

$$CH_4 + CO_2 \rightarrow 2H_2 + 2CO \quad \Delta H^\circ = 247\ \text{kJ/mol} \quad \Delta G^\circ = 170\ \text{kJ/mol} \qquad (8.R2)$$

$$CH_4 + 0.5O_2 \rightarrow 2H_2 + CO \Delta H^\circ = -36\ \text{kJ/mol} \Delta G^\circ = -87\ \text{kJ/mol} \qquad (8.R3)$$

Here, ΔH° and ΔG°, respectively, represent the standard enthalpy and Gibbs free energy changes of the reaction. Hereafter, H_2O is considered in the vapor phase. If H_2 is demanded for fuel cell application, the water gas shift (WGS) reaction (8.R4) is induced to enrich H_2 in the upgraded gas stream:

$$CO + H_2O \rightarrow H_2 + CO_2 \quad \Delta H^\circ = -41\ \text{kJ/mol} \quad \Delta G^\circ = -29\ \text{kJ/mol} \qquad (8.R4)$$

The role of nonthermal plasma can be uniquely distinguished into two categories depending on whether it is combined with an exothermic or endothermic system, which is discussed further in the following sections.

8.2.1 Exothermic System

8.2.1.1 CH_4 Partial Oxidation in Gliding Arc Discharge

One of the important roles of nonthermal plasma is to generate reactive species almost independently of the ambient temperature. Plasma-generated reactive species are used to initiate chemical reactions at much lower temperatures than conventional thermochemical reactions. Such peculiarity is particularly important in non-catalytic homogeneous partial oxidation of hydrocarbons [12] and plasma-assisted combustion [13]. The main driving force of an exothermic reaction is the heat released by partial oxidation of the initial feed. Although generation of nonthermal plasma consumes electrical energy, plasma itself is not necessarily an energy source in the exothermic system; therefore, the energy consumption of plasma-assisted partial oxidation is inevitably low when compared to an endothermic system. Reaction temperature, fuel/oxygen ratio, and plasma power can be varied to a large extent as long as plasma-generated reactive species induce a spontaneous chain reaction to sustain partial oxidation. Because plasma can lower the reaction temperature (seemingly, lowering the apparent activation energy) and create new reaction pathways, plasma is considered to possess a catalytic effect which is called *plasma catalysis*. However, the definition of *plasma catalysis* is not always clear. Because the generation of any type of plasma source is accompanied by heat generation. The individual contribution of radical injections, local heating, or a combination of these, it is not always well understood.

One of the most successful applications is the gliding arc discharge for the partial oxidation of hydrocarbon fuels. Figure 8.1 shows gliding arc discharges in various electrode configurations. A diverging two-blade electrode reactor is commonly used to generate relatively intense discharge channels (Fig. 8.1a) [14]. It is characterized as a two-dimensional transient arc plasma because discharge channels are generated

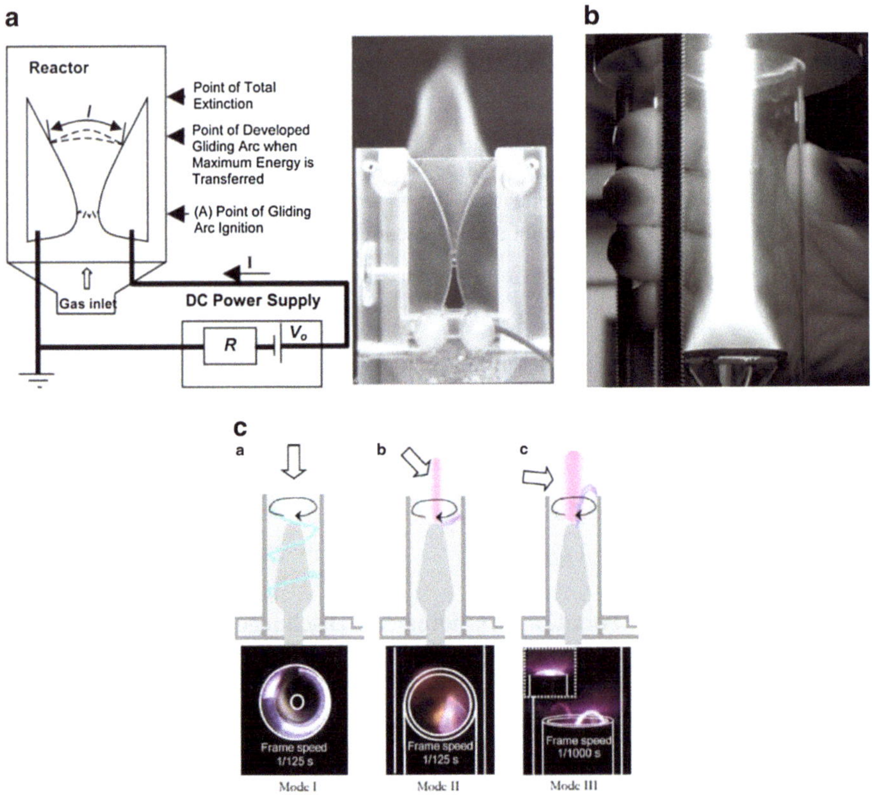

Fig. 8.1 Gliding arc type discharges in various configurations: (**a**) Diverging two-blade type (2D) [14], (**b**) Reverse vortex flow "tornado" type (3D) [14], (**c**) Rotating arc type [16]

in the planar geometry, and thus contact between discharge channels and the gas stream may be insufficient. In order to improve the interaction between the discharge channels and incoming gas flow, a three-dimensional configuration known as a reverse vortex flow, or tornado gliding arc discharge, was developed (Fig. 8.1b) [14]. The heat generated by the transient high-temperature arc column, as well as excited species, plays a key role. The principles of thermal and nonthermal characteristics of gliding arc discharge are investigated in detail by Fridman and co-workers [15]. A similar principle, but different electrode configuration, is named rotating arc discharge (Fig. 8.1c) [16]. Transient filamentary arc channels are literary rotating between diverging coaxial electrodes, improving the mixing and ionization of the reactant flow. Thermal and nonthermal characteristics of rotating arc discharges and their applications in fuel reforming are presented by Song and co-workers [17].

Gliding arc discharges have shown some of the best reforming performances for syngas production [7]. However, there are several challenges to overcome before widespread use in industry. First, relatively high temperatures under fuel-rich conditions produce coke and/or tar [18]. Second, the use of air as an oxidizer dilutes the output gas with N_2 by up to 50%, producing low calorific syngas [19]; alternatively, N_2 should be separated by post-plasma processes. The use of pure O_2 accompanies an additional energy penalty and a bulky air separation unit which is not compatible with a compact and low-cost reforming system. Pure O_2 from water electrolysis, obtainable from a power-to-gas system (water electrolysis), is an interesting option in future applications. Finally, full oxidation of fuels is likely to occur in the fuel/O_2 system. Syngas can be generated by the full oxidation of fuels, followed by endothermic steam/dry reforming as shown in Fig. 8.2. The principle is similar to a multistep reforming as depicted in the *in-direct route*: drastic improvement of energy efficiency may be hard to achieve unless CH_4 combustion is excluded.

8.2.1.2 Direct Conversion of CH_4 to CH_3OH in Microplasma Reactor

An alternative approach for gliding arc partial oxidation of CH_4 is a low-temperature and single-step CH_4 partial oxidation to methanol as indicated in *direct route* (Fig. 8.2). Direct conversion of methane to methanol can greatly reduce capital and operating costs of high-temperature, energy-intensive, multistep processes for syngas production [20, 21]:

$$CH_4 + 0.5O_2 \rightarrow CH_3OH_{\_liq} \quad \Delta H^{\circ} = -163 \ \ \mathrm{kJ/mol} \quad \Delta G^{\circ} = -116 \ \ \mathrm{kJ/mol} \tag{8.R5}$$

Although tremendous effort has been made on the direct CH_4 conversion to oxygenates in a homogeneous gas phase reaction and over heterogeneous catalysts [22–25], the yield for desired products was below the economic value [26]. More recently, atmospheric pressure nonthermal plasma is looked at as a viable synthesis

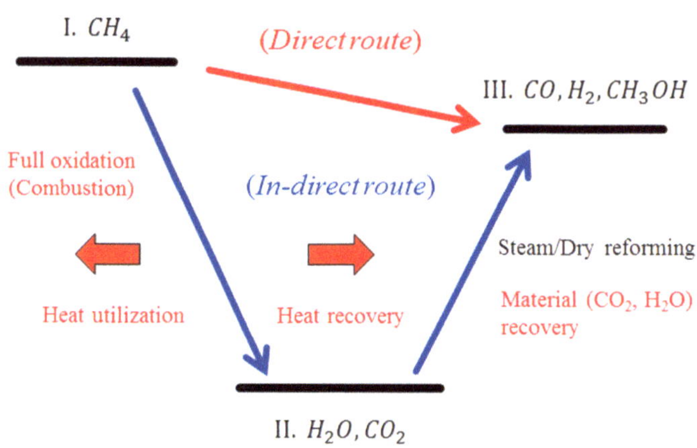

Fig. 8.2 Energy and material conversion pathways in CH_4 reforming: *direct* (I.-III.) and *in-direct* (I.-II.-III.) routes

method because high-energy electron impacts initiate CH_4 partial oxidation at low temperature, enabling a single-step CH_4 conversion into various oxygenates. Nevertheless, the one-pass yield for useful oxygenates was unsatisfactory [27–29]. Performing nonthermal discharge in a microreactor brings unconventional thermochemical conditions to materials processing, enabling better control over process parameters to selectively synthesize desirable products [30, 31]. DBD generated in a microreactor selectively produced methanol (CH_3OH), formaldehyde (HCHO), and formic acid (HCOOH) with a one-pass yield of more than 10% [32].

A schematic diagram of the microplasma reactor is presented in Fig. 8.3. Dielectric barrier discharge is generated in a thin glass tube (inner diameter, 1.5 mm, and

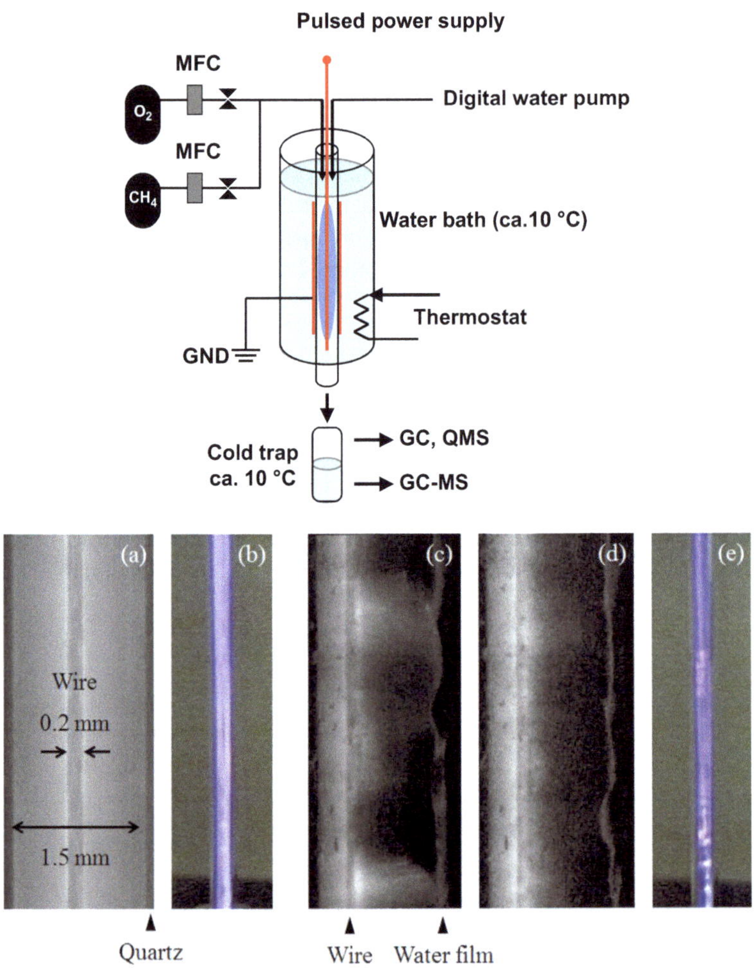

Fig. 8.3 A schematic diagram of the two-phase flow microplasma reactor and discharge images: (**a–b**) without water injection, (**c–e**) with water injection. The emission from discharge looks like (**b**) and (**e**) to the naked eye. Magnified images (**c**) and (**d**) show the water film on the wall. Conditions: 30 W and $CH_4 = O_2 = 20$ cm^3/min

length, 50 mm) equipped with an internal wire electrode (diameter, 0.2 mm). Gas breaks down with a nanosecond duration, and average gas temperature only slightly increases [33–35]. Meanwhile, high-energy electrons activate methane and oxygen, initiating CH_4 partial oxidation at room temperature; therefore, a stoichiometric flammable CH_4 and O_2 mixture (8.R5) was processed in the microplasma reactor without detonation. Moreover, distilled water was injected simultaneously from the top of the reactor, creating a water film on the inner wall of the tube. The microreactor configuration enhances the removal of heat generated by CH_4 partial oxidation. Therefore, ignition and propagation of flames, or full combustion, in CH_4/O_2 mixture are avoided. Moreover, the low-temperature operation enhances the condensation of liquid components onto the water film, realizing product separation from the oxygen containing reactive plasma. Product separation is important because organic oxygenates are much more reactive than CH_4 and readily decompose during oxidative methane activation. Figure 8.3a, b shows images of DBD without water injection. Although DBD consists of a number of filamentary discharges, the time- and space-averaged image is fairly uniform. In contrast, DBD is no longer uniform as a result of complex interaction with the water film (Fig. 8.3c–e). A high electric field creates small bumps on the water film due to the electrohydrodynamic interaction. The electric field is concentrated at these points, creating constricted and intense discharge channels.

The total selectivity of liquid components is shown in Fig. 8.4 with respect to CH_4 conversion [36]. The results are compared with ten from cited literature that used thermochemical methods [24, 25, 37–44]. This literature was selected mostly from the excellent review article given by Rasmussen and Glarborg [45]. A similar attempt was made by Casey and co-workers [26], comparing selected cited literature in terms of CH_3OH selectivity vs. CH_4 conversion. Although tremendous effort has been made toward direct CH_4 conversion to CH_3OH, there is no significant improvement between 1994 [26] and 2008 [45]. In the conventional thermochemical approach, relatively high-temperature thermal energy must be added to initiate methane partial oxidation, while a small amount of oxygen, typically 10% at most, was introduced so that successive oxidation of methanol is suppressed. As a result, CH_4 conversion beyond 10% was scarcely achieved. Although CH_4 conversion increased by increasing either the oxygen content or reaction temperature, selectivity for useful oxygenates dropped sharply. There is a clear trade-off relationship between CH_3OH selectivity and CH_4 conversion. Rasmussen and Glarborg studied CH_4 partial oxidation numerically, assuming fuel-rich, high-pressure (30–100 bar), and relatively low-temperature conditions (550–800 K). They concluded the highest one-pass methanol yield of 4.2% would be feasible at optimum conditions of 97.4 bar, 643.3 K, and the initial CH_4/O_2 ratio of 23.6 [45]. Theoretical prediction does a good job explaining experimental results reported so far, implying that the thermochemical method may not be a promising approach for direct conversion of methane to methanol at a high yield. In contrast, the total liquid selectivity (●) in the two-phase flow microplasma reactor decreases moderately with respect to CH_4

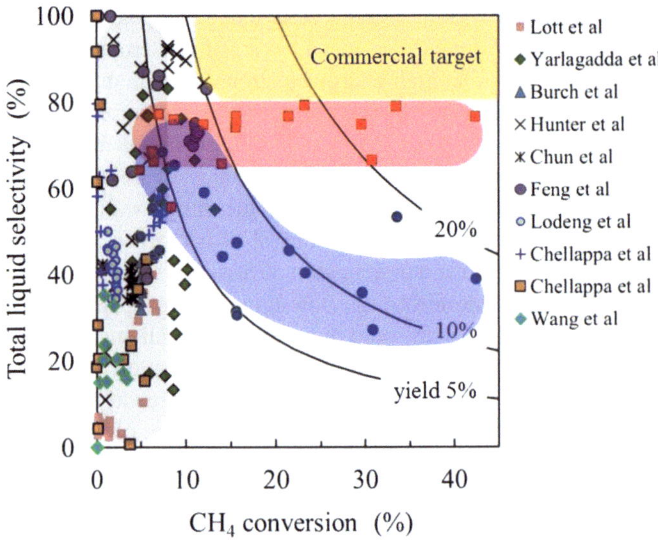

Fig. 8.4 Total liquid selectivity (● and ■) vs. CH_4 conversion. Conditions, O_2/CH_4, 0.5 and 1.0; reaction time, 130–530 ms. The three curves represent constant yield lines. Data is compared with cited literature values [24, 25, 37–44]

conversion. Although liquid selectivity was below 60%, a one-pass liquid yield with ca. 20% was obtainable. Assuming post DME (dimethyl ether, CH_3OCH_3) synthesis with the syngas obtained as by-product, the overall liquid selectivity could be enhanced up to 80% (■).

Atomic oxygen is considered as a key species to initiate CH_4 partial oxidation at low temperature [46]:

$$CH_4 + O \rightarrow CH_3 + OH \quad \Delta H^{\circ} = 11.7 \text{ kJ/mol} \quad E_a = 31.8 \text{ kJ/mol} \tag{8.R6}$$

ΔH° expresses standard enthalpy change and E_a is the activation energy [47]. O_2 dissociation by electron impact produces one ground-state atomic oxygen atom ($O(^3P)$) and an electrically excited one ($O(^1D)$) [48]:

$$O_2 + e \rightarrow O(^3P) + O(^1D) + e \tag{8.R7}$$

$$O_2 + e \rightarrow O(^3P) + O(^3P) + e \tag{8.R8}$$

However, based on the numerical simulation, neither $O(^3P)$ nor $O(^1D)$ abstract a H-atom from the strong C-H bond of CH_4 below 300 °C [49]. Presumably, 8.R6 is an endothermic reaction and the thermal energy (high-temperature environment) is needed to compensate the reaction enthalpy. Atomic oxygen can be a key species at elevated temperatures, but the results more closely resemble the total oxidation of CH_4, which contradicts low-temperature direct CH_4 conversion to CH_3OH. On the other hand, OH is apparently a desired oxidative agent because C−H dissociation proceeds via an exothermic reaction with a much smaller activation energy (8.R9) than 8.R6 [47]:

$$CH_4 + OH \rightarrow CH_3 + H_2O \quad \Delta H^{\circ} = -60.9 \ \text{kJ/mol} \quad E_a = 10.2 \ \text{kJ/mol} \tag{8.R9}$$

$$CH_4 + H \rightarrow CH_3 + H_2 \quad \Delta H^{\circ} = 3.35 \ \text{kJ/mol} \quad E_a = 36.6 \ \text{kJ/mol} \tag{8.R10}$$

$$CH_4 + M \rightarrow CH_3 + H + M \quad \Delta H^{\circ} = 440 \ \text{kJ/mol} \quad E_a = 446 \ \text{kJ/mol} \tag{8.R11}$$

C-H dissociation by atomic hydrogen (8.R10) is also possible when ΔH° and E_a are of the same levels as 8.R6. However, as the recombination of H and CH_3 (backward 8.R11) is preferable to 8.R10, CH_4 conversion would decrease which is against the expectation [49, 50]. CH_4 pyrolysis (8.R11) at low temperature is negligible because of the large activation energy and endothermic enthalpy.

The electron impact CH_4 dissociation is the key initiation pathway at room temperature (8.R12), and CH_3 is oxidized consecutively toward CH_3OH, HCHO, and HCOOH [49]. As predicted by the model reaction scheme (8.R12, 8.R13, 8.R14 and 8.R15), a large amount of methyl hydroperoxide (CH_3OOH) was detected in the plasma-treated liquid sample as the key intermediate species [51], implying that total liquid selectivity shown in Fig. 8.4 should further increase with appropriate posttreatment of the plasma-treated liquid.

As CH_4 conversion increases, partial oxidation of H_2 occurs preferentially to CH_4 oxidation, producing oxidative species such as H_2O_2, OH, and HOO that further accelerate H_2 partial oxidation [50]. Consequently, H_2 selectivity decreased markedly and CH_4 oxidation was decelerated: H_2 is clearly detrimental for efficient conversion of CH_4 to CH_3OH. Although nonthermal plasma creates unconventional reaction pathways at low temperature, it is hard to overcome trade-off relationship between CH_4 conversion and product selectivity:

$$CH_4 + e \rightarrow CH_3 + H + e \tag{8.R12}$$

$$CH_3 + O_2 \rightarrow CH_3OO \tag{8.R13}$$

$$CH_3OO + CH_4 \rightarrow CH_3OOH + CH_3 \tag{8.R14}$$

$$CH_3OOH \rightarrow CH_3OH, HCHO, HCOOH \tag{8.R15}$$

8.2.2 Endothermic System

Methane and CO_2/H_2O reforming is an endothermic reaction requiring high-temperature (>800 °C) thermal energy, which is supplied by the combustion of initial feed. It is truly detrimental because it directly leads to energy loss and releases CO_2 as well as NO_x: if the catalytic effect of nonthermal plasma lowers the reaction temperature and eliminates combustion of initial feed, it is beneficial for the greener use of a natural gas resource. Low-temperature waste heat can be used as an external heat source; however, unlike exothermic systems, the energy consumed by nonthermal plasma should cover a part of energy necessary for the endothermic reaction: nonthermal plasma is not only a radical source (reaction promotor) but also an energy source. It is noteworthy to mention that, in the case of a power-to-gas concept, energy consumption by nonthermal plasma is not necessarily low because excess renewable electricity needs to be converted to the chemical energy of syngas. The key issue is to pursue an appropriate combination of low-temperature thermal energy and electrical energy (energy consumed by plasma) in order to maximize energy and the material conversion efficiency.

8.2.2.1 Steam CH_4 Reforming in DBD/Catalyst Hybrid

A comprehensive analysis of plasma catalysis of CH_4 is presented by T Nozaki et al. [2]. Briefly, an energy efficiency of 10% was estimated for CH_4 conversion to C_2 hydrocarbons (mostly C_2H_6) by a simplified one-dimensional numerical analysis of the streamer-type discharge produced in pure CH_4 at ambient conditions (300 K and 101.3 kPa). It was also studied by the experiments, showing an energy efficiency of less than 1% was feasible: more than 90% of energy fed into the DBD was converted into heat and removed by water-cooled electrodes [34]. Another 10% was used for gas heating. In contrast, the energy efficiency of CH_4 conversion to syngas reached 50–80% when DBD was combined with a Ni/Al_2O_3 catalyst [52, 53]. The importance of the vibrational excitations of CH_4 and their interactions with heterogeneous catalysts was pointed out in the nonthermal plasma catalysis of steam methane reforming [54, 55].

The effect of radical injections on the reaction enhancement was kinetically analyzed by the Arrhenius plot method. The forward methane reaction rate based on the overall steam methane reforming using power-law kinetics is expressed as follows:

$$r = -\frac{d[CH_4]}{dt} = k[CH_4]^{\alpha}[H_2O]^{\beta} = k[CH_4]^{(0.85\sim1.4)}[H_2O]^{(-0.8\sim0)} \quad (8.1)$$

$$k = A \times exp\left(-\frac{E}{RT}\right) \quad (8.2)$$

Here, r denotes the forward methane reaction rate and k is the forward methane conversion rate constant for the overall SMR reaction. The reaction order for overall SMR on a nickel catalyst is $\alpha = 0.85\sim1.4$ in CH_4 and $\beta = -0.8\sim0$ in H_2O [56–59]. The phenomenological understanding of those values is that the rate-determining step is the activation of methane and the overall forward reaction rate

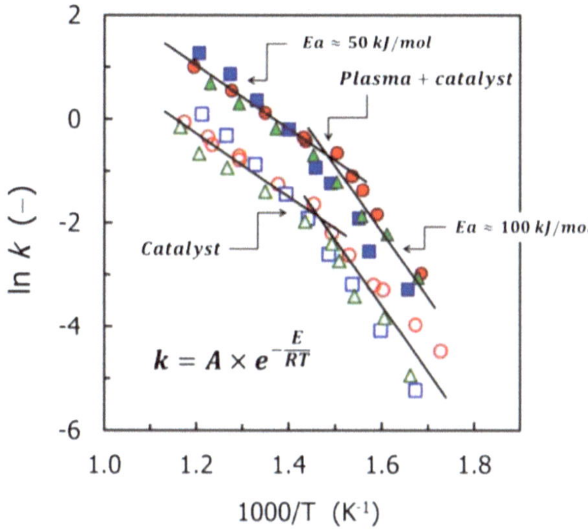

Fig. 8.5 An Arrhenius plot for the forward methane reaction rate constant. ○, ●, $GHSV = 18{,}000\ h^{-1}$, S/C = 1; □, ■, $GHSV = 18{,}000\ h^{-1}$, S/C = 3; Δ, ▲, $GHSV = 10{,}800\ h^{-1}$, S/C = 1, $P = 101.3$ kPa

is less dependent on H_2O [60]. Assume the overall rate constant is expressed in the Arrhenius form (Eq. 8.2), the results are shown in Fig. 8.5. An earlier study [61] presented detailed procedures to evaluate the reaction order (α and β), pre-exponential factor (A), and overall activation energy (E) under the influence of DBD.

The convex characteristics of the Arrhenius plot imply that overall SMR was in the reaction-limited regime when the catalyst temperature was lower than 460 °C. The diffusion-limited regime was clearly distinguished at this temperature. An analogous curve is obtained in the hybrid reaction where the reaction regime is separated at 420 °C. The overall activation energy in the reaction-limited regime is approximately 100 kJ/mol, which agrees well with the reported value [60], indicating that dissociative chemisorption of CH_4 over a Ni catalyst is the rate-determining step independent of DBD hybridization. The same trend is observed in the diffusion-limited regime, where only the pre-exponential factor was enhanced sevenfold, while the apparent activation energy was not influenced by the DBD. Figure 8.5 derives important conclusions and the hypothesis is as follows:

1. The overall activation energy was not influenced by DBD: the rate-determining step, i.e., C-H dissociation over catalyst, seems to be essentially unchanged by nonthermal plasma. The contribution of plasma-activated CH_4 is not clearly identified by the macroscopic observation.
2. Plasma-generated reactive species, such as OH, do not induce an alternative CH_4 activation pathway in the gas phase as the overall activation energy was unchanged. Therefore, the reaction mechanism of plasma-mediated catalysis could be explained by extending a Langmuir-Hinshelwood surface reaction mechanism.

3. The reaction order for CH_4 (α) was only slightly increased by DBD, while for H_2O (β), it was more than doubled [61].
4. The pre-exponential factor was clearly increased by DBD. One possible explanation is that plasma-activated H_2O removes adsorbed carbon species which regenerates active sites for subsequent CH_4 adsorption; thus the number of apparent active sites increases, but the CH_4 activation rate over individual sites would be kept unchanged.

8.2.2.2 Dry CH_4 Reforming in a DBD/Catalyst Hybrid

Dry methane reforming (DMR) has drawn keen attention as a viable CO_2 utilization technology as it may have one of the greatest commercial potentials [62, 63]. Various catalysts have been developed and reaction kinetics are well understood [64, 65]. However, a commercial installation of DMR is still difficult because the coke formation and catalyst deactivation problems have yet to be fully solved. An efficient supply of high-temperature heat is necessary; however the heat transport property of a fixed bed reactor is generally poor. Therefore, an excessively high-temperature operation is required which increases the energy penalty and deteriorates materials used for the reactor and catalysts. An increasing demand of DMR is therefore pursuing a new technology, and potential use of thermal and nonthermal plasma is highlighted. Dielectric barrier discharge is the most successful atmospheric pressure nonthermal plasma sources in industrial applications [66]. Fundamental physics and chemistry in DBD are well understood in relation to ozone synthesis. More recently, DBD and its combination with heterogeneous catalysts have been investigated for plasma-assisted dry methane reforming [3, 67–75]. Hereafter, plasma-assisted dry methane reforming over Ni/Al_2O_3 catalysts is focused based on our recent study [3, 74, 75].

8.3 Dry Methane Reforming over Heterogeneous Catalyst

8.3.1 Thermodynamics

Dry methane reforming is expressed by 8.R2. In addition, solid carbon (C) and H_2O are produced as undesired by-products:

$$\nu_1 CH_4 + \nu_2 CO_2 = \nu_3 H_2 + \nu_4 CO + \nu_5 C + \nu_6 H_2O \tag{8.R16}$$

ν_i expresses the stoichiometric coefficient for each component. Figure 8.6a, b, respectively, illustrates the equilibrium molar fraction for the six species (8.R16) with a stoichiometric mixture of $CH_4/CO_2 = 1$ at constant pressures of 100 kPa and 5 kPa. The CH_4 and CO_2 conversions are relatively high even below 500 °C; however, the selectivity for C and H_2O is much greater than that of syngas

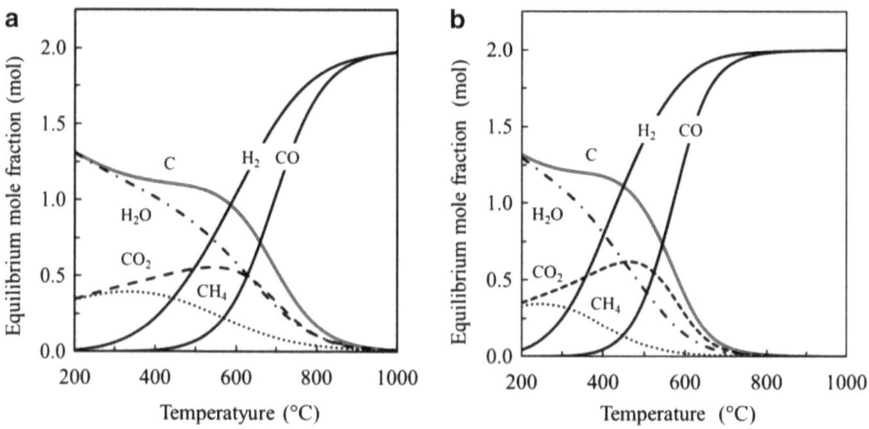

Fig. 8.6 Equilibrium composite for dry methane reforming at $CH_4/CO_2 = 1{:}1$. (**a**) Total pressure 100 kPa and (**b**) 5 kPa

(H_2 and CO), for C especially where selectivity is nearly 100% below 500 °C at a total pressure of 100 kPa. This is explained by the Boudouard equilibrium (8.R17); carbon formation is thermodynamically favored at low temperature:

$$CO + CO = C + CO_2 \quad \Delta H^{\circ} = -172 \quad \text{kJ/mol} \quad \Delta G^{\circ} = -120 \quad \text{kJ/mol} \tag{8.R17}$$

One of the major roles of nonthermal plasma is to lower the reaction temperature. Therefore, it is critically important to elucidate the reaction mechanism involving H_2O and C under the influence of nonthermal plasma. As shown in Fig. 8.6b, the production of syngas is shifted toward low temperatures at reduced pressure. Correspondingly, formation of H_2O and C shifts to the lower-temperature regime. Figure 8.7 highlights the C and CO mole fraction at different pressures. A low-pressure operation is thermodynamically beneficial for lowering reaction temperature with reduced coke formation. Dry methane reforming at reduced pressure is worth investigating as long as mass processing and thus low cost are guaranteed.

8.3.2 Reaction Kinetics Over Supported Catalyst

Dry methane reforming over supported metal catalyst on various support materials has been investigated extensively and excellent review articles are available [63–65]. The reaction mechanism is explained by the Langmuir-Hinshelwood (L-H) mechanism as schematically depicted in Fig. 8.8. Briefly, CH_4 is chemisorbed on metallic sites and almost irreversibly dehydrogenated towards the carbon atom (8.R18) [60]. Adsorbed carbon forms a solid solute with Ni and

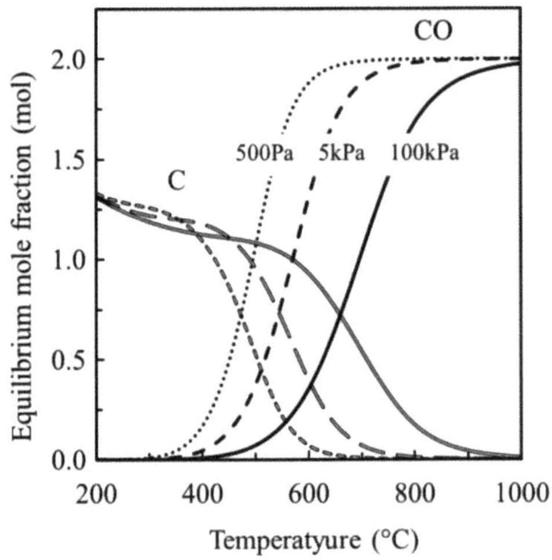

Fig. 8.7 Effect of pressure on C (solid carbon) and CO mole fraction

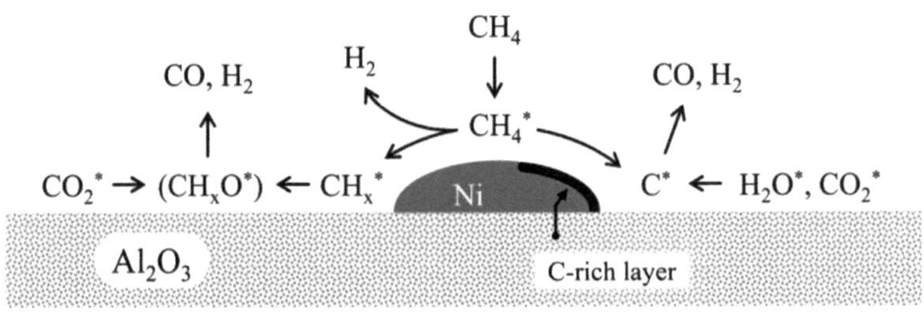

Fig. 8.8 A schematic representation of the Langmuir-Hinshelwood mechanism for dry methane reforming

may grow carbon filaments when the carbon concentration is saturated in Ni/NiC nanoparticles:

$$CH_4 \rightarrow CH_4^* \rightarrow CH_x^* \rightarrow C^* \leftrightarrow C \cdot Ni \rightarrow \text{carbon filament} \qquad (8.R18)$$

Independently, CO_2 is more likely to be adsorbed near the interface between the Ni catalyst and a metal oxide support such as Al_2O_3 [76]. Here, adsorbed species are denoted by $*$ in figures and equations. The key step is that CH_x^* is oxidized by CO_2-derived oxygen species to form CH_xO^* before complete dehydrogenation to C^* occurs. Successful production of CH_xO^* leads to syngas (CO and H_2) without coke formation [77]. However, successive dehydrogenation of CH_x^* to C^* is so fast that adsorbed carbon C^* becomes the abundant surface species [60].

Adsorbed carbon (C^*) readily combines with the transition metal catalyst and forms a core–shell like carbon-rich solid solute or molten thin layer on the surface of the catalyst nanoparticles [78]. Carbon diffusivity in the disordered molten layer is much faster than in the crystalline solid phase of the core region. Detailed kinetic analysis has been performed by Puretzky et al. within the scope of a carbon nanotube/nanofiber growth mechanism [79]. In the case of gas reforming, the surface carbon must be removed before it is saturated in the molten layer and grows carbon filaments. This phenomenon is known as *coking*, and it is purely detrimental because solid carbon blocks active sites and eventually clogs pellet pores. Although coke formation is observed frequently in CH_4 reforming, this process is not included in the scope of the L-H mechanism: a theoretical formulation of DMR and SMR in the L-H scheme is employed without coke formation.

Carbon oxidation is possible by a reverse Boudouard reaction (8.R19). However, this reaction is slow when compared to 8.R18: the overall carbon removal reaction is even more decelerated when the carbon diffusion within the catalyst particles becomes a rate-determining step [3]. The addition of H_2O vapor is known to be effective in avoiding coke formation, which is explained by thermodynamic considerations [78]. In a nonthermal plasma hybrid reaction, plasma-activated H_2O promotes carbon oxidation (8.R20) and overall syngas productivity is increased with reduced coke as discussed in Sect. 8.2.2.1. A similar principle is applicable to DMR [74]:

$$\mathrm{C} + \mathrm{CO_2} = \mathrm{CO} + \mathrm{CO} \quad \Delta H^{\circ} = 172\ \mathrm{kJ/mol} \quad \Delta G^{\circ} = 120\ \mathrm{kJ/mol} \tag{8.R19}$$

$$\mathrm{C} + \mathrm{H_2O} = \mathrm{CO} + \mathrm{H_2} \quad \Delta H^{\circ} = 131\ \mathrm{kJ/mol} \quad \Delta G^{\circ} = 91\ \mathrm{kJ/mol} \tag{8.R20}$$

8.3.3 Surface Reaction Model

8.3.3.1 Langmuir-Hinshelwood Mechanism

The overall reaction rate (r) for DMR is expressed by either Eq. 8.3 or Eq. 8.4 by taking into account 8.R16:

$$r = kP_{\mathrm{A}}{}^{n}P_{\mathrm{B}}{}^{m} \tag{8.3}$$

$$r = -\frac{1}{\nu_1}\frac{\mathrm{d}[\mathrm{CH_4}]}{\mathrm{d}t} = -\frac{1}{\nu_2}\frac{\mathrm{d}[\mathrm{CO_2}]}{\mathrm{d}t} = \frac{1}{\nu_3}\frac{\mathrm{d}[\mathrm{H_2}]}{\mathrm{d}t} = \frac{1}{\nu_4}\frac{\mathrm{d}[\mathrm{CO}]}{\mathrm{d}t} \tag{8.4}$$

Here, P expresses a partial pressure of a gaseous species; subscript A and subscript B represent CH_4 and CO_2, respectively; k is the overall reaction rate constant; n and m, respectively, express reaction order for CH_4 and CO_2. In many cases, reaction order takes the value of $n = 1$ and $m = 0$ as shown in Fig. 8.9 [65]. Wang et al. examined CO production rate as a representative overall reaction rate of DMR (r) and found it is proportional to the CH_4 partial pressure if coke formation is absent. In contrast, CO production rate is independent of CO_2 partial pressure when it is sufficiently high. This expression is similar to SMR [60, 61].

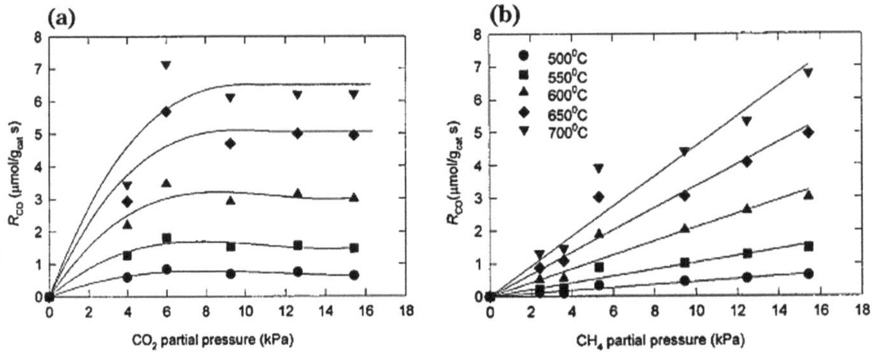

Fig. 8.9 Effect of CH_4 and CO_2 partial pressure on CO production rate, representing overall reaction order for DMR (Eq. 8.3) [65]

A heterogeneous L-H reaction is initiated with the adsorption of molecules onto the active sites of catalyst, followed by surface reactions and desorption to yield products. If CH_4 and CO_2 are adsorbed on the same active sites, the overall reaction rate is expressed by the following equation:

$$r = k_s n_A n_B \theta_A \theta_B = k_s n_A n_B \frac{K_A K_B P_A P_B}{(1 + K_A P_A + K_B P_B)^2} \quad (8.5)$$

Here, k_s expresses the overall surface reaction rate constant; θ is surface coverage which takes the value between 0 and 1; K is the equilibrium constant between adsorption and desorption; n is the total number of active sites per unit area. K and θ are correlated with the Langmuir isotherm where adsorption and desorption are equilibrated. For catalysts commonly used in DMR, CH_4 is selectively adsorbed on metallic sites and CO_2 absorbs somewhat selectively on the interface between the metallic catalyst and the metal oxide support. The mathematical expression becomes even more complex if the *WGS equilibrium*, reverse DMR, and type of catalyst are considered [76]. For simplicity, WGS and reverse DMR are not considered. Further assuming that CO_2 adsorbs only on metal-support interface, Eq. 8.5 is rearranged to Eq. 8.6:

$$r = k_s n_A n_B \frac{K_A K_B P_A P_B}{(1 + K_A P_A)(1 + K_B P_B)} \quad (8.6)$$

Assume CO_2 absorption is strong and equilibrated, i.e., $K_B P_B \gg 1$; Eq. 8.6 is approximated by Eq. 8.7:

$$r = k_s n_A n_B \frac{K_A P_A}{(1 + K_A P_A)} \propto P_B{}^0 \quad (8.7)$$

Equation 8.7 indicates that the reaction order is zeroth-order for P_B ($m = 0$ in Eq. 8.3); in other words, the overall DMR reaction rate is independent of the CO_2 partial pressure. You can further assume $K_A P_A \ll 1$, implying CH_4 is weakly bounded by the metallic sites. Eventually, Eq. 8.7 simplifies to Eq. 8.8:

$$r = k_s n_A n_B K_A P_A \propto P_A \tag{8.8}$$

The overall reaction rate is first-order in terms of CH_4 partial pressure. The overall reaction rate (r) is also proportional to n_A and n_B: catalysts having large surface area for both CH_4 and CO_2 adsorption are highly desired for the fast reaction.

There are two essential hypotheses: $K_A P_A \ll 1$ and $K_B P_B \gg 1$. A phenomenological understanding of the first hypothesis is that CH_4 is weakly absorbed on metallic sites and CH_4 dissociative absorption becomes the rate-determining step. The second hypothesis indicates that CO_2 absorption is relatively fast. Active sites are readily occupied by $CO_2{}^*$; therefore, CO_2 partial pressure is independent of the overall reaction rate (r). The overall surface reaction rate constant (k_s) is influenced by various elementary steps, and it is hard to identify individual contributions. It must be mentioned that Eq. 8.8 does not take coke formation kinetics into account. Meanwhile, plasma-activated H_2O and CO_2 play a critical role in oxidizing CH_4 fragments and suppressing coke formation. For a better understanding of plasma-enhanced catalysis of DMR, the kinetic analysis of the rate-determining step and for the coke formation behavior is discussed in detail in Sect. 8.5.1.

8.3.3.2 Precursor-Mediated Mechanism

Plasma-activated species are involved in the heterogeneous surface reactions. If molecules are excited primarily before impinging on the catalysts surface and being trapped at an active site, adsorption cannot be explained on the basis of thermal equilibrium. In other words, surface coverage (θ) is unable to be formulated using the equilibrium constant (K) within the scope of the Langmuir isotherm. This kind of adsorption process is categorized as a *precursor-mediated* mechanism as schematically presented in Fig. 8.10a. In the case of CH_4 activation, the chemisorption probability of vibrationally excited CH_4 is promoted exponentially with the internal energy [80]. If not the weakly bounded physiosorbed "hot CH_4" (precursor state admolecules) would lose internal energy and desorb. This process becomes an extremely important elementary step in plasma catalysis because vibrationally excited molecules are produced by low-energy electron impacts and easily populated in plasma [54, 81]. However, it must be pointed out that the vibrational excitation of CH_4 is overfocused in nonthermal plasma chemistry. As discussed previously, H_2O activation also plays an essential role because plasma-activated H_2O promotes oxidation of adsorbed carbon and regenerates active sites for the successive CH_4 adsorption. Otherwise, active sites are fully occupied by the adsorbed carbon and it is hard for further CH_4 conversion to take place. Likewise, we found experimentally that activated CO_2 has an ability to react with surface

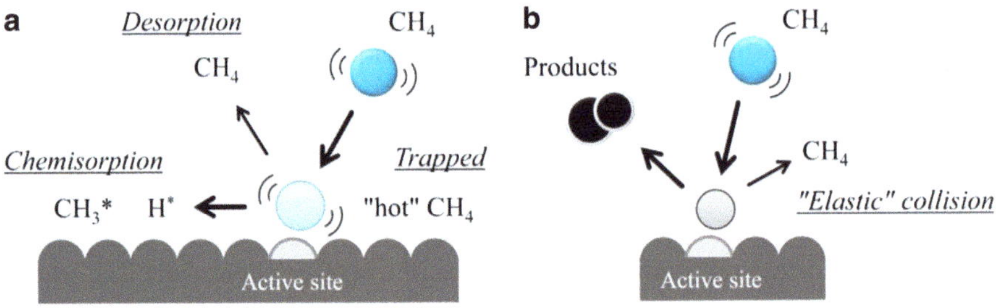

Fig. 8.10 A schematic representation of (**a**) precursor-mediated adsorption and (**b**) Eley–Rideal mechanism

carbon species. As discussed in Sect. 8.3.3.1, it is evident that both CH_4 and CO_2 must be equally excited by nonthermal plasma so that the overall reaction rate is enhanced by nonthermal plasma. Another important issue is the possibility of an Eley–Rideal mechanism. Active sites are not always vacant but mostly occupied by some kind of species. In that case, adsorbed species may have a direct interaction with incoming plasma-activated molecules which carries the internal energy beyond thermal equilibrium. Such a reaction should be explained by an Eley–Rideal mechanism (Fig. 8.10b), rather than precursor-type adsorption enhancement. The role of vibrationally excited species within the scope of Eley–Rideal mechanism has yet to be discussed in detail in plasma catalysis.

8.4 Plasma Hybrid Reaction

8.4.1 Packed-Bed DBD Reactor

A catalyst packed-bed DBD reactor is the most frequently used for hybrid operations. Figure 8.11 shows the hybrid reactor used for DMR in our study [75]. Briefly, the reactor consists of a 20 mm i.d. and 1.5 mm thick quartz tube, a 3 mm diameter centered electrode, and the external grounded electrode. The inner electrode was connected to a high-voltage power supply (Fig. 8.12). Although the voltage waveform is distorted from an ideal sine wave depending on the load impedance, it features compactness, low-cost, and a high-power output capacity (850 VA, $V_{pp} = -6\sim + 8$ kV, 13–14 kHz). The current waveform (Fig. 8.13a) exhibits spikelike current peaks at the rising and falling parts of the applied voltage, which is a standard characteristic of streamer-type DBD. Discharge power was measured by a voltage–charge (Lissajous) diagram [82] as shown in Fig. 8.13b. Spherical catalyst pellets (3 mm mean diameter, 12 wt.% Ni/Al_2O_3, Süd-Chemie) were packed for 40 mm length, and both ends of the catalyst bed were supported by metallic disks. The catalysts were reduced in a H_2/Ar flow (100/900 cm^3/min) at 600 °C for 60 min before the reforming reaction. CH_4, CO_2, H_2, and CO were quantitatively analyzed by a quadrupole mass spectrometer (QMS, Prisma; Pfeiffer Vacuum GmbH) after

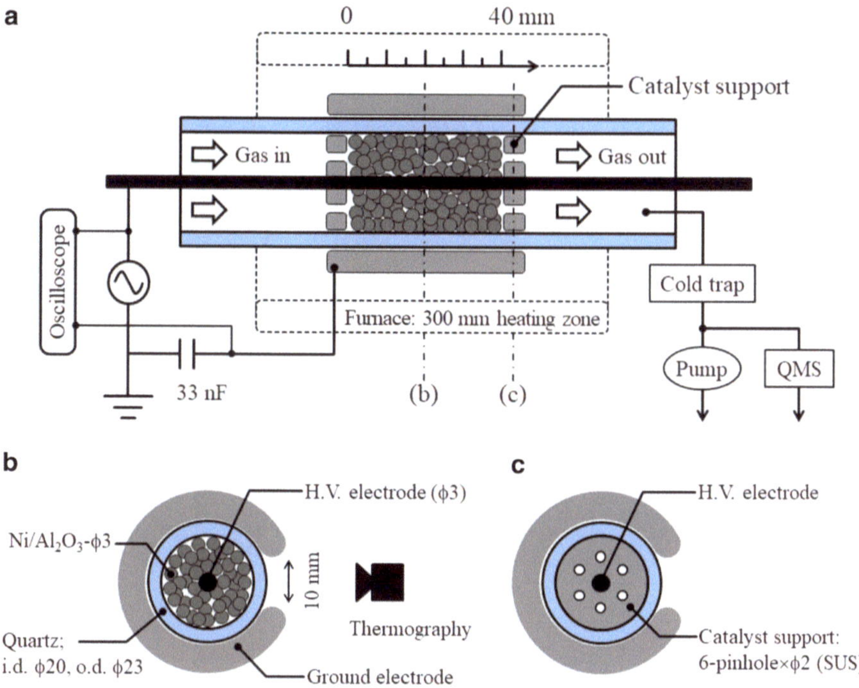

Fig. 8.11 Experimental setup: (**a**) overview of the packed-bed DBD reactor; (**b**, **c**) cross-sectional view at position (**b**) and (**c**)

H_2O removal by a cold trap (ca. 233 K). Figure 8.13c–e represents an overview of the catalyst bed, optical emission, and the temperature distribution: the catalyst temperature clearly decreased due to the endothermic reforming reaction. The gas temperature was estimated by optical emission spectroscopy of the CO(B-A) transition [83]. The rotational temperature of CO and the catalyst temperature measured by thermography were well matched as discussed in Sect. 8.4.1.2. The reactor pressure was evacuated until it reached 5 kPa so that DBD is generated stably even if carbon is deposited on the catalyst pellets. Streamer-type gas breakdown initiates at the pellet contacts; subsequently, a number of filamentary discharge channels propagate covering a broader area of the pellet surface when the electrical conductivity of catalysts is low [84]. Coke formation increases electrical conductivity of pellet surface and streamer propagation is suppressed substantially.

8.4.1.1 Basic Property of DBD

Energy consumption of plasma gas conversion is benchmarked by the *specific energy input* (*SEI*):

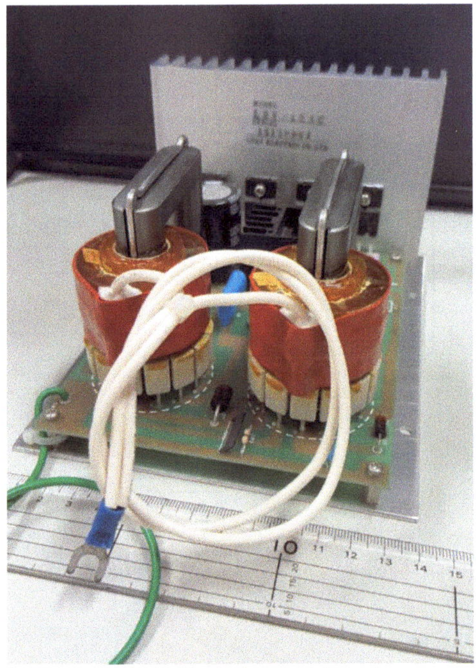

Fig. 8.12 The high-voltage power source used for DBD generation (Logy Electric, LHV-13 AC)

$$\mathrm{SEI} = C \times \frac{P}{F_{\mathrm{total}}} \quad (\mathrm{eV/molecule}) \tag{8.9}$$

C represents a conversion factor of the unit (cf. Appendix in Ref. [3]). The *SEI* is calculated unambiguously by the discharge power (*P*) and the total flow rate of feed gas (F_{total}) no matter what the reaction conditions would be. Therefore, performance levels of different conditions as well as different plasma reactors are classifiable on a *SEI* basis. According to the definition, the *SEI* expresses mean discharge energy consumption per unit volume of feed gas (e.g., J/cm^3), which is also interpreted as mean energy fed into a single molecule (eV/molecule). The *SEI* is used as an indication of the energy efficiency of plasma processes. For example, the *specific energy requirement* (*SER*) for an endothermic reaction is readily obtained from the reaction enthalpy: for dry methane reforming, *SER* = 247 kJ/mol = 2.56 eV/molecule (8.R2). Assuming the plasma catalysis of DMR is operated at *SEI* = 1 eV/molecule, about 40% of energy required for DRM can be supplied from plasma and 60% is from ambient thermal energy. This estimation is apparently different from energy efficiency (cf. Sect. 8.5.2); the point is that the *SEI* is preferably as small as possible to minimize the energy penalty of plasma process. When the *SEI* is greater than the *SER*, the energy and material conversion efficiency is inevitably low and such conditions must be avoided. For an exothermic system, the definition of energy efficiency and relationship with *SEI* is discussed in ref. [50].

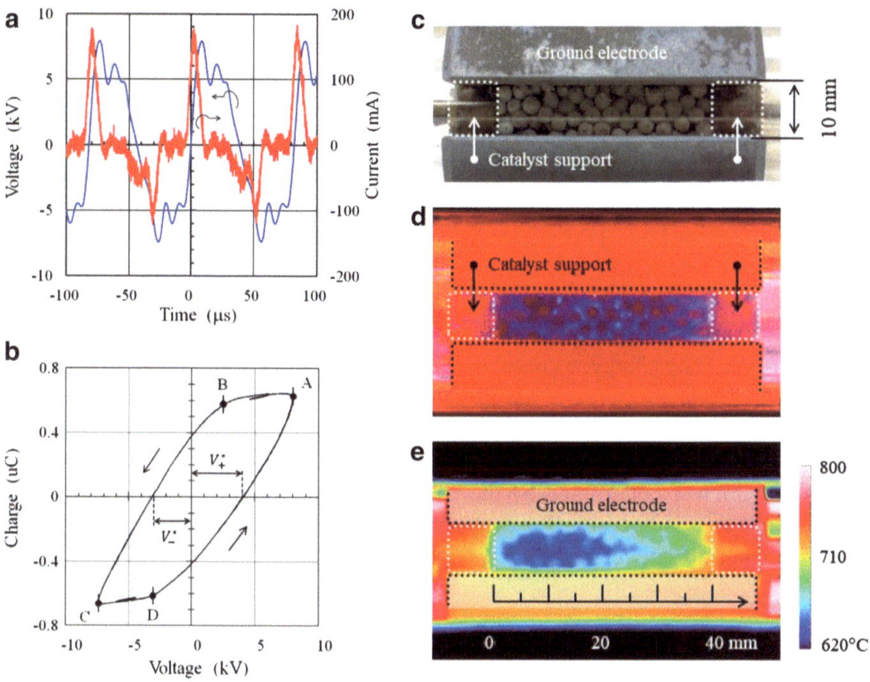

Fig. 8.13 (**a**) Voltage and current waveforms; (**b**) a voltage–charge (Lissajous) diagram; discharge period "B-C" and "D-A"; non-discharge period "A-B" and "C-D"; (**c**) packed-bed DBD reactor; (**d**) DBD in the catalyst bed; (**e**) temperature distribution of the catalyst bed. Conditions: total flow rate = 2000 cm ^{3}/min ($CH_4/CO_2 = 0.5$); $P = 89$ W; Frequency = 12 kHz; Pressure = 5 kPa

The *gas hourly space velocity* (*GHSV*) is used commonly to express contact time, showing how many times of volumetric gaseous materials are processed with respect to the total volume of packed materials per unit time (h^{-1}) at standard condition (25 °C and 101.3 kPa). Alternatively, *weight hourly space velocity* (*WHSV*) is used which is calculated from the mass flow rate of gaseous materials (g/h) divided by the total mass (weight) of catalyst materials packed into the reactor (g). *GHSV* and *WHSV* have the same unit (h^{-1}), which is apparently equivalent to the inverse of reaction time; therefore, for simplicity, we defined *GHSV* using the reactor volume (V) as an alternative parameter of reaction time:

$$\mathrm{GHSV} = (\mathrm{Reaction\ time})^{-1} = \frac{F_{\mathrm{total}}}{V} \quad \left(h^{-1}\right) \tag{8.10}$$

The product of the *GHSV* and *SEI* is equivalent to the power density:

$$\mathrm{GHSV} \times \mathrm{SEI} \propto \frac{P}{V} \ \left(\mathrm{W/cm^3}\right) \tag{8.11}$$

Power density is obtained by the discharge power and the reactor volume independently of the total gas flow rate. Assume the total gas flow rate is doubled; the energy density and reaction time (i.e., $GHSV^{-1}$) are halved simultaneously. Therefore, the conversion of initial feed may decrease by less than half. Such reaction behavior cannot be discussed by a power density basis. The individual contribution of *SEI* and *GHSV* must be taken into account for the appropriate analysis of reaction behavior.

Power density (W/cm^3) becomes the important parameter when designing a DBD reactor. DBD is characterized as a number of filamentary discharges with nanosecond durations. The charge accumulated on the dielectric barrier creates the reverse electric field, which suppresses the further development of weakly ionized streamer-type discharges into an intense arc. Therefore, the energy fed into a DBD reactor, or power density, inevitably has an upper limit in terms of the capacitance of dielectric materials which is determined by the electrode configuration. Discharge power is given by Manley's equation for the parallel-plate electrode configuration without packing materials between the gap [82]:

$$P = 4fV^*C_\mathrm{d}\left(V_\mathrm{p} - \left(1 + \frac{C_\mathrm{g}}{C_\mathrm{d}}\right)V^*\right) \ (W) \tag{8.12}$$

The f represents frequency; C_g and C_d, respectively, denote the capacitance of the gas gap and the dielectric material; V_p is the applied peak voltage and V^* expresses the effective gas breakdown voltage (cf. Fig. 8.13b). Assume $C_\mathrm{d} \gg C_\mathrm{g}$, Eq. 8.12 reduces to Eq. 8.13 showing power density:

$$\frac{P}{A \times d} \approx 4f\left(\frac{V^*}{d}\right)\left(\frac{C_\mathrm{d}}{A}\right)\left(V_\mathrm{p} - V^*\right) \ \left(\mathrm{W/cm^3}\right) \tag{8.13}$$

Here, V^*/d represents the mean electric field strength between the gas gap; C_d/A is the capacitance of the dielectric material per unit area. Power density is determined by the number of streamers generated per unit area and unit time; a typical number density of streamer in air DBD is 10^6 cm^{-2} s^{-1} [85]. Equation 8.13 does not display parameters indicating the electron density and electron energy per streamer, or the streamer number density. All these important properties of DBD are merged in V^* and not decoupled. In many cases, we do not tailor V^*; but we merely measure V^* which is determined as a consequence of complex discharge events. Because V^* is fairly independent of V_p and takes a fixed value under the given conditions, power density increases only slightly by a lone tuning knob, i.e., V_p. The total number of streamers per half cycle increases with V_p, but streamer number density, electron energy, and electron density are essentially unchanged. Meanwhile, frequency (f) has a much greater impact than V_p; power density increases linearly with operating frequency, enabling one to design a high-power density and compact DBD reactor.

Such peculiar characteristics of DBD have been reported by Kogelschatz in relation to ozone synthesis [66]. Later, Nozaki et al. demonstrated similar characteristics in CH_4-fed DBD by the comprehensive energy balance analysis [2, 34].

It must be mentioned the discharge property of DBD has a strong correlation with gas composition: the discharge property at the inlet (e.g., CH_4/CO_2) and the outlet (e.g., $CH_4/CO_2/H_2/CO/H_2O$) of the reactor must be different to some extent. Meanwhile, *SEI*, *GHSV*, and the power density (i.e., V^*) represent time- and space-averaged properties of DBD; overly detailed interpretation of plasma chemistry with macroscopic parameters might lead to inconsistent inferences.

8.4.1.2 Catalyst Temperature Measurement

Temperature measurements of the catalyst bed are critically important because heterogeneous reactions must be characterized by the surface temperature. However, the catalyst temperature has not been reported in many literatures; either electrode temperature or furnace temperature was monitored. We paid special attention to the catalyst temperature characterization: catalyst temperature was monitored by thermography and maintained constant by an electric furnace before the reaction so that the initial condition is known and fixed. Upon introducing the CH_4 and CO_2 mixture, the catalyst temperature decreased clearly because of the endothermic reaction, while it increased to some extent due to the heat generated by DBD. The catalyst temperature is modified in this way and the overall heat transfer from the furnace to the catalyst is biased from the initial conditions. Eventually, a steady-state catalyst temperature is established by negative feedback of the energy balance between (i) the endothermic reaction enthalpy per unit time, (ii) discharge power, and (iii) the overall heat transfer between the furnace and the catalyst bed; (i) and (iii) are temperature-dependent, while (ii) is a temperature-independent process. The gas temperature difference between the inlet and the outlet of the reactor was omitted in the analysis as it was such a minor effect. It is quite an engineering-related subject, but we have to carefully analyze the energy balance in relation to catalyst temperature in order for a truly beneficial comparison between plasma and thermal catalysis.

As can be seen in Fig. 8.13e, the individual catalyst pellet is identified in the thermal image, showing the outside layer of catalyst bed as the main part of the infrared emission. The transmittance of the quartz tube is generally higher than 90% when the wavelength is shorter than 3 μm, while the given thermography detected the infrared signal between 3.0 and 5.3 μm. It must be mentioned a thermography which detects long-wavelength (8–14 μm) measures the outside surface temperature of the quartz tube and individual catalyst pellets were unidentified by the thermal image. The thermography was calibrated by thermocouples without DBD under the reforming reaction. After steady state was established and making sure of a uniform temperature distribution by the thermal image, the emissivity of 0.82 provided the best correspondence between the thermocouples and thermography measurement. We confirmed that the emissivity (0.82) was unchanged for the reduced (Ni/Al_2O_3),

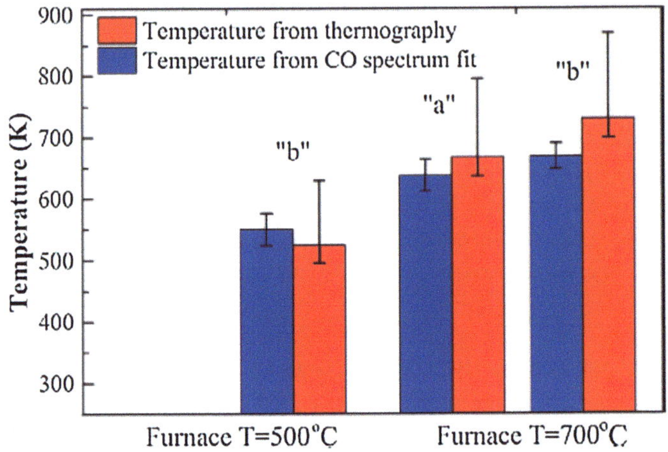

Fig. 8.14 A comparison of catalyst temperatures measured by the thermography and the plasma gas temperature by CO emission spectrum fitting [83]

oxidized (NiO/Al_2O_3), and coked ($C + Ni/Al_2O_3$) catalysts. Infrared emission from DBD via the vibrational relaxation of molecules was not taken into account.

The optical emission spectroscopy (OES) of the CO (B-A) transition was employed for gas temperature analysis during the reforming reaction [83]. Spectral fitting between 477 nm and 484 nm yielded the gas temperature close to the catalyst temperature with a smaller than 30 K difference (Fig. 8.14). Similar results were confirmed by the Boltzmann plot approach. DBD is characterized as a weakly ionized plasma with a nanosecond duration; temporal gas heating by the discharge propagation is several degrees higher than the ambient gas temperature [66], which is smaller than the measurement discrepancy between OES and thermography. In conclusion, temperature of catalyst pellets and plasma gas temperature are quasi-equilibrated regardless of heat absorption by endothermic reaction and heat generation by DBD.

8.4.2 Backward Reactions

In this section, the backward reaction of DMR is discussed. Nonthermal plasma primarily promotes forward DMR (8.R2). However, as CO and H_2 partial pressure increases under the presence of unreacted CO_2 as well as H_2O (cf. 8.R22.1), various side reactions occur simultaneously, and the individual reaction is influenced to some extent by nonthermal plasma. Important reactions are reverse DMR (8.R21) and the Sabatier reaction (8.R22):

$$2CO + 2H_2 = CH_4 + CO_2 \quad \Delta H^\circ = -247 \ \text{kJ/mol} \quad \Delta G^\circ = -170 \ \text{kJ/mol} \tag{8.R21}$$

$$CO_2 + 4H_2 = CH_4 + 2H_2O \quad \Delta H^\circ = -165 \ \text{kJ/mol} \quad \Delta G^\circ = -113 \ \text{kJ/mol} \tag{8.R22}$$

The Sabatier reaction is a combination of the reverse water gas shift (RWGS) (8. R22.1) and methanation reactions (8.R22.2):

$$CO_2 + H_2 = CO + H_2O \quad \Delta H^\circ = 41 \ \text{kJ/mol} \quad \Delta G^\circ = 29 \ \text{kJ/mol} \tag{8.R22.1}$$

$$CO + 3H_2 = CH_4 + H_2O \quad \Delta H^\circ = -206 \ \text{kJ/mol} \quad \Delta G^\circ = -142 \ \text{kJ/mol} \tag{8.R22.2}$$

The methanation reaction is identical with reverse steam methane reforming (8.R1). As shown in 8.R21 and 8.R22.2, CH_4 might be regenerated via the backward reactions of DMR and SMR, and the gross CH_4 conversion may be suppressed at some point of reaction. The backward reactions are possible over a Ni/Al_2O_3 catalyst: indeed, supported nickel catalysts are the most commonly studied materials in the catalytic hydrogenation of CO_2 [86]. Figure 8.15 compares the Gibbs free energy of the backward reactions. 8.R21 and 8.R22.2 are exothermic reactions and thermodynamically favored when the temperature is below ca. 600 °C. In the case of low-temperature CH_4 catalysis, CH_4 regeneration might be promoted by DBD under the presence of Ni/Al_2O_3 catalysts. Therefore, 8.R21 and 8.R22.2 were examined using CO_2 and H_2 as initial feed at various temperatures.

Figure 8.16 shows the reactant (CO_2, H_2) and product (CO, H_2O, CH_4) distribution obtained with and without DBD using a Ni/Al_2O_3 catalyst. The catalyst temperature was increased by a programmable electric furnace at a constant heating rate of 10 °C/min, while the catalyst temperature was recorded by thermography at

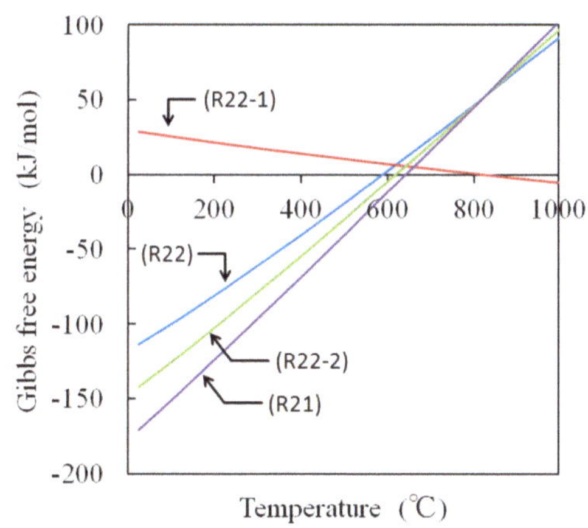

Fig. 8.15 The Gibbs free energy of reverse DMR (8.R21), Sabatier reaction (8.R22), RWGS (8.R22.1), and methanation reaction (8.R22.2)

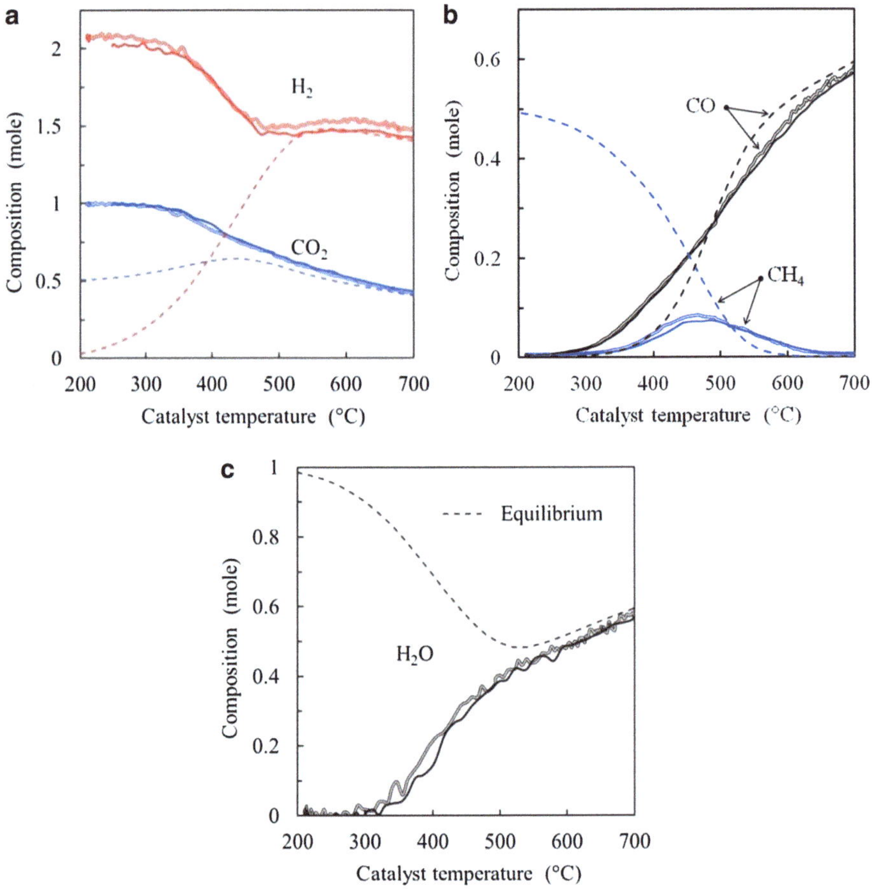

Fig. 8.16 The relationship between the *WGS equilibrium* and backward reactions. $GHSV = 15{,}000\ h^{-1}$, $H_2/CO_2 = 2$, total pressure = 5 kPa. Dotted line, equilibrium; solid line, with DBD; hollow line, without DBD. H_2O was estimated from the hydrogen balance (not a measured value)

1-min intervals. The stoichiometric H_2/CO ratio for reverse DMR (8.R21) is $H_2/CO = 1$, which is equivalent to the module ($M = 1$) with initial $H_2/CO_2 = 2$ (Eq. 8.14), because H_2 is consumed by RWGS (8.R22.1) and the same amount of CO is added to the system (see Ref. [11] page 119):

$$\text{When } H_2/CO_2 = 2, \quad M = \frac{[H_2] - [CO_2]}{[CO] + [CO_2]} = \frac{2-1}{0+1} = 1 \tag{8.14}$$

Likewise, the stoichiometric ratio of the methanation reaction (8.R22.2) is represented by $H_2/CO_2 = 4$, which leads to the equivalent $H_2/CO = 3$, or $M = 3$. Experiments with $H_2/CO_2 = 2$ and 4 ideally cover the reactions from

8.R21 to 8.R22.2. Steam (H_2O) was removed before QMS online analysis and not measured quantitatively, but H_2O selectivity was estimated from the hydrogen balance. The experimental results were compared with an equilibrium calculation. Solid carbon was not considered in the equilibrium calculation because the carbon balance among CH_4, CO, and CO_2 from the experiment was more than 95% and no carbon deposition was detected.

Reactions are generally slow at lower temperatures: there is a large offset between the equilibrium and experimental observation below 500 °C. As catalyst temperature increases, H_2 and CO_2 start to be consumed and CO is produced from ca. 300 °C via RWGS. Although a small amount of CH_4 was produced by either 8.R21 or 8.R22.2, CH_4 partial pressure is well below equilibrium when the catalyst temperature is lower than 500 °C. At high temperature (>600 °C), production of CH_4 is not favored thermodynamically (cf. Fig. 8.15) and the RWGS governs the gas composition. It deduces important conclusions: first, reactions between CO_2, CO, H_2, H_2O, and CH_4 are equilibrated (>500 °C) with the given H_2/CO_2 mixture as an initial feed. Moreover, there is little influence of DBD between 200 and 700 °C for these reactions. Hereafter, this characteristic is called the *WGS equilibrium*. Second, CH_4 regeneration is negligible between 200 and 700 °C. CH_4 conversion beyond equilibrium may be possible if the forward DMR reaction is promoted kinetically by nonthermal plasma. A similar trend was confirmed experimentally with an initial $H_2/CO_2 = 4$ (date is not shown).

8.4.3 Pulsed Reforming Spectrometry

We have developed pulsed reforming spectrometry which enables systematic analysis of DMR under the influence of nonthermal plasma. In order to overview the pulsed spectrometry, representative results are presented in Fig. 8.17 together with the optical emission spectroscopy data [3]. Pulsed CH_4 injection was employed for 1 min at 3-min intervals, while CO_2 was continuously supplied. Pulsed reforming was conducted over several cycles until the gas components and catalyst temperature reached a cyclic steady state.

Before the CH_4 injection ($t < 0.2$ min), Ni/Al_2O_3 catalysts would be almost fully adsorbed with CO_2^*. The CH_4 flow was turned on at $t = 0.2$ min and the CH_4 flowed over the CO_2-adsorbed catalyst pellets: at this moment, H_2 and CO are produced instantaneously, implying the CH_4 dehydrogenation on metallic sites, formation of CH_xO^*, and the desorption of H_2 and CO are sufficiently fast reactions. Moreover, the reaction between H_2^* and CO_2^* should produce CO which is governed by the *WGS equilibrium* as discussed in detail in Sect. 8.4.2. H_2 partial pressure monotonically decreased during the reforming period probably because the catalyst temperature decreases due to the endothermic nature of DMR. After 1 min into the reforming reaction, the CH_4 flow is turned off at $t = 1.2$ min and CO_2 only DBD was generated for 3 mins ($1.2 < t < 4.2$ min). Although CH_4 is absent, CO_2 is consumed by the reverse Boudouard reaction (8.R19) and CO partial pressure

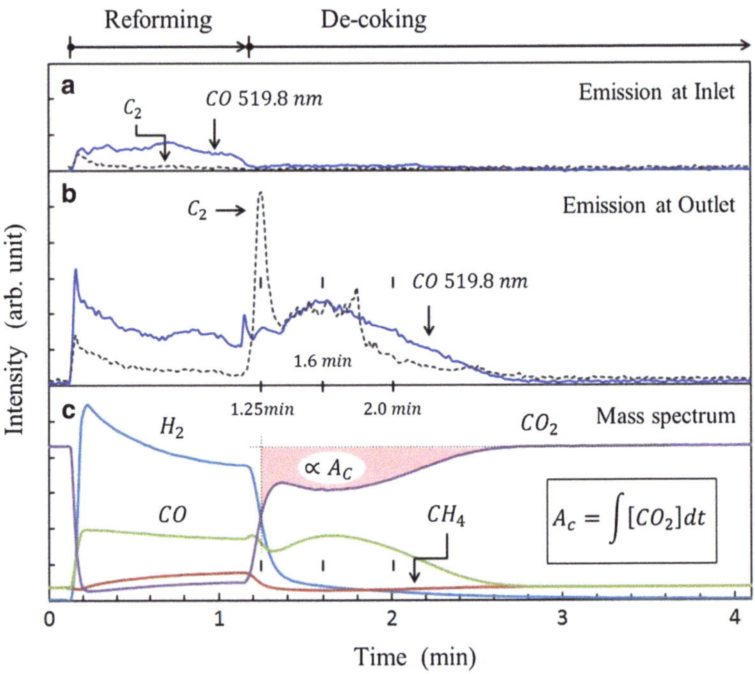

Fig. 8.17 The emission intensity of CO (518.3 nm) (solid blue line) and C_2 (dotted black line) at (**a**) the reactor inlet and (**b**) the outlet. (**c**) Mass spectrum profile for one cycle pulsed reforming. The emission profile is normalized by the emission intensity of C_2 peak at 1.25 min [3]. $CH_4/CO_2 = 2$, power 70 W

increases correspondingly. By integrating the CO_2 signal over the de-coking period, solid carbon formation during the 1-min reforming period is calculated quantitatively. Unlike the reforming reaction, carbon oxidation by CO_2^* is a slow reaction; therefore, the CO and CO_2 profiles change quite moderately and carbon removal requires ca. 1.4 min ($1.2 < t < 2.6$ min).

The time-dependent change of the line intensity of CO (519.8 nm) and the C_2 high-pressure Swan system (589 nm) are synchronized with the pulsed reforming. The emission profile is normalized by the intensity of the C_2 peak appearing at $t = 1.25$ min. At the reactor inlet (Fig. 8.17a), emission of CO is weak and almost absent in the de-coking period simply because CH_4 and CO_2 conversion is small. Emission from C_2 is negligibly small over the whole cycle at the reactor inlet. The CO profile at the reactor outlet (Fig. 8.17b) is well correlated with the CO mass spectrum profile over the cycle. This is because measurement by mass spectrometer directly reflects gas composition at the reactor outlet and the intensity of CO emission simply proportional to the CO partial pressure. In contrast, the C_2 emission profile is quite unique: C_2 emission intensity is weak during the reforming period and increased momentarily at $t = 1.25$ min, which is slightly after the CH_4 flow was turned off.

By comparing Fig. 8.17b, c, C_2 emission is observed only when CO exists. The intense C_2 peak at $t = 1.25$ min indicates that coke gasification by the reverse Boudouard reaction increase the CO partial pressure abruptly. Note, a sharp increase in CO at $t = 1.25$ min is not observed because CO is consumed to produce C_2. The C_2 emission is observed until $t = 2.6$ min under the presence of CO. The reaction mechanism from CO to C_2 is discussed in detail by S Kameshima et al. [3]. Although CO_2 DBD is generated between $2.6 < t < 4.2$ min, CO is not detected; thus the C_2 emission is absent. The specific energy input in this experiment is ca. 1 eV/molecule, and this value is too small to dissociate more than 1% of the CO_2 ($CO_2 + e \rightarrow CO + O + e$) in the gas phase [87]. The experimental observations lead to important conclusions: first, the excitation of CO_2, rather than dissociation, is the key reaction pathway for plasma hybrid heterogeneous conversion of CO_2 and CH_4. Second, heterogeneous oxidation of adsorbed carbon is well characterized by the gas-phase C_2 emission, which is utilized for a better understanding of the plasma-enhanced surface reaction mechanism.

8.4.4 *Other Reactions*

In this experiment, higher hydrocarbons, such as C_2H_6 and C_3H_8, were not synthesized. The higher hydrocarbons were presumably produced by DBD in the gas phase via dimerization of C_iH_j radicals; however, those species were fully converted to H_2 and CO at relatively high catalyst temperatures. The key influencing factor is the specific energy input. The production of higher hydrocarbons has been reported in packed-bed DBD reforming of CO_2/CH_4 [67–69] where the *SEI* was varied between 4 and 33 eV/molecule [67], 20–28 eV/molecule [68], and 1.4 eV/molecules [69]. An excessive energy input leads to the production of non-negligible amounts of higher hydrocarbons. Moreover, the complete conversion of higher hydrocarbons to syngas would not be possible at low catalyst temperatures (230 °C [67], unknown [68], 50–230 °C [69]). In contrast, the *SEI* in this experiment (1 eV/molecule) was sufficiently small and catalyst temperature (500–600 °C) is higher than that reported in the literature. In summary, H_2, CO, H_2O, and C were the major products, and the net production of higher hydrocarbons was absent.

8.5 Reaction Mechanism

8.5.1 *Kinetic Analysis*

In the thermal catalysis, the kinetic analysis is performed with either fixed CH_4 and variable CO_2 partial pressures or vice versa using N_2 or Ar as an inert balance. However, neither N_2 nor Ar dilution is suitable for plasma catalysis: excited N_2 and Ar are even more reactive than excited CH_4 and CO_2. Gas phase and surface chemistry would be biased by the unrecognized contribution of those species. Therefore, we must establish a reaction diagnostic method without a dilution gas. For this reason, multiple-pulsed DMR was performed by systematic variation of the CH_4/CO_2 ratio without dilution gas. CH_4 and CO_2 flow rates were controlled individually by programmable digital mass flow controllers. The CH_4/CO_2 ratio was initially set to 0.5. Every three-cycle operation, the CH_4/CO_2 ratio was incremented stepwise until $CH_4/CO_2 = 2$ consecutively, while the total flow rate was fixed at 2000 cm^3/min ($GHSV = 10{,}300\ h^{-1}$). The discharge power was 85–90 W where specific energy input was ca. 0.6 eV/molecule [74].

Reactant conversion and product yields are shown in Fig. 8.18. The definition for conversion and yield is provided in S Kameshima's research [74]. Figure 8.18a shows that the CH_4 conversion in the hybrid reaction was the same as that in the thermal reaction, implying that CH_4 dehydrogenation (8.R18) was not affected by DBD. Nevertheless, the H_2 yield was clearly increased by the hybrid reaction. A possible explanation is that the plasma-activated H_2O oxidizes adsorbed carbon (8. R20) to produce H_2. The reactivity of plasma-activated H_2O was studied by an Arrhenius plot analysis where reaction order for H_2O was doubled by DBD [61]. As a result, only the H_2 yield increased without increased CH_4 and CO_2 conversion when $CH_4/CO_2 < 0.67$. In contrast, the CO_2 conversion and the CO and H_2 yields were clearly promoted by DBD at $CH_4/CO_2 > 0.67$. Plasma-activated CO_2 and H_2O would promote surface reaction and increase CO and H_2 yield.

The experimental observation derives a tentative reaction scheme illustrated in Fig. 8.19. As discussed in Sect. 8.4.2, H_2O, CO, H_2, and CO_2 establish a *WGS equilibrium* with and without DBD. However, if plasma-activated H_2O^* promotes the surface reaction with adsorbed carbon (C^*), it creates additional pathways to syngas (8.R20). Similarly, the reaction between plasma-activated CO_2^* and C^* increases CO yield (8.R19). In fact, coke formation is suppressed by DBD (Fig. 8.18e): a similar result was reported in CH_4 steam reforming [53] and further confirmed in CH_4 dry reforming [75]. Coke (C^*) oxidation would proceed irreversibly and CO partial pressure increases rather selectively. Subsequently, due to the *WGS equilibrium*, product distribution is rearranged toward "CO_2 and H_2," producing more H_2 than CO when the temperature is well below 800 °C (cf. Fig. 8.15 and 8. R22.1). It was proven by the experiment that H_2 yield was enhanced much more than CO yield at 550 °C (Fig. 8.18c, d). Although production of C^* is detrimental for catalyst activity and lifetime, presence of C^* creates key pathways for emerging plasma-induced synergistic effect. As for CH_4 conversion, a synergistic effect was not observed clearly. Although experiments were carried out at fixed catalyst temperatures of ca. 550 °C, there was a possibility that the synergistic effect may occur at a different temperature range.

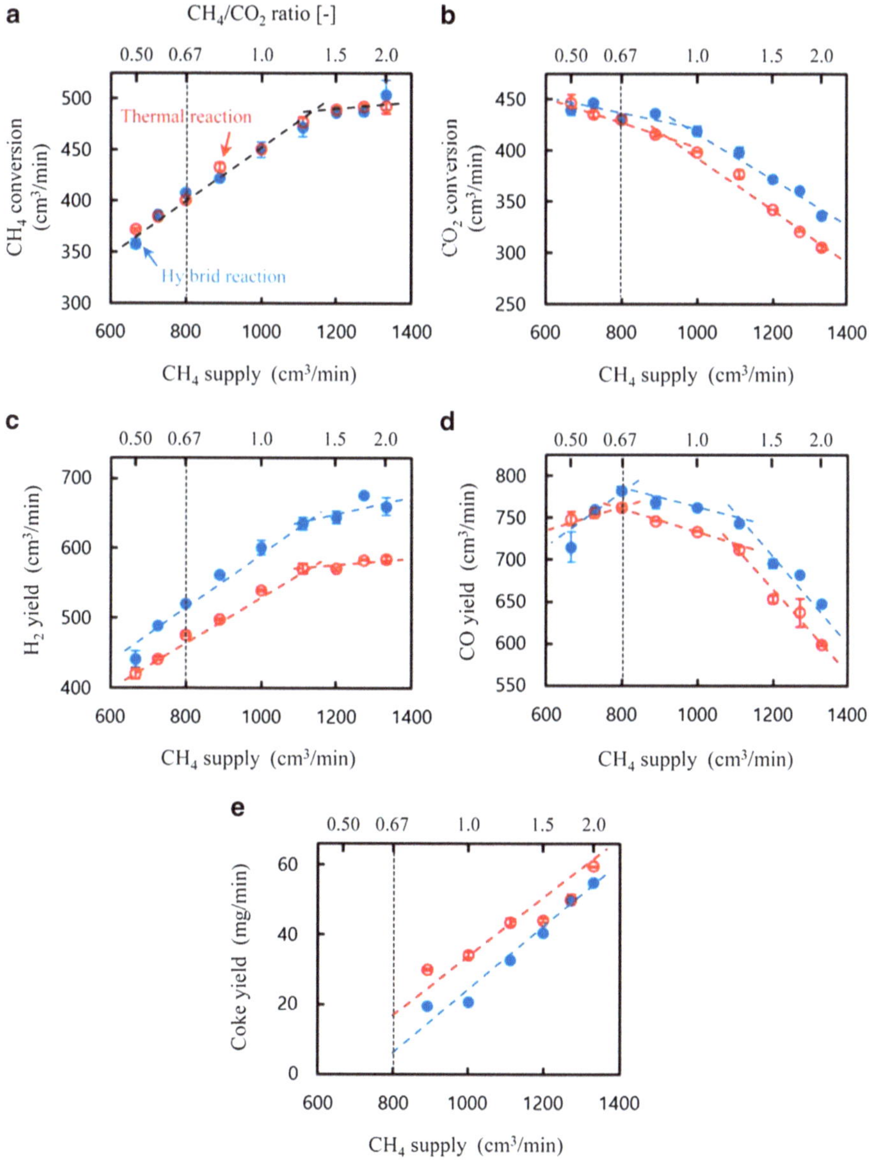

Fig. 8.18 The effect of the CH_4/CO_2 ratio on DMR at 550 °C: (**a**) CH_4 conversion; (**b**) CO_2 conversion; (**c**) H_2 yield; (**d**) CO yield; (**e**) coking rate. Open circle, thermal reaction (w/o DBD); solid circle, hybrid reaction

Fig. 8.19 A tentative reaction scheme of the DBD induced synergistic effect. Gasification of solid carbon by plasma-activated H_2O and CO_2 biased *WGS equilibrium* toward "CO_2 + H_2," yielding more H_2 than CO

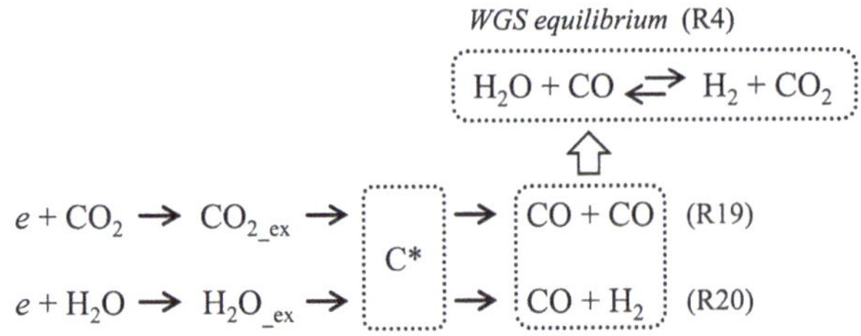

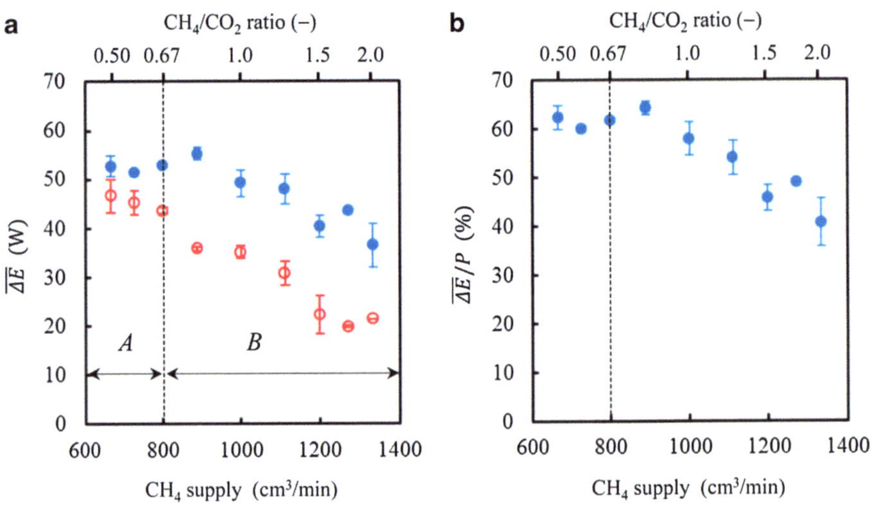

Fig. 8.20 (**a**) Mean energy absorption by DMR and (**b**) energy efficiency. Open circle, thermal reaction (w/o DBD); solid circle, hybrid reaction

8.5.2 Energy Efficiency

The average energy absorption by DMR ($\overline{\Delta E}$) over the reforming period ($0 < t < \tau$) and the energy efficiency (η) were calculated by the following equations:

$$\eta = \frac{\overline{\Delta E}}{P} \ (-) \tag{8.15}$$

$$\overline{\Delta E} = \frac{1}{\tau}\int_0^{\tau} \Delta E(t)dt \ (W) \tag{8.16}$$

$$\Delta E(t) = \nu_6 \Delta H^{\circ}_{\mathrm{H_2O|vapor}} + \nu_4 \Delta H^{\circ}_{\mathrm{CO}} - \nu_1 \Delta H^{\circ}_{\mathrm{CH_4}} - \nu_2 \Delta H^{\circ}_{\mathrm{CO_2}} \tag{8.17}$$

Here, P (W) is discharge power and ΔH° (J/mol) represents standard enthalpy of formation ($\Delta H^{\circ}_{\mathrm{C}} = \Delta H^{\circ}_{\mathrm{H_2}} = 0$ J/mol). The stoichiometric coefficient appearing in Eq. 8.17 (ν_{i}, mol/s) was obtained experimentally. Figure 8.20a shows that $\overline{\Delta E}$

increases by DBD, especially in region *B* where plasma-induced synergy clearly appears. According to the definition, enhancement of $\overline{\Delta E}$ is due to increased CO_2 conversion (the fourth term in Eq. 8.17) and H_2 selectivity (i.e., decreased H_2O selectivity, the first term). CH_4 conversion was not increased by DBD and net zero contribution to $\overline{\Delta E}$. The increased CO selectivity (the second term) is a negative contribution to the overall energy absorption ($\overline{\Delta E}$) because $\Delta H^{\circ}_{CO} < 0$. As discussed in Fig. 8.19, the increase in H_2 selectivity in the low-temperature plasma catalysis is much greater than that of CO, which is beneficial to increase energy efficiency. Energy efficiency reached 60–65% in region *A*; however, it decreases monotonically in region *B*, simply because CO_2 conversion decreases with CH_4/CO_2 ratio and the fourth term in Eq. 8.17 ($-\nu_2 H^{\circ}_{CO_2} > 0$) has a great impact on both $\overline{\Delta E}$ and η.

8.6 Conclusion and Future Outlook

Nonthermal plasma activation of CH_4 was overviewed, followed by a rather dedicated description of CH_4 and CO_2 reforming by DBD-enhanced heterogeneous catalysis. The nonthermal plasma hybrid reaction is explained based on the Langmuir-Hinshelwood (L-H) mechanism. The L-H mechanism consists of three steps: (I) adsorption of reactants, (II) surface diffusion/reaction, and (III) desorption of products. The rate-limiting step in CH_4 steam/dry reforming is the dissociation of the strong C-H bond over the metallic sites of the catalyst. By the simple kinetic analysis, vibrationally excited CH_4 is the most abundant and long-lived species produced by low-energy electron impact. Moreover, vibrationally excited CH_4 is known to promote dissociative chemisorption over the metal surface. In this sense, the role of nonthermal plasma in the initial stage of the L-H mechanism is quite clear. However, there is an inconsistency between macroscopic kinetic analysis and theoretical prediction. Based on our extensive study, activation of H_2O and CO_2 creates the key reaction pathways for the emerging synergistic effect. Meanwhile, reaction enhancement via CH_4 activation was not clearly identified in either steam or dry methane reforming. In addition to CH_4 activation, we need to study the vibrationally excited CO_2 and H_2O, as well as their interactions with the catalytic surface.

The role of nonthermal plasma on the second and third stage of the L-H kinetics is not well understood because nonthermal plasma has a limited interaction with the catalyst surface. For example, a plasma sheath is not generated clearly in the highly transient streamer-type discharge. Second, energy transfer from energetic ions to the surface is not expected in high-pressure conditions. Finally, nonthermal plasma is not generated in micro- and nanometer pores; excited species might be able to penetrate into catalyst pores of the order of 1–100 micrometers [88, 89]. We hypothesize that the plasma-induced nonthermal heating mechanism would play a key role for the promotion of overall surface reaction including diffusion of adsorbates and desorption of products. Charge recombination and association of radicals on the catalyst surface can release energy corresponding to 1–10 eV/molecule. If this energy is transferred directly to the admolecules, surface reaction/diffusion may be enhanced without increasing the macroscopic catalyst temperature. This reaction scheme would be explained by the Eley-Rideal mechanism rather than precursor-type adsorption enhancement (cf. Fig. 8.10). Alternatively, such an energy release may increase the local temperature on the micro-, nano-, and even molecular-scale and promotes surface chemistry. Although the plasma-induced energy transfer mechanism is commonly accepted in particle growth on the micrometer [90] and nanometer scale [91], it has yet to be investigated within the scope of plasma catalysis.

Kortshagen et al. estimated that the temperature of the free-standing silicon nanoparticles is likely to be equilibrated with the ambient gas temperature when particles grow beyond 10 nm [91], while the temperature of smaller particles (smaller than 5 nm) is elevated beyond the gas temperature. The heating and cooling events occur on the nanoparticle surface, while net temperature increase is determined by the gross energy gain divided by the volumetric heat capacity of particles. Inevitably, the surface to volume ratio of nanoparticles is the critically important parameter. In the case of supported catalysts, however, heat or energy readily escapes toward supporting materials. Therefore, the temperature of nanoparticle catalysts may be determined unambiguously by the temperature of supporting material no matter what their sizes are. A deep understanding of highly transient and nonequilibrium energy transfer via excited molecules, without macroscopic temperature change, needs to be achieved.

Exploring new types of catalysts, dedicated to plasma catalysis, is an important subject of research. This is one missing research thrust in the plasma catalysis community. In looking toward the successful catalyst design and selection of appropriate materials, materials science research should be supported by the deep understanding of plasma chemistry, diagnostics, and modeling of interfacial phenomena [92]. Performance of newly developed catalysts is benchmarked by the macroscopic diagnostics method [2, 3, 75]. These approaches should be integrated as a plasma catalysis science platform, leading to truly beneficial applications.

Acknowledgments This work is supported by the Japan Society for the Promotion of Science (JSPS) KAKENHI (JP16J09876). S.K. acknowledges JSPS for providing Research Fellowship for Young Scientists (DC1); Z.R. acknowledges financial support from the program of China Scholarships Council (No.201707040056). T.N. would like to thank Mr. Tinnapop Moonmuang (Mechanical Engineering, Chiang Mai University) for the experimental support.

References

1. Schiebahn, S., Grube, T., Robinius, M., Zhao, L., Otto, A., Kumar, B., Weber, M., & Stolten, D. (2013). In D. Stolten & V. Scherer (Eds.), *Transition to renewable energy systems* (pp. 813–847). Wiley-VCH.

2. Nozaki, T., & Okazaki, K. (2013). Non-thermal plasma catalysis of methane: Principles, energy efficiency, and applications. *Catalysis Today*, *211*, 29–38.
3. Kameshima, S., Tamura, K., Ishibashi, Y., & Nozaki, T. (2015). Pulsed dry methane reforming in plasma-enhanced catalytic reaction. *Catalysis Today*, *256*, 67–75.
4. Nozaki, T., & Okazaki, K. (2011). Innovative methane conversion technology using atmospheric pressure non-thermal plasma. *Journal of the Japan Petroleum Institute, 54*, 146–158.
5. Goede, A. P. H. (2015). CO_2-neutral fuels, EPJ Web of Conferences, *98*, 07002.
6. Gallagher, M. J., Jr., & Fridman, A. (2011). Plasma reforming for H_2-rich synthesis gas. *Fuel cells: Technologies for fuel processing, Chapter-8* (pp. 223–259). Elsevier.
7. Petitpas, G., Rollier, J.-D., Darmon, A., Gonzalez-Aguilar, J., Metkemeijer, R., & Fulcheri, L. (2007). A comparative study of non-thermal plasma assisted reforming technologies. *International Journal of Hydrogen Energy*, *32*, 2848–2867.
8. Whitehead, J. C. (2016). The known knowns, the known unknowns and the unknown unknowns. *Journal of Physics D: Applied Physics, 49*, 243001.
9. Lee, D. H., Kim, K.-T., Song, Y.-H., Kang, W. S., & Jo, S. (2013). Mapping plasma chemistry in hydrocarbon fuel processing processes. *Plasma Chemistry and Plasma Processing*, *33*, 249–269.
10. Nozaki, T., & Okazaki, K. (2012). Plasma enhanced C1-chemistry: towards greener methane conversion. *Green Processing and Synthesis*, *1*, 517–523.
11. Rostrup-Nielsen, J., & Christiansen, L. J. (2011). *Concepts in syngas manufacture* (1st ed.). Imperial College Press.
12. Gutsol, A., Rabinovich, A., & Fridman, A. (2011). Combustion-assisted plasma in fuel conversion. *Journal of Physics D: Applied Physics, 44*, 274001.
13. Ju, Y., & Sun, W. (2015). Plasma assisted combustion: Dynamics and chemistry. *Progress in Energy and Combustion Science*, *48*, 21–83.
14. Kalra, C. S., Gutsol, A. F., & Fridman, A. A. (2005). Gliding arc discharges as a source of intermediate plasma for methane partial oxidation. *IEEE Transactions on Plasma Science*, *33*, 32–41.
15. Mutaf-Yardimci, O., Saveliev, A. V., Fridman, A. A., & Kennedyd, L. A. (2000). Thermal and nonthermal regimes of gliding arc discharge in air flow. *Journal of Applied Physics*, *87*, 1632–1641.
16. Lee, D. H., Kim, K.-T., Cha, M. S., & Song, Y.-H. (2007). Optimization scheme of a rotating gliding arc reactor for partial oxidation of methane. *Proceedings of the Combustion Institute*, *31*, 3343–3351.
17. Lee, D. H., Kim, K.-T., Cha, M. S., & Song, Y.-H. (2010). Plasma-controlled chemistry in plasma reforming of methane. *International Journal of Hydrogen Energy*, *35*, 10967–10976.
18. Gutsol, A. (2010). *Handbook of combustion*. In M. Lackner et al. (Eds.), *New technologies* (Vol. 5, pp. 323–353). Weinheim: WILEY-VCH Verlag GmbH & Co. KGaA.
19. Rafiq, M. H., & Hustad, J. E. (2011). Synthesis gas from methane by using a plasma-assisted gliding arc catalytic partial oxidation reactor. *Industrial and Engineering Chemistry Research*, *50*, 5428–5439.
20. Periana, R. A., Mironov, O., Taube, D., Bhalla, G., & Jones, C. J. (2003). Catalytic, oxidative condensation of CH_4 to CH_3COOH in one step via CH activation. *Science*, *301*, 814–818.
21. York, A. P. E., Xiao, T., & Green, M. L. H. (2003). Brief overview of the partial oxidation of methane to synthesis gas. *Topics in Catalysis*, *22*, 345–358.
22. Bjorklund, M. C., & Carr, R. W. (2002). Enhanced methanol yields from the direct partial oxidation of methane in a simulated countercurrent moving bed chromatographic reactor. *Industrial and Engineering Chemistry Research*, *41*, 6528–6536.
23. Danen, W. C., Ferris, M. J., Lyman, J. L., Oldenborg, R. C., Rofer, C. K., Steit, G. E. (1991). Methane to methanol by direct partial oxidation. Preprints Petro. Chem. Div. by ASC, *36*, 166–171.

24. Feng, W., Knopf, F. C., & Dooley, K. M. (1994). Effects of pressure, third bodies, and temperature profiling on the noncatalytic partial oxidation of methane. *Energy & Fuels, 8*, 815–822.
25. Yarlagadda, P. S., Morton, L. A., Huntera, N. R., & Gesser, H. D. (1998). Direct conversion of methane to methanol in a flow reactor. *Industrial and Engineering Chemistry Research, 27*, 252–256.
26. Casey, P. S., McAllister, T., & Foger, K. (1994). Selective oxidation of methane to methanol at high pressures. *Industrial and Engineering Chemistry Research, 33*, 1120–1125.
27. Larkin, D. W., Lobban, L. L., & Mallinson, R. G. (2001). The direct partial oxidation of methane to organic oxygenates using a dielectric barrier discharge reactor as a catalytic reactor analog. *Catalysis Today, 71*, 199–210.
28. Okazaki, K., Kishida, T., Ogawa, K., & Nozaki, T. (2002). Direct conversion from methane to methanol for high efficiency energy system with exergy regeneration. *Energy Conversion and Management, 43*, 1459–1468.
29. Zhou, L. M., Xue, B., Kogelschatz, U., & Eliasson, B. (1998). Partial oxidation of methane to methanol with oxygen or air in a nonequilibrium discharge plasma. *Plasma Chemistry and Plasma Processing, 18*, 375–393.
30. Agiral, A., & Gardeniers, J. G. E. H. (2010). Microreactors with electrical fields. *Advances in Chemical Engineering, 38*, 37–102.
31. Lin, L., & Wang, Q. (2015). Microplasma: a new generation of technology for functional nanomaterial synthesis. *Plasma Chemistry Plasma Process, 35*, 925–962.
32. Nozaki, T., Hattori, A., & Okazaki, K. (2004). Partial oxidation of methane using a microscale non-equilibrium plasma reactor. *Catalysis Today, 98*, 607–616.
33. Nozaki, T., Unno, Y., Miyazaki, Y., & Okazaki, K. (2001). Optical diagnostics for determining gas temperature of reactive microdischarges in a methane-fed dielectric barrier discharge. *Journal of Physics D: Applied Physics, 34*, 2504–2511.
34. Nozaki, T., Miyazaki, Y., Unno, Y., & Okazaki, K. (2001). Energy distribution and heat transfer mechanisms in atmospheric pressure non-equilibrium plasmas. *Journal of Physics D: Applied Physics, 34*, 3383–3390.
35. Nozaki, T., Unno, Y., & Okazaki, K. (2002). Thermal structure of atmospheric pressure non-equilibrium plasmas. *Plasma Sources Science and Technology, 11*, 431–438.
36. Nozaki, T., Ağıral, A., Yuzawa, S., Gardeniers, J. G. E. H., & Okazaki, K. (2011). A single step methane conversion into synthetic fuels using microplasma reactor. *Chemical Engineering Journal, 166*, 288–293.
37. Lott, J. L., & Sliepcevich, C. M. (1967). Partial oxidation of methane at high pressures. *Industrial and Engineering Chemistry Process Design and Development, 6*, 67–74.
38. Burch, R., Squire, G. D., & Tsang, S. C. (1989). Direct conversion of methane into methanol. *Journal of the Chemical Society, Faraday Transactions, 85*, 3561–3568.
39. Hunter, N. R., Gesser, H. D., Morton, L. A., & Yarlagadda, P. S. (1990). Methanol formation at high pressure by the catalyzed oxidation of natural gas and by the sensitized oxidation of methane. *Applied Catalysis, 57*, 45–54.
40. Chun, J.-W., & Anthony, R. G. (1993). Catalytic oxidations of methane to methanol. *Industrial and Engineering Chemistry Research, 32*, 259–263.
41. Lødeng, R., Lindvåg, O. A., Søraker, P., Roterud, P. T., & Onsager, O. T. (1995). Experimental and modeling study of the selective homogeneous gas phase oxidation of methane to methanol. *Industrial and Engineering Chemistry Research, 34*, 1044–1059.
42. Chellappa, A. S., Fuangfoo, S., & Viswanath, D. S. (1997). Homogeneous oxidation of methane to methanol: Effect of CO_2, N_2, and H_2 at high oxygen conversions. *Industrial and Engineering Chemistry Research, 36*, 1401–1409.
43. Chellappa, A. S., & Viswanath, D. S. (1995). Partial oxidation of methane using ferric molybdate catalyst. *Industrial and Engineering Chemistry Research, 34*, 1933–1940.

44. Wang, X., Wang, Y., Tang, Q., Guo, Q., Zhang, Q., & Wan, H. (2003). MCM-41-supported iron phosphate catalyst for partial oxidation of methane to oxygenates with oxygen and nitrous oxide. *Journal of Catalysis, 217*, 457–467.
45. Rasmussen, C. L., & Glarborg, P. (2008). Direct partial oxidation of natural gas to liquid chemicals: chemical kinetic modeling and global optimization. *Industrial and Engineering Chemistry Research, 47*, 6579–6588.
46. Oshima, Y., Saito, M., Koda, S., & Tominaga, H. (1988). Partial oxidation of methane by laser-initiated chain reaction. *Chemistry Letters, 17*, 203–206.
47. NIST Chemistry WebBook. http://webbook.nist.gov/chemistry/
48. Eliasson, B., & Kogelschatz, U. (1991). Nonequilibrium volume plasma chemical processing. *IEEE Transactions on Plasma Science, 19*, 309.
49. Goujard, V., Nozaki, T., Yuzawa, S., Ağıral, A., & Okazaki, K. (2011). Selective conversion of methane to synthetic fuels using dielectric barrier discharge contacting liquid film. *Journal of Physics D: Applied Physics, 44*, 274010.
50. Nozaki, T., Abe, S., Moriyama, S., Kameshima, S., Okazaki, K., Goujard, V., & Ağıral, A. (2015). One step methane conversion to syngas by dielectric barrier discharge. *Japanese Journal of Applied Physics, 54*, 01AG01.
51. Nozaki, T., Goujard, V., Yuzawa, S., Moriyama, S., Ağıral, A., & Okazari, K. (2011). Selective conversion of methane to synthetic fuels using dielectric barrier discharge contacting liquid film. *Journal of Physics D: Applied Physics, 44*, 274010.
52. Nozaki, T., Tsukijihara, H., & Okazaki, K. (2006). Hydrogen enrichment of low-calorific fuels using barrier discharge enhanced Ni/γ-Al_2O_3 bed reactor: Thermal and nonthermal effect of nonequilibrium plasma. *Energy and Fuels, 20*, 339–345.
53. Nozaki, T., Fukui, W., & Okazaki, K. (2008). Reaction enhancement mechanism of the nonthermal discharge and catalyst hybrid reaction for methane reforming. *Energy and Fuels, 22*, 3600–3604.
54. Nozaki, T., Muto, N., Kado, S., & Okazaki, K. (2004). Dissociation of vibrationally excited methane on Ni catalyst: Part 1. Application to methane steam reforming. *Catalysis Today, 89*, 57–65.
55. Nozaki, T., Muto, N., Kado, S., & Okazaki, K. (2004). Partial oxidation of methane using a microscale non-equilibrium plasma reactor. *Catalysis Today, 89*, 67–74.
56. Lee, A. L., & Zabransky, R. F. (1990). Internal reforming development for solid oxide fuel cells. *Industrial and Engineering Chemistry Research, 29*, 766–773.
57. Rostyup-Nielsen, J. R., & Hansen, J.-H. B. (1993). CO_2-reforming of methane over transition metals. *Journal of Catalysis, 144*, 38–49.
58. Ahmed, K., & Foger, K. (2000). Kinetics of internal steam reforming of methane on Ni/YSZ-based anodes for solid oxide fuel cells. *Catalysis Today, 63*, 479–487.
59. Laosiripojana, N., & Assabumrungrat, S. (2005). Methane steam reforming over Ni/Ce–ZrO_2 catalyst: Influences of Ce–ZrO_2 support on reactivity, resistance toward carbon formation, and intrinsic reaction kinetics. *Applied Catalysis, A: General, 290*, 200–211.
60. Wei, J., & Iglesia, E. (2004). Isotopic and kinetic assessment of the mechanism of reactions of CH_4 with CO_2 or H_2O to form synthesis gas and carbon on nickel catalysts. *Journal of Catalysis, 224*, 370–383.
61. Nozaki, T., Tsukijihara, H., Fukui, W., & Okazaki, K. (2007). Kinetic analysis of the catalyst and nonthermal plasma hybrid reaction for methane steam reforming. *Energy and Fuels, 21*, 2525–2530.
62. Ravanchi, M. T., & Sahebdelfar, S. (2014). Carbon dioxide capture and utilization in petrochemical industry: potentials and challenges. *Applied Petrochemical Research, 4*, 63–77.
63. Shah, Y. T., & Gardner, T. H. (2014). Dry reforming of hydrocarbon feedstocks. *Catalysis Review: Science and Engineering, 56*, 476–536.
64. Gao, J., Hou, Z., Lou, H., & Zheng, X. (2011). Dry (CO_2) reforming. In D. Shekhawat, J. J. Spivey, & D. A. Berry (Eds.), *Fuel cells: Technologies for fuel processing* (pp. 191–221). Amsterdam: Elsevier.

65. Wang, S., & Lu, G. Q. (1999). A comprehensive study on carbon dioxide reforming of methane over Ni/γ-Al_2O_3 catalysts. *Industrial and Engineering Chemistry Research, 38*, 2615–2625.
66. Kogelschatz, U. (2003). Dielectric-barrier discharges: their history, discharge physics, and industrial applications. *Plasma Chemistry and Plasma Processing, 23*, 1–46.
67. Tu, X., & Whitehead, J. C. (2012). Plasma-catalytic dry reforming of methane in an atmospheric dielectric barrier discharge: Understanding the synergistic effect at low temperature. *Applied Catalysis B: Environmental, 125*, 439–448.
68. Tu, X., Gallon, H. J., Twigg, M. V., Gorry, P. A., & Whitehead, J. C. (2011). Dry reforming of methane over a Ni/Al_2O_3 catalyst in a coaxial dielectric barrier discharge reactor. *Journal of Physics D: Applied Physics, 44*, 274007.
69. Kraus, M., Eliasson, B., Kogelschatz, U., & Wokaun, A. (2001). CO_2 reforming of methane by the combination of dielectric-barrier discharges and catalysis. *Physical Chemistry Chemical Physics, 3*, 294–300.
70. Wang, Q., Yan, B.-H., Jin, Y., & Cheng, Y. (2009). Dry reforming of methane in a dielectric barrier discharge reactor with Ni/Al_2O_3 Catalyst: Interaction of catalyst and plasma. *Energy & Fuels, 23*, 4196–4201.
71. Long, H., Shang, S., Tao, X., Yin, Y., & Dai, X. (2008). CO_2 reforming of CH_4 by combination of cold plasma jet and Ni/γ-Al_2O_3 catalyst. *International Journal of Hydrogen Energy, 33*, 5510–5515.
72. Mahammadunnisa, S., Manoj Kumar Reddy, P., Ramaraju, B., & Subrahmanyam, C. H. (2013). Catalytic nonthermal plasma reactor for dry reforming of methane. *Energy Fuels, 27*, 4441–4447.
73. Pan, K. L., Chung, W. C., & Chang, M. B. (2014). Dry reforming of CH_4 with CO_2 to generate syngas by combined plasma catalysis. *IEEE Transactions on Plasma Science, 42*, 3809–3818.
74. Kameshima, S., Tamura, K., Mizukami, R., Yamazaki, T., & Nozaki, T. (2017). Parametric analysis of plasma-assisted pulsed dry methane reforming over Ni/Al_2O_3 catalyst. *Plasma Processes & Polymers, 14*, e1600096.
75. Kameshima, S., Mizukami, R., Yamazaki, T., Prananto, L. A., & Nozaki, T. (2018). Interfacial reactions between DBD and porous catalyst in dry methane reforming. *Journal of Physics D: Applied Physics, 51*, 114006.
76. Olsbye, U., Wurzel, T., & Mleczko, L. (1997). Kinetic and reaction engineering studies of dry reforming of methane over a Ni/La/Al_2O_3 catalyst. *Industrial and Engineering Chemistry Research, 36*, 5180–5188.
77. Osaki, T. (2015). Effect of nickel diameter on the rates of elementary steps involved in CO_2 reforming of CH_4 over Ni/Al_2O_3 catalysts. *Catalysis Letters, 145*, 1931–1940.
78. Jablonski, G. A., Geurts, F. W., Sacco, A., Jr., & Biederman, R. R. (1992). Carbon deposition over Fe, Ni, and Co foils from CO-H_2-CH_4-CO_2-H_2O, CO-CO_2, CH_4-H_2, and CO-H_2-H_2O gas mixtures: I. Morphology. *Carbon, 30*, 87–98.
79. Puretzky, A. A., Geohegan, D. B., Jesse, S., Ivanov, I. N., & Eres, G. (2005). In situ measurements and modeling of carbon nanotube array growth kinetics during chemical vapor deposition. *Applied Physics A: Materials Science & Processing, 81*, 223–240.
80. Dombrowski, E., Peterson, E., Sesto, D. D., & Utz, A. L. (2015). Precursor-mediated reactivity of vibrationally hot molecules: Methane activation on Ir(1 1 1). *Catalysis Today, 244*, 10–18.
81. Neyts, E. C. (2016). Plasma-surface interactions in plasma catalysis. *Plasma Chemistry and Plasma Processing, 36*, 185–212.
82. Falkenstein, Z., & Coogan, J. J. (1997). Microdischarge behaviour in the silent discharge of nitrogen - oxygen and water - air mixtures. *Journal of Physics D: Applied Physics, 30*, 817–825.
83. Du, Y., Tamura, K., Moore, S., Peng, Z., Nozaki, T., & Bruggeman, P. J. (2017). CO($B^1\Sigma^+ \rightarrow A^1\Pi$) Angstrom system for gas temperature measurements in CO_2 containing plasmas. *Plasma Chemistry and Plasma Processing, 37*, 29–41.
84. Kim, H.-H., Teramoto, Y., Sano, T., Negishi, N., & Ogata, A. (2015). Effects of Si/Al ratio on the interaction of nonthermal plasma and Ag/HY catalysts. *Applied Catalysis B: Environmental, 166–167*, 9–17.

85. Coogan, J. J., & Sappey, A. D. (1996). Distribution of OH within silent discharge plasma reactors. *IEEE Transactions on Plasma Science, 24*, 91–92.
86. Wang, W., & Gong, J. (2011). Methanation of carbon dioxide: an overview. *Frontiers of Chemical Science and Engineering, 5*, 2–10.
87. Aerts, R., Somers, W., & Bogaerts, A. (2015). Carbon dioxide splitting in a dielectric barrier discharge plasma: A combined experimental and computational study. *ChemSusChem, 8*, 702–716.
88. Kim, H.-H., Teramoto, Y., Negishi, N., & Ogata, A. (2015). A multidisciplinary approach to understand the interactions of nonthermal plasma and catalyst: A review. *Catalysis Today, 256*, 13–22.
89. Zhang, Y.-R., Van Laer, K., Neyts, E. C., & Bogaerts, A. (2016). Can plasma be formed in catalyst pores? A modeling investigation. *Applied Catalysis B: Environmental, 185*, 56–67.
90. Maurer, H. R., & Kersten, H. (2011). On the heating of nano- and microparticles in process plasmas. *Journal of Physics D: Applied Physics, 44*, 174029.
91. Kramer, N. J., Anthony, R. J., Mamunuru, M., Aydil, E. S., & Kortshagen, U. R. (2014). Plasma-induced crystallization of silicon nanoparticles. *Journal of Physics D: Applied Physics, 47*, 075202.
92. Neyts, E. C., Ostrikov, K., Sunkara, M. K., & Bogaerts, A. (2015). Synergistic effects at the nanoscale. *Chemical Reviews, 115*, 13408–13446.

Chapter 9
Plasma-Catalytic Conversion of Carbon Dioxide

Bryony Ashford, Yaolin Wang, Li Wang, and Xin Tu

9.1 Introduction

The emission of CO_2 is a pressing concern as its release into the atmosphere is a major source of global warming. As global temperatures rise due to the greenhouse effect and current technologies, such as carbon capture and storage (CCS) and a switch to renewables, fall short, expertise must be employed to find new, viable processes for the mitigation of CO_2. Focus is now on carbon dioxide utilization, as high-value chemicals and fuels can be produced, creating viable and sustainable processes. Current processes, however, such as thermal catalytic and electrochemical processes, require elevated temperatures and are not thermodynamically efficient, thus reducing their energy efficiency and feasibility. Plasma-catalytic processes have the potential to overcome these drawbacks due to their low-temperature operation and non-equilibrium characteristics which allow the high stability of the CO_2 molecule to be overcome without the need for large energy inputs. Alongside this, the catalyst acts to lower the activation barrier and enhance the selectivity to the required product. Interactions also occur between the catalyst and the plasma, creating synergy. Furthermore, quick start-up and shutdown enable plasma-catalytic processes to be used as a method of storing excess energy from renewable energy generation. A great number of reactions can potentially be carried

B. Ashford · Y. Wang · X. Tu (✉)
Department of Electrical Engineering and Electronics, University of Liverpool, Liverpool, UK
e-mail: xin.tu@liverpool.ac.uk

L. Wang
State Key Laboratory of Fine Chemicals, School of Chemical Engineering, Dalian University of Technology, Dalian, Liaoning, China

College of Environmental Sciences and Engineering, Dalian Maritime University, Dalian, Liaoning, China

Department of Electrical Engineering and Electronics, University of Liverpool, Liverpool, UK

X. Tu et al. (eds.), *Plasma Catalysis*, Springer Series on Atomic, Optical, and Plasma Physics 106, https://doi.org/10.1007/978-3-030-05189-1_9

out in a plasma-catalytic reactor, including CO_2 decomposition, dry reforming of methane and CO_2 hydrogenation; hence a great number of high-value products can be created (oxygenates, liquid hydrocarbons, syngas, etc.). Currently, a trade-off exists between high energy efficiency and high reactant conversion. This is due to conversion increasing with input power, which leads to a decrease in energy efficiency. A greater understanding of the plasma chemistry and the interactions between plasma and catalyst will further increase the viability of plasma-catalytic processes for the utilization of CO_2 on an industrial scale. The following chapter describes this process in detail for a number of different reactions and discusses recent advances in the area.

9.1.1 Carbon Dioxide Emission

Carbon dioxide is a major greenhouse gas and is largely responsible for the changes we are currently seeing in the climate. Since the industrial revolution, greenhouse gas emissions have risen, with a 75% global increase in greenhouse gases occurring since 1970. In 2010, 49×10^9 tonnes of CO_2 equivalent ($GtCO_2eq$) of greenhouse gases were emitted from anthropogenic sources, of which CO_2 sources currently make up the majority at 76% [1]. Industrial development and the release of carbon dioxide have always gone hand in hand, and with many countries still developing their industry and a rising global population, carbon dioxide emissions will continue to increase. As is well documented, a rise in CO_2 in the atmosphere leads to warming of the planet, causing a rise in sea level which can wipe out whole communities in low-lying areas and increasing the frequency of disasters such as tsunamis and forest fires. In recent years, these disasters have become more commonplace, resulting in societal values changing to reflect the growing urgency of the situation and forcing governments around the world to take action against climate change. This need for change culminated in the Paris Agreement, a legally binding document between 195 countries that aims to tackle climate change, aiming to limit the global temperature rise to well below 2 °C above pre-industrial levels. In order to fulfill targets set out in the agreement, a switch to renewable energy and a reduction in the release of greenhouse gases into the atmosphere are required, necessitating the design of novel technologies that enable this change while allowing society to prosper. The utilization of carbon dioxide from waste gas streams is one area which has the potential to fulfill these requirements, as CO_2 from fossil fuel and industrial processes accounts for 65% of total annual anthropogenic greenhouse gas emissions, as detailed in the Fifth Assessment Report of Intergovernmental Panel on Climate Change (IPCC) [1]. A significant reduction in carbon dioxide emissions can therefore be made by capturing CO_2 from waste gas streams and converting it into valuable fuels and chemicals.

9.1.2 Current Technologies to Reduce CO_2 Emission

There are several methods that can be used to convert carbon dioxide, including catalytic conversion, photocatalytic/photochemical processes, electrocatalytic/electrochemical processes, enzymatic/biochemical processes and plasma processes. Each of these methods results in a slight variation in the product created, with hydrocarbons, hydrogen and oxygenates forming via catalytic conversion, while carbon monoxide, hydrocarbons, syngas and oxygenates are the main products of plasma processes. Along with the need for high temperatures in thermal-catalytic processes, issues can be encountered with catalyst deactivation due to coking. Electrocatalytic processes show promise for producing methanol, formic acid and a variety of other organic chemicals. However, these processes have low thermodynamic efficiency. The least researched, yet predicted to be the most effective, is the plasma process [2, 3]. In comparison to the other processes, it is simple and fast: plasma has the potential to enable thermodynamically unfavorable chemical reactions (e.g. CO_2 dissociation) to occur at ambient conditions [3].

9.1.3 Carbon Dioxide Utilization Through Plasma Technology

Nonthermal plasma (NTP) can be operated at room temperature and atmospheric pressure while still generating highly active species and electrons, with mean electron energy between 1 and 10 eV. This electron energy is the optimum range for exciting molecular and atomic species and breaking chemical bonds. For CO_2 dissociation (9.R1) to occur in plasma, only 5.5 eV is required to break the $OC = O$ bond via stepwise vibrational excitation. Nonthermal plasma therefore shows great potential in the production of an efficient CO_2 utilization process, as it can overcome the stability of CO_2 without the need for the high temperatures required in thermal catalytic processes. Plasma technology is also advantageous over thermal processes as reaction rates are high and steady state is quickly reached [4]. This facilitates quick start-up and shutdown, a promising feature that enables plasma technology powered by renewable energy to act as efficient chemical energy storage through a localized or distributed system at peak grid times [5]. Different routes for CO_2 conversion have been explored using NTP (Fig. 9.1).

9.2 Plasma CO_2 Decomposition

$$2CO_2 \rightarrow 2CO + O_2 \tag{9.R1}$$

Decomposition of CO_2 into CO and O_2 using NTP has recently attracted significant interest as this reaction is almost impossible at low temperatures using

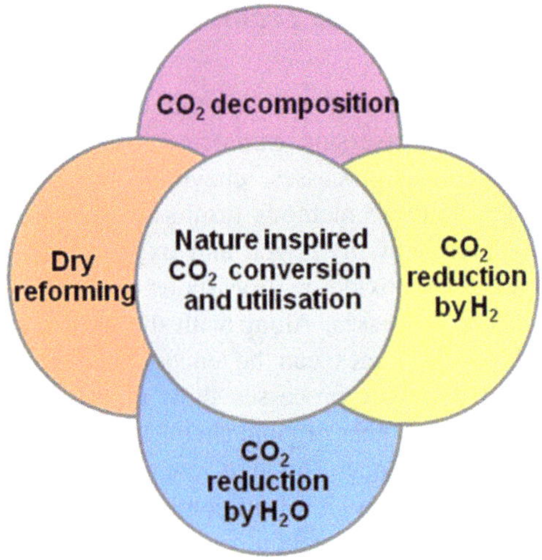

Fig. 9.1 Different routes for CO_2 conversion

conventional catalysis. Furthermore, the product CO is an important chemical feedstock for further synthesis of fuels and chemicals. Various plasma systems are reported to successfully convert CO_2 into CO and O_2 (9.R1), including glow discharge, where one study found a CO_2 conversion of 30% is achievable at an input voltage of 7 kV [6]; radio frequency discharge, which can achieve 90% CO_2 conversion at 1 kW [7]; and microwave discharge, in which a 100 W power input was accompanied by a 90% conversion [8]. A carrier gas, such as helium or argon, has previously been commonplace in the dissociation of CO_2 via plasma systems; however, this leads to an additional, undesired cost. Dielectric barrier discharge (DBD) reactors have been shown to successfully dissociate CO_2 in the absence of a carrier gas [9, 10], with one study achieving 30% conversion at a power density of 14.75 W/cm^3 [10].

In plasma, reactions mainly occur in the gas phase. Firstly, CO_2 is dissociated into CO and an oxygen atom. The oxygen atom created then either combines with another oxygen atom to form molecular oxygen (9.R2), or it reacts with CO_2 to form carbon monoxide and an oxygen molecule (9.R3):

$$O + O \rightarrow O_2 \tag{9.R2}$$

$$O + CO_2 \rightarrow O_2 + CO \tag{9.R3}$$

The production of carbon can also occur (9.R4), along with reverse CO_2 decomposition (9.R5):

$$CO + e \rightarrow C + O + e \tag{9.R4}$$

$$CO + O \rightarrow CO_2 \tag{9.R5}$$

CO_2 conversion mainly occurs via electronic dissociation, vibrational excitation and dissociative attachment [11]. A zero-dimensional chemical kinetics model has been developed to understand the reaction mechanisms in the plasma CO_2 dissociation [12]. Electron impact dissociation has been found dominant in CO_2 decomposition in DBD plasmas, leading to lower energy efficiency compared to microwave plasmas or gliding arc discharges, in which more energy-efficient vibrational excitation of CO_2 plays a key role in the decomposition of CO_2 [13–15]. A 1D model developed for an AC gliding arc reactor has shown that dissociation only occurs at the center of the arc where the gas, vibrational and electron temperatures, as well as the ionization degree, are at their maximums [13]. The dissociation of CO_2 in a gliding arc therefore only occurs significantly at the arc center. In a typical microwave plasma, gas temperature has shown to be an important factor in determining CO_2 conversion, with a higher temperature resulting in a greater conversion; as temperature increases, the reaction rate coefficients for heavy-particle dissociation also increase, along with the number of vibrationally excited species [16].

The reformation of CO_2 can be an issue in plasma reactors, decreasing the effective CO_2 conversion as CO recombines with oxygen species (O_2, O and O^{2-}) [17]. A novel approach to solving this problem was carried out by Mori and Tun, in which a DBD reactor was combined with a solid oxide electrolyser cell (SOEC) [17]. The SOEC was used to remove oxygen from the gas, hence restricting it from reacting with CO. The SOEC also contributed to CO_2 splitting [17]. When the two were used separately for this reaction, the SOEC process reached a maximum conversion of 3%, while the plasma-alone process had an optimum conversion of 15%. However, when combined, a synergistic effect occurred as the hybrid system reached a conversion of 93%, which was attributed to the SOEC removing oxygen from the system, as shown in Fig. 9.2 [17]. It is novel ideas such as this which can greatly increase the feasibility of plasma CO_2 decomposition on an industrial scale.

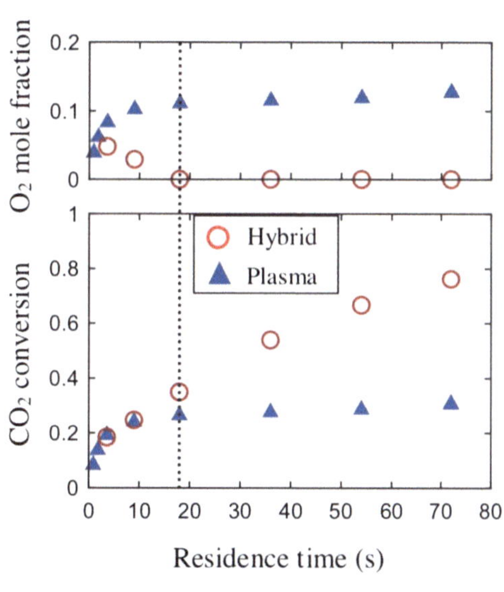

Fig. 9.2 CO_2 conversion and outlet O_2 mole fraction with residence time (plasma input power, 30 W; SOEC applied voltage, 7 V; SOEC current, 37–50 mA; CO_2 flow rate, 0.23–18 ml/min; pressure, 3 kPa; inside heater temperature, 800 °C; outside furnace temperature, 200 °C) [17]

9.2.1 *Influence of Process Parameters*

A design of experiments approach using a cylindrical DBD revealed that the main parameter which affects the energy efficiency of this process is the feed flow rate, while discharge power has the greatest influence on CO_2 conversion [18]. A high feed flow rate leads to a lower CO_2 conversion but is more energy efficient [18]. A high feed flow rate corresponds to a lower residence time, resulting in fewer interactions between the feed gases and the excited species, hence a lower conversion; however, a higher flow rate leads to lower specific energy input (SEI) (at constant plasma power) (Eqs. 9.1 and 9.2) and thus results in a more energy-efficient process (Eq. 9.3) (equations found in [19]) [12]:

$$\text{SEI}\ [\text{kJ/L}] = \left(\frac{\text{Power}\ [\text{kW}]}{\text{Flow}\ [\text{L/min}]}\right) \times 60\ [\text{s/min}] \tag{9.1}$$

$$\text{Power} = (1/T)\int_0^T (V(t) \times I(t))dt \tag{9.2}$$

$$E[\%] = \frac{\Delta H_{\text{r}}[\text{kJ/mol}] \times X_{\text{CO}_2}[\%]}{\text{SEI}[\text{kJ/L}] \times \text{molar volume}[\text{L/mol}]} \tag{9.3}$$

A trade-off therefore exists between energy efficiency and conversion, as energy efficiency decreases with increasing SEI, but conversion rises [12]. However, although SEI can remain constant when varying both plasma power and residence time simultaneously, a constant SEI does not necessarily result in the same values of CO_2 conversion and energy efficiency. In fact, it has been found that obtaining the same value of SEI using varying combinations of residence time and plasma power can lead to changes in reaction performance [12]. At high residence time and low plasma power, the maximum CO_2 conversion reached can be greater than when a low residence time and high power, but the same SEI, are used [12]. The effect of residence time on the conversion and energy efficiency is therefore greater than the effect of the plasma power. This is due to the length of time CO_2 stays within streamers (longer for a high residence time) being the major influencing factor on conversion, as opposed to high streamer intensity (as a result of high plasma power). At high SEIs, energy efficiency and conversion can therefore be increased simultaneously; however, energy efficiencies will be low even when maximized [12].

Discharge gap can play an important role in determining CO_2 conversion and energy efficiency. At constant SEI, an increase in discharge gap can lead to a decrease in energy efficiency and conversion if the increase is large enough to alter the streamer behavior [12, 20]. Smaller discharge gaps allow more streamers with higher peak currents to form; therefore, an increase in CO_2 dissociation occurs due to the greater effective plasma volume, while the increase in electron density resulting from the higher peak currents leads to a rise in electron impact reaction rates [12]. Alongside this, the average electron energy will be larger for smaller gaps; hence when collisions occur, more energy will be transferred [12].

The influence of discharge length has also been evaluated, with results showing that an increase in both discharge length and discharge power leads to an increased CO_2 conversion [20, 21]. This is due to higher discharge power resulting in an increase in the number of micro-discharges, thereby increasing the number of reaction channels for collisions to occur [20, 22]. However, this only appears to be the case up to a certain point, with one study finding an increase in applied power above 80 W results in a decreased conversion, possibly due to a change in discharge mode from surface discharge to filamentary discharge [19].

Dielectric material and frequency have been found to have no influence on the energy efficiency and CO_2 conversion [12, 20]. Alumina and quartz dielectrics were compared at various SEI values with results showing no significant differences in reaction performance between the two [12]. By contrast, Mei and Tu al reported that the thickness of dielectric materials affects the plasma conversion of CO_2 and energy efficiency using a DBD reactor [20]. Increasing the thickness of a quartz tube from 1.5 to 2.5 mm decreased the CO_2 conversion and energy efficiency of the plasma process by around 15% at a SEI of 120 kJ/L and a constant discharge gap of 2.5 mm [20]. In addition, they found that using a screw-type inner electrode in the DBD reactor significantly enhanced the conversion of CO_2 and energy efficiency compared to the reaction using a rod electrode [20]. The sharp edge of the screw-type electrode could distort the electric field and enhance the local electric field around the inner electrode and consequently generate more intensified filamentary discharge which can also be evidenced by increased amplitude and number of current pulses. This effect could generate more reaction channels for CO_2 conversion and makes a contribution to the enhanced reaction performance [20] (Fig. 9.3).

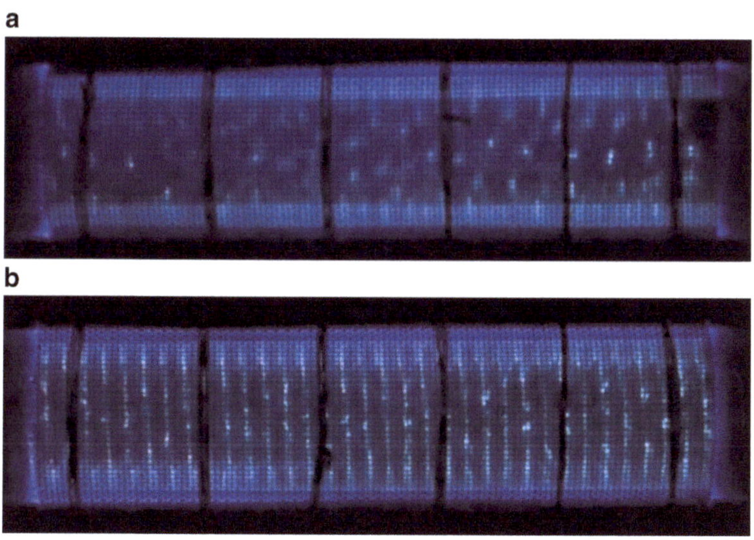

Fig. 9.3 Images of the CO_2 DBD plasma: (**a**) rod inner electrode; (**b**) screw-type inner electrode (discharge power, 40 W; discharge gap, 2.5 mm; discharge length, 100 mm; CO_2 feed flow rate, 25 mL/min; frequency, 9 kHz; outer electrode, stainless steel mesh) [20]

A diluent gas, such as helium or argon, can be used in the feed alongside CO_2. The addition of such gases can lead to a higher absolute CO_2 conversion in a DBD; however, if total feed rate is kept constant, the actual amount of CO_2 converted (effective CO_2 conversion) will be lower in the mixed feed (CO_2 + diluent) than for a pure CO_2 feed due to the decrease in CO_2 present in the feed [23]. The increase in absolute CO_2 conversion in the diluted feed can be explained by the lower breakdown voltage and the increased plasma formation due to an increase in plasma capacity; the addition of helium causes a more homogeneous plasma to form, while argon results in a greater micro discharge filament density [23]. As the threshold energies for inelastic collisions involving He and Ar are much higher than those for CO_2, and electron recombination is favored for CO_2^+ ions over Ar^+ and He^+, a greater fraction of plasma power will go into dissociating CO_2 and less will go into gas breakdown as this can occur at a lower voltage due to an increase in the electron mean free path [23]. As well as increasing conversion, this also leads to an increase in energy efficiency. However, as a large fraction of the input energy goes into exciting the diluent gas, the effective energy efficiency drops in comparison to the pure CO_2 feed [23].

A different effect results when N_2, an impurity present in many waste gas streams, is added to the pure CO_2 feed in DBD plasma. Below 50% N_2, the absolute CO_2 conversion increases as N_2 molecules enhance conversion due to the collision between N_2 metastable molecules and CO_2 resulting in CO_2 dissociation; thus, the effective conversion is tantamount to that of pure CO_2 as the increase in absolute CO_2 conversion cancels out the decrease in CO_2 in the feed [24]. However, above 50% N_2, a greater fraction of the input energy is transferred to N_2 molecules rather than going into CO_2 dissociation; hence effective CO_2 conversion and energy efficiency decrease [24]. The effect of N_2 addition to the feed differs in microwave plasma. Here, the effect on CO_2 conversion and energy efficiency is similar to that found when Ar or He is added to the feed in DBD plasma: absolute CO_2 conversion increases in comparison to pure CO_2 feed; however the reduction in CO_2 concentration results in a lower effective conversion and energy efficiency [25]. Absolute CO_2 conversion increases with a rise in N_2 concentration due to partial conversion of lower CO_2 vibrational levels into higher ones [25]. An important point to note is that on addition of N_2 at all concentrations to both microwave and DBD systems, harmful gases such as N_2O and NO_x are formed [24, 25].

9.2.2 *Influence of Packing and Catalytic Materials*

In current plasma systems, a trade-off exists between CO_2 conversion and energy efficiency [7, 26]. In order to solve this problem and hence create a feasible industrial process, further modification of the plasma system is required. One such modification is the addition of a catalyst into the plasma discharge, as research shows the hybrid plasma-catalytic process can result in higher CO_2 conversion while maintaining low energy consumption [19, 27]. The combination of plasma and

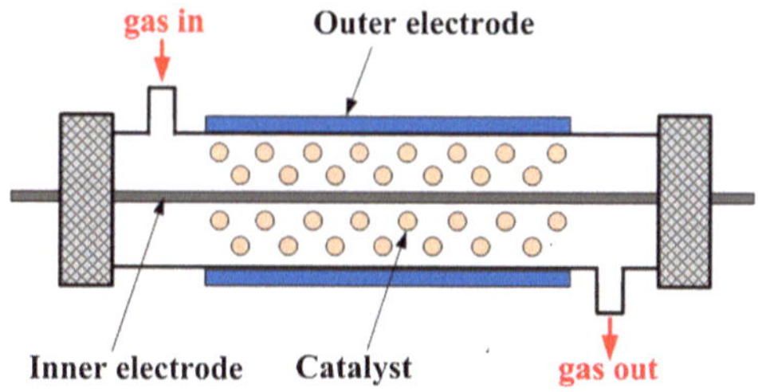

Fig. 9.4 Single-stage setup in a plasma-catalytic DBD reactor

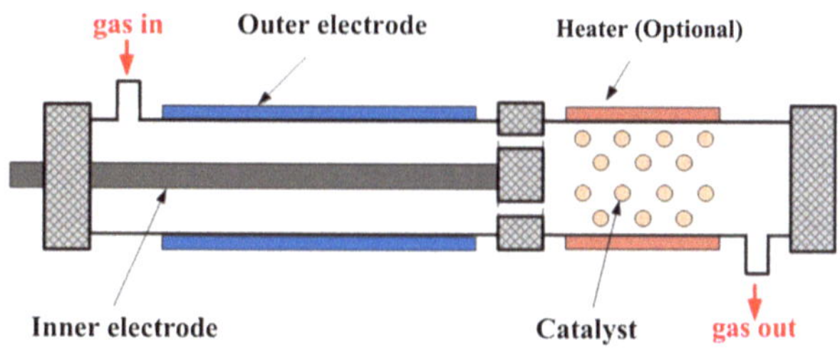

Fig. 9.5 Two stage setup in a plasma-catalytic DBD reactor

catalyst allows the beneficial aspects of each to be realized, along with the effect resulting from their interaction [28, 29]. This can lead to synergy in relation to conversion and efficiency, thus creating a more feasible process for the utilization of CO_2 on an industrial scale. When a catalyst is combined with plasma, the reactions that occur change from purely gas phase to a mixture of gaseous and heterogeneous [11]. The simplest method of combining plasma and catalyst in a single-stage setup is to do so in a DBD reactor (Fig. 9.4), as the catalyst can be placed directly into the discharge without the need for any adjustments in reactor geometry. In this setup, the catalyst is in direct contact with the plasma and can therefore interact with short-lived active species such as excited-state atoms and molecules, reactive radicals, photons and electrons.

In the two-stage setup (Fig. 9.5), a catalyst is placed downstream of the plasma discharge. The catalyst is not in direct contact with the plasma; hence it cannot interact with short-lived excited species, only with the exit gas which contains long-lived intermediates and, possibly, vibrationally excited species.

Catalysts can be incorporated into DBD reactors in a fully or partially packed-bed configuration [30, 31]. By mixing the catalyst and cheap packing material (e.g. glass beads, Al_2O_3 and quartz wool) before placing in the reactor, a packed-bed effect can also be realized without the need for high volumes of costly catalyst in addition to a catalytic effect. This setup also results in a quasi-homogeneous dispersion of the catalyst, which benefits the reaction performance as a greater number of CO_2 molecules will come into contact with the catalyst. The packing material can also lead to an increase in both CO_2 conversion and energy efficiency simultaneously. Zirconia beads with diameters in the range of 1.6–1.8 mm have been shown to increase CO_2 conversion by almost 100% while nearly doubling the energy efficiency in comparison to values obtained in the absence of any packing material [19].

The presence of a packing material will decrease the residence time of CO_2 molecules in the plasma when a fully packed bed is used. As mentioned in Sect.

9.2.2 on the effect of various process parameters, a decreased residence time results in a decrease in CO_2 conversion as there is less time for the feed gases and excited species to interact and hence for CO_2 splitting to occur. It would therefore be assumed that packing material would cause a reduction in CO_2 conversion. However, as CO_2 undergoes adsorption onto the surface of the packing material, this effect may be at least partially compensated for [22, 32].

Both catalysts and packing material will interact with the plasma. Fully packing $BaTiO_3$ and glass beads into the discharge gap have been shown to beneficially modify the discharge mode for the CO_2 decomposition reaction when used as a packing material, resulting in an increase in the average electric field and mean electron energy due to the formation of surface discharges alongside the typical filamentary discharges formed in the absence of a packed bed [22]. The extent of filamentary discharge formation is reduced in a packed-bed reactor as this type of discharge can only be formed within the gaps between the pellets and between the pellets and reactor wall, rather than throughout the whole discharge area; instead, surface discharges form at pellet contact points and across the pellet surfaces [11, 22, 32]. The increase in electron energy and local electric field occurs at the bead contact points as the external electric field causes them to become polarized [19]. Breakdown occurs more readily as the electron temperature is higher than in an empty reactor due to greater acceleration of the electrons in the enhanced electric field [19, 22]. This leads to a more efficient usage of the applied electrical energy which contributes to maximizing the energy efficiency and conversion [19].

The presence of a catalyst can increase both energy efficiency and CO_2 conversion simultaneously [33]. This is partly due to the electron temperature increasing when a catalyst is employed, even though the input power remains constant [33]. As mentioned above, polarization of the dielectric material occurs, enhancing the local electric field which increases the electron temperature [19, 22, 33]. At the contact points, the electron temperature has been shown to increase fourfold in comparison to the electron temperature in an empty reactor [33]. Consequently, there is a more efficient transference of energy from the applied electric power to the electrons in the form of heat [33]. As CO_2 splitting occurs through excitation and dissociation and electron impact ionization when the electrons transfer energy to CO_2, these processes are also more efficient, leading to an increase in energy efficiency in the presence of a catalyst [33]. Recent works have shown that the combination of DBD plasma with a photocatalyst ($BaTiO_3$ and TiO_2) using a partial catalyst packing configuration significantly enhanced the conversion of CO_2 and energy efficiency by up to 250% at low temperatures (~150 °C) compared to the plasma conversion of CO_2 in the absence of a catalyst, as shown in Fig. 9.6 [29]. The presence of the catalyst pellets in the part of the discharge gap has been found to induce plasma physical effects, such as the enhanced local electric field by 10% due to the polarization of the catalytic materials which increases the electron temperature and produces more energetic electrons and reactive species. More importantly, this work has demonstrated that energetic electrons generated by the plasma have acted as the main driving force to activate both photocatalysts for CO_2 conversion, making a major contribution to the enhanced CO_2 conversion and energy efficiency, while the

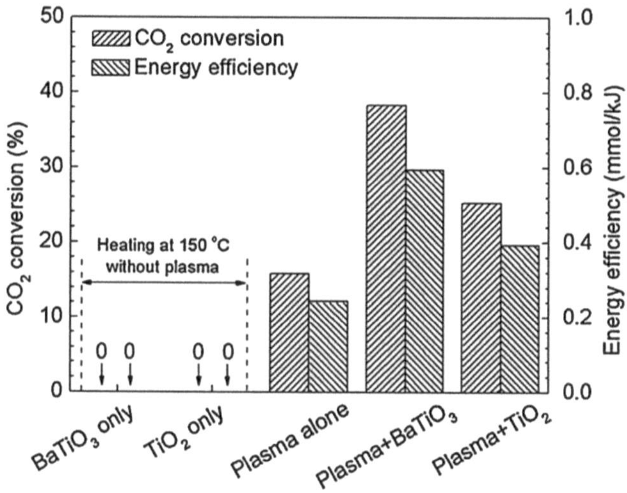

Fig. 9.6 Effect of photocatalysts ($BaTiO_3$ and TiO_2) on plasma conversion of CO_2 at a SEI of 28 kJ/L [29]

UV emission generated by the plasma has played a minor role in the activation of these photocatalysts for CO_2 conversion [29]. Thus, this exceptional synergistic effect resulting from the integration of DBD with photocatalysis for CO_2 conversion at low temperatures can be attributed to both the physical effect induced by the presence of the catalyst in the discharge and the dominant photocatalytic surface reaction driven by energetic electrons from the CO_2 discharge [29].

Furthermore, breakdown voltage is also affected by the packing material. In comparison to the breakdown voltage measured in a DBD reactor in the absence of a packing material (3.43 kV), the presence of glass beads has been shown to roughly halve this value (1.56 kV), while the use of $BaTiO_3$ has been shown to decrease breakdown voltage to less than a third (1.03 kV) [22]. Additionally, the decrease in electron density caused by the use of a dielectric material promotes oxygen radical recombination which in turn impedes the recombination of CO and O to form CO_2 [11, 34]. The reaction performance (CO_2 conversion and energy efficiency) is therefore enhanced due to the physical changes resulting from the presence of the packing material, with CO_2 conversion and energy efficiency increasing by up to 75% for a packed bed of $BaTiO_3$ (albeit not simultaneously) in comparison to no packing material [22]. However, not all packing material gives the same effect; quartz wool is very porous and interacts strongly with plasma, leading to the formation of intense filamentary discharges [22]. A summary of the energy efficiency achieved in different plasma-catalytic systems can be seen in Table 9.1. Also included are the energy efficiencies achieved without the use of a catalyst, in order to ascertain the effect of the catalyst. Clearly, different types of plasma system result in varying energy efficiencies (defined as (Eq. 9.3)) (Table 9.1). Generally, gliding arc and microwave plasmas attain higher energy efficiencies than DBDs;

Table 9.1 Comparison of CO_2 conversion and energy efficiency using different atmospheric pressure plasma sources

Plasma type	Packing materials/catalysts	Maximum CO_2 conversion			Maximum energy efficiency			Ref
		SEI (kJ/L)	CO_2 conversion (%)	Energy efficiency (%)	SEI (kJ/L)	CO_2 conversion (%)	Energy efficiency (%)	
DBD	–	120	27.2	2.8	24	20.0	10.4	[18]
DBD	–	229.0	34.0	1.9	4.3	3.1	8.0	[12]
DBD	–	297.6	46.6	2.0	1.6	2.3	17.7	[35]
DBD	–	60.0	18.2	1.7	20.0	9.6	3.8	[36]
GAD	–	15.4	17.4	14.1	9.8	15.2	19.3	[37]
DBD	–	22.2	24.3	12.6	11.7	16.0	15.7	[38]
DBD	–	240.0	30.0	1.6	45.1	14.0	3.9	[10]
Corona	–	80	10.9	1.7	5.2	3.1	7.5	[39]
Microwave	NiO/TiO_2 (Ar plasma treated)	–	–	–	30	41.3	17.2	[40]
DBD	$BaTiO_3$	–	–	–	28	38.3	16.6	[29]
Packed-bed DBD	CaO	75.8	41.9	5.7	45.5	32.9	7.1	[41]
Packed-bed DBD	ZrO_2	240	42.3	4.6	36	9.6	9.6	[19]
Packed-bed DBD	$CaTiO_3$	52.9	20.5	4.8	32.4	15.8	6.1	[34]
Packed-bed DBD	$BaTiO_3$	60.0	28.2	5.9	24.0	13.7	7.1	[22]

however, DBDs are more easily combined with a catalyst and are more suited to industrial-scale applications.

The size and form of the packing material or catalyst can affect the reaction performance. If bead sizes are too small, the reaction performance can decrease in comparison to the reaction without packing. This is due to the decreased residence time, as well as the number of contact points being too great for the applied electrical energy to cause significant enhancement of the local electric field and the electron energy [19]. In one study that used a DBD reactor, CO_2 conversion generally increased with an increase in ZrO_2 bead size range (from 0.9–1 mm beads to 2–2.24 mm beads) at 80 W input power and various flow rates [19]. However, in another case, 0.18–0.25 mm beads of γ-Al_2O_3, MgO and CaO all resulted in a higher CO_2 conversion in comparison to larger beads with size range 0.25–0.42 mm [11]. The decrease in conversion for the larger beads was attributed to an increased void fraction causing a drop in the electric field strength and discharge area. As can be seen from just these two results with conflicting conclusions, the interactions between plasma and packing material are far from simple and many factors will contribute to reaction performance, making it difficult to predict the outcome for a particular system.

Another aspect which must be taken into account is the role the catalyst or packing material plays in the splitting of the CO_2 molecule via adsorption of the molecule onto the beads. Two types of adsorption can occur: chemical (occurs with materials such as CaO and MgO) and physical (such as for γ-Al_2O_3). Chemisorption leads to a higher CO_2 conversion as CO_2 molecules adsorbed this way more readily decompose [11]. Chemisorption is affected by the acid-base properties of the packing material, with high basicity materials leading to a greater CO_2 conversion as they aid adsorption. This is because highly basic metals are more easily reduced.

The number of surface oxygen vacancies present in the catalyst is a highly important factor for determining CO_2 conversion [29, 42], as dissociative electron attachment is facilitated by oxygen vacancies; hence a high number of vacancies can lead to an increase in CO_2 conversion (Fig. 9.7). Recently, Mei and his co-workers have shown that the presence of oxygen vacancies on the surface of $BaTiO_3$ and TiO_2 photocatalysts contributes to the enhanced CO_2 conversion in comparison to the plasma reaction without a catalyst [29]. They found that more oxygen vacancies were formed in $BaTiO_3$ than in TiO_2, resulting in the higher CO_2 conversion using $BaTiO_3$ in the plasma-catalytic conversion of CO_2 [29]. Various pretreatments of catalysts can also be used for the synthesis of a catalyst with a large number of oxygen vacancies. One study used plasma pretreatment, in which three different gases (CO_2, Ar and O_2) were used to treat NiO/TiO_2 catalysts [42]. Both O_2 and CO_2 pretreatments failed to result in a catalyst with high affinity for CO_2 decomposition; however, the Ar pretreatment led to a catalyst that increased both the energy efficiency and CO_2 conversion by a factor of 2 in comparison to the plasma-alone process. This difference in reaction performance between the catalysts prepared using different gases was attributed to the increase in the number of oxygen vacancies in the Ar-treated catalyst. This is because dissociative electron attachment occurs at the oxygen vacancy sites as CO_2 is adsorbed more easily here than on

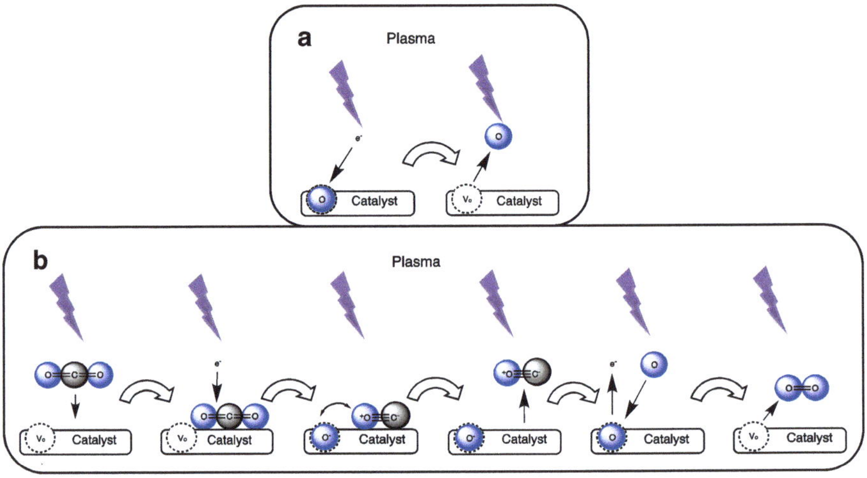

Fig. 9.7 (**a**) Generation of oxygen vacancies (V_o) at the catalyst surface via bombardment with plasma-generated electrons; (**b**) CO_2 dissociation mechanism via dissociative electron attachment at the catalyst surface

defect-free sites [42]. CO_2 can then be dissociated to CO and O^-, through the transient CO_2^- ion, due to electrons from the plasma colliding with the molecule. The CO molecule is then desorbed from the active site and the O^- ion releases an electron as it fills the oxygen vacancy on the catalyst surface [42]. Furthermore, the oxygen vacancies are regenerated, preventing any decrease in catalytic activity [42]. This could be due to a gaseous oxygen atom in the plasma recombining with surface-adsorbed O_2, according to (9.R6). As oxygen atoms in the plasma can also be in excited states, the recombination process may occur to an even greater extent due to this enhanced energy state [42]:

$$O_{ads} + O(g) \rightarrow O_2(g) \tag{9.R6}$$

9.3 Plasma Dry Reforming of CH_4 with CO_2

$$CO_2 + CH_4 \rightarrow 2CO + 2H_2 \tag{9.R7}$$

Dry reforming of methane with CO_2 (9.R7) has the benefit of utilizing two greenhouse gases in the form of different sources (e.g. landfill gas, biogas and shale gas) in a single process. This process usually produces syngas, a mixture of hydrogen and carbon monoxide, alongside other valuable chemicals and fuels. Syngas is a vital chemical intermediate that can be used to produce a variety of chemicals and fuels, including via the Fischer-Tropsch process. Higher hydrocarbons, such as C_2H_2, C_2H_4, C_2H_6 and C_3H_8, can also form from the dry reforming reaction, although the concentration of these is always much smaller than that of syngas [43–45].

Very recently, Wang et al. have developed a water electrode DBD plasma reactor for the direct, one-step reforming of CH_4 with CO_2 into higher-value liquid fuels and chemicals (e.g. acetic acid, methanol, ethanol and formaldehyde) with high selectivity at room temperature (30 °C) and atmospheric pressure (Fig. 9.8) [46]. The total selectivity to oxygenates was approximately 50–60%, with acetic acid being the major liquid product at 40.2% selectivity. Two possible reaction pathways could contribute to the formation of acetic acid in this process. CO can react with a CH_3 radical to form an acetyl radical (CH_3CO) with a low energy barrier of 28.77 kJ/mol, followed by recombination with OH to produce acetic acid with no energy barrier. Direct coupling of CH_3 and carboxyl radicals (COOH) could also form acetic acid based on density functional theory (DFT) modeling. A few groups have also found the formation of trace oxygenates (e.g. alcohols and acids) as by-products of syngas production in plasma dry reforming of methane. In a DBD reactor, acetic, formic, butanoic and propanoic acids have all successfully been formed, along with methanol and ethanol [43]. There are a number of pathways through which formic acid and propanoic acid could form. The most likely pathway is the addition of CO to an ethyl radical, although an ethyl radical may also couple with a carboxyl radical; furthermore, the carboxyl radical could couple with a hydrogen radical to form formic acid. The carboxyl radicals required for acid formation are thought to result from the addition of CO and OH [43].

Carbon nanomaterials are also possible by-products of the dry reforming reaction (Fig. 9.9). Multiwall carbon nanotubes and spherical carbon nanoparticles have been formed in a gliding arc reactor [4]. These are important by-products as carbon

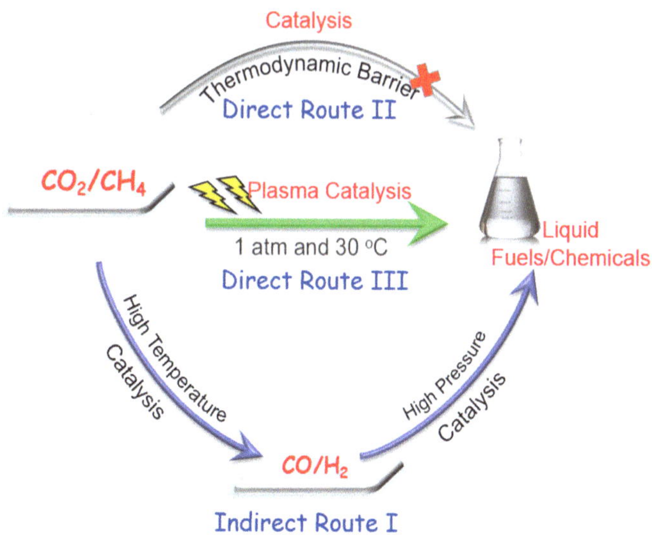

Fig. 9.8 Direct and indirect dry reforming of methane to liquid fuels and chemicals [46]

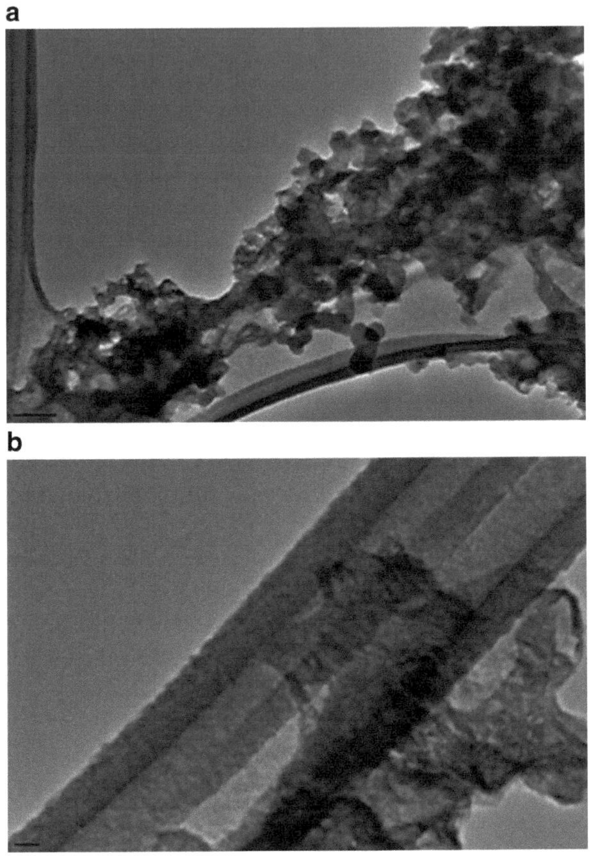

Fig. 9.9 TEM images of carbon nanomaterials formed in the plasma dry reforming of CH_4 and CO_2 using a gliding arc plasma: (**a**) spherical carbon nanoparticles and (**b**) carbon nanotubes [4]

nanotubes have a variety of applications; this could prove a sustainable and energy-efficient method for their production if further research is focused in this area [4].

9.3.1 Influence of Process Parameters

The influence of a wide range of operating parameters on the energy efficiency and conversions of reactants has been investigated using different plasma systems. The reactant conversion and the H_2/CO molar ratio, along with the product yields and selectivities, are affected by the molar ratio of CO_2/CH_4 in the feed [47, 48]. An increase in CO_2 content in the feed leads to a rise in both energy efficiency and total conversion [5, 49, 50]. This is because CO_2 and CH_4 decomposition occur via electron impact dissociation, forming O and H atoms, respectively. These atoms in turn react with each other to form OH, thus limiting the recombination of CH_3 and H and increasing CH_4 conversion [5].

One important factor that determines reaction performance is the number of micro-discharge filaments that come into contact with the gas molecules in the DBD, with a greater number of lower-energy filaments resulting in an increase in energy efficiency and conversion [5]. It has been reported that duty cycle affects CO_2 and CH_4 conversion in a DBD, with conversion increasing with duty cycle, when a sinusoidal voltage with square wave modulation is applied [43]. Pulsed plasma has been shown to effectively enhance the performance of the dry reforming process, with an increase in pulse frequency or applied peak voltage leading to higher total conversion [48, 51].

An increase in plasma power results in higher total conversion as a greater number of higher-energy electrons are formed, which can go on to initiate reactions [5]. However, input power can also affect product distribution, with a greater concentration of carbon powder and water forming at higher power. The use of a longer residence time can also increase total conversion, as the SEI is increased; however, it bears almost no impact on CO and H_2 selectivities [5, 48]. The SEI term incorporates both power and residence time, and if both parameters are adjusted but the SEI remains constant, the change in total conversion will be negligible [5]. However, an increase in SEI leads to higher CO_2 and CH_4 conversions in a variety of plasmas (gliding arc, DBD, corona), but has the opposite effect on the energy efficiency, resulting in a trade-off between the two [5].

Optical emission spectroscopic diagnostics has been used to understand the formation of a wide range of reactive species generated in the reforming process [27], while plasma chemical kinetic modeling has been used to understand the underlying plasma chemistry and reaction pathways of the dry reforming process [45]. The latter model demonstrates how selectivity towards different products can be achieved through manipulation of the residence time due to the spatially averaged densities of some molecules continually increasing with residence time, while others peak at a certain value [45].

In order to maximize reactant conversion and energy efficiency, the process parameters must be optimized. This is a difficult task as each plasma system will have a different set of optimal parameters and numerous experiments must be carried out to realize these. Alternatively, a modeling approach can be used. Some models have now shown good agreement with experimental results, even though they are much simplified versions of the actual plasma chemistry. As mentioned previously, a trade-off exists between energy efficiency and reactant conversion; hence a middle ground must be found that results in a viable process.

9.3.2 Catalytic Reforming Versus Plasma-Catalytic Reforming

As shown by 9.R7, dry reforming is a highly endothermic process. As such, high temperatures are required (>1000 °C) in the thermal process to overcome the stability of both the CO_2 and CH_4 molecules. A catalyst can be used to convert the reactants at lower temperatures; however, reactions temperatures must still be above 700 °C [47, 52]. Carbon deposition can also occur on the catalyst surface, leading to catalyst deactivation [47, 52]. Both these issues incur cost, as the energy input must be high and catalysts must be replaced periodically. In order to overcome these issues, catalysts can be combined with NTP. In the plasma-catalytic system, the benefits of using NTP which can overcome the high stability of the CO_2 and CH_4 molecules at low temperature and atmospheric pressure, along with those of catalysis (reduction in activation barrier, increased selectivity and conversion), can be realized simultaneously. This can increase reaction performance [53, 54], as well as reducing costs. Furthermore, interactions occur between the catalyst and the plasma which can lead to a synergistic effect in terms of conversion and energy efficiency [47, 55]. However, coking can still be an issue in plasma-catalytic systems, although this may be dampened through removal by excited hydrogen species [53]. One method to overcome this is to flow pure CO_2 through the reactor to remove the deposited carbon [53]. The oxidation of carbon by CO_2 occurs much faster in a DBD reactor than a thermal one [53].

Thermal-catalytic and plasma-catalytic dry reforming differ in the reagent conversion ratio. Conversions of both CO_2 and CH_4 should be equal due to the stoichiometry of the reaction; however, as a result of the reverse water-gas shift reaction, in the thermal process, a higher conversion of CO_2 occurs in comparison to CH_4. This differs to the plasma reaction, in which the CH_4 conversion is higher. This can be attributed to the prevalence of gas-phase reactions that lead to the dissociation of the CH_4 molecule (9.R8, 9.R9, and 9.R10), as well as the production of CO_2 via 9.R11 [47]:

$$CH_4 + e \rightarrow CH_3 + H + e \tag{9.R8}$$

$$CH_4 + e \rightarrow CH_2 + H_2 + e \tag{9.R9}$$

$$CH_4 + e \rightarrow CH + H + H_2 + e \tag{9.R10}$$

$$CO_2{}^{+} + CH_4 \rightarrow CH_4{}^{+} + CO_2 \tag{9.R11}$$

As mentioned above, in the plasma-catalytic dry reforming reaction, plasma reactions occur in the gas phase to dissociate CH_4 (9.R8, 9.R9, 9.R10, and 9.R11) and CO_2 (9.R12) [51]:

$$CO_2 + e \rightarrow CO + O + e \tag{9.R12}$$

Active species created in the plasma can also adsorb onto the catalyst surface, from where they can form CO and H_2 products [55]:

$$CH_x + \square \rightarrow (CH_x)_{ad} \tag{9.R13}$$

$$H + \square \rightarrow H_{ad} \tag{9.R14}$$

$$O + \square \rightarrow O_{ad} \tag{9.R15}$$

$$(CH_x)_{ad} + O_{ad} \rightarrow CO_{ad} + xH_{ad} \quad (9.R16)$$

$$CO_{ad} \rightarrow CO + \square \quad (9.R17)$$

$$2H_{ad} \rightarrow H_2 + \square \quad (9.R18)$$

where □ is an empty adsorption site present on the catalyst surface. The simultaneous occurrence of both plasma and surface reactions in plasma-catalytic systems at low temperature can lead to synergy in terms of product selectivities and reactant conversions, which cannot occur in the catalytic process at low temperature due to insufficient energy input. Desorption of the species on the surface of the catalyst can also occur more readily in the plasma process as the plasma can affect the catalyst properties [55]. Thus, the plasma-catalytic process is more beneficial due to increased CO_2 and CH_4 conversion and product selectivities and a reduction in coking at lower energy input than the thermal catalytic process.

9.3.3 *Influence of Catalyst*

In the thermal catalytic dry reforming process, supported metal catalysts have been used widely, with non-noble metals being prevalent due to their low cost and wide availability. Catalysts with high activity for the thermal catalytic process have been used as the starting point for the plasma-catalytic process. Supported metal catalysts with a core and shell structure have been investigated as this allows the active sites to be uniformly distributed, while deactivation due to carbon deposition and sintering is kept to a minimum as a result of the strong interactions between the core and shell [56]. Zeolite 3A [57], NaX and NaY [58, 59], Ni/γ-Al_2O_3 [27, 30, 47, 60–62], Co/γ-Al_2O_3 [47], Mn/γ-Al_2O_3 [47], Ag/Al_2O_3 [63], Pd/Al_2O_3 [63, 64], Cu/Al_2O_3 [47, 64], Fe/Al_2O_3 [65], La_2O_3/γ-Al_2O_3 [66], $LaNiO_3$ [67], Cu-Ni/Al_2O_3 [55] and $LaNiO_3$@SiO_2 [56, 68] catalysts have all been tested in the plasma-catalytic process, with Ni/γ-Al_2O_3 being the most commonly used. More recently, K-, Mg- and Ce-promoted Ni/γ-Al_2O_3 catalysts have also been evaluated in plasma-catalytic dry reforming of methane and CO_2 [69]. However, the scope of catalysts available that are active for the thermal process has only just been touched upon for plasma dry reforming.

In the plasma-catalytic system, the catalyst structure can be influenced by the plasma, while the presence of the catalyst can affect the discharge properties [70, 71]. As a result of the interactions between the plasma and the catalyst, a synergistic effect can result, whereby the reaction performance is greater than the sum of the plasma-alone and the purely catalytic processes [27, 55]. The formation of radicals in the plasma can change the catalysts reaction mechanism as these species are adsorbed onto the catalyst surface, while adsorbed vibrational excited species can facilitate CO_2 and CH_4 dissociation through dissociative adsorption due to their high internal energies [55, 70]. It may also be possible that the catalytic activity is improved due to the charged particles on the catalyst surface and the applied voltage acting similarly to the electrochemical promotion of catalysis [70]. Other plasma influences include changing the lattice structure of the catalyst due to the transference of thermal energy from ions and electrons causing thermal perturbation which can increase coking resistance and catalyst activity, as well as structural changes due to particle bombardments which lead to changes in catalyst pores, active metals and promoters [70]. One such change that can occur is in the crystallinity of the active metal and support due to variation in valence state, resulting in an increase in oxygen affinity which leads to greater conversion of CH_4 [70].

The changes discussed this far are a result of the plasma changing the properties of the catalyst. As mentioned above, the catalyst can also affect the plasma properties [53, 70]. Catalysts change the discharge mode to a combination of micro-discharge and surface discharge [71]. An increase in the local electron density due to energy concentrating in the gaps between the catalyst pellets results in an increase in reactions occurring in both the plasma and at the catalyst surface as the electric field is enhanced [70]. The dielectric constant of the catalyst also affects the plasma, with a high dielectric constant catalyst resulting in an increase in the plasma electric field [53, 70, 71]. High dielectric catalysts, such as ferroelectrics, can increase the production of syngas in the dry reforming reaction [70]. The catalyst properties, such as dielectric constant, and geometry are therefore highly important in determining reaction performance [54, 72]; thus, the reaction performance can be optimized through selection of catalyst.

Packing geometry can also influence the interactions between the plasma and the catalyst; in a DBD reactor, partially packing a Ni/γ-Al_2O_3 catalyst into the discharge gap results in an enhancement in reaction performance in comparison to a fully packed reactor [27, 30]. This is because the discharge in the partially packed reactor retains the strong filamentary discharge, whereas the reduction in discharge volume in the fully packed reactor changes the discharge mode to surface discharge and spatially limited micro-discharge [27].

In a DBD reactor, the use of a Ni/γ-Al_2O_3 catalyst has been shown to enhance the conversion of CH_4, along with the yield of CO and H_2, in comparison to the plasma-alone process; however, the CO_2 conversion decreased slightly upon addition of the catalyst [47]. This Ni/γ-Al_2O_3 catalyst resulted in higher H_2 and CO yields and CH_4 conversion than the Co/γ-Al_2O_3, Cu/γ-Al_2O_3 and Mn/γ-Al_2O_3 catalysts that were also tested in this reactor, with a maximum CH_4 conversion of 19.6% being achieved at a flow rate of 50 ml/min and 7.5 W discharge power [47]. Although the energy efficiency of the plasma reaction is not always increased by the addition of a catalyst, such as is the case when Co/γ-Al_2O_3 or Cu/γ-Al_2O_3 catalysts are used, both Ni/γ-Al_2O_3 and Mn/γ-Al_2O_3 catalysts were found to enhance the energy efficiency [47].

Energy efficiency is higher in gliding arc discharge in comparison to other types of discharge, and catalysts can increase this still further [54]. The use of a NiO/Al_2O_3 catalyst, placed in the afterglow of the discharge in a gliding arc reactor, was found to increase energy efficiency by over 20% in comparison to that achieved using plasma only [54]. H_2 yield, along with CO_2 and CH_4 conversions, was also increased. The concentration of active metal was found to influence reaction performance, as a 33wt% NiO/Al_2O_3 catalyst resulted in a decrease in reaction performance in comparison to an 18wt% NiO/Al_2O_3 catalyst, while a smaller catalyst diameter was found to be beneficial [54].

The addition of dopants and use of bimetallic catalysts have also been studied. Zhang et al. investigated the effect of varying the concentrations of Cu and Ni in various Cu-Ni/γ-Al_2O_3 catalysts and found that the 12 wt% Cu-12 wt % Ni/γ-Al_2O_3 catalyst gave the optimum results for both CH_4 and CO_2 conversion (Fig. 9.10) [55]. This catalyst also achieved the maximum selectivity to CO of 75%. However, this selectivity was also achieved when using the 5 wt% Ni-12 wt% Cu/γ-Al_2O_3 catalyst, whereas the maximum selectivity to H_2 was achieved with 16 wt% Ni-12 wt % Cu/γ-Al_2O_3 and 20 wt% Ni-12 wt% Cu/γ-Al_2O_3 catalysts [55].

Another factor which must be taken into consideration is the catalyst support, as the support, along with the interactions between it and the active metal, can affect the reaction performance. A study completed by Mei et al. investigated the use of a Ni catalyst supported on γ-Al_2O_3, TiO_2, MgO and SiO_2 [73]. The results of this experiment concluded that the γ-Al_2O_3 support was most beneficial on the reaction performance, giving the highest CO_2 (26.2%) and CH_4 (44.1%) conversions, as well as the maximum achieved energy efficiency and highest yields of CO and H_2. This was attributed to the increased reducibility of the Ni/γ-Al_2O_3 catalyst and the number of stronger basic sites present at its surface (which facilitate CO_2 chemisorption and activation), along with its higher specific surface area and greater dispersion of smaller NiO particles [73]. Carbon deposition also occurred to a lower extent on this catalyst, as the increase in CO_2 chemisorption and activation may have resulted in adsorbed CO_2 undergoing gasification by surface-adsorbed oxygen [73]. Weaker interactions between the catalyst and support are favorable as this increases the reducibility of the catalyst, increasing its activity [27].

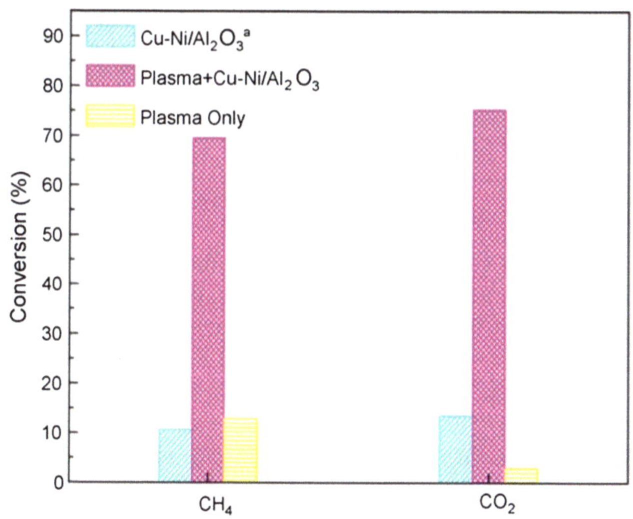

Fig. 9.10 Conversion of CH_4 and CO_2 in the dry reforming reaction using three different processes (plasma only, catalysis only and plasma catalysis) [55]

9.4 Plasma CO_2 Hydrogenation

A particularly significant route being developed for CO_2 conversion is CO_2 hydrogenation, which has a lower thermodynamic limitation compared to direct CO_2 decomposition and dry reforming of methane. Carbon dioxide can be hydrogenated in plasma via reaction with hydrogen at atmospheric pressure, thus avoiding the use of high pressure required by conventional thermal catalytic processes. CO_2 methanation (9.R20) and the reverse water-gas shift (RWGS) reaction (9.R19) prevail when hydrogen is reacted with CO_2 in the plasma process [74]; however, it is also possible to produce higher-value chemicals and fuels such as methanol and ethanol [75]. The main barrier to this process is the source of hydrogen. In order for this process to be both economically viable and sustainable, hydrogen must be produced using a low-cost, environmentally friendly and sustainable process. Currently, coal gasification and steam reforming of methane dominate the production pathways of H_2, leading to the emission of CO_2 [76]. Due to this, CO_2 conversion using hydrogen must convert a greater amount of CO_2 than hydrogen production pathways generate [76]. In thermal catalysis, the production of methane from H_2 and CO_2 is not considered a viable method of fuel production due to the low energy per unit volume and high H_2 consumption [77]. As plasma processes have the potential to convert large amounts of CO_2 at high energy efficiencies, interest is increasing into the development of these systems. In comparison to thermal CO_2 hydrogenation which requires high temperature and high pressure (30–300 bar), plasma systems operate at room temperature and pressure, hence increasing their viability. If plasma systems can be combined with a sustainable source of hydrogen, such as from water splitting using renewable energy, or indeed be used to split the hydrogen source in situ, this could prove a vital pathway for CO_2 mitigation.

9.4.1 *CO_2 Hydrogenation to CO*

$$CO_2 + H_2 \rightarrow CO + H_2O \tag{9.R19}$$

The conversion of CO_2 in this reaction has been found to rise as the H_2 content in the feed is increased. A higher ratio (H_2/CO_2) can also increase CO selectivity and yield, with a ratio of 4:1 resulting in a slight increase in CO selectivity and a threefold increase in CO yield in comparison to a feed ratio of 1:1 in a DBD [74]. Furthermore, this increase in feed ratio also results in a rise in CO production efficiency.

The selectivity to CO has been found to increase with a rise in total flow rate [78]. This is most likely due to the decreased residence time associated with an increase in flow rate resulting in the recombination of CO and O being suppressed, along with the further hydrogenation of CO to form hydrocarbons. This hypothesis is supported by the decrease in selectivity to CH_4 that occurs as the total flow rate is raised, as detailed below.

In order to enhance the production of CO, a catalyst can be added into the reactor. In a DBD reactor, it was found that the addition of a Mn/γ-Al_2O_3 catalyst leads to an increase (of 114%) in CO yield in comparison to the plasma-alone process, as well as an increase (of 116%) in CO production efficiency [74].

9.4.2 CO_2 Hydrogenation to CH_4

$$CO_2 + 4H_2 \rightarrow CH_4 + 2H_2O \tag{9.R20}$$

In this process, a higher H_2 content in the feed in comparison to CO_2 is desirable as this increases the conversion of CO_2. This has been determined both experimentally [74, 79] and through the use of a 1D fluid model [76]. A 3:1 ratio (H_2/CO_2) is optimal for enhancing CH_4 yield [74]. Optimizing the total flow rate can also maximize the CH_4 selectivity and CO_2 conversion. A very low total flow rate can lead to reverse reactions occurring, reforming CO from CH_4 according to 9.R21, due to the longer residence time increasing the interactions between the CO_2 hydrogenation products and the excited species in the plasma [78]:

$$CH_4 + H_2O \rightarrow CO + 3H_2 \tag{9.R21}$$

As the total flow rate is increased, the rate of formation of CH_4 becomes larger than the rate of the reverse reaction, leading to an increase in CH_4. At high total flow rates however, the residence time is too low for reagent gases to interact with excited plasma species, decreasing the overall production of CH_4 as both the forward and reverse reactions occur to a lesser extent [78].

For DBD plasmas, the use of an alumina reactor instead of quartz is beneficial on reaction performance due to the enhanced relative dielectric permittivity coefficient of alumina [79].

Addition of a magnetic field can enhance CO_2 conversion, increasing the CH_4 selectivity by over 10% at a discharge power of 30 W while also tripling the energy efficiency of the process [80]. This study however employed low pressure (200 Pa), reducing simplicity of design and requiring extra energy input, thus detracting from the benefits of NTP. The function of the magnetic field is to prevent electrons from diffusing to the downstream region. In the downstream region, CH_4 is produced but also decomposed through electron impact dissociation (Eq. 9.4), leading to a decrease in the yield of CH_4 [80]. Magnetized electrons cannot travel out of the magnetic field; hence the downstream recombination reaction is suppressed, while CH_4 production can still proceed via reactions involving neutral radicals as these are not magnetized. The magnetic field also increases the electron density, due to confinement of the electrons in the magnetic field, which leads to an increase in the CO_2 decomposition reaction as well as the energy efficiency [80].

An increase in power input generally results in a higher selectivity to CH_4 due to the increase in power density this results in [79, 80]. However, it has been found that at high-power input (>160 W), energy is transferred to the electrodes through heating rather than being used for plasma production, resulting in no further increase in CH_4 selectivity [80]. A rise in voltage also causes an increase in plasma density, which again leads to an increase in CH_4 selectivity. In radio frequency (RF) discharge, the relationship between CH_4 concentration and voltage can be expressed as [78]:

$$[CH_4] \propto (V)^N \ (N \geq 2) \tag{9.4}$$

Hence at higher voltages, the production efficiency of CH_4 increases. A study into the CH_4 production dependence on the repetition frequency of a low-pressure RF discharge has shown that CH_4 yield increases with repetition frequency [78]. The production of CO increases linearly with repetition frequency (up to its saturation point) due to an increase in the number of electrons as CO is formed via electron impact dissociation of CO_2 [78]. CH_4 is formed via reaction between CO and H; hence there is a power law relationship between the formation of CH_4 and the repetition frequency (and number of electrons) [78].

A smaller discharge gap is beneficial on the CO_2 conversion and CH_4 selectivity due to the rise in input power density caused by enhancement in the electric field [80]. In fact, a smaller discharge gap can achieve the same CH_4 selectivity at a lower input power than when using a larger discharge gap [80]. A reduction in discharge gap can also increase the production efficiency of the system. It must be noted here that for the case of [80], a decrease in discharge gap also resulted in an increase in magnetic field.

As expected, an increase in power input results in a decreased energy efficiency [80]. However high-power inputs give rise to larger conversions. Catalysts have therefore been employed as a method to combat this trade-off. The use of $Mn/\gamma\text{-}Al_2O_3$ and $Cu/\gamma\text{-}Al_2O_3$ catalysts in a coaxial DBD reactor has been found to increase CO_2 conversion as well as the energy efficiency of both CH_4 and CO production in comparison to the plasma-alone process [74]. The $Cu/\gamma\text{-}Al_2O3$ catalyst was found most beneficial for the production of CH_4 as this catalyst achieved the highest yield and selectivity to CH_4, while the maximum CO_2 conversion was achieved using the $Mn/\gamma\text{-}Al_2O_3$ catalyst. However, the $Mn/\gamma\text{-}Al_2O_3$ catalyst resulted in a decrease in CH_4 selectivity in comparison to the process in the absence of a catalyst. The decrease in CO_2 conversion attained when using the $Cu/\gamma\text{-}Al_2O_3$ catalyst in comparison to the $Mn/\gamma\text{-}Al_2O_3$ catalyst may be attributed to the increased prevalence of the water gas shift reaction in the presence of this catalyst as Cu catalysts are often used for catalyzing this reaction; hence the apparent CO_2 conversion will be reduced [74]. It is therefore important to select a catalyst that will suppress the water gas shift reaction and simultaneously increase the CO_2 conversion and the selectivity to CH_4.

The combination of plasma and catalysis allows the CO_2 methanation reaction to occur at much lower temperatures than those required in the purely catalytic process [81]. Below 250 °C, negligible CO_2 conversion occurs for the catalytic process; however, when combined with plasma, the CO_2 conversion reaches 80%, with 90% selectivity to CH_4, for the addition of a Ce-Zr supported Ni catalyst in a DBD reactor [81]. This is due to the creation of excited species in the plasma, which generate new pathways for CO_2 dissociation; hence the reaction is not limited by the rate of CO_2 dissociation at the catalyst surface as it is in the purely catalytic process [82]. The use of nickel-containing hydrotalcite catalysts has also shown promise in the plasma-catalytic CO_2 methanation reaction, with a CO_2 conversion of 80% and selectivity to CH_4 of nearly 100% having been achieved in a DBD reactor [83]. It is thought that the high number of low- and medium-strength basic sites is responsible for the high activity of this catalyst, as when promoted with metals containing high-strength basic sites (Ce and Zr), the conversion and yield decrease and no other important morphological changes could be identified [83].

9.4.3 *Pathways of CO and CH_4 Formation*

In order for methane to form, CO_2 is first dissociated to CO. Reaction between CO and H_2 can then occur to form CH_4. However, the oxygen radical produced in the dissociation of CO_2 will compete to react with hydrogen, forming water [84].

The dominating pathway for production of CO occurs via electron impact dissociation of CO_2 (9.R12). At low CO_2 concentrations in the feed, electron impact dissociation is also the main pathway for the dissociation of H_2 (9.R22); however, at high CO_2 concentrations, H_2 is mostly consumed through reaction with H_2O^+ and H_3O^+ [76]:

$$H_2 + e \rightarrow H_2{}^* \rightarrow H + H \tag{9.R22}$$

Dissociation of CO can also occur; however, this reaction is highly endothermic; hence it occurs to a much lesser extent [78].

The net loss rate of CO_2 remains constant at all inlet concentrations, but the net loss rate of H_2 varies. At high H_2 concentrations, the net loss rate is high as there is more H_2 in the feed; as the H_2 concentration is decreased, the net loss rate follows the same trend [76]. The net loss rate of H_2 is higher than that of CO_2, with this effect being more pronounced at high H_2 concentrations.

Downstream of the reactor, electron energies have decreased and are usually insufficient for dissociation reactions. Recombination reactions prevail (9.R23 and 9.R24) as these exothermic reactions only require low-energy electrons [78]:

$$CO + 3H_2 \rightarrow CH_4 + H_2O \tag{9.R23}$$

$$CO + 2H_2 \rightarrow CH_3OH \qquad (9.R24)$$

However, the dissociation of CH_4 can occur via electron impact dissociation (9. R25) [80], reducing the obtained yield:

$$CH_4 + e \rightarrow CH_x + (4 - x)H + e(x = 1, 2, 3) \qquad (9.R25)$$

The water gas shift (WGS) reaction (9.R26) can also occur, decreasing the apparent CO_2 and H_2 conversions [74]:

$$CO + H_2O \rightarrow CO_2 + H_2 \qquad (9.R26)$$

In this simplified summary of reaction pathways, it can therefore be seen that the formation of CH_4 occurs via CO formation, which forms as a result of CO_2 dissociation. Methanol is also a possible product. Unwanted reactions also occur in the plasma system, such as the water gas shift reaction and the dissociation of CH_4; hence these reactions must be suppressed to ensure high yields of CO and CH_4.

9.4.4 CO_2 Hydrogenation to Liquid Fuels

CO_2 hydrogenation to liquid fuels (e.g. methanol, ethanol and dimethyl ether (DME)) is one of the most attractive routes for CO_2 conversion and utilization (Fig. 9.11). Significant efforts have been concentrated on CO_2 hydrogenation to methanol (9.R27) using heterogeneous catalysis at high pressures [85]. CH_3OH is a valuable fuel substitute and additive and is also a key feedstock for the synthesis of other higher-value chemicals. In addition, methanol is considered a promising hydrogen carrier, suitable for storage and transportation [85]:

$$CO_2 + 3H_2 \rightarrow CH_3OH + H_2O \qquad (9.R27)$$

Cu-based catalysts have attracted considerable interest for catalytic hydrogenation of CO_2 to methanol, owing to the excellent activity of metallic Cu for this reaction. Extensive efforts have also been devoted to modifying the structure of Cu-based catalysts using various supports (Al_2O_3, ZnO, ZrO_2, SiO_2, Nb_2O_5, Mo_2C and carbon materials, etc.), promoters (Zn, Zr, Ce, Ga, Si, V, K, Ti, B, F and Cr) and preparation methods [85–87].

Up until now, very limited research has concentrated on CO_2 hydrogenation using nonthermal plasmas, either with or without a catalyst [88–91]. The majority of this research reports CO as the dominant chemical, with CH_4 formed as a minor product and no CH_3OH detected [80–82]. In the late 1990s, Eliasson and co-workers investigated CO_2 hydrogenation to CH_3OH using a DBD plasma reactor [92]. However, only trace amounts of CH_3OH were produced, with a maximum CH_3OH yield

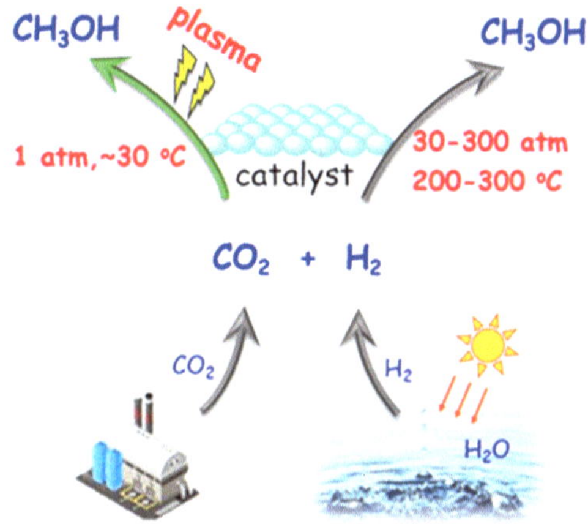

Fig. 9.11 Scheme of CO_2 hydrogenation to methanol [75]

of 0.2% obtained at atmospheric pressure (1 bar), a relatively high plasma power of 400 W, a total flow rate of 250 ml/min and a H_2/CO_2 molar ratio of 3:1. They also found that packing a $Cu/ZnO/Al_2O_3$ catalyst (a commercial methanol synthesis catalyst) in the discharge increased the methanol yield (from 0.1 to 1.0%), methanol selectivity (from 0.4 to 10.0%) and CO_2 conversion (from 12.4% to 14.0%) at a higher pressure (8 bar) under similar operating conditions [92]. However, the methanol yield and selectivity were still significantly lower than those reported in catalytic CO_2 hydrogenation processes. The formation of trace CH_3OH in plasma CO_2 reduction was also reported using a radio frequency impulse discharge at low pressures (1–10 torr) [78]. Very recently, Wang et al. developed a specially designed water electrode DBD reactor for the highly selective hydrogenation of CO_2 to methanol at room temperature (30 °C) and atmospheric pressure [75]. They found that the methanol production was strongly dependent on the structure of the DBD reactor; the DBD reactor with a special water electrode design and a single dielectric showed the highest reaction performance in terms of the conversion of CO_2 and methanol yield (Fig. 9.12) [75].

The combination of the plasma with $Cu/\gamma\text{-}Al_2O_3$ or $Pt/\gamma\text{-}Al_2O_3$ catalyst significantly enhanced the CO_2 conversion and methanol yield compared to the plasma hydrogenation of CO_2 without a catalyst. The maximum methanol yield of 11.3% and methanol selectivity of 53.7% were achieved over the $Cu/\gamma\text{-}Al_2O_3$ catalyst with a CO_2 conversion of 21.2% in the plasma-catalytic process, while no reaction occurred at ambient conditions without using plasma [75]. Compared to catalytic CO_2 hydrogenation to methanol, which has been carried out using a wide range of catalysts, very limited catalysts that are active for thermal catalytic process have been examined for plasma hydrogenation of CO_2. In addition, the production of dimethyl ether from plasma CO_2 hydrogenation was reported using an atmospheric pressure surface discharge with a CO_2 conversion of 15% and a H_2/CO_2 molar ratio of 1:1 [91].

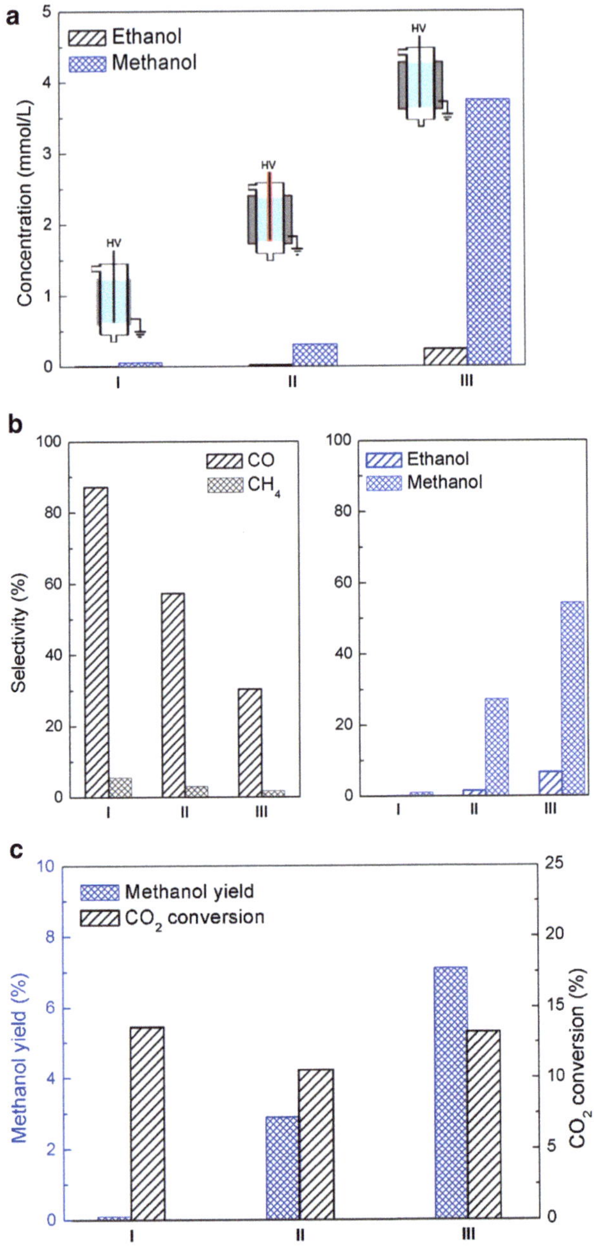

Fig. 9.12 Influence of reactor structure (reactor I, II and III) on plasma hydrogenation of CO_2 process: (**a**) concentration of oxygenates; (**b**) selectivity of gas and liquid products; (**c**) methanol yield and CO_2 conversion (reaction pressure 1 atm, H_2/CO_2 molar ratio 3:1) [75]

9.5 Plasma CO_2 Conversion with Water

Carbon dioxide can be hydrogenated with water to produce syngas [93, 94]. Limited research exists into this method of splitting CO_2. A higher H_2O content in the feed appears to be beneficial on the production of H_2, although the study citing this only uses ratios between 10:50 and 50:50 (H_2O/CO_2); hence it is unclear if further increasing the H_2O content will continue the trend [93]. The opposite is true for the production of CO [93].

SEI can affect product yields, with maximum H_2 yields occurring at low SEI [93]. A rise in the feed flow rate results in H_2 yields decreasing; this is expected due to the decrease in residence time and the relationship between SEI and flow rate discussed previously. The reduction in H_2 yield due to high SEI and/or increased flow rate can be attributed to the occurrence of the reverse water gas shift reaction, as the yield of CO remains constant [93]:

Syngas is not the only possible product of plasma CO_2 hydrogenation with water: methane can also be produced [94, 95]. Although the current obtainable methane concentration is low (ppm) and reported energy efficiencies are well below feasible, the proof of concept is important. Methane production from CO_2 and water using NTP could potentially provide a one-step process for creating a useful energy source from a sustainable source of hydrogen.

The addition of a catalyst to this reaction can remarkably increase the yield of methane, as dissociative adsorption of H_2, CO (formed via plasma gas phase reactions) and CO_2 occurs at the catalyst surface, enabling the hydrogenation of carbon species via this mechanism as opposed to through plasma gas-phase reactions alone [95]. The use of a NiO/Al_2O_3 catalyst is beneficial for the production of methane as this catalyst facilitates the hydrogenation of CO [94]. Furthermore, the use of a reduced Ni catalyst can facilitate the production of carbon nanofibers through plasma-assisted chemical vapor decomposition of methane (Fig. 9.13) [94].

The reduction of CO_2 using water is possible using electrochemical processes [96] as well as through photoreduction [97]. These processes produce methanol, an important chemical intermediate [96]. To the best of the authors' knowledge, this reaction has not been carried out using plasma processes; however, the production of methanol has been successful in a DBD (albeit in very small quantities) when using hydrogen in the feed [92]. The addition of a $CuO/ZnO/Al_2O_3$ catalyst has also been shown to increase the methanol yield by a factor of 10 [92]. As H_2O can be successfully split into H_2 in plasma, the production of methanol from CO_2 and H_2O is theoretically possible. The main competing reaction for the formation of methanol is the production of methane [92].

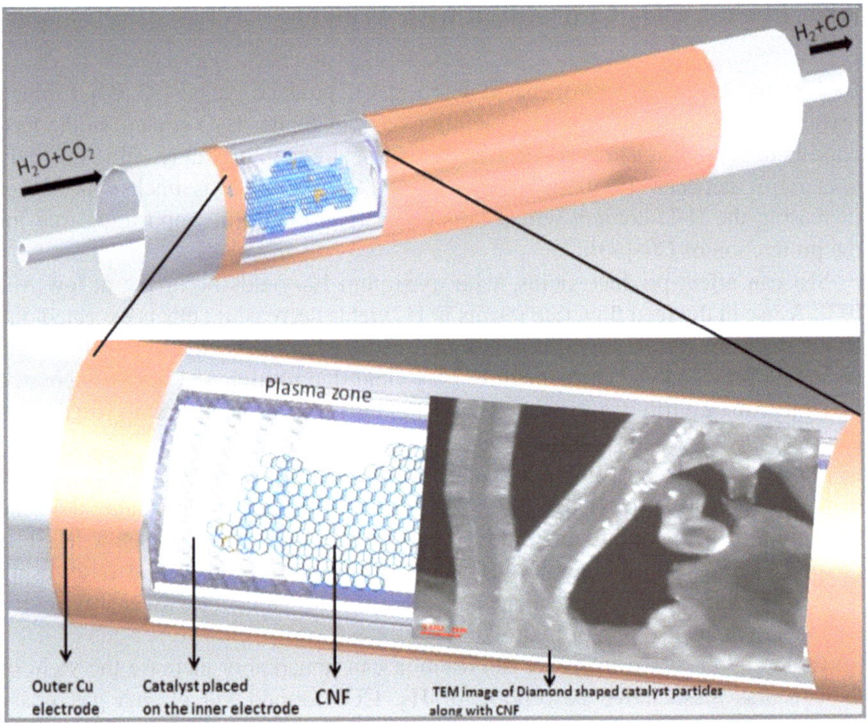

Fig. 9.13 Formation of CNF in the plasma conversion of CO_2 with water over Ni/Al_2O_3 catalyst [94]

9.6 Summary and Future Outlook

Plasma-based technologies for the conversion of CO_2 into value-added fuels and chemicals show great potential due to the ability of nonthermal plasma to break bonds in the highly stable CO_2 molecule while operating at room temperature and pressure. Plasma systems therefore have an advantage over thermal processes, which require high-temperature inputs; hence plasma conversion of CO_2 could prove much more feasible on an industrial scale. However, a trade-off between energy efficiency and CO_2 conversion currently exists in the plasma process as conversion increases when energy input is raised, which also causes a decrease in energy efficiency. Initial research has shown that this problem can be overcome by modifying the plasma system, such as by combining the plasma discharge with a catalyst; however, further research is required to promote the simultaneous increase of energy efficiency and conversion. Once plasma processes can concurrently operate at high conversion and energy efficiency, they will become a front runner in green technologies for the conversion of carbon dioxide.

Plasma chemistry is highly complex, and although much research is being conducted into plasma modeling, the models being used are greatly simplified versions of the reactions taking place [16, 93, 98]. Newer models can also leave their predecessors redundant; therefore, further study is required to produce comprehensive models for a variety of plasma processes [99]. When it comes to plasma catalysis, the chemistry becomes even more complex due to the interactions occurring between the plasma and catalyst. The number of different catalysts that can be employed in plasma processes, along with variations in catalyst preparation method, loading amount, pretreatment, etc., makes it tricky to use a 'one-model-fits-all' approach. If a greater understanding of plasma reactions and the interactions between plasma and catalyst can be realized, a more comprehensive model could be produced. In situ experimental techniques, such as in situ infrared spectroscopy, can lead to a greater understanding of plasma-assisted surface reactions, for example, plasma adsorption or desorption on the surface of the catalyst. Using techniques such as this could help to create such a model, drastically reducing the time required to optimize a process and leading to selection of the optimum catalyst without the need for numerous laborious experiments.

The potential exists to produce more complex carbon-based liquid products using plasma. A variety of liquids can currently be produced in small quantities, such as formaldehyde, acetic acid and methanol, as well as ethylene and C_4 hydrocarbons [100, 101]. However, the selectivity to many of these products still needs to be improved for the process to be viable. The selection of an appropriate catalyst that can increase the selectivity to the required product is therefore required for progress to be made in this area.

Thermal catalytic techniques are currently used to produce liquid hydrocarbons such as DME from carbon dioxide and hydrogen. For the production of DME, high temperatures (240–270 °C) and pressures (3 MPa) are required [102–104]. If NTP at atmospheric pressure can be used instead, the energy input can be drastically reduced. In order to produce hydrocarbons directly from CO_2, bifunctional catalysts are required [105]. Novel catalysts and new reactor setups may help offset the need for high pressures in a plasma reactor. Much more research is needed to create plasma processes that produce liquid hydrocarbons currently only produced by other (non-plasma) techniques, but if successful these processes could transform the chemical and energy industries.

The scope and potential of plasma processes for the utilization of CO_2 are therefore vast. These processes reduce the concentration of CO_2 in our atmosphere and allow the chemical storage of energy which can be transferred to the system from renewable energy sources at peak times. As well as producing fuels, valuable chemicals can also be formed. A greater understanding of the plasma chemistry, both through modeling of plasma and coupling with other techniques such as catalysis, as well as further insight into synthesizing a catalyst which will create synergy when combined with plasma [106], will allow this field to expand. Alongside this, further research into the conversion of CO_2 in feed gases mixed with other gases from industrial waste streams could also be beneficial for creating large-scale plasma processes for industrial applications [24].

Acknowledgments The authors acknowledge financial support from the UK EPSRC SUPERGEN Hydrogen & Fuel Cell (H2FC) Hub (EP/J016454/1), EPSRC SUPERGEN Bioenergy Challenge II Programme (EP/M013162/1), and EPSRC Impact Acceleration Account (IAA). We acknowledge the funding from the European Union's Horizon 2020 Research and Innovation Programme under the Marie Sklodowska-Curie Action (Grant Number 823745).

References

1. IPCC. (2014). *Climate Change 2014: Mitigation of Climate Change*. Contribution of Working Group III to the Fifth assessment report of the intergovernmental panel on climate change, New York.
2. Snoeckx, R., & Bogaerts, A. (2017). Plasma technology: A novel solution for CO_2 conversion. *Chemical Society Review, 46*, 5805–5863.
3. Ashford, B., & Tu, X. (2017). Non-thermal plasma technology for the conversion of CO_2. *Current Opinion in Green and Sustainable Chemistry, 3*, 45–49.
4. Tu, X., & Whitehead, J. C. (2014). Plasma dry reforming of methane in an atmospheric pressure AC gliding arc discharge: Co-generation of syngas and carbon nanomaterials. *International Journal of Hydrogen Energy, 39*, 9658–9669.
5. Snoeckx, R., Zeng, Y. X., Tu, X., & Bogaerts, A. (2015). Plasma-based dry reforming: Improving the conversion and energy efficiency in a dielectric barrier discharge. *RSC Advances., 5*, 29799–29808.
6. Wang, J., Xia, G., Huang, A., Suib, S. L., Hayashi, Y., & Matsumoto, H. (1999). CO_2 decomposition using glow discharge plasmas. *Journal of Catalysis, 185*, 152–159.
7. Spencer, L. F., & Gallimore, A. D. (2011). Efficiency of CO_2 dissociation in a radio-frequency discharge. *Plasma Chemistry and Plasma Processing, 31*, 79–89.
8. Tsuji, M., Tanoue, T., Nakano, K., & Nishimura, Y. (2001). Decomposition of CO_2 into CO and O in a microwave-excited discharge flow of CO_2/He or CO_2/Ar mixtures. *Chemistry Letters, 1*, 22–23.
9. Aerts, R. (2014). *Experimental and computational study of dielectric barrier discharges for environmental applications*. Belgium: University of Antwerp.
10. Paulussen, S., Verheyde, B., Tu, X., De Bie, C., Martens, T., Petrovic, D., Bogaerts, A., & Sels, B. (2010). Conversion of carbon dioxide to value-added chemicals in atmospheric pressure dielectric barrier discharges. *Plasma Sources Science and Technology, 19*, 034015.
11. Duan, X., Hu, Z., Li, Y., & Wang, B. (2015). Effect of dielectric packing materials on the decomposition of carbon dioxide using DBD microplasma reactor. *American Institute of Chemical Engineers, 61*, 898–903.
12. Aerts, R., Somers, W., & Bogaerts, A. (2015). CO_2 splitting in a dielectric barrier discharge plasma: A combined experimental and computational study. *ChemSusChem, 8*, 702–716.
13. Wang, W. Z., Berthelot, A., Kolev, S., Tu, X., & Bogaerts, A. (2016). CO_2 conversion in a gliding arc plasma: 1D cylindrical discharge mode. *Plasma Sources Science and Technology, 25*, 065012.
14. Silva, T., Britun, N., Godfroid, T., & Snyders, R. (2014). Optical characterization of a microwave pulsed discharge used for dissociation of CO_2. *Plasma Sources Science and Technology, 23*, 025009.
15. Pietanza, L. D., Colonna, G., D'Ammando, G., Laricchiuta, A., & Capitelli, M. (2015). Vibrational excitation and dissociation mechanisms of CO_2 under non-equilibrium discharge and post-discharge conditions. *Plasma Sources Science and Technology, 24*, 042002.
16. Berthelot, A., & Bogaerts, A. (2016). Modeling of plasma-based CO_2 conversion: Lumping of the vibrational levels. *Plasma Sources Science and Technology, 25*, 045022.
17. Mori, S., & Tun, L. L. (2017). Synergistic CO_2 conversion by hybridization of dielectric barrier discharge and solid oxide electrolyser cell. *Plasma Processes and Polymers, 14*, 1600153.

18. Mei, D. H., He, Y. L., Liu, S. Y., Yan, J. D., & Tu, X. (2016). Optimisation of CO_2 conversion in a cylindrical dielectric barrier discharge reactor using design of experiments. *Plasma Processes and Polymers, 13*, 544–556.
19. Van Laer, K., & Bogaerts, A. (2015). Improving the conversion and energy efficiency of carbon dioxide splitting in a zirconia-packed dielectric barrier discharge reactor. *Energy Technology, 3*, 1038–1044.
20. Mei, D., & Tu, X. (2017). Conversion of CO_2 in a cylindrical dielectric barrier discharge reactor: Effects of plasma processing parameters and reactor design. *Journal of CO_2* Utilization, 19, 68–78.
21. Winanti, W. S., Purwanto, W. W., & Bismo, S. (2014). Decomposition of carbon dioxide in the three-pass flow dielectric barrier discharge plasma reactor. *International Journal of Technology, 1*, 1–11.
22. Mei, D., Zhu, X., He, Y. L., Yan, J. D., & Tu, X. (2015). Plasma-assisted conversion of CO_2 in a dielectric barrier discharge reactor: Understanding the effect of packing materials. *Plasma Sources Science and Technology, 24*, 015011.
23. Ramakers, M., Michielsen, I., Aerts, R., Meynen, V., & Bogaerts, A. (2015). Effect of argon or helium on the CO_2 conversion in a dielectric barrier discharge. *Plasma Processes and Polymers, 12*, 755–763.
24. Snoeckx, R., Heijkers, S., Wesenbeeck, K., Lenaerts, S., & Bogaerts, A. (2016). CO_2 conversion in a dielectric barrier discharge plasma: N_2 in the mix as helping hand or problematic impurity? *Energy & Environmental Science, 9*, 999–1011.
25. Heijkers, S., Snoeckx, R., Kozak, T., Silva, T., Godfroid, T., Britun, N., Snyders, R., & Bogaerts, A. (2015). CO_2 conversion in a microwave plasma reactor in the presence of N_2: Elucidating the roles of vibrational levels. *The Journal of Physical Chemistry C, 119*, 12815–12828.
26. Butterworth, T., Elder, R., & Allen, R. (2016). Effects of particle size on CO_2 reduction and discharge characteristics in a packed bed plasma reactor. *Chemical Engineering Journal, 293*, 55–67.
27. Tu, X., & Whitehead, J. C. (2012). Plasma-catalytic dry reforming of methane in an atmospheric dielectric barrier discharge: Understanding the synergistic effect at low temperature. *Applied Catalysis B: Environmental, 125*, 439–448.
28. Neyts, E. C., & Bogaerts, A. (2014). Understanding plasma catalysis through modelling and simulation—A review. *Journal of Physics D: Applied Physics, 47*, 224010.
29. Mei, D., Zhu, X., Wu, C., Ashford, B., Williams, P. T., & Tu, X. (2016). Plasma-photocatalytic conversion of CO_2 at low temperatures: Understanding the synergistic effect of plasma-catalysis. *Applied Catalysis B: Environmental, 182*, 525–532.
30. Tu, X., Gallon, H. J., Twigg, M. V., Gorry, P. A., & Whitehead, J. C. (2011). Dry reforming of methane over a Ni/Al_2O_3 catalyst in a coaxial dielectric barrier discharge reactor. *Journal of Physics D: Applied Physics, 44*, 274007.
31. Mei, D., & Tu, X. (2017). Atmospheric pressure non-thermal plasma activation of CO_2 in a packed-bed dielectric barrier discharge reactor. *ChemPhysChem, 18*, 3253–3259.
32. Ray, D., & Subrahmanyam, C. (2016). CO_2 decomposition in a packed bed DBD plasma reactor: Influence of packing materials. *RSC Advances, 6*, 39492.
33. Bogaerts, A., Kozak, T., van Laer, K., & Snoeckx, R. (2015). Plasma-based conversion of CO_2: Current status and future challenges. *Faraday Discussions, 183*, 217–232.
34. Yu, Q., Kong, M., Liu, T., Fei, J., & Zheng, X. (2012). Characteristics of the decomposition of CO_2 in a dielectric packed-bed plasma reactor. *Plasma Chemistry and Plasma Processing, 32*, 153–163.
35. Belov, I., Paulussen, S., & Bogaerts, A. (2016). Appearance of a conductive carb.onaceous coating in a CO_2 dielectric barrier discharge and its influence on the electrical properties and the conversion efficiency. *Plasma Sources Science and Technology, 25*, 015023.
36. Duan, X., Li, Y., Ge, W., & Wang, B. (2015). Degradation of CO_2 through dielectric barrier discharge microplasma. *Greenhouse Gases: Science and Technology, 5*, 131–140.

37. Indarto, A., Yang, D. R., Choi, J. W., Lee, H., & Song, H. K. (2007). Gliding arc plasma processing of CO_2 conversion. *Journal of Hazardous Materials, 146*, 309–315.
38. Ozkan, A., Dufour, T., Bogaerts, A., & Reniers, F. (2016). How do the barrier thickness and dielectric material influence the filamentary mode and CO_2 conversion in a flowing DBD? *Plasma Sources Science and Technology, 25*, 045016.
39. Xu, W., Li, M. W., Xu, G. H., & Tian, Y. L. (2004). Decomposition of CO_2 using DC corona discharge at atmospheric pressure. *Japanese Journal of Applied Physics, 43*, 8310–8311.
40. Chen, G. X., Georgieva, V., Godfroid, T., Snyders, R., & Delplancke-Ogletree, M.-P. (2016). Plasma assisted catalytic decomposition of CO_2. *Applied Catalysis B: Environmental, 190*, 115–124.
41. Duan, X. F., Hu, Z. Y., Li, Y. P., & Wang, B. W. (2015). Effect of dielectric packing materials on the decomposition of carbon dioxide using DBD microplasma reactor. *AIChE Journal, 61*, 898–903.
42. Chen, G., Georgieva, V., Godfroid, T., Snyders, R., & Delplancke-Pgletree, M. P. (2016). Plasma assisted catalytic decomposition of CO_2. *Applied Catalysis B: Environmental, 190*, 115–124.
43. Martini, L. M., Dilecce, G., Guella, G., Maranzana, A., Tonachini, G., & Tosi, P. (2014). Oxidaton of CH_4 by CO_2 in a dielectric barrier discharge. *Chemical Physics Letters, 593*, 55–60.
44. De Bie, C., Martens, T., van Dijk, J., Paulussen, S., Verheyde, B., Corthals, S., & Bogaerts, A. (2011). Dielectric barrier discharges used for the conversion of greenhouse gases: Modeling the plasma chemistry by fluid simulations. *Plasma Sources Science and Technology, 20*, 024008.
45. De Bie, C., van Dyke, J., & Bogaerts, A. (2015). The dominant pathways for the conversion of methane into oxygenates and syngas in an atmospheric pressure dielectric barrier discharge. *The Journal of Physical Chemistry C, 119*, 22331–22350.
46. Wang, L., Yi, Y., Wu, C., Guo, H., & Tu, X. (2017). One-step reforming of CO_2 and CH_4 to high-value liquid chemicals and fuels at room temperature by plasma-driven catalysis. *Angewandte Chemie International Edition, 56*, 13679–13683.
47. Zeng, Y., Zhu, X., Mei, D., Ashford, B., & Tu, X. (2015). Plasma-catalytic dry reforming of methane over γ-Al_2O_3 supported metal catalysts. *Catalysis Today, 256*, 80–87.
48. Nguyen, H. H., Nasonova, A., Nah, I. W., & Kim, K. S. (2015). Analysis on CO_2 reforming of CH_4 by corona discharge process for various process variables. *Journal of Industrial and Engineering Chemistry, 32*, 58–62.
49. Wang, Q., Yan, B. H., Jin, Y., & Cheng, Y. (2009). Investigation of dry reforming of methane in a dielectric barrier discharge reactor. *Plasma Chemistry and Plasma Processing, 29*, 217–228.
50. Bo, Z., Yan, J., Li, X., Chi, Y., & Cen, K. (2008). Plasma assisted dry methane reforming using gliding arc gas discharge: Effect of feed gases proportion. *International Journal of Hydrogen Energy, 33*, 5545–5553.
51. Nguyen, H. H., & Kim, K. S. (2015). Combination of plasmas and catalytic reactions for CO_2 reforming of CH_4 by dielectric barrier discharge process. *Catalysis Today, 256*, 88–95.
52. Scapinello, M., Martini, L. M., Dilecce, G., & Tosi, P. (2016). Conversion of CH_4/CO_2 by a nanosecond repetitively pulsed discharge. *Journal of Physics D: Applied Physics, 49*, 075602.
53. Kameshima, S., Tamura, K., Ishibashi, Y., & Nozaki, T. (2015). Pulsed dry methane reforming in plasma-enhanced catalytic reaction. *Catalysis Today, 256*, 67–75.
54. Abd Allah, Z., & Whitehead, J. C. (2015). Plasma-catalytic dry reforming of methane in an atmospheric pressure AC gliding arc discharge. *Catalysis. Today, 256*, 76–79.
55. Zhang, A. J., Zhu, A. M., Guo, J., Xu, Y., & Shi, C. (2010). Conversion of greenhouse gases into syngas via combined effects of discharge activation and catalysis. *Chemical Engineering Journal, 156*, 601–606.
56. Zheng, X. G., Tan, S. Y., Dong, L. C., Li, S. B., & Chen, H. M. (2015). Silica-coated $LaNiO_2$ nanoparticles for non-thermal plasma assisted dry reforming of methane: Experimental and kinetic studies. *Chemical Engineering Journal, 265*, 147–156.

57. Gallon, H. J., Tu, X., & Whitehead, J. C. (2012). Effects of reactor packing materials on H2 production by CO_2 reforming of CH_4 in a dielectric barrier discharge reactor. *Plasma Processes and Polymers, 9*, 90–97.
58. Eliasson, B., Liu, C. J., & Kogelschatz, U. (2000). Direct conversion of methane and carbon dioxide to higher hydrocarbons using catalytic dielectric-barrier discharges with zeolites. *Industrial & Engineering Chemistry Research, 39*, 1221–1227.
59. Zhang, K., Kogelschatz, U., & Eliasson, B. (2001). Conversion of greenhouse gases to synthesis gas and higher hydrocarbons. *Energy Fuels, 15*, 395–402.
60. Mahammadunnisa, S., Reddy, P. M. K., Ramaraju, B., & Subrahmanyam, C. (2013). Catalytic nonthermal plasma reactor for dry reforming of methane. *Energy & Fuels, 27*, 4441–4447.
61. Mei, D. H., Liu, S. Y., & Tu, X. (2017). CO_2 reforming with methane for syngas production using a dielectric barrier discharge plasma coupled with γ-Al_2O_3 catalysts: Process optimization through response surface methodology. *Journal of CO_2* Utilization, 21, 314–326.
62. Wang, Q., Cheng, Y., & Jin, Y. (2009). Dry reforming of methane in an atmospheric pressure plasma fluidized bed with Ni/γ-Al_2O_3 catalyst. *Catalysis Today, 148*, 275–282.
63. Sentek, J., Krawczyk, K., Mlotek, M., Kalczewska, M., Kroker, T., Kolb, T., Schenk, A., Gericke, K. H., & Schmidt-Szalowski, K. (2010). Plasma-catalytic methane conversion with carbon dioxide in dielectric barrier discharges. *Applied Catalysis B: Environmental, 94*, 19–26.
64. Kroker, T., Kolb, T., Schenk, A., Krawczyk, K., Mlotek, M., & Gericke, K. H. (2012). Catalytic conversion of simulated biogas mixtures to synthesis gas in a fluidized bed reactor supported by a DBD. *Plasma Chemistry and Plasma Processing, 32*, 565–582.
65. Krawczyk, K., Mlotek, M., Ulejczyk, B., & Schmidt-Szalowski, K. (2014). Methane conversion with carbon dioxide in plasma-catalytic system. *Fuel, 117*, 608–617.
66. Pham, M. H., Goujard, V., Tatibouet, J. M., & Batiot-Dupeyrat, C. (2011). Activation of methane and carbon dioxide in a dielectric-barrier discharge-plasma reactor to produce hydrocarbons – Influence of La_2O_3/γ-Al_2O_3 catalyst. *Catalysis Today, 171*, 67–71.
67. Goujard, V., Tatibouet, J. M., & Batiot-Dupeyrat, C. (2011). Use of a non-thermal plasma for the production of synthesis gas from biogas. *Applied Catalysis A: General, 353*, 228–235.
68. Zheng, X. G., Tan, S., Dong, L., Li, S., & Chen, H. (2009). $LaNiO_3$@SiO_2 core-shell nanoparticles for the dry reforming of CH_4 in the dielectric barrier discharge plasma. *International Journal of Hydrogen Energy, 39*, 11360–11367.
69. Zeng, Y. X., Wang, L., Wu, C. F., Wang, J. Q., Shen, B. X., & Tu, X. (2018). Low temperature reforming of biogas over K-, Mg- and Ce-promoted Ni/Al_2O_3 catalysts for the production of hydrogen rich syngas: Understanding plasma-catalytic synergy. *Applied Catalysis B: Environmental, 224*, 469–478.
70. Chung, W. C., & Chang, M. B. (2016). Review of catalysis and plasma performance on the dry reforming of CH_4 and possible synergistic effects. *Renewable and Sustainable Energy Reviews, 62*, 13–31.
71. Zheng, X. G., Tan, S. Y., Dong, L. C., Li, S. B., Chen, H. M., & Wei, S. A. (2015). Experimental and kinetic investigation of the plasma catalytic dry reforming of methane over perovskite $LaNiO_3$ nanoparticles. *Fuel Processing Technology, 137*, 250–258.
72. Montoro-Damas, A. M., Brey, J. J., Rodriguez, M. A., Gonzalez-Elipe, A. R., & Cotrino, J. (2015). Plasma reforming of methane in a tunable ferroelectric packed-bed dielectric discharge reactor. *Journal of Power Sources, 296*, 268–275.
73. Mei, D., Ashford, B., He, Y. L., & Tu, X. (2017). Plasma-catalytic reforming of biogas over supported Ni catalysts in a dielectric barrier discharge reactor: Effect of catalyst supports. *Plasma Processes and Polymers, 14*, e1600076.
74. Zeng, Y., & Tu, X. (2016). Plasma-catalytic CO_2 hydrogenation at low temperatures. *IEEE Transactions on Plasma Science, 44*, 405–411.
75. Wang, L., Yi, Y., Guo, H., & Tu, X. (2017). Atmospheric pressure and room temperature synthesis of methanol through plasma-catalytic hydrogenation of CO_2. *ACS Catalysis, 8*, 90–100.

76. De Bie, C., van Dijk, J., & Bogaerts, A. (2016). CO_2 hydrogenation in a dielectric barrier discharge plasma revealed. *The Journal of Physical Chemistry C, 120*, 25210–25224.
77. Saeidi, S., Amin, N. A. S., & Rahimpour, M. R. (2014). Hydrogenation of CO_2 to value-added products – A review and potential future developments. *Journal of CO_2* Utilization, 5, 66–81.
78. Kano, M., Satoh, G., & Iizuka, S. (2012). Reforming of carbon dioxide to methane and methanol by electric impulse low-pressure discharge with hydrogen. *Plasma Chemistry and Plasma Processing, 32*, 177–185.
79. Mora, E. Y., Sarmiento, A., & Vera, E. (2016). Alumina and quartz as dielectrics in a dielectric barrier discharges DBD system for CO_2 hydrogenation. *Journal of Physics: Conference Series, 687*, 012020.
80. Arita, K., & Iizuka, S. (2015). Production of CH_4 in a low-pressure CO_2/H_2 discharge with magnetic field. *Journal of Chemical Engineering and Materials Science, 3*, 69–77.
81. Nizio, M., Albarazi, A., Cavadias, S., Amouroux, J., Galvez, M. E., & Da Costa, P. (2016). Hybrid plasma-catalytic methanation of CO_2 at low temperature over ceria zirconia supported Ni catalysts. *International Journal of Hydrogen Energy, 41*, 11584–11592.
82. Jwa, E., Lee, S. B., Lee, H. W., & Mok, Y. S. (2013). Plasma-assisted catalytic methanation of CO and CO_2 over Ni-zeolite catalysts. *Fuel Processing Technology, 108*, 89–93.
83. Nizio, M., Benrabbah, R., Krzak, M., Debek, R., Motak, M., Cavadias, S., Galvez, M. E., & Da Costa, P. (2016). Low temperature hybrid plasma-catalytic methanation over Ni-Ce-Zr hydrotalcite-derived catalysts. *Catalysis Communications, 83*, 14–17.
84. Ponduri, S. (2016). Understanding CO_2 containing non-equilibrium plasma: Modelling and experiments. PhD Thesis, *Eindhoven: Technische Universiteit Eindhoven, Netherlands.*
85. Wang, W., Wang, S. P., Ma, X. B., & Gong, J. L. (2011). Recent advances in catalytic hydrogenation of carbon dioxide. *Chemical Society Reviews, 40*, 3703–3727.
86. Porosoff, M. D., Yan, B., & Chen, J. G. (2016). Catalytic reduction of CO_2 by H_2 for synthesis of CO, methanol and hydrocarbons: Challenges and opportunities. *Energy & Environmental Science, 9*, 62–73.
87. Liu, P., Yang, Y. X., & White, M. G. (2013). Theoretical perspective of alcohol decomposition and synthesis from CO_2 hydrogenation. *Surface Science Reports, 68*, 233–272.
88. Zou, J. J., & Liu, C. J. (2010). Utilization of carbon dioxide through nonthermal plasma approaches. *Carbon Dioxide as Chemical Feedstock*, M. Aresta (Ed.), pp 267–290. Wiley VCH.
89. Amouroux, J., Cavadias, S., & Doubla, A. (2011). Carbon dioxide reduction by non-equilibrium electrocatalysis plasma reactor. *IOP Conference Series: Materials Science and Engineering, 19*, 012005.
90. Zeng, Y. X., & Tu, X. (2017). Plasma-catalytic hydrogenation of CO_2 for the cogeneration of CO and CH_4 in a dielectric barrier discharge reactor: Effect of argon addition. *Journal of Physics D: Applied Physics, 50*, 184004.
91. Hayashi, N., Yamakawa, T., & Baba, S. (2016). Effect of additive gases on synthesis of organic compounds from carbon dioxide using non-thermal plasma produced by atmospheric surface discharges. *Vacuum, 80*, 1299–1304.
92. Eliasson, B., Kogelschatz, U., Xue, B., & Zhou, L. M. (1998). Hydrogenation of carbon dioxide to methanol with a discharge-activated catalyst. *Industrial & Engineering Chemistry Research, 37*, 3350–3357.
93. Chen, G., Silva, T., Georgieva, V., Godfroid, T., Britun, N., Snyders, R., & Delplancke-Ogletree, M. P. (2015). Simultaneous dissociation of CO_2 and H_2O to syngas in a surface-wave microwave discharge. *International Journal of Hydrogen Energy, 40*, 3789–3796.
94. Mahammadunnisa, S., Reddy, E. L., Ray, D., Subrahmanyam, C., & Whitehead, J. C. (2013). CO_2 reduction to syngas and carbon nanofibres by plasma-assisted in situ decomposition of water. *International Journal of Greenhouse Gas Control, 16*, 361–363.
95. Hoeben, W. F. L. M., van Heesch, E. J. M., Beckers, F. J. C. M., Boekhoven, W., & Pemen, A. J. M. (2015). Plasma-driven water assisted CO_2 methanation. *IEEE Transactions on Plasma Science, 43*, 1954–1958.

96. Keets, K., Morris, A., Zeitler, E., Lakkaraju, P., & Bocarsly, A. (2010). *Catalytic conversion of carbon dioxide to methanol and higher order alcohols at a photoelectrochemical interface*. Proceedings in Solar Hydrogen and Nanotechnology V. SPIE Solar Energy + Technology, San Diego (Vol. 7770, article ID: 77700R).
97. Zbudniewek, K., Goralski, J., & Rynkowski, J. (2012). Studies on TiO_2/SiO_2 and $Pd/TiO_2/SiO_2$ catalysts in photoreduction of CO_2 with H_2O to methanol. *Russian Journal of Physical Chemistry A, 86*, 2057–2062.
98. Aerts, R., Somers, W., & Bogaerts, A. (2014). CO_2 splitting in a dielectric barrier discharge plasma: A combined experimental and computational study. *ChemSusChem, 8*, 702–716.
99. Bogaerts, A., Wang, W., Berthelot, A., & Guerra, V. (2016). Modeling plasma-based CO_2 conversion: crucial role of the dissociation cross section. *Plasma Sources Science and Technology, 25*, 055016.
100. Gómez-Ramírez, A., Rico, V. J., Cotrino, J., González-Elipe, A. R., & Lambert, R. M. (2014). Low temperature production of formaldehyde from carbon dioxide and ethane by plasma-assisted catalysis in a ferroelectrically moderated dielectric barrier discharge reactor. *ACS Catalysis, 4*, 402–408.
101. Pham, M. H., Goujard, V., Tatibouët, J. M., & Batiot-Dupeyrat, C. (2011). Activation of methane and carbon dioxide in a dielectric-barrier discharge-plasma reactor to produce hydrocarbons—Influence of $La_2O_3/\gamma\text{-}Al_2O_3$ catalyst. *Catalysis Today, 171*, 67–71.
102. Frusteri, F., Bonura, G., Cannilla, C., Drago Ferrante, G., Aloise, A., Catizzone, E., Migliori, M., & Giordano, G. (2015). Stepwise tuning of metal-oxide and acid sites of CuZnZr-MFI hybrid catalysts for the direct DME synthesis by CO_2 hydrogenation. *Applied Catalysis B: Environmental, 176–177*, 522–531.
103. Witoon, T., Permsirivanich, T., Kanjanasoontorn, N., Akkaraphataworn, C., Seubsai, A., Faungnawakij, K., Warakulwit, C., Chareonpanich, M., & Limtrakul, J. (2015). Direct synthesis of dimethyl ether from CO_2 hydrogenation over Cu–ZnO–ZrO_2 hybrid catalysts: effects of sulfur-to-zirconia ratios. *Catalysis Science & Technology, 5*, 2347–2357.
104. Zhang, Y., Li, D., Zhang, Y., Cao, Y., Zhang, S., Wang, K., Ding, F., & Wu, J. (2014). V-modified CuO–ZnO–ZrO_2/HZSM-5 catalyst for efficient direct synthesis of DME from CO_2 hydrogenation. *Catalysis Communications, 55*, 49–52.
105. Centi, G., & Perathoner, S. (2009). Opportunities and prospects in the chemical recycling of carbon dioxide to fuels. *Catalysis Today, 148*, 191–205.
106. Chen, G., Georgieva, V., Godfriod, T., Snyders, R., & Delplancke-Ogletree, M. P. (2016). Plasma assisted catalytic decomposition of CO_2. *Applied Catalysis B: Environmental, 190*, 115–124.

Chapter 10
Plasma-Catalytic Reforming of Alcohols

Dae Hoon Lee

10.1 Introduction

Methanol and ethanol are one of the most important basic materials for chemical industries. Methanol is used as a raw material in a wide range of fields such as synthetic detergents, photosensitive materials, synthetic resins, ethers, esters, dyes, and adhesives in the chemical industry and as a test and disinfectant in the medical field [1–4].

In recent years, as environmental problems and a lack of petroleum resources are expected, the need for various types of nontraditional energy sources such as renewable energy is increasing. Since a typical alternative energy source is biomass-derived fuels, which consist of alcohols, attempts have been made to utilize alcohol, especially methanol, for various energy sources and as chemical raw materials in this background [5–7].

This movement is largely driven by (1) the use of alcohol itself as a fuel, (2) the production of hydrocarbons such as olefins and paraffins through chemical synthesis and the production of synthetic oils therefrom, (3) and the production of hydrogen via reforming.

A typical case of using the alcohol itself is as a fuel to directly produce electricity, such as in a DMFC (direct methanol fuel cell) of which the principles of operation are schematically shown in Fig. 10.1 [8].

In recent years, biofuels have been used for transportation in the sense of utilizing renewable energy. Among possible biofuel resources, bioethanol extracted from

D. H. Lee (✉)
Plasma Engineering Laboratory, Korea Institute of Machinery and Materials, Yuseong-Gu, Daejeon, South Korea

Department of Environment & Energy Engineering, University of Science and Technology, Yuseong-Gu, Daejeon, South Korea
e-mail: dhlee@kimm.re.kr

X. Tu et al. (eds.), *Plasma Catalysis*, Springer Series on Atomic, Optical, and Plasma Physics 106, https://doi.org/10.1007/978-3-030-05189-1_10

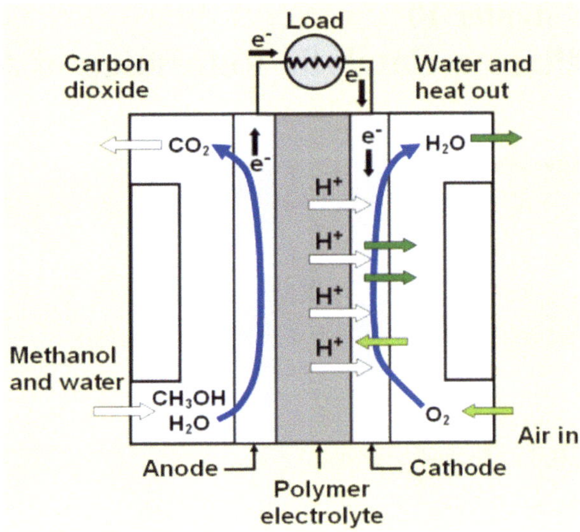

Fig. 10.1 Schematic principle of a DMFC in operation [8]

biomass fermentation is the most widely used among them [9, 10]. In the United States, 96% of renewable fuel additives are bioethanol [11].

Recently, a variety of applications for alcohols have been suggested through chemical synthesis. The most representative field is that of the MTO (methanol to olefin) process for synthesizing olefins from methanol. Figure 10.2 shows a simplified process flow diagram commercialized by UOP [12, 13].

$$2CH_3OH \rightarrow CH_3OCH_3 + H_2O \rightarrow C_2H_4 + 2H_2O \tag{10.1}$$

Olefins, especially ethylene, are the most basic chemicals in the petrochemical industry. Diesel-related technologies are being developed to replace existing naphtha-based processes, in relation to ethylene production to prepare for rising oil prices and future declines in crude oil production. Ethylene produced through the MTO process itself is used as a raw material for various polymers, as well as in the manufacturing process of synthetic fuels [14] (MTG (methanol to gasoline) process, Eq. (10.2)). These processes have been commercialized by ExxonMobil et al. Fig. 10.3 shows the process flow for the MTG process [15, 16].

$$\begin{aligned} &\text{Methanol} \rightarrow \text{DME(Dimethyl Ether)} \rightarrow \text{Light Olefin} \rightarrow C_5 + \text{Olefin} \\ &\quad \rightarrow \text{Paraffin/Naphthenes/Aromatics(Gasoline)} \end{aligned} \tag{10.2}$$

Alcohol, DME, and the like have merit in that the hydrogen/carbon ratio is much higher than that of conventional hydrocarbons, and thus a high hydrogen yield can be obtained. Based on this merit, various researches and developments have been made, in order to produce synthesis gas and hydrogen through a variety of industrial and chemical routes, such as those based on reforming and direct decomposition reactions.

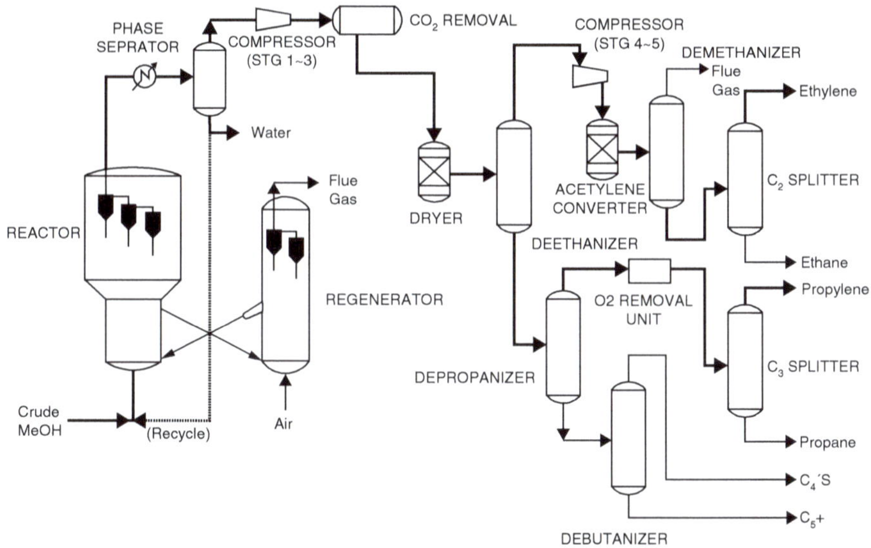

Fig. 10.2 Schematic of a simplified process flow for the UOP/Hydro MTO process [13]

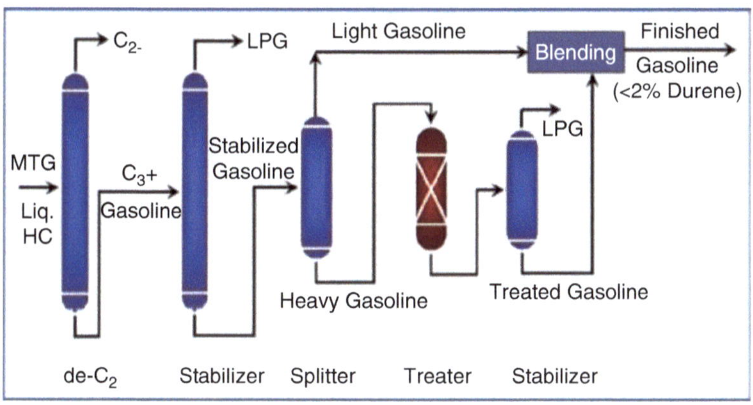

Fig. 10.3. A schematic of an MTG reactor section showing the process flow [15]

This chapter discusses reforming/direct decomposition reactions, especially ones utilizing the plasma process, among various other alcohol-related processes.

10.2 Reforming and Decomposition of Alcohol

Alcohol reforming/decomposition reactions for hydrogen production can be classified into four reaction types: direct decomposition, partial oxidation, steam reforming, and oxidative steam reforming (OSR). Each reaction has different characteristics such as the composition of the reactants, the composition of the products, and the thermal properties (endothermic, exothermic). These reaction characteristics also change the efficiency and meaning of the reaction depending on which catalyst is used [17–21].

- Alcohol decomposition reaction

$$CH_3OH \rightarrow 2H_2 + CO \tag{10.3}$$

$$C_2H_5OH \rightarrow CH_4 + CO + H_2 \tag{10.4}$$

$$C_2H_5OH + 2H_2 \rightarrow 2CH_4 + H_2O \text{ (hydrogenlysis)} \tag{10.5}$$

$$C_2H_5OH \rightarrow C_2H_4 + H_2O \text{ (dehydration)} \tag{10.6}$$

- Partial oxidation

$$CH_3OH + 1/2O_2 \rightarrow 2H_2 + CO_2 \tag{10.7}$$

$$C_2H_5OH + 3/2O_2 \rightarrow 2CO_2 + 3H_2 \tag{10.8}$$

- Steam reforming

$$CH_3OH + H_2O \rightarrow 3H_2 + CO_2 \tag{10.9}$$

$$C_2H_5OH + 3H_2O \rightarrow 2CO_2 + 6H_2 \tag{10.10}$$

- Oxidative steam reforming

$$CH_3OH + (1-n)H_2O + 0.5O_2 \rightarrow (3-n)H_2 + CO_2 \quad 0 < n < 1 \tag{10.11}$$

$$C_2H_5OH + (3-2n)H_2O + nO_2 \rightarrow 2CO_2 + (6-2n)H_2 \quad 0 < n < 0.5 \tag{10.12}$$

In the case of alcohol reforming, steam reforming has been reported most frequently. In the case of the OSR reaction, a thermally neutral reaction can be formed according to the composition of the reactants, and thus the reaction has recently attracted a great deal of attention [18].

The reforming of alcohol has been realized through (1) catalytic, (2) plasma, and (3) plasma-catalytic systems. Traditionally, the reforming reaction of alcohols has been realized through the catalytic reaction. However, as the technique of low-temperature plasma generation at atmospheric pressure condition has expanded, the application of plasma for alcohol reforming/decomposition reaction has also been tried. The catalytic system mainly utilizes the principle of increasing the reactivity of the surface reaction on the catalyst surface by depending on the thermal equilibrium, while the plasma system utilizes the nonequilibrium collisional process. In this case, a product having a composition different from the thermal equilibrium can be obtained.

Particularly, in the case of low-temperature conditions, the amount of the by-products produced by the catalytic reaction increases, whereas the by-products of the plasma reaction do not change much with changes in temperature. On the other hand, in the case of the plasma process, one disadvantage is that there is a low density of the excited active species, which can initiate the reaction. This is due to the threshold temperature condition of the electrons which can cause bond breakage in the alcohol structure. This disadvantage is directly related to the cost of the reaction. Plasma-catalyzed reactions have been studied in recent years to overcome these shortcomings of both catalyst and plasma systems and to obtain synergy. The characteristics of each reaction are introduced below.

10.2.1 Catalytic Process

The catalytic process has been carried out using metal catalysts such as Cu or elements from group 8 to 10. The factors affecting the catalytic reaction are as follows: (1) the synthesis method (coprecipitation, homogeneous precipitation, hydrothermal synthesis, impregnation, wet impregnation, and so on), (2) the promoter type and composition ratio (ZnO, Zn, Mn, CeO_2, Cr_2O_3, etc.), and (3) the reduction temperature of the catalyst. Table 10.1 shows the changes in the performance for reforming depending on the type of promotor when a Cu catalyst is used [22].

Since the catalytic reaction is basically dependent on the thermal equilibrium state, intermediate conversions can occur during the reforming/decomposition reaction of the alcohol with a catalyst, particularly in the case of a low-temperature reaction. Dehydration and dehydrogenation especially are faster than the slow steam reforming reaction, so there is a possibility for kinetic control of the reforming reaction itself at low-temperature conditions [23].

For the reforming reaction of alcohols, especially in the case of the steam reforming reaction, the reaction path varies depending on the ratio between water and alcohol. Ethanol is more affected by this reaction condition due to (1) an increase of water gas shift (WGS) reaction resulting in a decrease of CO and increase of the ratio of CO_2/H_2 and (2) a reduction of by-products by the shift of the equilibrium conditions.

It is still controversial as to how the alcohol degradation/reforming reaction proceeds on the catalyst. Various reaction paths have been reported that are dependent on the Cu surface conditions involving Cu^0, Cu^+, or a Cu^0-Cu^+ pair, and it is known that the intermediate species vary depending on the active sites of these catalysts. In the case of methanol, representative oxygenate intermediate species that can be formed on the transition metal surface are shown below [24] (Fig. 10.4).

Ethanol produces additional aldehyde-type intermediates as shown in Fig. 10.5 [23].

It has been reported that the hydroxyl end of alcohol and the O atom on the surface enhance the adsorption process. A positive valence at the Cu site formed by the O atom at the surface promotes adsorption through the electron donating O lone pair.

Table 10.1 The influence of the type of promotor on the performance of different Cu-based catalysts for methanol steam reforming [22]

Catalyst	T (°C)	X_{MeOH} (%)	Activity (μmol $H_2/(g_{cat} \cdot s)$)	Y_{CO} (%)	S_{CO} (%)
Cu/ZnO	260	75	–	0.0073	–
$Cu/ZnO/Al_2O_3$	260	79	–	0.0073	–
$Cu/ZnO/ZrO_2/Al_2O_3$	260	92	–	0.0010	–
Cu/Zn/Al	250	39	–	–	0.6
Cu/Zn/Zr/Al	250	61	–	–	0.4
Cu/SiO_2	300	50	–	–	0.15
$ZnO/Cu/SiO_2$	300	75	–	–	0.3
$Cu/Zn/Al_2O_3$ (commercial)	300	57	–	–	1.1
$Cu/ZnO/Al_2O_3$ (commercial)	270	59.0	–	–	–
$Cu/ZnO/Al_2O_3$	270	89.2	–	–	0.92
$Cu/ZnO/ZrO_2/Al_2O_3$	270	92.4	–	–	0.97
$Cu/ZnO/CeO_2/ZrO_2/Al_2O_3$	270	89.4	–	–	0.85
$Cu/ZnO/CeO_2/Al_2O_3$	270	79.3	–	–	0.66
Cu/ZnO	308	90	–	0.0014	–
$Cu/ZnO/Al_2O_3$	305	90	–	0.0011	–
$Cu/ZnO/ZrO_2$	295	90	–	0.0005	–
$Cu/ZnO/ZrO_2/Al_2O_3$	278	90	–	0.00045	–
Cu/Zn/Al	260	76	203	0.011	–
Cu/Zn/Ce/Al	260	90	244	0.0006	–
$Cu/Zn/Al_2O_3$	300	100	185	0.012	–
$Cu/Cr/Al_2O_3$	300	63	117	0.011	–
$Cu/Zr/Al_2O_3$	300	44	81	0.0075	–
$Cu-Cr_2O_3$	240	–	28	–	5
Cu-ZnO	240	–	24	–	7
Cu-CoO	240	–	17	–	14
CuO/CeO_2	250	80	75	–	–
$CuO/CeO_2/ZrO_2$	250	–	97	–	–
Cu/CeO_2	260	91	135	–	2.3
Cu/ZnO	260	67	99	–	0.9
Cu/Zn(Al)O	260	58	86	–	0.8
Cu/Al_2O_3	260	22	32	–	0.4
Cu-Mn-O	240	99	–	–	3.1
Cu-Ce-O	240	37	–	–	0.8
CuMn-spinel	213	77	–	–	0.2
Non-spinel CuMn	214	61	–	–	0.1
CuZn	250	39	–	0.0011^2	–
Pd/CuZn	250	47	–	0.003^2	–

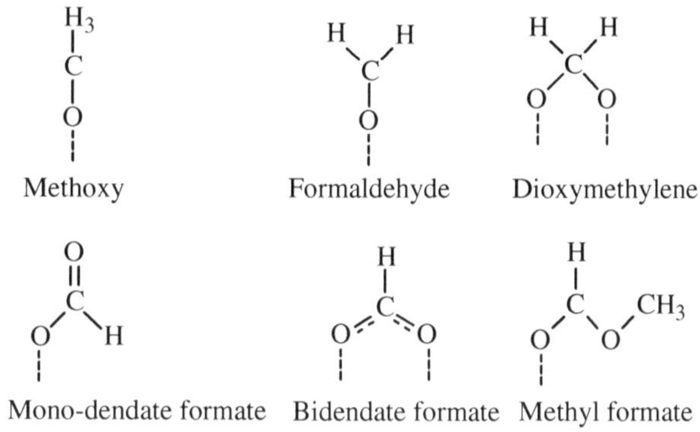

Fig. 10.4 Molecular structure of surface intermediate species of methanol [23]

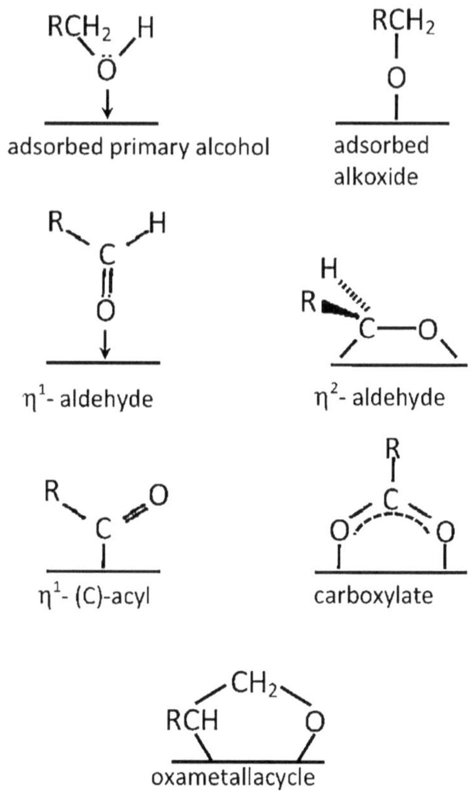

Fig. 10.5 Molecular structure of surface intermediate species of ethanol [24]

The electron-donating property by O at the surface is one of the characteristics that appears in the surface discharge conditions. One of the typical phenomena that occurs as a result of this discharge is the generation of active hydroxyl/oxygen species on the surface. Considering this point, it is expected that the loop connecting the discharge and the catalytic reaction can be utilized for enhancement of the reaction.

10.2.2 Plasma Process

In the case of low-temperature plasma such as dielectric barrier discharge (DBD) and corona discharge, the temperature of the heavy species is not high. Therefore, the temperature of the bulk gas volume is at most a few hundred K, which is lower than that required to obtain a thermally driven reaction, but the electron temperature is relatively high in the range of several eV or above. On the other hand, in the case of a high-temperature plasma such as an arc discharge, the temperature of the heavy species is relatively high, so the possibility of a thermally driven reaction becomes high. However, the temperature of the electrons is relatively low when compared with low-temperature plasma and the relative amount of electron collision-driven reactions is reduced. The bond break energy required to initiate the alcohol reforming/decomposition reaction requires a high energy of about 7–8 eV. Therefore, in the case of an alcohol decomposition/reforming reaction, low-temperature plasma sources have been preferred. Typical plasma sources applied for the alcohol reforming reaction are corona discharge [25, 26], DBD [27, 28], gliding arc [29, 30], rotating arc [31, 32], hollow cathode plasma [33, 34], and spark discharge [35, 36].

The ease of reactor construction, discharge stability, and reforming characteristics depend on the plasma source used. Table 10.2 shows a comparison of selected plasma sources [37].

On the other hand, in the case of the plasma reaction, since the energetic species are statistically distributed due to the nature of the plasma generation, the selectivity of the reaction product is relatively low compared to that of the catalyst. Recently, plasma catalysis has been studied to compensate the disadvantages of catalyst and plasma methods and to obtain synergy.

Examples of the plasma source being actually applied to alcohol reforming/decomposition are shown in Table 10.3.

10.3 Reaction Mechanism and Pathway

The start of a typical catalytic reaction begins with the adsorption of the reactants. At this time, various adsorbent species exist depending on the kind and condition of the catalyst. Although there are still controversial debates on these intermediate species, several representative reaction paths have been identified.In the case of plasma, there are various types of collision-based radical products early in the reaction. Since the bond breakage energies at different bonding sites are different, the reaction is initiated with different intermediate species depending on the energy of the colliding electrons.

On the other hand, new considerations arise in the case of alcohol decomposition/reforming by plasma catalysis. In the case of plasma catalysis, gas phase and surface reactions are combined, resulting in a more complex reaction path. To understand this, it is necessary to understand the reaction characteristics in plasma catalysis. In

Table 10.2 Comparison of reforming performance of different plasma sources [37]

Type	Compactness	Stability	Reliability	Reforming efficiency	Description
Gliding arc discharge	++	++	++	+	Excellent reformer, while the resistance time of reactant should be extended further for mixing better and reacting longer
Microware discharge	—	+	+	+	High economic cost, a high conversion rate, and good H_2 selectivity, while it has low H_2 production rate and is limited by the extra energy requirement for a vacuum device
Dielectric barrier discharge	++	++	++	+	Low H_2 yield in the products, suitable for combining with catalysts
Corona discharge	+	+	+	+	Limited by power input
Glow discharge	++	+	+	—	Low pressure and conversion rate for reforming process
Radio frequency	—	+	+	—	Very low pressure and high frequencies; it usually is used to produce diamond
Arc	++	+	+	+	Short lifetime of electrodes, high conversion rate, and low SER

"—" means poor. "+" means moderate. "++" means good

the case of the plasma catalysis combined reaction, discharge occurs in the space where the catalyst is packed. In this case, the discharge characteristic is basically a combination of discharge in the gas phase and discharge on the catalyst surface.

10.3.1 Reaction Pathway in the Catalytic Process

The reaction pathway in a typical catalytic reaction is determined by the adsorption pattern. Adsorption patterns of adsorbed species formed by alcohol species are determined depending on the kind of catalyst (Fig. 10.6) [47–50]. Figure 10.6 shows the adsorption route of methanol along the adsorption surface, and Fig. 10.7 shows the adsorption species of ethanol on the transition metal surface [51–53].

The catalytic process in methanol and ethanol is driven mostly through a Langmuir-Hinshelwood mechanism [54] where the adsorption stage determines the following reaction steps. As a result, control of molecular adsorption and

Table 10.3 Configuration of plasma reactors applied for alcohol reforming and decomposition

Source type	Reactor Configuration	Reference
Arc	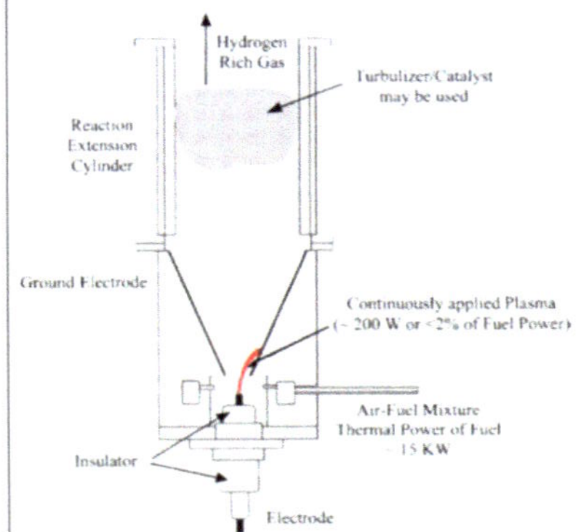 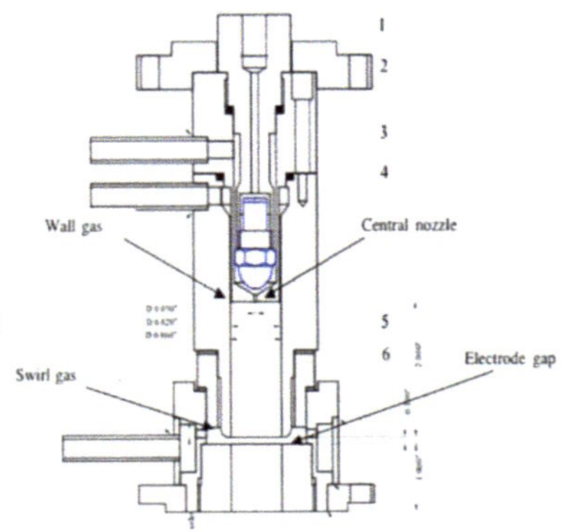	[38]

Gliding arc	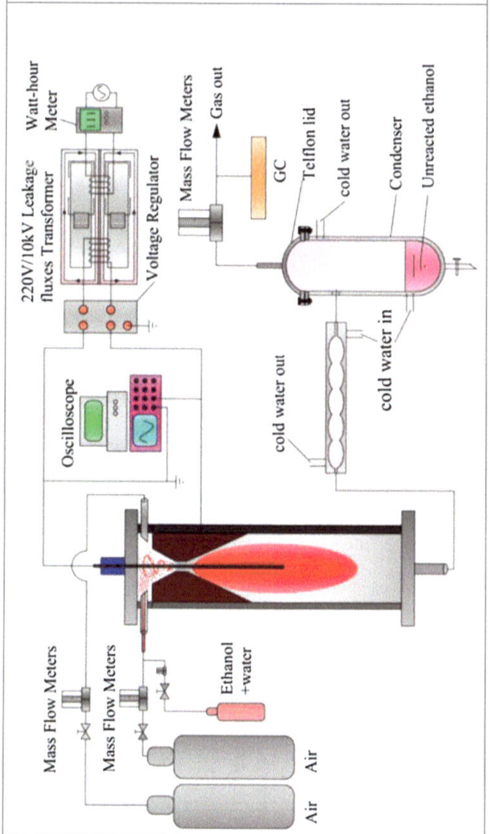	[39]

(continued)

Table 10.3 (continued)

Source type	Reactor Configuration	Reference
MW	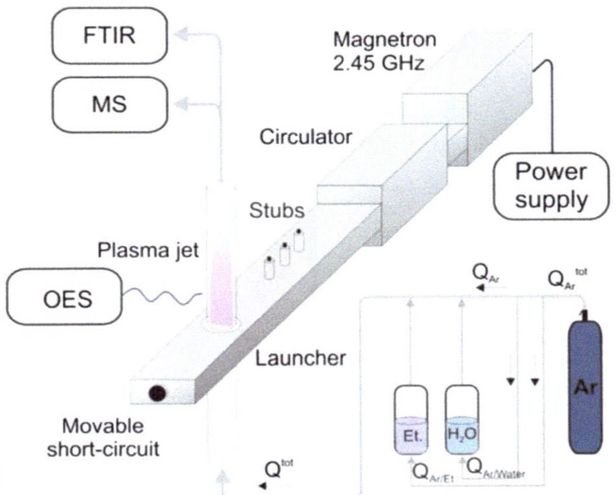	[40]

DBD	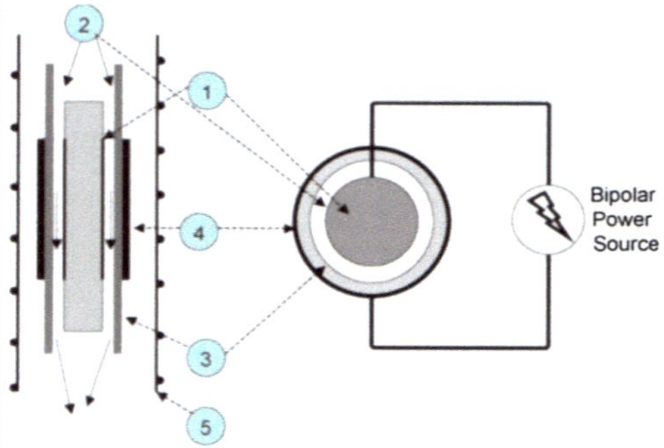	[41]
Corona	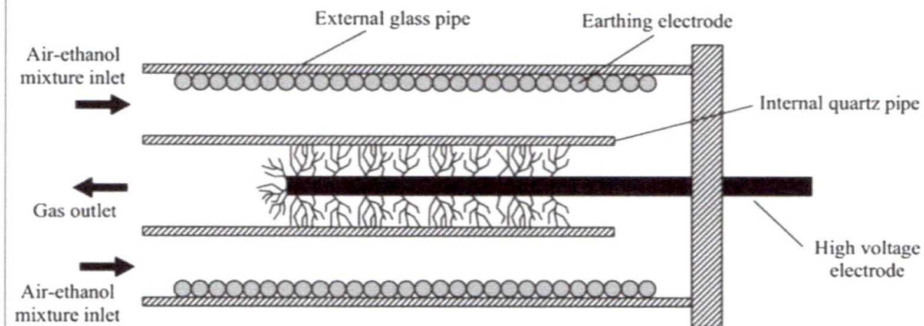	[42]

(continued)

Table 10.3 (continued)

Source type	Reactor Configuration	Reference
Pulse corona	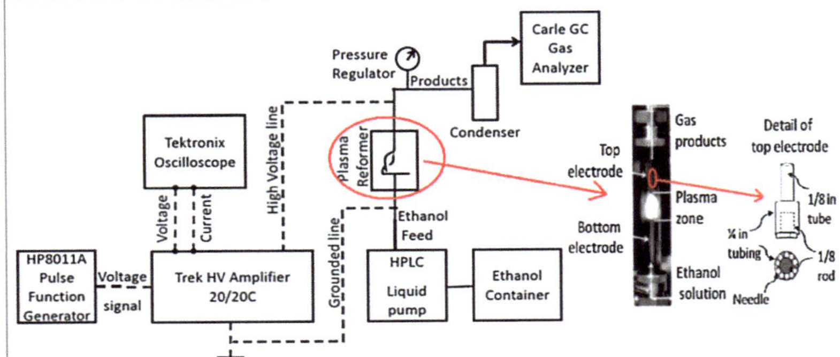	[43]

Glow discharge plasma electrolysis	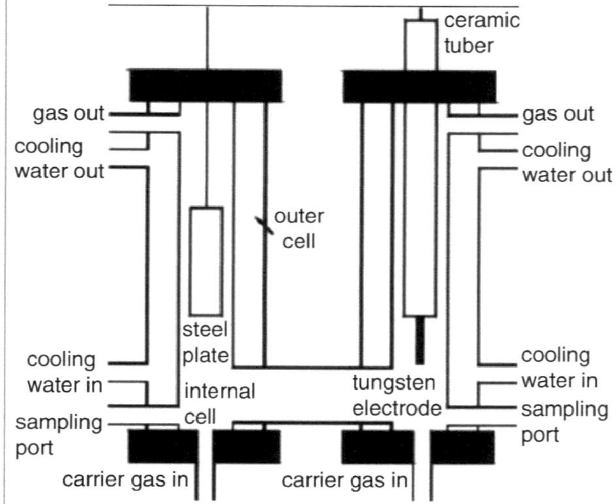	[44]

(continued)

Table 10.3 (continued)

Source type	Reactor Configuration	Reference
Glow discharge (atmospheric pressure)	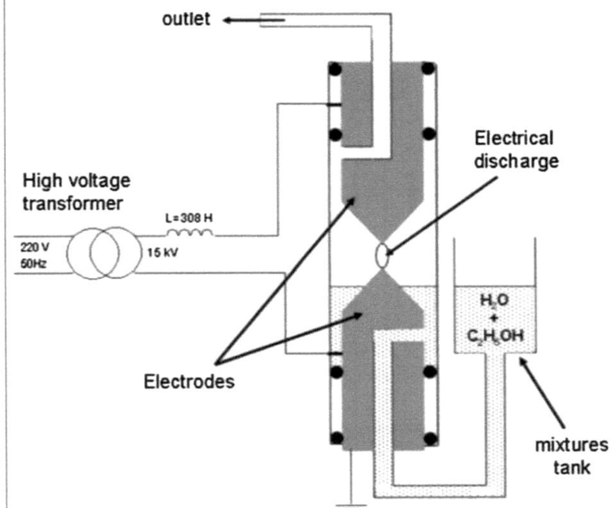	[45]

Spark (discharge in a gas channel with a liquid wall)	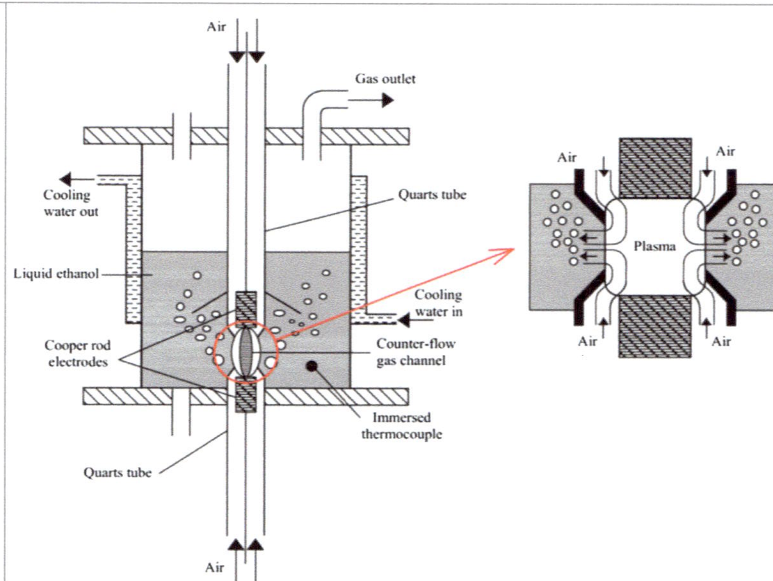 	[46]

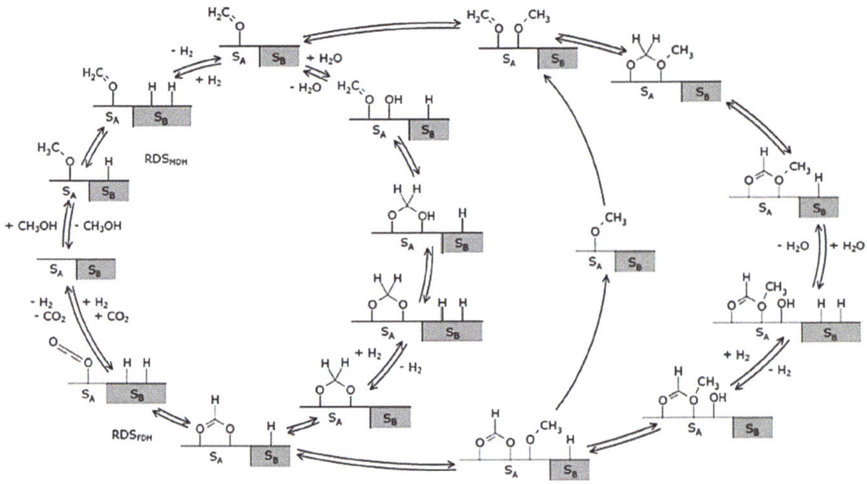

Fig. 10.6 Catalytic cycle of methanol steam reforming according to the different reactive surfaces S_A and S_B [51]

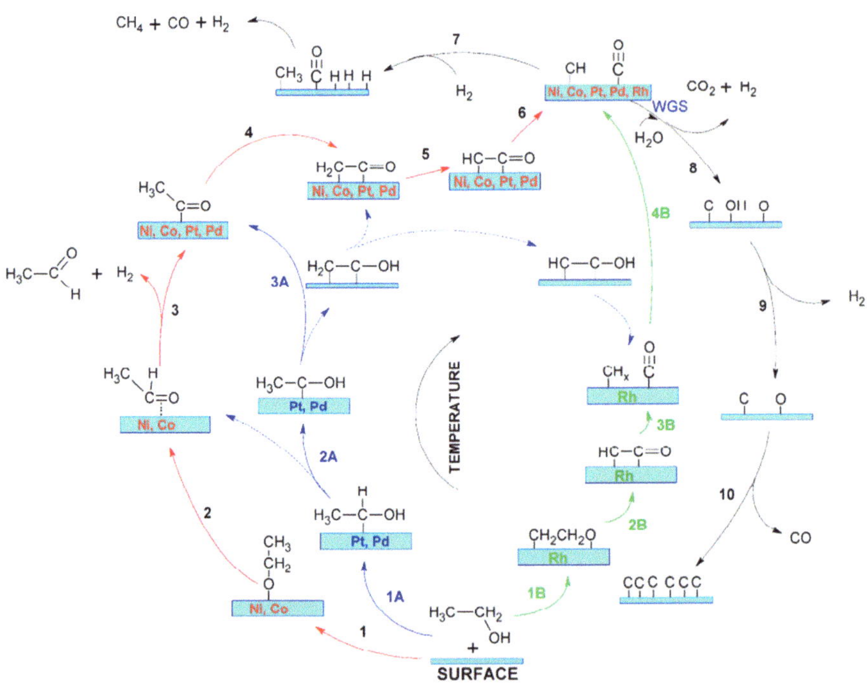

Fig. 10.7 Ethanol adsorption according to different transition metal surfaces [52, 53]

generation of intermediate species are the main route to control the reaction. Here, control of molecular adsorption is basically a thermally controlled process, and there is possibly less chance to control the adsorption itself. Generation of intermediate species can be controlled by modification of the catalytic surface, and enhancement of catalytic reforming can be obtained by modification of the catalyst and support materials.

This is the point where plasma can affect the catalytic process. Plasma can produce different types of species before adsorption including excited states and radicals. Also, depending on the energy state of the adsorbed species, the path for the generation of intermediate species can possibly be changed. Representation of such a change can be obtained by an Eley-Rideal-type reaction between surface intermediate and gas-phase excited species [55]; because of that, plasma generating reaction pathways should also be understood to control the catalyst reaction pathways.

10.3.2 Reaction Pathway in the Plasma Process

The plasma reaction path is basically that of a reaction by collision-generated species. The bond breakage energy of the various chemical bonds contained in the alcohol molecules, and the energy distribution of the electrons, and excited species produced by the plasma determine the kind of primary radicals generated in the initial collision and thus the different reaction paths that are formed (Fig. 10.8).

The type of primary radicals determined by the collision is determined by the energy of the colliding electrons. In the case of methanol, the threshold energy level for each radical formation in the primary radical reaction through decomposition is as follows [57–60].

$$CH_3OH + e \rightarrow CH_3O + H + e \qquad 7.95\ \text{eV} \tag{10.13}$$

$$CH_3OH + e \rightarrow CH_3 + OH + e \qquad 8.28\ \text{eV} \tag{10.14}$$

$$CH_3OH + e \rightarrow CH_2OH + H + e \qquad 8.28\ \text{eV} \tag{10.15}$$

Although there is no significant difference, the distribution of primary radicals can be different due to the variance of each threshold energy level. (Depending on the value of the median energy, the density difference between each primary radical can affect the reaction.)

In this case, the product composition changes depending on what the primary radical is. For example, when the density of CH_3 is relatively high among the primary radicals, as shown in reaction path of Fig. 10.9 (P3), the amount of gas phase products such as CH_4 and C_2H_6 can be relatively increased [61].

P1 is similar with that observed in typical catalytic process and results in the production of CO or reformed synthesis gas, while P2 still has possibility of producing C2 species (C_2H_2, C_2H_4) that cannot be observed in typical catalytic reforming processes.

In the case of ethanol, the reaction pathway with C_2H_5O as the primary radical can proceed as shown in Fig. 10.10 [56]. Pathways for other primary radicals are also possible and provide a path for the generation of gas-phase products other than H_2

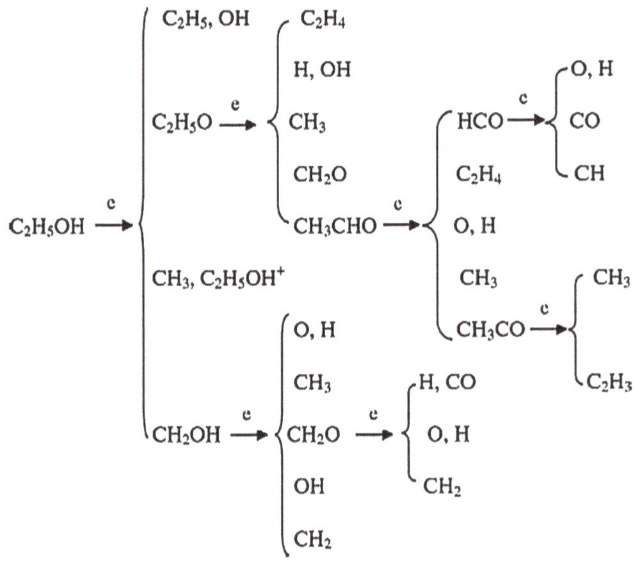

Fig. 10.8 Radical cascade from ethanol by a collisional reaction [56]

Fig. 10.9 Reaction paths depending on the primary radicals [61]

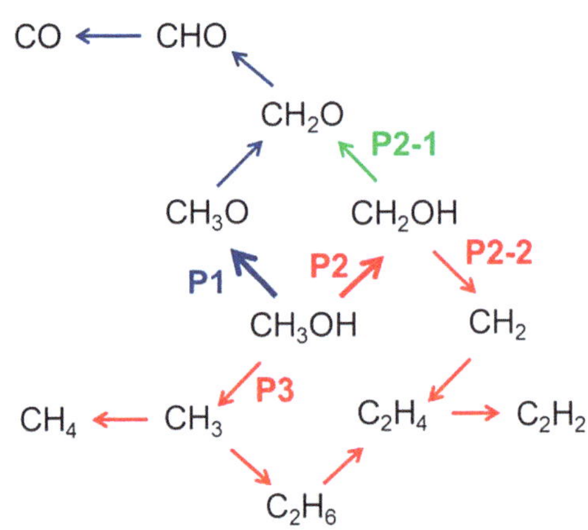

and CO, but the production of H_2CO is the main reaction pathway resulting in synthesis gas.

10.3.3 *Reaction Pathway in the Plasma-Catalytic Process*

Plasma catalysis can be thought of as a combination of a general gas-phase discharge and surface reactions in a catalyst, but there are phenomena that appear when two

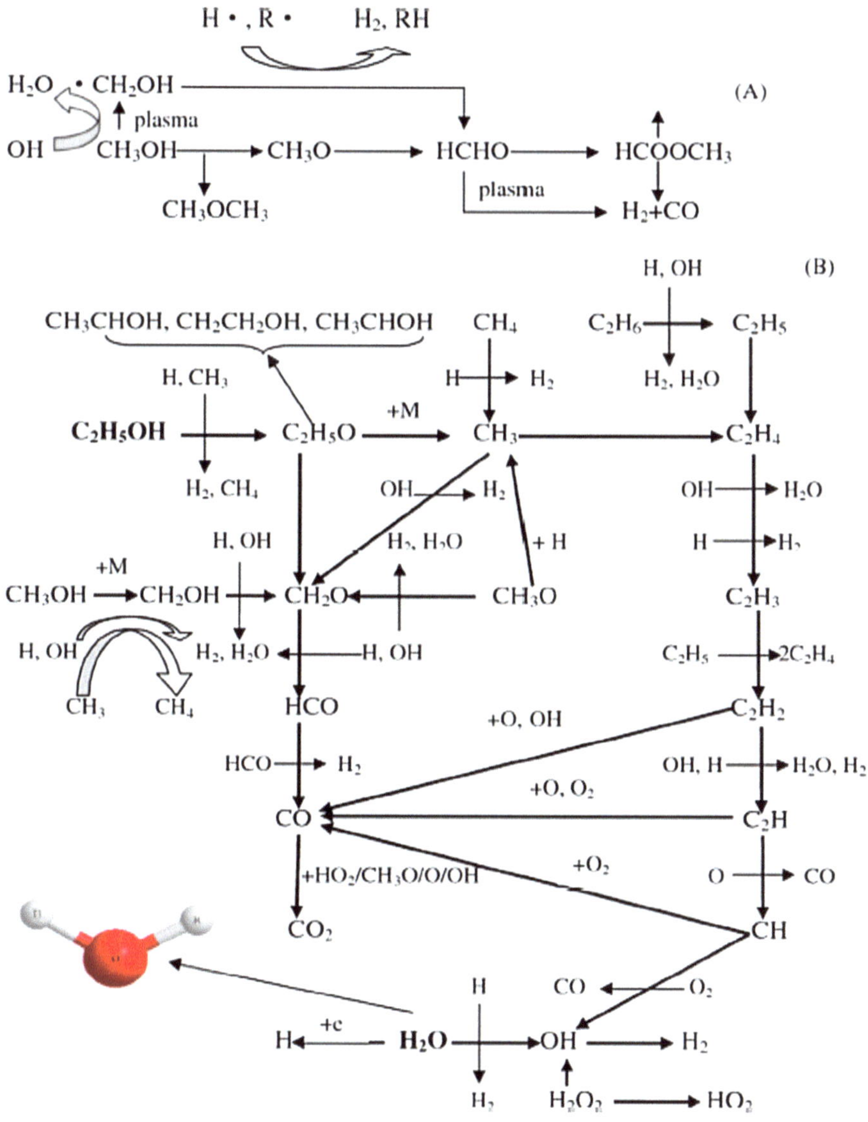

Fig. 10.10 Radical reaction paths for the plasma reforming of alcohol [56]

processes are combined that are more than a simple summation. Detailed aspects of the various interactions of the plasma and the catalyst can be found in the appropriate chapter. In this chapter, we will only mention some points that should be considered responsive.

The first is that the discharge itself in the gas phase occurs preferentially in the adjacent region of the catalyst. As shown in Fig. 10.11, a change in the electric field

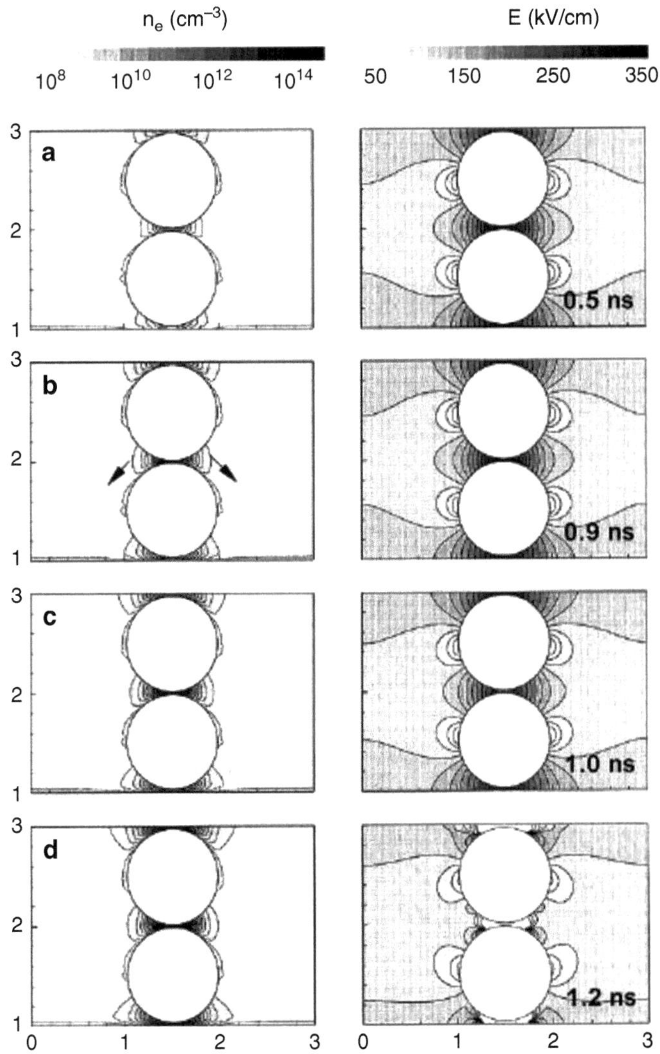

Fig. 10.11 Distribution of electron density (left) and electric field strength (right) inside a double-barrier DBD reactor packed with spherical ferroelectric materials [62]

distribution produced by the presence of the catalyst, whether a metal-supported catalyst or an oxide-supported catalyst, occurs, and the strongest streamer is found in the vicinity of the contact point between the catalysts [62]. Apart from the physical properties of the catalyst itself (whether it is conductive or in the oxide state), the size and shape of the catalyst (or support like alumina) in the catalyst-packed reactor are a major factor in affecting the electric field distribution.

The second is that the discharge pattern on the catalyst surface varies depending on the conditions of the catalyst (Fig. 10.12, [63]). In this case, the streamer and surface discharge can be changed by different amounts of charge accumulation on the surface, which is determined by the conductivity and amount of the catalyst [64]. Once the energy states of the produced electrons and active species are different, the primary reaction products of the gaseous reactants in contact with them will also be different (mainly radical intermediates).

The reaction path in the plasma-catalytic reaction includes both that of the plasma and the catalytic reaction. It is important to note that the alcohol molecules experience the discharge firstly in the state of gas phase. Considering this, in plasma catalysis, it is helpful to understand the phenomenon of reaction by considering the reaction characteristics in the 1. adsorption step, 2. reaction step, and 3. desorption step separately.

1. Adsorption step

When the alcohol molecules are inside the reaction volume, the reaction progresses in different ways depending on whether the molecule experienced the discharge before adsorption. Molecules that have not undergone discharge (without gas-phase collision) are adsorbed onto the catalyst like an ordinary catalytic reaction pathway, while molecules experiencing discharge approach the surface of the

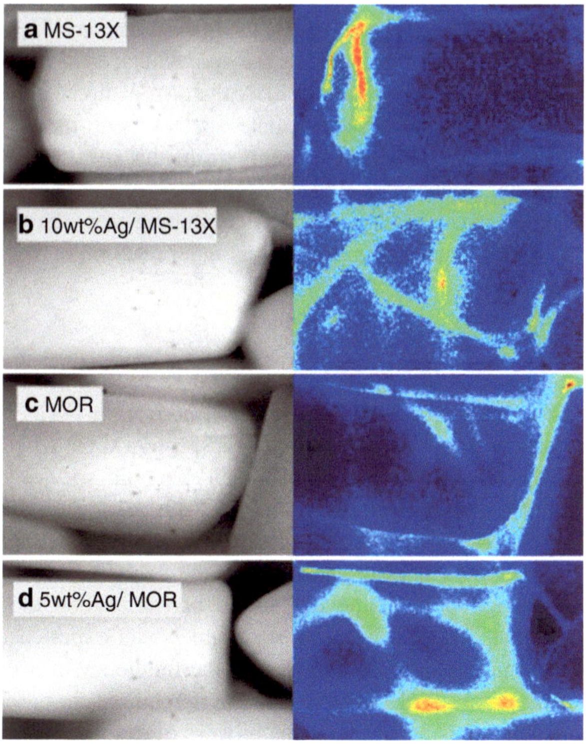

Fig. 10.12 Surface streamers propagating on the surface of various catalysts: (**a**) MS-13X, (**b**) 10 wt% Ag/MS-13X, (**c**) mordenite (MOR), (**d**) 5 wt % Ag/MOR. An AC 17 kV of 50 Hz was applied. The ICCD gate time was 40 ms. The diameter of the catalyst pellet is 1.6 mm [63]

catalyst in an excited or radical state. There is debate as to whether the excited state can enhance the adsorption or not.

For example, in the case of vibrationally excited methane, there have been reports of increased adsorption of methane on the Ni surface [65–67]. Considering that the sticking probability varies significantly depending on the crystal orientation or incident angle of the approaching species on the adsorption surface, if the catalyst is not in a precisely aligned crystal orientation, the increase in adsorption due to plasma will be limited [68]. The synergistic effect of plasma on the adsorption process still requires more investigation to be fully understood.

2. Reaction step

Plasma appears to occur on the surface of the catalyst, but since the discharge does not actually occur within the micropores in which the catalyst is present, the transport process is required for the species formed by the plasma to reach the surface of the catalyst. Gas-phase transport and surface transport can be considered, and the gas phase transport has been reported to occupy a significant portion of transport mechanism [55].

Since discharge does not occur on the surface of the catalyst, the generation of primary species by the plasma and the reaction of adsorbed species have different time scales and do not occur concurrently. Therefore, the most important feature of the plasma-catalyzed reaction in the reaction step is that the reaction path is of an Eley-Rideal type and thus relatively increased compared to the existing catalytic reaction or a Langmuir-Hinshelwood mechanism [55, 69, 70].

If there is a discharge in the space adjacent to the catalyst surface, the active radicals formed by the plasma can form a direct bond with the surface-adsorbed species. When the intermediate species is formed through an Eley-Rideal-type reaction, the reaction can begin from the specific intermediate step of a typical reaction mechanism starting from the adsorption of the molecular state. If the omitted reaction step is a rate-determining step, the plasma-catalyzed reaction can obtain a synergy effect that has a higher reaction rate than the conventional catalytic reaction.

3. Desorption step

The desorption process is basically a process in which adsorbed species participate and are controlled by thermal activation [71]. That is, there is relatively no room for the plasma to affect the desorption stage when compared to the adsorption or reaction steps. The collision of energetic species is likely to intervene in the reaction step but not likely to affect the desorption process. However, it has been suggested that the energy of intermediate species made by plasma is higher than that of other intermediate species and, the excited state may lead to different rates of desorption beyond the desorption barrier. When the low-temperature plasma is generated, the local region of streamers can experience heating, which may affect desorption. However, considering the frequency of occurrence of streamers and the area, a more in-detail investigation is required, in order to decide whether or not a discharge has an effect on the process of desorption.

10.4 Evaluation of the Process

Plasma is essentially a nonequilibrium phenomenon. This means that there is a possibility that the thermal efficiency of energy distribution in the system is not high. The onset of a discharge or breakdown process requires a high electric field and a following sufficiently high-energy supply. Therefore, it is always necessary to consider the cost of the plasma process. Generally, in the chemical process, the process cost due to plasma generation is calculated based on the amount of electric power consumed for plasma generation. Several parameters can be defined depending on the reactants or product conditions in order to take the power requirements into account. Firstly, one needs to evaluate the efficiency of the reaction itself. In the course of defining the efficiency of reaction, the concept of conversion, selectivity, and yield is introduced as below:

$$\text{Conversion }(\%) = \frac{\text{mole of }(R)\text{ in reactant} - \text{mole of }(R)\text{ in product}}{\text{mole of }(R)\text{ in reactant}} \times 100(\%) \tag{10.16}$$

$$\text{Selectivity }(\mathrm{H_2})(\%) = \frac{2}{2n+2} \times \frac{\text{mole of}(\mathrm{H_2})\text{ in product}}{\text{mole of }(R)\text{ in reactant} - \text{mole of }(R)\text{ in product}} \times 100(\%) \tag{10.17}$$

$$\text{Selectivity }(\mathrm{CO})(\%) = \frac{1}{n} \times \frac{\text{mole of}(CO)\text{in product}}{\text{mole of }(R)\text{ in reactant} - \text{mole of }(R)\text{ in product}} \times 100(\%) \tag{10.18}$$

$$\text{Yield }(A) = \text{Conversion} \times \text{Selectivity}(A) \tag{10.19}$$

Where (R) = Alcohol species, in the case of $C_nH_{2n+1}OH$ decomposition

Conversion is a parameter of how much of a reactant is converted through the reaction and is obtained by a relative comparison of the amount of a species in the reactants and products. Selectivity is a parameter of how selectively reactants are used for a particular reaction. The selectivity is for defining the chemical efficiency of the reaction or as a criterion for determining how much the total reaction was induced by the desired reaction without any other side reactions.

Yield is the value of conversion times selectivity, and it is a parameter for how efficiently the production of the product is achieved, relative to the reactants put into the reaction, apart from those participating in or not in desired reaction.

Conversion, selectivity, and yield all are parameters measuring the completeness of a reaction. However, these parameters do not provide information on the efficiency of the reaction in terms of energy cost. Therefore, additional parameters must be defined to evaluate the energy cost of the reaction process. Commonly used parameters are as follows [72, 73].

SEI: Specific energy input. It is a parameter generally used for the consideration of how much energy is used to run the reaction regardless of the reaction's chemical efficiency (without considering selectivity). It is often used when considering the cost of the plasma generation used to initiate a reaction. It is defined as the energy supplied to the reactor versus the reactant flow rate. The unit is J/L or J/mole.

SER: Specific energy requirement. It is a parameter that can compare the power consumption or energy input based on a specific product. Regardless of the amount of reactants, it is defined as the value of power supplied to produce a unit quantity of specific product. It is defined as the energy supplied to the unit mass/volume of specific product. J/L and J/kg are usually used as the units.

For the definition of thermal efficiency, in general, the ratio of total enthalpy of product to that of reactant is often used as shown in Eq. (10.20):

$$\eta_{\mathrm{th}}(\%) = \frac{\sum_{ip} H_{fi}^{0} \times Y_i}{\sum_{ir} H_{fi}^{0} \times X_i} \times 100(\%) \tag{10.20}$$

In the case of the plasma reaction, the power requirement for plasma generation is considered in terms of energy cost, and the thermal efficiency can be defined as (total enthalpy of product)/(required power + total enthalpy of reactant) as in Eq. (10.21):

$$\eta_{\mathrm{th}}(\%) = \frac{\sum_{ip} H_{fi}^{0} \times X_i}{\mathrm{Power} + \sum_{ir} H_{fi}^{0} \times X_i} \times 100(\%) \tag{10.21}$$

Compared with the existing definition of thermal efficiency, the thermal efficiency of the plasma process decreases because the power supplied to the system is included in the denominator. Strictly to say, it is more appropriate to change the "power" in the denominator of Eq. (10.21) to be "$\eta \times P$," by defining the conversion efficiency η to be the parameter for how much of the electric power for the generation of plasma is converted to heat or thermal energy:

$$\eta_{\mathrm{th}}(\%) = \frac{\sum_{ip} H_{fi}^{0} \times X_i}{\eta\,\mathrm{Power} + \sum_{ir} H_{fi}^{0} \times X_i} \times 100(\%) \tag{10.22}$$

As clearly shown in Eq. (10.22), the application of plasma results in a decrease in efficiency. In other words, despite the fact that the application of plasma is such a factor that lowers the efficiency, to gain the benefits of plasma application, the product should have a higher total enthalpy of the product or plasma application should provide higher selectivity of the specific product species compared to that of the typical catalytic reaction.

In general, the catalytic reaction is controlled by the equilibrium according to the temperature, so that the best of the yield and enthalpy of the product are predictable values, and in the case of the alcohol reforming reaction, the yield of the product is close to 100% at a sufficiently high-temperature condition. In other words, it can be understood that the benefits of plasma application in alcohol reforming process should be found at low temperatures at which condition the yield and efficiency of the existing catalytic reaction are low.

10.5 Other Considerations

10.5.1 Benefits of Using Alcohols

In the case of the alcohol reforming reaction at room temperature, the liquid-phase reaction must proceed, and in the case of the reaction using only plasma, there is an energy cost due to the vaporization process which is required. Therefore, having the reaction at a temperature condition in which the gaseous state is possible would be advantageous for the plasma generation and reaction. However, since the equilibrium reforming reaction proceeds at a relatively low temperature (200–250 °C) in the presence of a catalyst, there is not much difference between the temperature required for vaporizing the reactant and the temperature for securing the activity of the catalyst. That is, the region (temperature range) where the efficiency increase can be obtained through the application of the plasma is not so large. Therefore, it is advantageous to use the plasma-catalyzed method rather than just plasma alone in terms of thermal efficiency [74–76]. However, besides the advantage of obtaining the activity of the alcohol reactant under a relatively low temperature instead of the conventional catalytic activation temperature, there is a limited range of conditions to get the benefits of plasma application.

Therefore, the benefits of plasma application should be considered not in viewpoint of process cost or thermal efficiency, but in terms of securing yield in relative low-temperature conditions. In addition to factors other than yield, the fact that the catalyst can be regenerated and that a high reactivity can be obtained even under the conditions of deactivated catalyst should be considered as the basis of plasma application.

10.5.2 Reforming Catalysts

In general, transition metals, especially Cu-based catalysts, are widely used for the reforming of alcohols. The catalytic performance of Cu catalysts can be improved by using promoters such as ZrO_2, ZnO-CNTs, Zn, and SiO_2 [77–80]. The effect of synthesis conditions and the relative composition of promoters on the catalytic performance have been investigated [81].

However, in the plasma-catalytic reaction, especially the reaction proceeding under conditions when the activity of the catalyst is not high, a catalyst useful for generating plasma other than Cu may be more advantageous for the efficiency of the whole reaction. In general plasma catalysis, a conductive metal catalyst is advantageous for surface discharge, while an oxide catalyst having dielectric properties is advantageous for the generation of a strong streamer. As a consequence, consideration should be given in the course of catalyst selection whether the plasma effect should be utilized, or catalytic reaction will be important in the target process [82, 83]. However, other oxygenates can also be used for hydrogen production such as DME, butanol, and IPA (isopropyl alcohol) using nonthermal plasmas [84–90].

10.5.3 Recommended Resources

Lots of publications can be found regarding the plasma or plasma-catalytic-driven reforming/decomposition of alcohol. Actually the history of catalytic alcohol reforming is rather long and there are numbers of extensive review articles. Suggested review articles are listed below.

- Meng, N. I., Leung, D. Y. C., Leung, M. K. H. (2007). A review on reforming bio-ethanol for hydrogen production. *International Journal of Hydrogen Energy*, *32*(15), 3238–3247.
- Chen, F., Huang, X., Cheng, D., Zhan, X. Hydrogen production from alcohols and ethers via cold plasma: A review. *International Journal of Hydrogen Energy*. *39*(17), 9036–9046.
- Palo, D. R., Dagle, R. A., Holladay, J. D. (2007). Methanol steam reforming for hydrogen production. *Chemical Reviews*, *107*(10), 3992–4021.

Most of the contents in this chapter focused on methanol and ethanol. However, there are other oxygenates that can also be used for hydrogen production such as DME, butanol, and IPA (isopropyl alcohol). Plasma can be used for the reforming of these oxygenates and publications on these topics can also be found [84–90].

10.6 Summary and Suggestions

The decomposition/reforming process of alcohol using plasma is introduced. Catalytic reactions, plasma reactions, and plasma-catalytic reactions are introduced for reaction paths of general reforming reactions. The reaction pathway is relatively well known for the methanol/ethanol reforming reaction, and the reaction by the plasma is easier to understand than the typical reforming/decomposition of gas-phase hydrocarbon species.

Unlike methane and other hydrocarbon species, alcohol has a low activation energy, and the reforming/decomposition temperature of the catalytic reaction is relatively low in the range of 200–300 °C. Therefore, the advantage of decomposition/modification of methanol/ethanol at a relatively lower temperature by using plasma is not remarkable. Actually, the gas-phase plasma reaction itself requires temperature of above 80 °C for the vaporization. The advantages of lowering the reaction temperature from that of the catalytic reaction are therefore not much; in the case of plasma catalysis reactions, this means the advantage on lowering the activation barrier by plasma application is not so high. The advantages of plasma application should be found in that it is possible to produce a reaction path different from that of the existing catalytic reaction. The gaseous radicals generated by the plasma can produce products which cannot be obtained in the conventional catalytic reaction, either independently of the catalytic reaction or by reacting with adsorbent species on the catalytic surface.

From this point of view, in particular, in the reforming/decomposition reaction of alcohols using only plasma, a difference from conventional catalyst-based processes can be obtained in the case of applying a low-temperature plasma to the reaction in a liquid phase. The liquid-phase environment can inhibit thermal oxidation of intermediate species in reforming reaction and can produce oxygenated hydrocarbon species such as aldehydes instead of a general reforming product such as hydrogen and carbon monoxide. This can be the most unique process characteristic that the plasma process can bring.

Methanol itself is an intermediate and is highly utilized in various industrial processes. As in the case of MTO, the reaction route generated from the syngas is mostly used to generate methanol. The methanol reforming reaction, which is a reverse reaction thereof, is not used much except for the case of DMFC. Nonetheless, studies on the methanol/ethanol reforming by plasma can provide important insights into the study of various hydrocarbon species based on the work done on low-temperature oxidation and generation of oxygenated hydrocarbons.

References

1. Olah, G. A. (2005). Beyond oil and gas: The methanol economy. *Angewandte Chemie International Edition, 44*, 2636–2639.
2. Sheldon, R., Arends, I. W. C., Arends, E., & Dijksman, A. (2000). New developments in catalytic alcohol oxidations for fine chemicals synthesis. *Catalysis Today, 57*, 157–166.
3. Calkins, W. H. (1984). Chemicals from methanol. *Catalysis Reviews Science and Engineering, 26*, 347–358.
4. Kaeding, W. W., & Butter, S. A. (1980). Production of chemicals from methanol: I. low molecular weight olefins. *Journal of Catalysis, 61*, 155–164.
5. Lamy, C., Lima, A., LeRhun, V., Delime, F., Coutanceau, C., & Léger, J. (2002). Recent advances in the development of direct alcohol fuel cells (DAFC). *Journal of Power Sources, 105*, 283–296.

6. Hill, J., Nelson, E., Tilman, D., Polasky, S., & Tiffany, D. (2006). Environmental, economic, and energetic costs and benefits of biodiesel and ethanol biofuels. *Proceedings of the National Academy of Sciences, 103*, 11206–11210.
7. Dillon, R., Srinivasan, S., Aricò, A. S., & Antonucci, V. (2004). International activities in DMFC R&D: Status of technologies and potential applications. *Journal of Power Sources, 127*, 112–126.
8. Bguyen, V. L., Cao, M. T., Masayuki, N., & Michitaka, O. (2012) Heat treatment - conventional and novel applications, Chapter 11, INTECH, ISBN 978-953-51-0768-2, 2012 under.
9. Balat, M., & Balat, H. (2009). Recent trends in global production and utilization of bio-ethanol fuel. *Applied Energy, 86*, 2273–2282.
10. http://www.consumerenergycenter.org/transportation/afvs/ethanol.html
11. Farrell, A. E., Plevin, R. J., Turner, B. T. Jones, A. D., O'Hare, M., & Kammen, D. M. (2006). Ethanol can contribute to energy and environmental goals. *Science, 311*, 506–508.
12. Tian, P., Wei, Y., Ye, M., & Liu, Z. (2015). Methanol to olefins (MTO): From fundamentals to commercialization. *ACS Catalysis, 5*, 1922–1938.
13. Coal to chemicals: Is it coal's time again? *Nexant/ChemSystems MultiClient Report*, January 2006.
14. Stöcker, M. (1999). Methanol-to-hydrocarbons: Catalytic materials and their behavior. *Microporous and Mesoporous Materials, 29*, 3–48.
15. Hindman, H. (2013). Methanol to gasoline technology. The Twenty-third International Offshore and Polar Engineering Conference. International Society of Offshore and Polar Engineers.
16. Methanol to Gasoline (MTG): Production of Clean Gasoline from Coal. Exon Mobil Research and Engineering. Available from http://www.exxonmobil.com/Apps/Refining Technologies/files/sellsheet_09_mtg_brochure.pdf.
17. Peppley, B. A., Armphlett, J. C., Kearns, L. M., & Mann, R. F. (1999). Methanol–steam reforming on $Cu/ZnO/Al_2O_3$. Part 1: The reaction network. *Applied Catalysis A: General, 179*, 21–29.
18. Velu, S., Suzuki, K., Okazaki, M., Kapoor, M. P., Osaki, T., & Ohashi, F. (2000). Oxidative steam reforming of methanol over CuZnAl (Zr)-oxide catalysts for the selective production of hydrogen for fuel cells: Catalyst characterization and performance evaluation. *Journal of Catalysis, 194*, 373–384.
19. Deo, G., & Wachs, I. E. (1994). Reactivity of supported vanadium oxide catalysts: The partial oxidation of methanol. *Journal of Catalysis, 146*, 323–334.
20. Alejo, L., Lago, R., Peña, M. A., & Fierro, J. L. G. (1997). Partial oxidation of methanol to produce hydrogen over Cu-Zn-based catalysts. *Applied Catalysis A: General, 162*, 281–297.
21. Ueno, A., Onishi, T., & Tamaru, K. (1971). Reaction intermediates in methyl alcohol decomposition on ZnO. *Transactions of the Faraday Society, 67*, 3585–3589.
22. Sá, S., Silva, H., Bandão, L., Sousa, J. M., & Mendes, A. (2010). Catalysts for methanol steam reforming—A review. *Applied Catalysis B: Environmental, 99*, 43–57.
23. Mattos, L. V., Jacobs, G., Davis, B. H. & Noronha, F. B. (2012). Production of hydrogen from ethanol: Review of reaction mechanism and catalyst deactivation. *Chemical Reviews, 112*, 4094–4123.
24. Yong, S. T., Ooi, C. W., Chai, S. P., & Wu, X. S. (2013). Review of methanol reforming-Cu-based catalysts, surface reaction mechanisms, and reaction schemes. *International Journal of Hydrogen Energy, 38*, 9541–9552.
25. Chang, J.-S., Lawless, P. A., & Yamamoto, T. (1991). Corona discharge processes. *IEEE Transactions on Plasma Science, 19*, 1152–1166.
26. Tabrizchi, M., Khayamian, T., & Taj, N. (2000). Design and optimization of a corona discharge ionization source for ion mobility spectrometry. *Review of Scientific Instruments, 71*, 2321–2328.
27. Kogelschatz, U. (2003). Dielectric-barrier discharges: Their history, discharge physics, and industrial applications. *Plasma Chemistry and Plasma Processing, 23*, 1–46.
28. Massines, F., Ségui, P., Gherardi, N., Khamphan, C., & Ricard, A. (2003). Physics and chemistry in a glow dielectric barrier discharge at atmospheric pressure: Diagnostics and modelling. *Surface and Coatings Technology, 174*, 8–14.

29. Czernichowski, A. (1994). Gliding arc: Applications to engineering and environment control. *Pure and Applied Chemistry, 66*, 1301–1310.
30. Fridman, A., Nester, S., Kennedy, L. A., Saceliev, A., & Mutaf-Yardimci, O. (1999). Gliding arc gas discharge. *Progress in Energy and Combustion Science, 25*, 211–231.
31. Lee, D., Kim, T., Cha, M., & Song, Y. (2007). Optimization scheme of a rotating gliding arc reactor for partial oxidation of methane. *Proceedings of the Combustion Institute, 31*, 3343–3351.
32. Lee, D., Kim, T., Cha, M., & Song, Y. (2010). Plasma-controlled chemistry in plasma reforming of methane. *International Journal of Hydrogen Energy, 35*, 10967–10976.
33. Schoenbach, K. H., El-Habachi, A., Shi, W., & Ciocca, M. (1997). High-pressure hollow cathode discharges. *Plasma Sources Science and Technology, 6*, 468.
34. Schoenbach, K. H., Verhappen, R., Tessnow, T., & Peterkin, F. E. (1996). Microhollow cathode discharges. *Applied Physics Letters, 68*, 13–15.
35. Meek, J. M. (1940). A theory of spark discharge. *Physical Review, 57*, 722.
36. Loeb, L. B., & Meek, J. M. (1940). The mechanism of spark discharge in air at atmospheric pressure. I. *Journal of Applied Physics, 11*, 438–447.
37. Du, C. M., Mo, J. M., & Li, H. X. (2014). Renewable hydrogen production by alcohols reforming using plasma and plasma-catalytic technologies: Challenges and opportunities. *Chemical Reviews, 115*, 1503–1542.
38. Bromberg, L., Cohn, D. R., Rabinovich, A., Alexeev, N., Samokhin, N., Hadidi, K., & Kaaja, R. (2006). Plasma Science and Fusion Center, JA-02-30.
39. Du, C., Li, H., Zhang, L., Wang J., Huang D., Xiao M., Cai J., Chen Y., Yan H., & Xiong Y. (2012). Hydrogen production by steam-oxidative reforming of bio-ethanol assisted by Laval nozzle arc discharge. *International Journal of Hydrogen Energy, 37*, 8318–8329.
40. Bundaleska, N., Tatarova, E., Dias, F. M., & Ferreira, C. M. (2014). Steam reforming of ethanol into hydrogen-rich gas using microwave Ar/water "tornado"–Type plasma. *International Journal of Hydrogen Energy, 39*, 5663–5670.
41. Sarmiento, B., Brey, J. J., Viera, I. G., González-Elipe, A. R., Cotrino, J., & Rico, V. J. (2007). Hydrogen production by reforming of hydrocarbons and alcohols in a dielectric barrier discharge. *Journal of Power Sources, 169*, 140–143.
42. Paterkowski, W., Parus, W., & Kalisiak, S. (2009). Destructive oxidation of ethanol in the corona discharge reactor. *Polish Journal of Chemical Technology, 11*, 57–62.
43. Hoang, T. Q., Zhu, X. l., Lobban, L. L., & Mallinson, R. G. (2011). Effects of gap and elevated pressure on ethanol reforming in a non-thermal plasma reactor. *Journal of Physics D: Applied Physics, 44*, 274003.
44. Yan, Z. C., Li, C., & Lin, W. H. (2009). Hydrogen generation by glow discharge plasma electrolysis of methanol solutions. *International Journal of Hydrogen Energy, 34*, 48–55.
45. Aubry, O., Met, C., Khacef, A., & Cormier, J. M., (2005). On the use of a non-thermal plasma reactor for ethanol steam reforming. *Chemical Engineering Journal, 106*, 241–247.
46. Chernyak, V. Y., Olszewski, S. V., Yukhymenko, V. V., Solomenko E. V., Prysiazhnevych, I. V., Naumov, V. V., Levko, D. S., Shchedrin, A. I., Ryabtsev, A. V., Demchina, V. P., Kudryavtsev, V. S., Martysh, E. V., & Verovchuck, M. A. (2008). Plasma-assisted reforming of ethanol in dynamic plasma-liquid system: Experiments and modeling. *IEEE Transactions on Plasma Science, 36*, 2933–2939.
47. Jiang, C. J., Trimm D.L., Wainwright, M.S. & Cant, X.W. (1993). Kinetic mechanism for the reaction between methanol and water over a Cu-ZnO-Al_2O_3 catalyst. *Applied Catalysis A: General, 97*, 145–158.
48. Jiang, C. J., Trimm D. L., Wainwright, M. S., & Cant, X. W., (1993). Kinetic study of steam reforming of methanol over copper-based catalysts. *Applied Catalysis A: General, 93*, 245–255.
49. Peppley, B. A., Amphlett, J. C. C., Kenrns, L. M., & Mann, F. R. (1999). Methanol–steam reforming on Cu/ZnO/Al_2O_3. Part 1: The reaction network. *Applied Catalysis A: General, 179*, 21–29.

50. Takezawa, N., & Iwasa, N. (1997). Steam reforming and dehydrogenation of methanol: Difference in the catalytic functions of copper and group VIII metals. *Catalysis Today, 36*, 45–56.
51. Frank, B., Jentoft, F. C., Soerijanto, H., Schlogl, R., & Schomacker, R. (2007). Steam reforming of methanol over copper-containing catalysts: Influence of support material on microkinetics. *Journal of Catalysis, 246*, 177–192.
52. Zanchet, D., Santos, J. B. O., Damyanova, S., Gallo, J. M. R., & Bueno, J. M. C. (2015). Toward understanding metal-catalyzed ethanol reforming. *ACS Catalysis, 5*, 3841–3863.
53. Mavrikakis, M., & Barteau, M. A. (1998). Oxygenate reaction pathways on transition metal surfaces. *Journal of Molecular Catalysis A: Chemical, 131*, 135–147.
54. Baxter, R. J., & Hu, P. (2002). Insight into why the Langmuir–Hinshelwood mechanism is generally preferred. *The Journal of Chemical Physics, 116*, 4379.
55. Kim, T., Lee, D. H., Jo, S., Pyun, S. H., Kim, K., & Song Y. H. (2016). Mechanism of the accelerated reduction of an oxidized metal catalyst under electric discharge. *ChemCatChem, 8*, 685–689.
56. Li, H. (2012). Master Thesis, Sun Yat-Sen University, Ethanol reforming using nonthermal arc discharges.
57. Lee, D. H., & Kim, T. (2014). Effect of catalyst deactivation on kinetics of plasma-catalysis for methanol decomposition. *Plasma Processes and Polymers, 11*, 455–463.
58. Levko, D., Shchedrin, A., Chernyak, V., Olszewski, S., & Nedybaliuk, O. (2011). Plasma kinetics in ethanol/water/air mixture in a 'tornado'-type electrical discharge. *Journal of Physics D: Applied Physics, 44*, 145206.
59. Levko, D., Shchedrin, A., Naumov, V., Chernyak, V., Yukhymenko, V., Prysiazhnevych, I., & Olszewski, S. (2008). Modeling of plasma-assisted conversion of liquid ethanol into hydrogen enriched syngas in the nonequilibrium electric discharge plasma-liquid system. *arXiv preprint arXiv, 0809.0968*.
60. Shchedrin, A. I., Levko, D. S., Ryabtsev, A. V., Chernyak, V. Ya., Yukhimenko, V. V., Ol'shevskiy, S. V., Prisyazhnevich, I. V., Solomenke, E. V., Naumov, V. V., Demchina, V. P., & Kudryavtsev, V. S. (2008). Plasma's kinetics in discharge in mixture of air, water and ethanol steams and the questions of alternative fuel. *Вопросы атомной науки и техники*.
61. Lee, D. H., & Kim, T. (2014). Effect of catalyst deactivation on kinetics of plasma-catalysis for methanol decomposition. *Plasma Processes and Polymers, 11*, 455–463.
62. Kang, W. S., Park, J. M., Kim, Y., & Hong, S. H. (2003). Numerical study on influences of barrier arrangements on dielectric barrier discharge characteristics. *IEEE Transactions on Plasma Science, 31*, 504–510.
63. Kim, H. H., Teramoto, Y., Negishi, N., & Ogata, A. (2015). A multidisciplinary approach to understand the interactions of nonthermal plasma and catalyst: A review. *Catalysis Today, 256*, 13–22.
64. Jo, S., Kim, T., Lee, D. H., kang, W. S., & Song, Y. H. (2014). Effect of the electric conductivity of a catalyst on methane activation in a dielectric barrier discharge reactor. *Plasma Chemistry and Plasma Processing, 34*, 175–186.
65. Crim, F. F. (2008). Chemical dynamics of vibrationally excited molecules: Controlling reactions in gases and on surfaces. *Proceedings of the National Academy of Sciences, 105*, 12654–12661.
66. Halonen, L. (2001). Reactivity of vibrationally excited methane on nickel surfaces. *The Journal of Chemical Physics, 115*, 5611.
67. Shirazi, M., Neyts, E. C., & Bogaerts, A. (2017). DFT study of Ni-catalyzed plasma dry reforming of methane. *Applied Catalysis B: Environmental, 205*, 605–614.
68. Juaristi, J. I., Alducin, M., Diez Muiño, R., Busnengo, H. F., & Salin, A. (2008). Role of electron-hole pair excitations in the dissociative adsorption of diatomic molecules on metal surfaces. *Physical Review Letters, 100*, 116102.
69. Kim, T., Lee, Dae H., Jo, S., Pyun, S. H. & Kim, K. T. (2016). Mechanism of the accelerated reduction of an oxidized metal catalyst under electric discharge. *ChemCatChem, 8*, 685–689.

70. Kratzer, P., & Brenig, W. (1991). Highly excited molecules from Eley-Rideal reactions. *Surface Science, 254*, 275–280.
71. King, D. A. (1975). Thermal desorption from metal surfaces: A review. *Surface Science, 47*, 384–402.
72. Bromberg, L., Cohn, D. R., Rabinovich, A., & Alexeev, N. (1999). Plasma catalytic reforming of methane. *International Journal of Hydrogen Energy, 24*, 1131–1137.
73. Petitpas, G., Rolloer, J.-D., Darmon, A., Gonzalez-Aguilar, J., Metkemeijer, R., & Fulcheri, L. (2007). A comparative study of non-thermal plasma assisted reforming technologies. *International Journal of Hydrogen Energy, 32*, 2848–2867.
74. Whitehead, J. C. (2010). Plasma catalysis: A solution for environmental problems. *Pure and Applied Chemistry, 82*, 1329–1336.
75. Durme, J. V., Dewulf, J., Leys, C., & Langenhove, H. V. (2008). Combining non-thermal plasma with heterogeneous catalysis in waste gas treatment: A review. *Applied Catalysis B: Environmental, 78*, 324–333.
76. Neyts, E. C., & Bogaerts, A. (2014). Understanding plasma catalysis through modelling and simulation—A review. *Journal of Physics D: Applied Physics, 47*, 224010.
77. Oguchi, H., Kanai, H., Utani, K., Matsumura, Y., & Imamura, S. (2005). Cu_2O as active species in the steam reforming of methanol by CuO/ZrO_2 catalysts. *Applied Catalysis A: General, 293*, 64–70.
78. Yang, H.-M., & Liao, P.-H. (2007). Preparation and activity of Cu/ZnO-CNTs nano-catalyst on steam reforming of methanol. *Applied Catalysis A: General, 317*, 226–233.
79. Jones, S. D., & Hagelin-Weaver, H. E. (2009). Steam reforming of methanol over CeO_2-and ZrO_2-promoted Cu-ZnO catalysts supported on nanoparticle Al_2O_3. *Applied Catalysis B: Environmental, 90*, 195–204.
80. Matsumura, Y., & Ishibe, H. (2009). Selective steam reforming of methanol over silica-supported copper catalyst prepared by sol–gel method. *Applied Catalysis B: Environmental, 86*, 114–120.
81. Li, D., Li, X., & Gong, J. (2016). Catalytic reforming of oxygenates: State of the art and future prospects. *Chemical Reviews, 116*, 11529–11653.
82. Ogata, A., Kim, H. H., Futamura, S., Kushiyama, S., & Mizuno, K. (2004). Effects of catalysts and additives on fluorocarbon removal with surface discharge plasma. *Applied Catalysis B: Environmental, 53*, 175–180.
83. Tu, X., Gallon, H. J., & Whitehead, J. C. (2011). Electrical and spectroscopic diagnostics of a single-stage plasma-catalysis system: effect of packing with TiO_2. *Journal of Physics D: Applied Physics, 44*, 482003.
84. Song, L., Li, X., & Zheng, T. (2008). Onboard hydrogen production from partial oxidation of dimethyl ether by spark discharge plasma reforming. *International Journal of Hydrogen Energy, 33*, 5060–5065.
85. Zou, J.-J., Zhang, Y.-p., & Liu, C.-J. (2007). Hydrogen production from partial oxidation of dimethyl ether using corona discharge plasma. *International Journal of Hydrogen Energy, 32*, 958–964.
86. Xing-hu, L. I. (2010). Production of hydrogen-rich gas by plasma reforming of dimethyl ether. *Journal of Fuel Chemistry and Technology, 38*, 201–206.
87. Pan, Y.-X., Han, Y., & Liu, C.-J. (2007). Pathways for steam reforming of dimethyl ether under cold plasma conditions: A DFT study. *Fuel, 86*, 2300–2307.
88. Zou, J.-J., Zhang, Y.-P., & Liu, C.-J. (2007). Hydrogen production from dimethyl ether using corona discharge plasma. *Journal of Power Sources, 163*, 653–657.
89. Jiménez, M., Rincón, R., Marinas, A., & Calzada, M. D. (2013). Hydrogen production from ethanol decomposition by a microwave plasma: Influence of the plasma gas flow. *International Journal of Hydrogen Energy, 38*, 8708–8719.
90. Demidiouk, V., & Chae, J. O. (2005). Decomposition of volatile organic compounds in plasma-catalytic system. *IEEE Transactions on Plasma Science, 33*, 157–161.

Chapter 11
Plasma Catalysis: Challenges and Future Perspectives

J. Christopher Whitehead

Plasma catalysis has been demonstrated as a promising alternative to thermal catalysis for environmental clean-up and the synthesis of platform chemicals and fuels from different feedstocks at low temperatures. There have been considerable and increasing research activities in this emerging and interdisciplinary field in recent years. However, plasma catalysis, particularly using a single-stage configuration, is a very complex process involving both gas-phase reactions driven by the plasma and plasma-assisted surface reactions. A number of challenges need to be addressed to achieve significant advancement in this field and the full potential of this emerging technology.

11.1 Scientific and Engineering Challenges

11.1.1 Experimental Challenges

In general terms, most of our existing information about plasma catalysis has come from experiments performed under a variety of conditions in which the products were measured in the output gas stream and the catalyst might be characterized before and after processing. Such gas-phase analyses are restricted to stable species (which might include some by-products and very long-lived intermediates such as ozone) but contain no information about short-lived intermediates. In addition, there is always a possible complication that reactions unconnected with the studied plasma catalysis process might occur after processing either in the tubing or the collection vessel but before sampling. Similarly, analysis of the surface of the catalyst after the experiment

J. C. Whitehead (✉)
School of Chemistry, The University of Manchester, Manchester, UK
e-mail: j.c.whitehead@manchester.ac.uk

X. Tu et al. (eds.), *Plasma Catalysis*, Springer Series on Atomic, Optical, and Plasma Physics 106, https://doi.org/10.1007/978-3-030-05189-1_11

will only yield information about the strongly bound or long-lived products or intermediates and may become contaminated post experiment but before analysis.

The need is to be able to routinely measure not just the final products but also the intermediates, i.e. we need in situ measurements to extend our ex situ ones giving us information that can aid the formation and validation of accurate reaction mechanisms. These measurements will be performed both in the gas-phase and on the catalyst surface. They can be of time-averaged concentrations for these species, but even more valuable would be to be able to measure their time evolution and know how they vary in different parts of the reactor. Also we would like to identify the species in the boundary layer just above the surface as well as what is absorbed onto the surface. Knowing the variation in concentration of all the species in the gas-phase on moving away from the surface and also in the direction of the gas flow will yield valuable information, i.e. we need to obtain 2- or 3-D maps of the gas-phase species in the reactor. As an example of what can be achieved, the reader could refer to the 2-D imaging of OH and toluene in an atmospheric gliding arc study of the destruction of toluene in air using a planar laser beam and detected the spatially resolved laser-induced fluorescence [1]. There is no catalyst in that experiment, but one can easily imagine the extension to plasma catalysis. An alternative is to use some of the variety of laser absorption or fluorescence techniques where the laser beam can be moved in the reactor relative to the catalyst surface. Teramoto et al. [2] give an example of measuring OH radical profiles in humid air plasma as a function of distance from an electrode. Temporal information can also be obtained using a pulsed plasma system and undertaking the detection after the pulse with a variable delay. This can be applied to absorption and fluorescence systems using pulsed lasers to give information about ground-state species and also for excited species using time resolved UV/visible or infrared emission.

Analysis of species on the catalyst surface can be achieved in situ and in real time using some infrared techniques such as diffuse reflectance infrared Fourier transform spectroscopy (DRIFTS), attenuated total reflection (ATR) and absorption [3–5]. These techniques have a sensitivity for species located on and a short distance into the catalyst surface (of the order of microns). These techniques have all been used in a plasma environment, and the DRIFTS and absorption measurement have already been applied successfully to plasma–catalytic systems. They are capable of time-resolved measurement with a resolution of about 1 s and have been combined with parallel gas-phase measurements [3].

11.1.2 Theoretical Calculations and Modeling Challenges

Theoretical and computational methods currently play an important role in the understanding and development of plasma catalysis, and their role will continue to increase. At various levels, they can provide an understanding based on an assumed model, which is inevitably approximate to some extent but can be used to replicate and extend the results of experiments and to provide information to inform the design of new experiments and processes. The experimental information obtained from in situ measurements on the gaseous and surface species can be used in conjunction with quantum mechanical calculations of the reaction pathways in terms of their energetics and configurations coupled with modeling and simulations of the dynamics on such reaction pathways to gain an understanding of the reaction dynamics in these plasma catalytis systems. The calculations can make predictions about the nature of intermediates, the time profiles of the evolution and decay of the different species that can be compared with experiments, thereby validating the assumptions made. Inevitably, the calculations may be of reduced dimensionality for computational convenience, and we can expect to see that restriction disappearing as computational power and the associated software develop [6–10]. Considerable progress is being made at the moment in the use of modeling to explain and evaluate the electrical effects of discharges with packing [11–13] including the effect of different gas mixtures [12]. These calculations are forerunners of models for the effect of packing in plasma catalysis. Models are being developed for different plasma types such as dielectric bed discharges, gliding arc and microwave systems [14–16] that are stepping stones towards the incorporation of a catalyst.

Hensen et al. have developed a microkinetic methodology for describing the mechanism of catalytic reactions that makes use of density functional theory (DFT) quantum calculations of the surface reaction potential energies for the intermediates and calculates the kinetics without any assumptions of the critical reaction pathway incorporating the dynamic changes of the surface during the reaction. This has been applied to the Fischer–Tropsch process [17, 18]. For the purposes of describing a plasma–catalytic process, it would be necessary to incorporate in a dynamic manner the effect of the plasma on the binding of species to the surface and on the nature of the transition state. Kim et al. [19] have presented a kinetic method to elucidate the key steps involved in the plasma–catalytic processing of the dry reforming of methane with CO_2 by using experimental results at a range of temperatures as input. They can separate the gas-phase reactions taking place from those on the Ni surface. In contrast to thermal catalysis for this reaction that shows Arrhenius behavior, the plasma catalytic systems exhibit a non-Arrhenius temperature dependence. A value for the energy barrier for plasma catalysis obtained from the relationship between rate constant and input power is 20 kJ/mol to be compared with the activation energy for thermal catalysis of 70 kJ/mol. They attribute the reduction to an energetically favorable interaction between plasma-generated vibrationally excited CH_4 and the Ni sites on the catalyst.

11.1.3 Engineering Challenges

Compared with the developments and progress that are being made by experiment and theory, there is comparatively little progress being made with scaling up plasma catalytis systems to the industrial scale. Many experiments are now focusing on the energy efficiency of the processing and attempting to benchmark results with competing techniques. These benchmark parameters are important in allowing different plasma catalysis experiments to be compared among themselves in terms of the effectiveness of different plasma types, different catalysts and other system variables and as the means of comparing plasma catalysis with other competing technologies. An excellent example of this approach can be found in the recent work of Snoeckx and Bogaerts for the conversion of CO_2 [20].

When considering possible scale-up and industrial applications of plasma catalysis, a whole life cost analysis must be performed to determine if the process will be economically competitive [21]. Unless a process can be developed which has 100% selectivity for the desired product, there can be additional costs associated with separating a range of products such as the use of pressure swing adsorption or membranes that require additional capital and energy input. A promising recent development has been the hybridization of a dielectric barrier discharge (DBD) reactor with a solid oxide electrolyser cell (SOEC) for the conversion of CO_2 into CO and O_2 where the SOEC removes oxygen from the reactor as it is formed, preventing the back reaction of CO and O_2 to reform CO_2 allowing complete conversion of the CO_2 [22].

Many of the catalysts that are being used in the present experiments involve expensive and scarce metals, and effort must be applied to using catalysts that are inexpensive. It is also necessary to evaluate the long-term performance of the catalyst in a plasma environment to demonstrate that it is at least as good as in the thermal experiment. Interestingly, most of the catalysts used in plasma catalysis are identical to the ones used in the analogous thermal experiment. We need to break free of this mind-set and use the accumulated knowledge of our studies to identify the specific characteristics of a successful catalyst in plasma catalysis and go for the goal of designing catalysts specifically for use in a plasma catalysis system. This way we might achieve better conversions, selectivities and higher energy efficiencies.

11.2 The Future

Despite at least 25 years of worldwide research into plasma catalysis, the vast majority of experiments are of the laboratory scale and are probing the mechanism of plasma catalysis. Considerable progress is being made in gaining an understanding of the intricacies of the mechanism of plasma catalysis, and this is significantly assisted by the theoretical and computational studies. There has been an evolution of emphasis from exhaust gas remediation and NO_x removal, through volatile organic compounds (VOCs) destruction to the current focus on energy-related topics such as methane reforming for syngas and hydrogen production, oxidative conversion of methane, gas-to-liquid conversion of hydrocarbons and reutilization of carbon dioxide and its implications for energy storage. It is vital for the research to remain relevant as it gives it impact and helps to secure funding. However, there will come a time when all this promise must be realized and blue skies research must be scaled up, be commercialized and enter the hard commercial world. When will plasma catalysis make this next step? It has to be before goodwill towards continued funding goes cold. The question to reflect upon is what we now need to do that we have not yet done to make this step. We need to identify and exploit the unique features of plasma catalysis processing (e.g. simplified construction, portability, rapid response, low-temperature operation, etc.). Consideration must be given to application into niche areas of low throughput versus competition with existing high-volume processes. Much to consider over the next 25 years!

References

1. Gao, J., Zhu, J., Ehn, A., Aldén, M., & Li, Z. (2017). In-situ non-intrusive diagnostics of toluene removal by a gliding arc discharge using planar laser-induced fluorescence. *Plasma Chemistry and Plasma Processing, 37*, 433–450.
2. Teramoto, Y., Kim, H. H., Ogata, A., & Negishi, N. (2014). Measurement of OH ($X^2\Sigma$) in immediate vicinity of dielectric surface under pulsed dielectric barrier discharge at atmospheric pressure using two geometries of laser-induced fluorescence. *Journal of Applied Physics, 115*, 133302.
3. Jia, Z., Wang, X., Thevenet, F., & Rousseau, A. (2017). Dynamic probing of plasma-catalytic surface processes: Oxidation of toluene on CeO_2. *Plasma Processes and Polymers, 14*, 1600114.
4. Klages, C.-P., Hinze, A., & Khosravi, Z. (2013). Nitrogen plasma modification and chemical derivatization of polyethylene surfaces - an in situ study using FTIR-ATR spectroscopy. *Plasma Processes and Polymers, 10*, 948–958.
5. Stere, C. E., Anderson, J. A., Chansai, S., Delgado, J. J., Goguet, A., Graham, W. G., Hardacre C., Taylor S. F. R., Tu, X., Wang, Z., & Yang, H. (2017). Non-thermal plasma activation of gold-based catalysts for low-temperature water-gas shift catalysis. *Angewandte Chemie, International Edition, 56*, 5579–5583.
6. Neyts, E. C., & Bogaerts, A. (2014). Understanding plasma catalysis through modelling and simulation-a review. *Journal of Physics D: Applied Physics, 47*, 224010.
7. Somers, W., Bogaerts, A., Van Duin, A. C. T., Huygh, S., Bal, K. M., & Neyts, E. C. (2013). Temperature influence on the reactivity of plasma species on a nickel catalyst surface: An atomic scale study. *Catalysis Today, 211*, 131–136.
8. Tennyson, J., Rahimi, S., Hill, C., Tse, L., Vibhakar, A., & Akello-Egwel, D. (2017). QDB: A new database of plasma chemistries and reactions. *Plasma Sources Science & Technology, 26*, 055014.
9. Bogaerts, A., De Bie, C., Eckert, M., Georgieva, V., Martens, T., Neyts, E., & Tinck, S. (2010). Modeling of the plasma chemistry and plasma-surface interactions in reactive plasmas. *Pure and Applied Chemistry, 82*, 1283–1299.
10. Aerts, R., Tu, X., Van Gaens, W., Whitehead, J. C., & Bogaerts, A. (2013). Gas purification by nonthermal plasma: A case study of ethylene. *Environmental Science & Technology, 47*, 6478–6485.
11. Koen Van, L., & Annemie, B. (2016). Fluid modelling of a packed bed dielectric barrier discharge plasma reactor. *Plasma Sources Science and Technology, 25*, 015002.
12. Zhang, Y., Wang, H. Y., Jiang, W., & Bogaerts, A. (2015). Two-dimensional particle-in cell/ Monte Carlo simulations of a packed-bed dielectric barrier discharge in air at atmospheric pressure. *New Journal of Physics, 17*, 12.
13. Van Laer, K., & Bogaerts, A. (2015). Improving the conversion and energy efficiency of carbon dioxide splitting in a zirconia-packed dielectric barrier discharge reactor. *Energy Technology, 3*, 1038–1044.

14. De Bie, C., Martens, T., van Dijk, J., Paulussen, S., Verheyde, B., Corthals, S., & Bogaerts, A. (2011). Dielectric barrier discharges used for the conversion of greenhouse gases: modeling the plasma chemistry by fluid simulations. *Plasma Sources Science and Technology, 20*, 024008.
15. Heijkers, S., Snoeckx, R., Kozák, T., Silva, T., Godfroid, T., Britun, N., Snyders, R., & Bogaerts, A. (2015). CO_2 conversion in a microwave plasma reactor in the presence of N_2: Elucidating the role of vibrational levels. *Journal of Physical Chemistry C, 119*, 12815–12828.
16. Cleiren, E., Heijkers, S., Ramakers, M., & Bogaerts, A. (2017). Dry reforming of methane in a gliding arc plasmatron: towards a better understanding of the plasma chemistry. *ChemSusChem, 10,* 4025-4036.
17. van Santen, R. A., Markvoort, A. J., Filot, I. A. W., Ghouri, M. M., & Hensen, E. J. M (2013). Mechanism and microkinetics of the Fischer-Tropsch reaction. *Physical Chemistry Chemical Physics, 15*, 17038–17063.
18. Filot, I. A. W., van Santen, R. A., & Hensen, E. J. M. (2014). The optimally performing Fischer-Tropsch catalyst. *Angewandte Chemie-International Edition, 53*, 12746–12750.
19. Kim, J., Go, D. B., & Hicks, J. C. (2017). Synergistic effects of plasma-catalyst interactions for CH_4 activation. *Physical Chemistry Chemical Physics, 19*, 13010–13021.
20. Snoeckx, R., & Bogaerts, A. (2017). Plasma technology - a novel solution for CO_2 conversion? *Chemical Society Reviews, 46*, 5805-5863.
21. Hessel, V., Cravotto, G., Fitzpatrick, P., Patil, B. S., Lang, J., & Bonrath, W. (2013). Industrial applications of plasma, microwave and ultrasound techniques: Nitrogen-fixation and hydrogenation reactions. *Chemical Engineering and Processing, 71*, 19–30.
22. Mori, S., Matsuura, N., Tun, L. L., & Suzuki, M. (2016). Direct synthesis of carbon nanotubes from only CO_2 by a hybrid reactor of dielectric barrier discharge and solid oxide electrolyser cell. *Plasma Chemistry and Plasma Processing, 36*, 231–239.

MIX
Papier aus verantwortungsvollen Quellen
Paper from responsible sources
FSC® C105338

If you have any concerns about our products,
you can contact us on
ProductSafety@springernature.com

In case Publisher is established outside the EU,
the EU authorized representative is:
Springer Nature Customer Service Center GmbH
Europaplatz 3, 69115 Heidelberg, Germany

Printed by Libri Plureos GmbH
in Hamburg, Germany